The Guide to Cooking Schools

Eleventh Edition

ShawGuides
NEW YORK

Inquiries concerning this book should be addressed to: Editor, ShawGuides, P.O. Box 1295, New York, New York 10023, Phone: (212) 799-6464, Fax: (212) 724-9287, E-mail: info@shawguides.com, URL: www.shawguides.com.

Please note that the information herein has been obtained from the listed cooking schools and organizations and is subject to change. The editor and publisher accept no responsibility for inaccuracies. Schools should be contacted prior to sending money and/or making travel plans.

Library of Congress Catalog Card Number 88-92516
ISSN 1040-2616
ISBN 0-945834-25-X

Printed in the United States of America by
R. R. Donnelley & Sons Company

How to Choose a Cooking School

Since the first edition of this directory was published, in 1989, the interest in food and cooking has grown steadily as have the number of educational opportunities for both professional and amateur cooks. This eleventh annual edition is the largest to date, containing over 1,000 career and recreational programs offered by colleges, universities, private culinary schools, foodservice establishments, resorts, cookware shops, travel companies, and individual professionals worldwide. More than 200 programs are new to this edition.

If you're considering a career in the culinary arts or seeking advanced training, section one (pages 1-178) provides detailed descriptions of 401 career and continuing education programs. These range in length from a few days or weeks for specialized certificate programs to four years or more for college degree programs that combine a culinary and academic curriculum. The last part of this section (pages 166-178) lists 94 American Culinary Federation-approved on-the-job apprenticeships.

If you're an amateur cook who wants to learn new recipes and techniques and gain a better understanding of food and cooking, section two (pages 181-350) describes 584 recreational programs, including classes, courses, and vacations that feature cooking lessons and food-related excursions.

Additional sections provide information about 94 wine courses taught by schools and individuals recognized by the American Wine Society and Society of Wine Educators (pages 303-320), 21 food and wine organizations (pages 373-378), and 48 food and wine publications (pages 387-390).

Consider the following points in selecting a culinary career program or a cooking vacation.

CHOOSING A CAREER SCHOOL

1. How long is the program? Career programs range from a few weeks to four years or more. Curricula for programs of a year or less consist primarily of culinary courses that prepare you for an entry level position. College degree programs include general education courses and electives that provide a more well-rounded education. The 3-year apprenticeship program sponsored by the American Culinary Federation (ACF) offers paid on-the-job training in a foodservice establishment and the opportunity to earn a college degree.

2. Is it affordable? Tuition ranges from a few hundred dollars at community colleges to over $10,000 per program or year at trade schools that offer a specialized curriculum. If cost is an obstacle, inquire about scholarships or loans, which are offered by many schools and some culinary organizations (see page 394).

3. What are the scheduling options? If you're unable to attend classes full-time, consider programs that permit you to enroll part-time or offer flexible schedules.

4. How qualified is the faculty? Instructor credentials should include certification by the American Culinary Federation (see page 373), college degree, and/or industry experience.

5. Is the school accredited? A school in operation for five years or more should be accredited. Colleges are accredited by one of six regional associations, private and trade schools by three organizations (see page 398). The ACF accredits 93 post-secondary institution programs in culinary arts and foodservice management (see page 381).

6. Is real-world experience part of the program? Some schools have student-staffed foodservice facilities on-campus where students are required to work as part of the program. Others offer intern- or externships in an off-campus setting.

7. What courses, textbooks, and course materials are provided? Has the school's curriculum adapted to today's healthier lifestyles with emphasis on fresh ingredients, nutrition, and a variety of international cuisines. Do they offer specialized courses in the subjects that interest you?

8. What kind of job offers can you expect? Will the school's placement office be able to find you a position in the setting you desire? Obtain the names of graduates and contact them to determine whether the school met their expectations for training and placement.

CHOOSING A COOKING VACATION

1. What can you expect to learn? Are the dishes appealing and suited to your expertise? Will you be learning the how's and why's of cooking, rather than just following recipes?

2. Will classes be demonstration or hands-on? Hands-on classes are necessary for learning techniques. Demonstrations are appropriate for experienced cooks and those who prefer observing to participating. Most vacation programs combine both.

3. What are the cooking and lodging facilities like? For hands-on classes will you have your own work station and utensils? Are appliances modern and in good working condition? Is the space large enough for everyone to move about comfortably? For demonstrations, is there an overhead mirror and is seating close enough that you'll be able to see clearly? Is lodging part of a chain or rated by a recognized travel guide? Do the rooms have private baths?

4. What are the qualifications of the instructor? If the teacher has written a cookbook, obtain a copy to determine whether the recipes appeal to you. If the teacher is a chef, will the recipes be adapted to a home kitchen? Request copies of some of the recipes that will be prepared and speak with the instructor to get a sense of his or her teaching style and communication skills.

5. What is scheduled during non-cooking time? Some vacation programs emphasize cooking over other activities, some offer a few classes with more time devoted to sightseeing, visiting food-related sites, shopping, dining out, or at leisure. Obtain a detailed itinerary so you'll know what to expect.

6. What is covered by the cost? The cost always covers classes and the meals prepared, usually covers sightseeing, most other meals, and ground transportation, and sometimes covers lodging and airfare. Find out what your payment covers and how much you should budget for the rest.

7. Request the names of recent participants and contact them. Did the program meet their expectations, does it offer the features you desire, would they would recommend it?

Although we strive to make each listing accurate, changes do occur. For updates and new listings, check our web site – http://www.shawguides.com – which contains the unabridged contents of this directory, updated weekly and accessible at no charge.

We thank the school and organization directors for their cooperation. We also appreciate the help provided by Stacey Shane-Nusbaum, Brian Shannon, and Patty Shannon.

May you find pleasure and success in all your culinary endeavors.

ShawGuides

Contents

1

Career and Professional Programs

ALABAMA

BISHOP STATE COMMUNITY COLLEGE

Mobile/Year-round

This two-year college offers a 6-quarter 90 credit-hour certificate/114 credit-hour AAS degree in Commercial Food Service; 2-semester+summer term/74 credit-hour AAS degree in Commercial Food Service. Established in 1963. Accredited by ACFEI, SACS. Calendar: semester. Curriculum: culinary, core. Admission dates: August, January. Total enrollment: 25. 100% of applicants accepted. 75% receive financial aid. 25% enrolled part-time. Student to teacher ratio 10:1. 100% of graduates obtain employment within six months. Facilities: Fully equipped kitchen.

Courses: Nutrition, commercial food service, purchasing, menu planning, terminology, dairy products, meats, vegetables, fruits, sauces, poultry, seafood, baking, garde manger, food and beverage management, cake decorating.

Faculty: Two certified chefs.

Costs: Tuition is $50 per credit-hour in-state, $100 per credit-hour out-of-state. Admission requirements: High school diploma or GED. Scholarships: yes. Loans: no.

Contact: Levi Ezell, Director, Commercial Food Service, Bishop State Community College-Carver Campus, 414 Stanton St., Mobile, AL 36617 US; 334-473-8692, Fax 334-473-7915, E-mail cfs@bscc.cc.al.us, URL http://www.bscc.cc.al.us.

FAULKNER STATE COMMUNITY COLLEGE

Gulf Shores/Year-round

This two-year college offers a 1-year/86 quarter-hour certificate and 2-year/126 quarter-hour AAS degree in Culinary Arts. Program started 1994. Accredited by SACS, ACFEI. Calendar: quarter. Curriculum: culinary, core. Admission dates: September, January, March, June. Total enrollment: 65. 25 each admission period. 90% of applicants accepted. 80% receive financial aid. 10% enrolled part-time. Student to teacher ratio 1:15. 100% of graduates obtain employment within six months. Facilities: New 18,000-square-foot state-of-the-art building.

Courses: Food safety and production, baking and pastry, food styling. Externship: 2,100 hours. Year-round: sanitation, nutrition, personnel management.

Faculty: Gerhard Brill, CEC, Nancy Hartly, CEC, Andy Camardella, CEC, Ron Koetter, CEC, CCE, CMB, AAC, Program Coordinator.

Costs: $32.50 per quarter-hour in-state, $65 out-of-state; no fees. Admission requirements: High school diploma or GED & entrance exam. Scholarships: yes. Loans: yes.

Contact: Ron Koetter, Program Coordinator, Faulkner State Community College, 3301 Gulf Shores Pkwy., Gulf Shores, AL 36542 US; 334-968-3104, Fax 334-968-3120, E-mail rkoetter@faulkner.cc.al.us, URL http://www.faulkner.cc.al.us.

JEFFERSON STATE COMMUNITY COLLEGE

Birmingham/Year-round

This two-year college offers a 34-hour Culinary Apprentice Option leading to an AAS degree (requires 6,000-hour on-the-job internship), 28-hour Food Service Management option. Curriculum: culinary.

Courses: Food preparation, meal management, baking, garde manger, beverage management.

Contact: George White, Jefferson State Community College, 2601 Carson Rd., Birmingham, AL 35215-3098 US; 205-853-1200, Fax 205-856-6070.

LAWSON STATE COMMUNITY COLLEGE

Birmingham/September-May

This college offers a 21-month certificate in Culinary Arts. Program started 1949. Accredited by SACS. Admission dates: Quarterly. Student to teacher ratio 23:1. 50% of graduates obtain employment within six months.

Faculty: 2 full-time.

Costs: Tuition $39 per credit-hour in-state, $78 per credit-hour out-of-state. Admission requirements: High school diploma or equivalent and admission test.

Location: The 2,000-student 50-acre suburban campus is 100 miles from Detroit.

Contact: Deborah Harris, Lawson State Community College, Commercial Food Preparation, 3060 Wilson Rd., SW, Birmingham, AL 35221-1717 US; 205-925-2515.

TRENHOLM STATE TECHNICAL COLLEGE

Montgomery/Year-round

This public post-secondary occupational education institution offers a 42-hour (12-month) certificate in Culinary Arts, 64-hour (24-month) associate degree in Culinary Arts Applied Technology, 6000-hour Chef Apprenticeship-Management option, 74-hour Hospitality Management option. Accredited by COE. Calendar: semester. Curriculum: culinary, core. Admission dates: January, June, August. Total enrollment: 101. 50+ each admission period. 80% of applicants accepted. 50% receive financial aid. 10% enrolled part-time. Student to teacher ratio 8:1. 100% of graduates obtain jobs within six months. Facilities: Fully equipped kitchen with electric and gas equipment, classrooms, formal and informal dining rooms, conference room.

Courses: Food production, catering, garde manger, nutrition, menu design, table service, restaurant/hotel/motel management. Apprenticeship and training at approved sites.

Faculty: American Culinary Federation certified chefs. The college has served as the home practice site for the USA Culinary Olympic Team, Southeast.

Costs: $42 per semester cr-hour, approximately $300 per semester for uniforms, books, insurance. Admission requirements: 17 yrs of age, high school diploma or GED. Scholarships: yes. Loans: yes.

Contact: Mary Ann Campbell, CEC, CCE, Director of Culinary Arts, Trenholm State Technical College, Culinary Arts, Hospitality Management, 1225 Air Base Blvd., P.O. Box 9039, Montgomery, AL 36108 US; 334-262-4728, Fax 334-832-2433, E-mail chefcampbell@mindspring.com.

WALLACE STATE COMMUNITY COLLEGE

Hanceville/Year-round

This two-year college offers an 18-month diploma in Commercial Food Technology, 24-month degree in Commercial Foods. Program started 1979. Accredited by SACS. Calendar: semester. Curriculum: culinary, core. Admission dates: August, January, June. Total enrollment: 18. 100% of applicants accepted. 80-90% receive financial aid. 5% enrolled part-time. Student to teacher ratio 15:1. 98% of graduates obtain employment within six months. Facilities: Classroom and lab.

Faculty: 2 full-time.

Costs: Tuition is $1,100-$1,300 in-state. Admission requirements: High school diploma or equivalent.

Contact: Donna Jackson, Dept. Chair, Wallace State Community College, Commercial Foods & Nutrition, Box 2000, Hanceville, AL 35077 US; 256-352-8227, Fax 256-352-8228.

ALASKA

UNIVERSITY OF ALASKA – FAIRBANKS

Fairbanks/August-April

This university offers a 2-year certificate and 2-year AAS degree in Culinary Arts. Program started 1979. Accredited by NASC, ACCSCT. Calendar: semester. Curriculum: culinary. Admission dates: Fall, spring. Total enrollment: 30-45. 85% of applicants accepted. Student to teacher ratio 8-10:1. 95% of graduates obtain employment within six months.

Courses: Externship provided.

Faculty: 3 full-time, 7 part-time.

Costs: Annual tuition in-state $74 per credit hour, out-of-state $224 per credit hour. Admission requirements: High school diploma or equivalent.

Contact: Frank U. Davis, CCE, CEC, University of Alaska, Tanana Valley Campus, 510 Second Ave., Fairbanks, AK 99701 US; 907-474-5196, Fax 907-474-7335, URL http://www.uaf.edu.

ARIZONA

ARIZONA WESTERN COLLEGE

Yuma/August-June

This state-supported community college offers a 1-year (2-semester, 24-credit) program. Program started 1996. Accredited by NCA. Calendar: semester. Curriculum: culinary. Admission dates: August, January. Total enrollment: 20. 15 each admission period. 100% of applicants accepted. 50% receive financial aid. 20% enrolled part-time. Student to teacher ratio 15:1. 50% of graduates obtain employment within six months. Facilities: Fully-equipped kitchen/lab and dining room.

Courses: Classic European-style preparation. Interns do voluntary placement under contract with local country club.

Faculty: One full-time registered dietitian with master's degree, 2 part-time apprenticeship-trained chefs with 25 years experience.

Costs: $1,400 for the year. Admission requirements: Open enrollment. Placement tests in reading & math, application, personal essay. Scholarships: yes. Loans: yes.

Contact: Nancy Meister, Coordinator, Culinary Arts Program, Arizona Western College, Culinary Arts Program, PO Box 929, Yuma, AZ 85364 US; 520-344-7779, Fax 520-344-7730, E-mail aw_meistern@rocky.awc.cc.az.edu, URL http://www.awc.cc.az.edu.

ART INSTITUTE OF PHOENIX – SCHOOL OF CULINARY ARTS

(See display ad page 8)

Phoenix/Year-round

This private school specializing in the creative and applied arts offers an 18-month AAS degree in Culinary Arts (24-months in the evening), certification in Sanitation and Safety and Nutrition. Program started 1996. Accredited by ACCSCT. Calendar: quarter. Curriculum: core. Admission dates: October, January, April, July. Total enrollment: 97. 40 each admission period. 90% of applicants accepted. 70% receive financial aid. 7% enrolled part-time. Student to teacher ratio 18:1. 100% of graduates obtain employment within six months. Facilities: 3 production kitchens for garde manger, basic skills, regional cuisine, nutritional cooking, baking and pastry, catering; dining lab.

Courses: Include culinary skills, food production, regional cuisine, nutritional cooking, baking and pastry, catering, a la carte cooking, nutrition, hospitality, facilities and design, purchasing and cost control, general education.

Faculty: Chef Director Bill Sy, MBA, CEC, Chef Walter Leible, Certified Master Chef, Terry Barkley, Anthony Rea CEC, Linda Marcinko, Matt Baer, Eric Watson, Jim Diamond, Roger Gerard, Peter Leitner. Also Robin DeBell MS, RD, Steven Durham.

Costs: $230 per credit hour, $3,680 quarterly. Application fee $50, administrative fee $100, supply kit $595, lab fee $250 per quarter. Total tuition and fees for program $24,325. Admission requirements: High school diploma or equivalent, 150-word essay, interview. Scholarships: yes. Loans: yes.

Location: The 62,000-square-foot campus is in North Phoenix, near Rte. I-17, buses, shopping, dining, housing.

Contact: Timothy Dengler, Asst. Director of Admissions, Art Institute of Phoenix-School of Culinary Arts, 2233 W. Dunlap Ave., Phoenix, AZ 85021 US; 800-474-2479 x102, Fax 602-216-0439, E-mail denglert@aii.edu, URL http://www.aii.edu.

MARICOPA SKILL CENTERS

Phoenix/Year-round

This community college division offers 14- to 27-week certificates in Cook's Apprentice, Kitchen Helper, Baker's Helper, Pantry Goods Maker. Program started 1964. Accredited by NCA.

Curriculum: culinary. Admission dates: Any Monday. Total enrollment: 25. 100% of applicants accepted. 80% receive financial aid. Student to teacher ratio 7:1. 80% of graduates obtain employment within six months. Facilities: Commercial kitchen.

Courses: Entry level.

Faculty: 3 full-time: Dan Bochicchio, CWC, and 2 assistants.

Costs: $2,635 tuition, $200 lab fee. Admission requirements: At least age 16. Scholarships: yes.

Location: Center of Valley of Sun, a metropolitan area.

Contact: Barbara Lacy, Public Relations, Maricopa Skill Centers, 1245 E. Buckeye Rd., Phoenix, AZ 85034 US; 602-238-4300, Fax 602-238-4307, E-mail lacybarbara@gwc.maricopa.edu, URL http://www.maricopa.edu/msc/.

PHOENIX COLLEGE

Phoenix/February-May, August-December

This two-year college offers associate degree and 16-week certificate of completion programs in Culinary Studies and Foodservice Administration. Established in 1972. Accredited by North Central Accrediting. Calendar: semester. Curriculum: culinary, core. Admission dates: August and January. Total enrollment: 38. 40 each admission period. 98% of applicants accepted. 20% receive financial aid. 60% enrolled part-time. Student to teacher ratio 18:1. 95% of graduates obtain employment within six months. Facilities: 5,000-square-feet of teaching facilities, the latest equipment, on-site 40-seat restaurant.

Courses: Include nutrition, menu planning, commercial baking, garde manger, professional cooking, French, American regional, and international cuisines, food purchasing and management, customer service, sanitation, safety. Banquet food and beverage service externship. Cont. ed. courses available.

Faculty: Chefs Scott Robinson, Steve Slansky, Michael Whelan, Guentar Haub, Ray Vicencio.

Costs: ~$3,200. Admission requirements: GED, previous transcripts, application. Scholarships: yes. Loans: yes.

Location: Downtown Phoenix.

Contact: Scott Robinson, Coordinator, Phoenix College, 1202 W. Thomas Rd., Phoenix, AZ 85013 US; 602-285-7901, Fax 602-285-7700, E-mail robinson@pc.maricopa.edu, URL http://www.pc.maricopa.edu.

PIMA COMMUNITY COLLEGE

Tucson/September-May

This state-supported college offers a 2-year certificate, 2-year AAS degree in Culinary Arts. Program started 1972. Accredited by NCA. Admission dates: January, May, August. Total enrollment: 50. 22 each admission period. 100% of applicants accepted. 92% enrolled part-time. Student to teacher ratio 18:1. 90% of graduates obtain employment within six months. Facilities: 1 kitchen, 2 classrooms.

Courses: Gourmet cooking, garde manger, baking. 4-year degree continuation at Northern Arizona University.

Faculty: 11 full-time and part-time.

Costs: Annual tuition in-state $32 per credit hour, out-of-state $55 per credit hour. Other fees $20 -$40 per semester. Off-campus housing $200-$600 per month. Admission requirements: High school diploma and 1 year food service experience.

Contact: John P. Dailey, Dept. Chair, Pima Community College, Hospitality, 1255 N. Stone Ave., Tucson, AZ 85703 US; 800-860-7462 x6283, Fax 520-884-6201. Additional phones: 520-748-4500/884-6283, URL http://www.pima.edu.

SCOTTSDALE COMMUNITY COLLEGE

Scottsdale/August-May

This state-supported college offers a 9-month certificate, 2-year AAS degree in Culinary Arts.

Scottsdale
Culinary
Institute
Professional culinary training
based on classical French
methods and contemporary
trends. Accelerated
15-month A.O.S. Degree,
paid externship, and hands-on
curriculum in small classes.
Housing and Placement
Assistance. American Culinary
Federation accredited.
Financial Aid if qualified.
Please call to request a
catalog and arrange a tour.
8100 East Camelback Road • Scottsdale, Arizona 85251
(602) 990-3773 • (800) 848-2433
Visit our Web site @ www.chefs.com/culinary/

Program started 1985. Accredited by NCA. Calendar: semester. Curriculum: culinary, core. Admission dates: August. Total enrollment: 30. 30 each admission period. 60% of applicants accepted. 0% enrolled part-time. Student to teacher ratio 7:1. 90% of graduates obtain employment within six months. Facilities: 1 kitchen, 2 classrooms, 1 student-run dining room.

Courses: Hospitality management, culinary principles, menu planning, hot foods, bakery/pastry, garde manger; student-operated restaurant.

Faculty: 2 full-time, Dominic O'Neill and Sarah Labensky; 2 part-time.

Costs: Annual tuition in-state $1,600, out-of-state $5,400, includes course fee of $250 per semester. Admission requirements: High school diploma and 1 year food service experience. Scholarships: yes. Loans: yes.

Location: The 10,000-student 160-acre campus is on Salt River-Pima Indian Community land in the greater Phoenix area.

Contact: Dominic O'Neill, Director, Scottsdale Community College, Culinary Arts Program, 9000 E. Chaparral Rd., Scottsdale, AZ 85250 US; 602-423-6244, Fax 602-423-6200, URL http://www.sc.maricopa.edu.

SCOTTSDALE CULINARY INSTITUTE

Scottsdale/Year-round *(See display ad page 5)*

This private institution offers an accelerated 15-month 78-credit-hour AOS degree in Culinary Arts and Sciences and Restaurant Management. Established in 1986. Accredited by ACCSCT, ACFEI. Curriculum: culinary. Admission dates: January, February, April, May, July, August, October, November. Total enrollment: 360. 40 each admission period. 70% of applicants accepted. 70% receive financial aid. Student to teacher ratio 15:1. 97% of graduates obtain employment within six months. Facilities: Include 5 modern, full-service kitchens, bakery, meat fabrication shop, student-run Mobil 3-star L'Ecole restaurant.

Courses: Emphasis is classic French techniques and principles of Escoffier. Final 12 weeks are a national paid externship program; 80% of positions become permanent. Continuing education for professionals.

Faculty: 17 full-time, 2 part-time. Founder/Director Elizabeth Leite developed a commercial foods curriculum at Scottsdale Voc-Tech Institute. Her staff of American- and European-trained professionals are selected for their teaching skills.

Costs: Tuition is $16,150. One-time $860 fee covers uniforms, knives, textbooks. $25 fee accompanies application. Housing adjoins the campus. Admission requirements: High school diploma or equivalent, application, essay, 2 letters of recommendation. Scholarships: yes. Loans: yes.

Location: Phoenix metropolitan area.

Contact: Scottsdale Culinary Institute, Admissions, 8100 E. Camelback Rd., Scottsdale, AZ 85251 US; 602-990-3773/800-848-2433, Fax 602-990-0351, URL http://www.chefs.com/culinary/.

TUCSON CULINARY ALLIANCE

Tucson/Year-round

This professional job bank, mentoring service, provider of assorted customized foodservice training offers Pima Community College's 2-year AAS degree in Hotel/Restaurant Management and Cook Certification with the ACFEI. Program started 1993. Accredited by U.S. Dept. of Labor Bureau of Apprenticeship Training. Calendar: semester. Curriculum: culinary, core. Admission dates: Open. Total enrollment: 30-45 (maximum). 5-10 per year each admission period. 90% of applicants accepted. 25% receive financial aid. 45% enrolled part-time. Student to teacher ratio 1:1. 100% of graduates obtain employment within six months. Facilities: All o.j.t./apprenticeship training accrues within the apprentice's place of employment and Pima Community College.

Courses: Depends on apprentice's needs and goals. Apprenticeship training is paid, on-the-job program.

Faculty: Advisory committee includes Program Director Robert Shell, CCC, CCE, Lorraine Adler, EOD, CEC, CCE, Cathy Ochoa, CEC, Dave Perkins, CEPC.

Costs: Total cost is $1,800-$2,500. Admission requirements: 18 years old, high school diploma or equivalent, pass pre-apprenticeship courses. Scholarships: yes. Loans: yes.

Location: 120 miles south of Phoenix.

Contact: Robert B. Shell, CCC, CCE, Program Director, Tucson Culinary Alliance, 3124 E. Pima St., Bldg. B, Tucson, AZ 85716 US; 520-327-3594 (8-11 am).

ARKANSAS

OZARKA TECHNICAL COLLEGE CULINARY ARTS PROGRAM

Melbourne/August-May

This public 2-year college offers a 9-month technical certificate program, 2-year AAS General Technology degree option. Established in 1975. Accredited by NCA. Calendar: semester. Curriculum: culinary, core. Admission dates: August, January. Total enrollment: 15. 6 each admission period. 100% of applicants accepted. 80% receive financial aid. 10% enrolled part-time. Student to teacher ratio 15:1. 80% of graduates obtain employment within six months. Facilities: Commercial kitchen, new culinary lab under construction.

Courses: Food safety, basic principles and techniques of food preparation, dining room service and catering, baking, advanced topics. Student-managed Restaurant Nights on campus.

Faculty: 1 instructor.

Costs: Tuition $888 per year, fees $132 per year. Admission requirements: High school graduate or GED. Scholarships: yes. Loans: yes.

Location: ~100 miles north of Little Rock.

Contact: Richard Tankersley, Chef, Ozarka Technical College Culinary Arts Program, PO Box 10, Melbourne, AR 72556 US; 870-368-7371/800-821-4335, Fax 870-368-4733, URL http://www.ozarka.tec.ar.us.

CALIFORNIA

AMERICAN RIVER COLLEGE

Sacramento/Year-round

This two-year college offers a 1-year/39 unit certificate and 2-year AA degree in Culinary Arts/Restaurant Management. Calendar: semester. Curriculum: culinary, core.

Courses: Food theory and preparation, professional cooking, baking and pastry, cost control, purchasing, advertising and sales, management and supervision, beverage operations, dining room management. 360-hour internship off campus.

Costs: Tuition is $13 per unit in-state, $138 per unit out-of-state.

Contact: Susan Barry, Program Coordinator, American River College, 4700 College Oak Dr., Sacramento, CA 95841 US; 916-484-8656, Fax 916-484-8880, E-mail barrys@arc.losrios.cc.ca.us, URL http://www.arc.losrios.cc.ca.us.

ART INSTITUTE OF LOS ANGELES

(See display ad page 8)

Los Angeles/Year-round

This college offers a 77-week (2-year) AS degree in Culinary Arts. Program started 1997. Accredited by ACCSCT. Calendar: quarter. Curriculum: core. Admission dates: July, October, January, April. Total enrollment: 20. 20-40 each admission period. 98% of applicants accepted. 75% receive financial aid. Student to teacher ratio 20:1. 90% of graduates obtain employment within six months. Facilities: 2 kitchens, classrooms.

Courses: Basic cooking skills, food production, garde manger, a la carte, baking, nutrition, sanitation, purchasing and cost control, dining room and menu management, wines and spirits, facilities design, 24 hours of general ed. 198-hour externship is required in the 7th quarter.

Faculty: 2 full-time: Richard Battista, CEC, CEPC, CCE and Cliff Chapman, CCE.

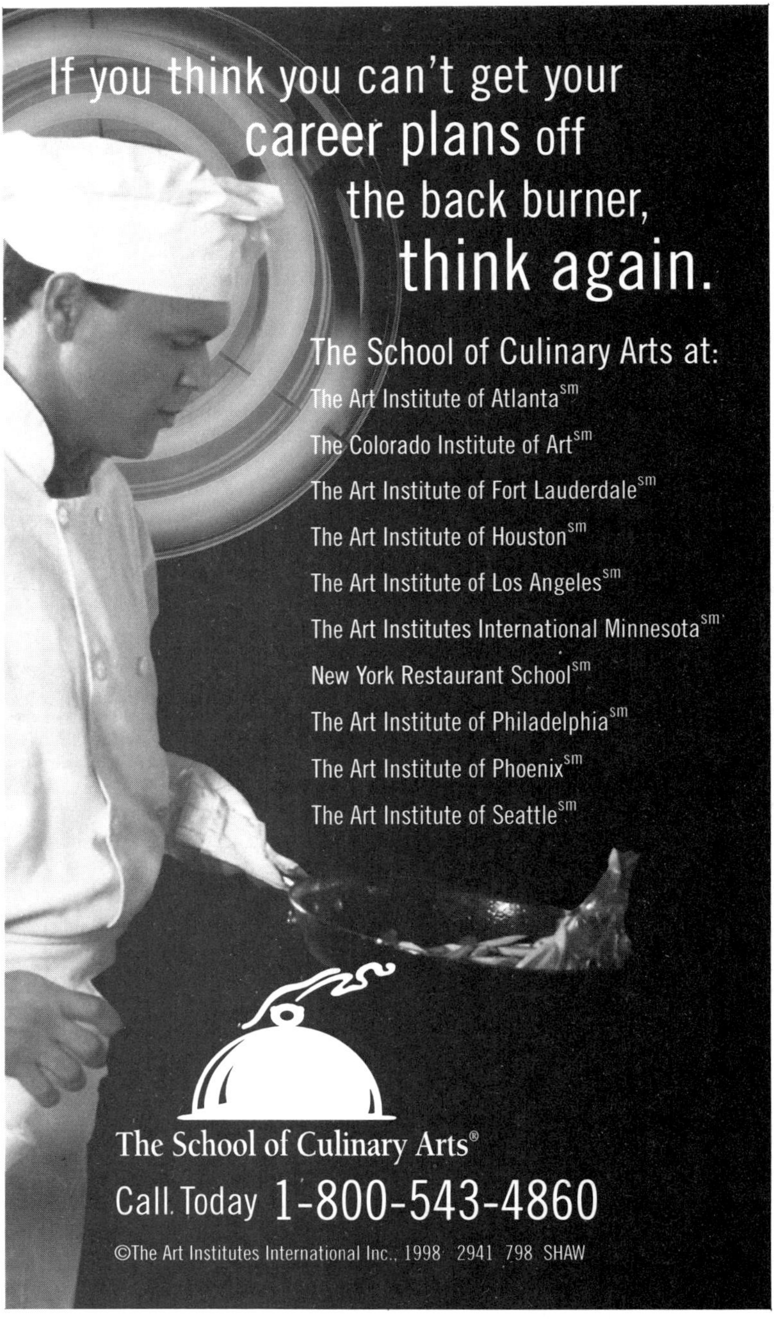

Costs: $4,144 per quarter. Application fee $50, administrative fee $100, lab fee $300, starting supply kit $660. Scholarships: yes. Loans: yes.

Location: The 38,000-square-foot campus is near the beach in Santa Monica/Los Angeles.

Contact: Mary Jo Placka, The Art Institute of Los Angeles, 2900-31st St., Santa Monica, CA 90405 US; 888-646-4610/310-752-4700, Fax 310-752-4708, E-mail plackam@aii.edu, URL http://www.aii.edu.

BERINGER VINEYARD'S SCHOOL FOR AMERICAN CHEFS

St. Helena/January

This nonprofit school offers an annual 2-week post-graduate course. Program started 1989. Total enrollment: 8. 8 each admission period. 25-30% of applicants accepted. Student to teacher ratio 8:1. Facilities: Beringer Vineyards' Culinary Arts Center: one kitchen for practicing and one for public dinner preparation.

Courses: Tailored to student requests. Required courses are wine and food pairing, wine tasting, and elementary viticulture.

Faculty: School Director Madeleine Kamman was born in Paris, taught French cuisine in the US for over 30 years, authored 9 cookbooks, and hosts a PBS TV series.

Costs: None. Tuition, ingredients, and wines are provided by Beringer Vineyard. Off-campus lodging averages $250 per week. Admission requirements: American citizen, age 21 with high school diploma or equivalent, 2 years experience as working chef.

Location: The Napa Valley, 90 minutes from San Francisco.

Contact: Allison Simpson, Administrator, Beringer Vineyards, PO Box 111, St. Helena, CA 94574 US; 707-963-7115 x2162, Fax 707-963-1735.

CABRILLO COLLEGE

Aptos/August-May

This community college offers 30-unit certificate and 60 unit AS degree programs in Culinary Arts and Hospitality Management. Program started 1972. Calendar: semester. Curriculum: core. Admission dates: August, January. Total enrollment: 200+. 150-200 each admission period. 100% of applicants accepted. 70% enrolled part-time. 100% of graduates obtain employment within six months. Facilities: Restaurant kitchen, quantity foods kitchen, bake shop, lecture/demonstration room, student-run restaurant.

Courses: Include culinary arts, baking and pastry arts, cake decorating, garde manger, chocolate.

Faculty: 2 full-time 4 part-time.

Costs: Annual tuition for full-time students $464 in-state, $4,820 out-of-state. Off-campus housing available. Admission requirements: High school diploma or equivalent. Scholarships: yes.

Location: 100 miles south of San Francisco, on the coast.

Contact: Eric Carter, Director of Culinary Arts, Cabrillo College, Culinary Arts & Hospitality Management, 6500 Soquel Dr., Aptos, CA 95003 US; 408-479-5749, URL http://www.cabrillo.cc.ca.us. Admissions: 408-479-6213.

CALIFORNIA CULINARY ACADEMY

(See also page 186) (See display ad page 10) **San Francisco/Year-round**

This proprietary institution offers a full-time 18-month AOS degree in Culinary Arts, 30-week certificate in Baking and Pastry Arts, professional weekend certificates in Baking and Pastry and Basic Skills. Salinas campus: Basic Professional Culinary Skills certificate. Established in 1977. Accredited by ACFEI, ACCSCT. Curriculum: culinary. Admission dates: Every two weeks except December. Total enrollment: 700 in degree program. 25 degree, 22 certificate each admission period. 85% of applicants accepted. 75% receive financial aid. Student to teacher ratio 20-25:1. 93% of graduates obtain employment within six months. Facilities: 14 production kitchens, auditorium, 2 restaurants, confiseries, butcher and seafood prep kitchens, garde manger, baking/pastry kitchens.

Courses: Sequential learning curriculum for degree program covers food preparation and pre-

sentation, global cuisine, baking and pastry, nutrition, wine, menu, facilities planning; certificate program covers baking and pastry, chocolate, candies, decorating, pastillage. 3-month externship for degree students. 200+ courses/year for professionals and novices.

Faculty: Full-time staff of 30 chefs averaging 15 years experience, 3 maitres d'hotel, 6 part-time, guest professionals.

Costs: Approximately $27,000 ($12,000) for degree (certificate) program, which includes meals, uniforms, textbooks, equipment and accident insurance. Application fee $35. Approximate cost of housing is $5,200. Admission requirements: High school diploma or equivalent; industry experience recommended. Scholarships: yes. Loans: yes.

Location: Near San Francisco's Civic Center, two blocks from City Hall.

Contact: Sandra Weber, Director of Admissions, California Culinary Academy, Admissions, 625 Polk St., San Francisco, CA 94102 US; 415-771-3536/800-BAY-CHEF, Fax 415-771-2194, URL http://www.baychef.com.

CHAFFEY COLLEGE

Alta Loma/Year-round

This public 2-year community college offers a 1-year certificate in Culinary Arts and 2-year associate degree in Culinary Arts and Food Service Management. Established 1986. Accredited by WASC. Calendar: semester. Curriculum: culinary. Admission dates: January, June, August. Total enrollment: 52. 20 each admission period. 100% of applicants accepted. 85% receive financial aid. 80% enrolled part-time. Student to teacher ratio 18:1. 98% of graduates obtain employment within six months.

Courses: Management. Internship required prior to certificate.

Faculty: 2 full- and 17 part-time.

Costs: $12 per unit, books $150 per semester. Admission requirements: High school diploma preferred. Scholarships: yes. Loans: yes.

Contact: D. Suzanne Johnson, Dept. Chair, Chaffey College-Hotel & Food Service Management, 5885 Haven Ave., Alta Loma, CA 91737-3002 US; 909-941-2711, Fax 909-466-2831, E-mail sjohnson@chaffey.cc.ca.us, URL http://www.chaffey.cc.ca.us.

CITY COLLEGE OF SAN FRANCISCO

San Francisco/August-May

This college offers a 4-semester AS degree in Hotel and Restaurant Operations, Award of Achievement and ACFEI certificate. Program started 1935. Accredited by WASC, ACFEI. Calendar: semester. Curriculum: culinary, core. Admission dates: August, January. Total enrollment: 200. 86 each admission period. 90% of applicants accepted. Student to teacher ratio 20:1. 90% of graduates obtain employment within six months. Facilities: 4 kitchens and classrooms, student-run restaurant.

Courses: Elementary and advanced foods, bake shop, advanced pastry, meat analysis, garde manger, and general education. 240-hour externship.

Faculty: 12 full- and 4 part-time.

Costs: Annual tuition in-state $390, out-of-state $3,720. Admission requirements: Age 18, or high school diploma. Scholarships: yes. Loans: yes.

Contact: Lynda Hirose, Advisor/Placement Counselor, City College of San Francisco, Hotel and Restaurant, 50 Phelan Ave., San Francisco, CA 94112 US; 415-239-3152, Fax 415-239-3913, URL http://www.hills.ccsf.cc.ca.us:9878/~hotelrst/index.html. Additional contact: Patrick Wille, Dept. Chair.

COLLEGE OF THE DESERT

Palm Desert/Year-round

This two-year college offers a 20-unit certificate in Basic Culinary Arts, 62-unit AA degree in Culinary Management. Calendar: semester. Curriculum: culinary, core.

Courses: Principles of cooking, baking, pantry, operations management.

Costs: $13 per unit in-state, $140.25 per unit out-of-state.

Contact: Steve Beno, Professor of Culinary Arts, College of the Desert, 43500 Monterey Ave., Palm Desert, CA 92260 US; 760-346-8041, Fax 760-341-8678, URL http://www.desert.cc.ca.us.

COLLEGE OF THE SEQUOIAS

Visalia/August-May

This two-year college offers an 11- to 13-unit certificate in Food Service, 28-32 unit certificate in Food Service Management, 20 unit certificate in Dietetic Service Supervisor. Calendar: semester. Curriculum: culinary. Admission dates: August and January. Total enrollment: 100. 50 each admission period. 100% of applicants accepted. 60% enrolled part-time. Student to teacher ratio 20:1. Facilities: Commercial food lab.

Courses: Commercial food, food service management, nutrition.

Faculty: 3 full- and 4 part-time instructors (R.D.s and chef).

Costs: Enrollment fee is $12 per unit for all students, non-resident tuition is $118 per unit. Loans: yes.

Contact: Barbara Reynolds, Consumer/Family Division Chair, College of the Sequoias, 915 S. Mooney Blvd., Visalia, CA 93277 US; 209-730-3717, Fax 209-730-3894, E-mail barbarar@giant.sequoias.cc.ca.us, URL http://www.sequoias.cc.ca.us.

COLUMBIA COLLEGE

Sonora/August-May

This California community college offers a 2-year AS degree in Culinary Arts/Hospitality Management; 10 culinary certificates. Program started 1977. Accredited by WASC. Calendar: semester. Curriculum: culinary. Admission dates: August, January. Total enrollment: 125. 30 each admission period. 99% of applicants accepted. 40% receive financial aid. 40% enrolled part-time. Student to teacher ratio 10-15:1. 90% of graduates obtain employment within six months. Facilities: 2 kitchens, 4 classrooms, 3-star restaurant.

Courses: Cooking, baking, pastry, restaurant desserts, wines, bartending, garde manger, service, restaurant management. 38 hours of culinary courses required for graduation. Off-campus 3-star restaurant and hotel for 2nd-year students.

Faculty: 2 full-time, 7 part-time. Qualifications: full-time: lifetime teaching credential and industry experience; part-time: A.S. degree + 6 years industry experience.

Costs: Annual tuition in-state $312, out-of-state $3,144. Health and student fees $30 per semester. On-campus housing: 180 spaces; average cost is $300 per month. Average off-campus housing cost $500 per month. Admission requirements: Admission test. Scholarships: yes. Loans: yes.

Location: The 3,000-student campus is 120 miles from San Francisco.

Contact: Gene Womble, Hospitality Management Program Coordinator, Columbia College, Hospitality Management, 11600 Columbia College Dr., Sonora, CA 95370 US; 209-588-5135, Fax 209-588-5316, E-mail ChefGeneColumbiaCollege@juno.com, URL http://www.ccc-infonet.edu/~naomi/columbia/hospmgmt.html.

CONTRA COSTA COLLEGE

San Pablo

This college offers a 2-year certificate. Program started 1962. Admission dates: August, January. Total enrollment: 75-100. 90% of applicants accepted. Student to teacher ratio 20-25:1. 90% of graduates obtain employment within six months.

Faculty: 3 full-time, 1 part-time.

Costs: $13/unit enrollment fee in-state, no tuition; $125 per unit tuition + $13 per unit enrollment fee out-of-state. Admission requirements: Admission test.

Contact: Steve Cohen, Contra Costa College, Culinary Arts, 2600 Mission Bell Dr., San Pablo, CA 94806 US; 510-235-7800.

THE CULINARY INSTITUTE OF AMERICA AT GREYSTONE

(See also pages 80 and 234) (See display ad page 81) **St. Helena/Year-round**

The CIA's Greystone campus, a center for continuing education of food and wine professionals, offers fundamental to advanced 3- to 5-day and 2- to 21-week intensives for foodservice professionals, and a 30-week Baking and Pastry Arts certificate program. Established in 1995. Accredited by Council of Private and Post-Secondary Vocational Education of State of Cal. Curriculum: culinary. Admission dates: On-going. Student to teacher ratio 18:1. Facilities: 15,000-square-foot open teaching kitchens which include Bonnet Cooking Suites and Bongard Hearth ovens, 125-seat Ecolab Theatre, on-campus vineyards and gardens.

Courses: Professional cooking, professional baking and pastry, garde manger, skill development, baking and pastry arts certificate program, wine, career discovery, special subjects, corporate programs. Other than certificate program, Greystone is exclusively devoted to continuing education.

Faculty: 13 full- and part-time instructors, visiting instructors include chef/owners of fine restaurants, also drawn from the 125 instructors at the Hyde Park, NY, campus.

Costs: Tuition ranges from $695 for 30 hours of instruction to $15,400 for the Baking and Pastry Arts certificate program. 18-room guest house on campus. Rooms available by the week. Admission requirements: For cooking production classes, a minimum of 6 months experience in a professional kitchen.

Location: One-and-a-half hours north of San Francisco, in the Napa Valley.

Contact: Holly Briwa, Education Office Manager, The Culinary Institute of America, Education Dept., 2555 Main St., St. Helena, CA 94574 US; 800-333-9242/707-967-0600, Fax 707-967-2410, URL http://www.ciachef.edu.

CYPRESS COLLEGE

Cypress/Year-round

This college offers a 1-year certificate, 2-year AS degree in Food Service Management, Hotel Operations, and Culinary Arts. Program started 1975. Accredited by WASC. Calendar: semester. Curriculum: culinary, core. Admission dates: August, January. Total enrollment: 140. 60 each admission period. 90% of applicants accepted. 45% receive financial aid. 55% enrolled part-time. Student to teacher ratio 16:1. 85% of graduates obtain employment within six months. Facilities: 1 kitchen, 4 classrooms, student-run dining room.

Courses: Basic food production, advanced cooking techniques, quantity food production, international gourmet foods, dining room service, food and beverage costing, kitchen management, menu planning and design, kitchen planning design, baking and pastry, pantry skills. 255-hour salaried externship.

Faculty: 3 full-time, 5 part-time.

Costs: Tuition in-state $13 per unit, out-of-state $114 per unit. Lab fees: $5 per lab. Average off-campus housing cost $250-$750 per month. Admission requirements: High school diploma or equivalent. Scholarships: yes. Loans: yes.

Location: The 16,200-student campus is 35 minutes from downtown Los Angeles.

Contact: David Schweiger, Department Coordinator, Cypress College, Hospitality Management/Culinary Arts, 9200 Valley View, Cypress, CA 90630 US; 714-826-2220 x208, Fax 714-527-8238, URL http://www.cypress.cc.ca.us.

DIABLO VALLEY COLLEGE

Pleasant Hill/Year-round

This college offers a program in Culinary Arts, Baking and Patisserie, Restaurant Management and Hotel Administration. Program started 1971. Accredited by ACFEI, WASC. Calendar: semester. Curriculum: core. Admission dates: August, January. Total enrollment: 750. 50 each admission period. 100% enrolled part-time. Student to teacher ratio 24:1. 100% of graduates obtain employment within six months. Facilities: Include a fully-equipped food production kitchen, demonstration laboratory, 130-seat open-to-the-public restaurant.

Courses: Advanced food preparation, catering, garde manger, menu planning, costing, nutrition, California Cuisine and baking. One semester externship at local hotels and restaurants.

Faculty: 5 full-time, 14 part-time. Qualifications: BA degrees and 7 years industry experience. Includes Jack Hendrickson, Dept. Chair Linda Sullivan, Nader Sharkes, Robert Eustes, Paul Bernhardt.

Costs: In-state tuition for first semester is $225. Fees and deposits: $13 per unit for residents, $127 per unit for non-residents, $135 per unit for international students. Average off-campus housing cost is $500 per month. Admission requirements: High school diploma or equivalent. Scholarships: yes. Loans: yes.

Location: The 23,000-student 100-acre suburban campus is off Contra Costa Blvd. and the 680 Freeway.

Contact: Jack Hendrickson, Department Chair, Diablo Valley College, Hotel and Restaurant Management Dept., 321 Golf Club Rd., Pleasant Hill, CA 94523 US; 925-685-1230 x555, Fax 925-825-8412, URL http://www.dvc.edu.

EPICUREAN SCHOOL OF CULINARY ARTS

Los Angeles/Year-round *(See also page 188)*

This private school offers a 9-month Professional Chef certificate program (Pro Chef I and Pro Chef II). Established in 1985. Admission dates: Year-round. Total enrollment: 15. 50 each admission period. 100% of applicants accepted. 100% enrolled part-time. Student to teacher ratio 15:1. Facilities: Teaching kitchen with 5 work stations.

Courses: Classic French and contemporary cuisines, breads and pastries, a variety of other topics and specialties.

Faculty: 4 part-time instructors are CIA and CCA graduates. Includes Karen Umland and Carol Cotner.

Costs: $2,800 for Pro Chef I and II.

Contact: Epicurean School of Culinary Arts, 8759 Melrose Ave., Los Angeles, CA 90069 US; 310-659-5990, Fax 310-659-0302, E-mail epicureans@aol.com.

GLENDALE COMMUNITY COLLEGE

Glendale/August-May

This college offers 2-year certificates in Culinary Arts, Restaurant Management, Hotel Management, and Dietary Services. Program started 1974. Accredited by State. Calendar: semester. Curriculum: culinary. Admission dates: August. Total enrollment: 338. 95% of applicants accepted. 50% enrolled part-time. Student to teacher ratio 35:1. 80-85% of graduates obtain employment within six months.

Courses: Restaurant and cost control management, quantity foods and purchasing. Other required courses: wine and beverages, catering, baking, dining room service, international cooking, and nutrition. Externships.

Faculty: 1 full-time, 5 part-time. Qualifications: BS or MS degree, at least 6 years experience.

Costs: Annual tuition in-state $12 per unit, out-of-state $138 per unit. Admission requirements: High school diploma or equivalent.

Contact: Yeimei Wang, Prof. of Food & Nutrition and Coordinator, Glendale Community College, Culinary Arts Dept., 1500 N. Verdugo Rd., Glendale, CA 91208 US; 818-240-1000 x5597, Fax 818-549-9436, URL http://www.glendale.cc.ca.us.

GROSSMONT COLLEGE

El Cajon/Year-round

This two-year college Regional Occupational Program offers a 1-year certificate in Culinary Arts, 2-year associate degree in Culinary Arts. Program started 1988. Calendar: semester. Curriculum: culinary, core. Admission dates: August and January. Total enrollment: 150. 150 each admission period. 100% of applicants accepted. 35% receive financial aid. 40% enrolled part-time. Student to teacher ratio 25:1. 95% of graduates obtain jobs within six months. Facilities: Classroom, lab, cafeteria.

Courses: Basic knowledge, health/nutrition, professional demeanor and abilities. 1-semester externship with certificate, 2-semester certificate with associate degree.

Faculty: 9 instructors: 3 executive chefs (2 Culinary Olympic Gold Medal Winners/teammembers), 2 professors (master's degree), 4 instructors (master's degree).

Costs: $12 per unit in-state $180 per semester, $121 per unit for out-of-state $1,815 per semester. Admission requirements: Ability to read and write. Loans: yes.

Location: 15 minutes east of San Diego.

Contact: Cathie Robertson, Professor, Coordinator of Culinary Arts Program, Grossmont College, 8800 Grossmont College Dr., El Cajon, CA 92020 US; 619-644-7327/7550, Fax 619-644-7190, E-mail croberts@mail.gcccd.cc.ca.us.

JUDY PECK PRINDLE "THE ART OF FOOD STYLING" SEMINARS

Los Angeles/October

Food stylist Judy Peck Prindle offers a 3-day weekend seminar that covers the role of the food stylist in TV and skills and techniques of preparing food for the camera. Total enrollment: 30.

Faculty: Judy Peck Prindle has worked for Kraft, General Foods, Pillsbury, General Mills, and over 20 other food companies.

Costs: $900 includes workbook, breakfasts, lunches.

Location: Ms. Prindle's photography studio.

Contact: Judy Peck Prindle, Judy Peck Prindle "The Art of Food Styling" Seminars, 106 N. Mansfield Ave., Los Angeles, CA 90036 US; 213-939-7009, Fax 213-939-4219.

LANEY COLLEGE

Oakland/Year-round

This community college offers a 2-year AA degree in Culinary Arts, 2-year certificate in Retail Baking. Program started 1948. Accredited by WASC. Admission dates: August, January. Total enrollment: 200. 60 each admission period. 80% of applicants accepted. 70% receive financial aid. 10% enrolled part-time. Student to teacher ratio 12:1. 100% of graduates obtain employment within six months. Facilities: Include 7 kitchens and classrooms, a student-run restaurant, retail bakery.

Faculty: 5 full-time, 4 part-time.

Costs: Tuition in-state $12 per unit, out-of-state $138 per unit.

Contact: Wayne Stoker, Culinary Arts Co-Dept. Chair, Laney College, Culinary Arts Dept., 900 Fallon St., Oakland, CA 94607 US; 510-464-3407, Fax 510-464-3240, URL http://laney.peralta.cc.ca.us. Additional contact: Cleo Ross.

LOS ANGELES CULINARY INSTITUTE, INC.

(See display ad page 16)

Encino/Year-round

This private school offers a 12-month accelerated diploma program in Culinary Arts, 3-month certificate courses in Cooking, Baking, and Pastry. Established in 1991. Calendar: semester. Curriculum: culinary. Admission dates: January, February, April, May, July, August, October, November. Total enrollment: 60-80. 12 each admission period. 90% of applicants accepted. Student to teacher ratio 12:1. 95-98% of graduates obtain employment within six months. Facilities: Up-to-date, well-equipped kitchens and an open-to-the-public restaurant.

Courses: 48 weeks of instruction and 6-week internship. Combines European-style apprenticeship with academic studies.

Faculty: 10 full-time. Includes Chef Raimund Hofmeister, Certified Master Chef.

Costs: $16,995, which includes tools, uniforms, books. $25 nonrefundable application fee, $100 enrollment fee. Admission requirements: High school diploma or equivalent. Scholarships: yes.

Contact: Howard Broockner, Director of Admissions, Los Angeles Culinary Institute, Inc., 17401 Ventura Blvd., Encino, CA 91316 US; 888-343-5224, Fax 818-380-1139, URL http://www.laci.com.

LOS ANGELES MISSION COLLEGE

Sylmar/Year-round

This two-year college offers a 2-year/60-64 unit AAS degree in Culinary Arts. Calendar: semester.

Costs: $13 per unit in-state, $128 per unit out-of-state.

Contact: Eloise Cantrell, Los Angeles Mission College, 13356 Eldridge Ave., Sylmar, CA 91342 US; 818-364-7600 x7611, Fax 818-364-7755, URL http://www.lamission.cc.ca.us.

LOS ANGELES TRADE-TECHNICAL COLLEGE

Los Angeles/September-May

This college offers a 1-year certificate, 2-year AA degree in Culinary Arts and Professional Baking. Program started 1941. Accredited by WASC. Calendar: semester. Curriculum: culinary, core. Admission dates: August, January. Total enrollment: 250. 75 each admission period. 75% of applicants accepted. 40% receive financial aid. Student to teacher ratio 25:1. 80% of graduates obtain employment within six months. Facilities: Include 2 kitchens, 6 classrooms.

Courses: Chef training, professional baking.

Faculty: 8 full-time. Qualifications: AA degree, ACF certification, industry experience.

Costs: Annual tuition in-state $150+, out-of-state $250+. $600 for tools, uniforms, books. Scholarships: yes. Loans: yes.

Contact: Carole Lung, Assoc. Dean, Los Angeles Trade-Technical College, Culinary Arts Dept., 400 W. Washington Blvd., Los Angeles, CA 90015; 213-744-9480, Fax 213-748-7334, lungck@laccd.cc.ca.us.

NAPA VALLEY COOKING SCHOOL

St. Helena/Year-round *(See also page 191) (See display ad page 17)*

Napa Valley College offers a 14-month certificate (Professional Training for Fine Restaurants) that consists of 9 months of school and 5 months of externship. Established in 1996. Curriculum: culinary. Admission dates: August. Total enrollment: 14-16. 14-16 each admission period. 75% of applicants accepted. 50% receive financial aid. Student to teacher ratio 14:1. 100% of graduates obtain employment within six months. Facilities: Modern, light teaching kitchen.

Courses: Basic to advanced techniques, pastry and baking, and food and wine education. Special emphasis on skills for entry and advancement in fine restaurants. Salaried externship in a Napa Valley restaurant the final semester.

Faculty: Northern California chef-instructors. Guest lecturers include area chefs, growers, specialty food producers, viticulturists, and winemakers. Includes Executive Chef, George Torassa, ACF, and Chef-Instructor Eric Lee.

Costs: Approximately $12,000. Average off-campus housing cost is $500+ per month. Admission requirements: High school diploma or equivalent & industry experience recommended. Loans: yes.

Location: Napa Valley, 75 minutes from San Francisco.

Contact: Eric Lee, Coordinator, Napa Valley Cooking School, 1088 College Ave., St. Helena, CA 94574; 707-967-2900 x2930, Fax 707-967-2909, URL http://www.napacomed.org/cookingschool.

ORANGE COAST COLLEGE

Costa Mesa/August-May

This college offers a 1-year certificate, 2-year AA degree. Program started 1964. Accredited by WASC, ACFEI. Calendar: semester. Curriculum: culinary. Admission dates: January, August. Total enrollment: 350. 100-125 each admission period. 100% of applicants accepted. 40% enrolled part-time. Student to teacher ratio 15:1. 100% of graduates obtain employment within six months. Facilities: Full-service cafeteria (seats 300), 80-seat restaurant, full-service bakery.

Faculty: 15 full-time.

Costs: Tuition in-state $120 per year, out-of-state $102 per unit. Admission requirements: High school diploma or equivalent. Scholarships: yes. Loans: yes.

Location: 25,000-student suburban campus in southern California.

Contact: Bill Barber, Program Coordinator, Orange Coast College, Hospitality Dept., 2701 Fairview Blvd., Box 5005, Costa Mesa, CA 92628-5005 US; 714-432-5835, Fax 714-432-5609.

OXNARD COLLEGE

Oxnard/August-May

This college offers a 1-year certificate, 2-year degree. Program started 1985. Accredited by WASC. Admission dates: August, January. Total enrollment: 75-125. 30 each admission period. 100% of applicants accepted. Student to teacher ratio 20:1. 95% of graduates obtain jobs within six months.

Faculty: 1 full-time, 5 part-time.

Costs: Tuition in-state $12 per unit, out-of-state $117 per unit.

Contact: Frank Haywood, Oxnard College, Hotel & Restaurant Management, 4000 S. Rose Ave., Oxnard, CA 93033 US; 805-986-5869, Fax 805-986-5865, URL http://www.oxnard.cc.ca.us.

RICHARDSON RESEARCHES, INC.

Hayward/March-April, June-July, October

This product development and research company for the confectionery food industry offers 8 to 10 five-day certificate courses per year. Established in 1976. 18 each admission period. Student to teacher ratio 9:1. Facilities: The company's 4,000-square-foot professional facility, which has 3 kitchen areas, a lecture room, and a specially-equipped laboratory.

Courses: Theoretical and practical aspects of confectionery and chocolate technology: Chocolate Technology, Confectionery Technology, Continental Chocolates, and Lite/Sugar-Free/Reduced Calorie and No Sugar Added.

Faculty: Terry Richardson, a graduate of the London Borough Polytechnic in Confectionery and Chocolate Technology, worked for major companies and holds several patents for new products and processes. Margaret Knight has a BS in Food Science.

Costs: Tuition ranges from $1,395-$1,495.

Location: 30 minutes from the San Francisco Airport and 20 minutes from the Oakland Airport.

Contact: Richardson Researches, Inc., 23449 Foley St., Hayward, CA 94545 US; 510-785-1350, Fax 510-785-6857, E-mail info@richres.com, URL http://www.richres.com.

SAN FRANCISCO STATE UNIVERSITY (SFSU)

San Francisco/Year-round

This state university continuing education department offers a 6-week professional cooking series, wine and hospitality courses. Program started 1997. Curriculum: culinary. Facilities: SFSU's food laboratories and classrooms.

Courses: Professional cooking and baking.

Faculty: Chef Christopher Sheldon, executive chef of SFSU's Vista Room, a graduate of the California Culinary Academy with 15+ years professional experience.

Costs: $870 plus food fee.

Contact: Mary Pieratt, Program Director, San Francisco State University College of Extended Learning, 1600 Holloway Ave., San Francisco, CA 94132 US; 415-338-1533, E-mail maryp@sfsu.edu, URL http://www.cel.sfsu.edu.

SAN JOAQUIN DELTA COLLEGE

Stockton/August-May

This college offers a 1-semester certificate, 3-4 semester certificate in Basic and Advanced Culinary Arts, 2-semester certificate in Dietetic Services Supervisor. Program started 1979. Accredited by WASC. Calendar: semester. Admission dates: Rolling. Total enrollment: 60. 20-30 each admission period. 100% of applicants accepted. 40% receive financial aid. 40% enrolled part-time. Student to teacher ratio 15:1. 90% of graduates obtain employment within six months. Facilities: 2 kitchens, 2 classrooms and a student-run restaurant.

Courses: Culinary arts, baking, restaurant operations, nutrition, menu planning, food purchasing, and catering. 1-semester salaried externship.

Faculty: 2 full-time, 3 part-time. Qualifications: Master's degree. Includes Char Britto, RD, FADA, and John Britto, CEC.

Costs: Tuition in-state $13 per unit, out-of-state $115 per unit. Students supply equipment, supplies. Housing is $400 per month. Admission requirements: High school graduate or age 18. Scholarships: yes.

Location: North of San Francisco.

Contact: Constance Smith, Division Chairperson, San Joaquin Delta College, Culinary Arts Dept., 5151 Pacific Ave., Stockton, CA 95207 US; 209-954-5516, Fax 209-954-5600, E-mail hhill@sjdccd.cc.ca.us, URL http://www.sjdccd.cc.ca.us/fchs/culinaryarts/c.a.toc.html.

SANTA BARBARA CITY COLLEGE

Santa Barbara/August-May

This college offers a 2-year certificate, 2-year AS degree in Culinary Arts and Restaurant-Hotel Management. Program started 1970. Accredited by WASC, ACFEI. Calendar: semester. Curriculum: culinary. Admission dates: Fall, spring. Total enrollment: 120. 50 each admission period. 80% of applicants accepted. 60% receive financial aid. 0% enrolled part-time. Student to teacher ratio 10-15:1. 100% of graduates obtain employment within six months. Facilities: Include 6 kitchens and classrooms, a gourmet dining room, coffee shop, bake shop, lecture/lab room, cafeteria, and snack shop.

Courses: International cuisine, wines, bar management, production service, nutrition, meat analysis, garde manger, baking, and restaurant ownership. Paid internships.

Faculty: 3 full-time, 6 part-time, and 11 lab teaching assistants.

Costs: Annual tuition in-state $500, out-of-state $3,360. Average off-campus housing cost $350 per month. Admission requirements: High school diploma or equivalent. Scholarships: yes. Loans: yes.

Location: The 12,000-student suburban campus is 90 miles from Los Angeles.

Contact: John Dunn, Chairperson, Santa Barbara City College, Hotel/Restaurant & Culinary Dept., 721 Cliff Dr., Santa Barbara, CA 93109-2394 US; 805-965-0581 x2457, Fax 805-963-7222.

SANTA ROSA JUNIOR COLLEGE

Santa Rosa

This college offers a 1-year certificate. Accredited by WASC. Total enrollment: 25-40. Student to teacher ratio 100. 100% of graduates obtain employment within six months.

Faculty: 3 full-time, 10 part-time.

Costs: Tuition in-state $12 per unit, out-of-state $121 per unit + $12 per unit enrollment fee.

Contact: Harriett Lewis, Santa Rosa Junior College, Consumer & Family Studies Dept., 1501 Mendocino Ave., Santa Rosa, CA 95401 US; 707-527-4395.

SHASTA COLLEGE

Redding

This college offers a 1-year certificate, 2-year AA program. Program started 1975. Total enrollment: 125. Student to teacher ratio 25:1. 100% of graduates obtain employment within six months.

Faculty: 1 instructor.

Costs: Annual tuition in-state $12 per unit, out-of-state $116 per unit. Admission requirements: High school diploma or equivalent and admission test.

Contact: Kathleen Kistler, VP, Academic and Student Affairs, Shasta College, Culinary Arts, 1155 N. Old Oregon Tr., PO Box 496006, Redding, CA 96049-6006 US; 530-225-4600, Fax 530-225-4841, E-mail kistler@shasta.cc.ca.us, URL http://www.shasta.cc.ca.us.

SOUTHERN CALIFORNIA SCHOOL OF CULINARY ARTS

South Pasadena/Year-round *(See also page 193) (See display ad page 19)*

This private school offers a 15-month full-time Professional Culinary Arts diploma, and 15-week Advanced Professional Cooking and Advanced Professional Baking diplomas. Established in 1994. Accredited by ACICS. Calendar: trimester. Curriculum: culinary. Admission dates: Fall, spring, summer. Total enrollment: 140. 54 each admission period. 90% of applicants accepted. 94% of graduates obtain employment within six months. Facilities: 5 kitchens for garde manger, baking/pastry, hot food production; learning resource center, classrooms.

Courses: Include Introduction to Culinary Arts, Garde Manger, Baking and Pastry, Hot Food Production, Sanitation, Nutrition, Supervision, Purchasing and Food Service Principles. Externship assisted. Continuing education: classes available year round.

Faculty: 6 full-time. Includes Director Chef Christopher F. Becker, Executive Chef Instructor Robert Danhi, Head Pastry Chef Instructor Leslie Bilderback.

Costs: Tuition is $23,332. Other purchases and fees $1,540. Off-campus lodging averages $625 per month. Admission requirements: High school diploma or equivalent and entrance test required. Scholarships: yes. Loans: yes.

Location: The 12,000-square-foot school facility is in an historic suburb, 12 minutes from downtown Los Angeles.

Contact: Pamela Ramirez, Admissions Director, Southern California School of Culinary Arts, 1420 El Centro St., S. Pasadena, CA 91030 US; 888-900-CHEF, Fax 626-403-8494, E-mail scsca@earthlink.net, URL http://www.scsca.com.

TANTE MARIE'S COOKING SCHOOL

(See also page 194) (See display ad above) **San Francisco/Year-round**

This small private school offers 6-month certificate programs in culinary arts (full-time) and pastry (part-time), nonvocational evening and weekend courses, culinary travel programs, cooking vacations, cooking parties. Established in 1979. Accredited by IACP. Curriculum: culinary. Admission dates: April, September. Total enrollment: 24-28. 12-14 each admission period. 95% of applicants accepted. 50% enrolled part-time. Student to teacher ratio 12:1. 95% of graduates obtain employment within six months.

Courses: Culinary courses cover basic French techniques, breads and pastries, desserts, ethnic cuisines, food purchasing and handling, menu planning, and taste refinement. 4-week internship in a local quality restaurant, bakery, or hotel.

Faculty: Founder Mary Risley studied at Le Cordon Bleu and La Varenne; Catherine Pantsios, former chef/owner of Zola's; Cathy Burgett, former pastry chef of Compton Place Hotel.

Costs: Tuition is $14,500 for the 6-month culinary certificate course, $5,800 for the 6-month part-time pastry course. Local apartments are available. Admission requirements: High school diploma. Scholarships: yes. Loans: yes.

Location: On San Francisco's Telegraph Hill, near Fisherman's Wharf and public transportation.

Contact: Peggy Lynch, Administrator, Tante Marie's Cooking School, 271 Francisco St., San Francisco, CA 94133 US; 415-788-6699, Fax 415-788-8924, URL http://www.tantemarie.com.

UCLA EXTENSION, HOSPITALITY/FOODSERVICE MANAGEMENT

Los Angeles/Year-round *(See also page 195)*

This self-supported continuing higher education institution offers sequenced certificate courses in Cooking, Pastry and Baking, Catering. Offered in the evening, the sequences take ~2 years to complete and consist of 54-hour 6-unit courses conducted over 12-week quarters. Calendar: quarter. Curriculum: culinary, core. Admission dates: Fall, spring. Total enrollment: Varies. 45 each admission period. 75% of applicants accepted. 20% receive financial aid. Student to teacher ratio 15:1. Facilities: Local commercial kitchens.

Courses: 17-course professional designation in Culinary Arts, 11-course certificate in Cooking, 10-course certificate in Pastry and Baking, 14-course certificate in Catering, certificate in Vintage. Professional Cooking and Catering require 200 hours, Professional Pastry and Baking requires 140 hours. Nonvocational courses are also offered.

Faculty: Local restaurant chefs, culinary specialists, and graduates of the CIA and CCA.

Costs: Professional designation in Culinary Arts $15,800, certificate in Cooking $11,000, certificate in Pastry and Baking $8,900, certificate in Catering $10,600. Certificate enrollment fee $175. Students provide uniforms and tools. Admission requirements: Successful completion of Introduction to Culinary Arts, a 12-wk 6-unit course. Scholarships: yes. Loans: yes.

Contact: Yvette De La Cruz, Administrator Certificate Programs, UCLA Extension, Business Management, 10995 Le Conte Ave., Room 515, Los Angeles, CA 90024-0901 US; 310-206-1578, Fax 310-206-7249, E-mail ydelacruz@unex.ucla.edu, URL http://www.unex.ucla.edu.

WESTLAKE CULINARY INSTITUTE

Westlake Village/Year-round *(See also page 190, Let's Get Cookin')*

This private school offers a 24-session professional series, 6-session baking series, 3-session catering series, certificate granted upon completion. Established in 1984. Curriculum: culinary. Admission dates: Variable. Total enrollment: 36. 12 each admission period. 80% of applicants accepted. 5% receive financial aid. 100% enrolled part-time. Student to teacher ratio 12:1. 90% of graduates obtain employment within six months. Facilities: 1,000-square-foot combination demonstration/participation facilities, cookware store.

Courses: Include classical food preparation, skills, methods, techniques and presentation, international cuisines, menu costing and planning.

Faculty: 25-member guest and regular faculty. Includes owner/director, Phyllis Vaccarelli, CCP and Cecilia DeCastro, CCP who coordinates professional programs for UCLA Extension.

Costs: $2,250 for the professional series, $495 for the baking series, $225 for the catering series. Admission requirements: Commitment, written application, basic skills, attitude.

Location: North of Malibu, 30 minutes from Los Angeles.

Contact: Phyllis Vaccarelli, Owner/Director, Let's Get Cookin', 4643 Lakeview Canyon Rd., Westlake Village, CA 91361 US; 818-991-3940, Fax 805-495-2554.

COLORADO

COLORADO INSTITUTE OF ART – SCHOOL OF CULINARY ARTS

Denver/Year-round *(See display ad page 8)*

This private school offers a 21-month AAS degree in Culinary Arts, 39-month BA degree in Culinary Management. Program started 1994. Accredited by ACCSCT. Calendar: quarter. Curriculum: core. Admission dates: January, April, July, October. Total enrollment: 400. 100 each admission period. 98% of applicants accepted. 75% receive financial aid. 15% enrolled part-time.

Student to teacher ratio 20:1. 90% of graduates obtain employment within six months. Facilities: Include 5 kitchens, classrooms, full dining facility.

Courses: Basic skills, baking and pastry, food production, garde manger, a la carte, dining room, sanitation, nutrition, management, cost control, wines and spirits, facilities design, and general education courses. 1-quarter internship required. Continuing education provided.

Faculty: 16 full-time, 14 part-time. Qualifications: professional certification, BA degree, and 20 years experience.

Costs: Quarterly tuition $3,600. Application fee $50, tuition deposit $100. Other fees $250 per quarter lab fee, $600 supply kit. Admission requirements: High school diploma or equivalent and essay. Scholarships: yes. Loans: yes.

Location: 3 miles from downtown Denver.

Contact: Barbara Browning, V.P., Director of Admissions, Colorado Institute of Art, 200 E. 9th Ave., Denver, CO 80203 US; 800-275-2420, Fax 303-860-8520, E-mail cia.adm@www.cia.aii.edu, URL http://www.aii.edu.

COLORADO MOUNTAIN CULINARY INSTITUTE

(See display ad above) **Keystone, Vail/Eagle Valley/Year-round**

Colorado Mountain College with Keystone Resort and five resorts in Vail and Beaver Creek offers a 3-year program: AAS in Culinary Arts, ACF Certificate of Apprenticeship. Established in 1993. Accredited by NCA. Calendar: trimester. Curriculum: culinary, core. Admission dates: June and August (application deadline April and May). Total enrollment: 45-75. 12-15 each admission period. 40% of applicants accepted. 60% receive financial aid. Student to teacher ratio 6-15:1. 100% of graduates obtain employment within six months. Facilities: Include 10 full-service kitchens and classrooms and 25 food and beverage outlets. Semester rotation through the various properties.

Courses: 6,000 hours of structured work experience combined with 850 hours of classroom lecture required for graduation. 6,000-hour ACF apprenticeship.

Faculty: Includes Doug Schwartz CEC, Chris Wing CEC, Chris Ryback CEC, Bob Burden CEC, John Goldfarb, CEC, Larry Pirner, Alisa Mathews, Julie Anne Lichtiege.

Costs: Annual tuition in-district $1,300-$1,668, in-state $2,340, out-of-state $4,350-$4,500 (tuition assistance for out-of-state students), fees $750. On-campus housing $280 per month, off-campus housing $350-$450 per month. Admission requirements: Personal interview or conference call. Scholarships: yes.

Location: Includes Keystone Resort, The Lodge & Spa at Cordillera, Sonnenalp Resort of Vail, Beaver Creek Lodege, Vail Cascade Club, Chateau Vail.

Contact: Doug Schwartz or John Goldfarb, Directors of Culinary Education, Colorado Mountain Culinary Institute, Admissions, PO Box 10,001, Glenwood Springs, CO 81602 US; 800-621-8559, Fax 970-947-8385, E-mail JoinUs@coloradomtn.edu, URL http://www.coloradomtn.edu.

COOK STREET SCHOOL OF ADVANCED CULINARY ARTS

Denver/Year-round

This private school offers a 20-week full-time accelerated culinary certificate program including 4 weeks of study in France and Italy. Program started 1999. Calendar: trimester. Curriculum: culinary. Admission dates: February, June, October. Total enrollment: 72. 24 each admission period. Student to teacher ratio 6:1. Facilities: Teaching kitchen with the latest equipment.

Courses: A technical, educational and cultural culinary foundation, with emphasis on classical French techniques. 4 weeks of study in France and Italy.

Faculty: Instructors with teaching and industry experience in hotels, restaurants and pastry kitchens.

Costs: $16,000 plus clothing, books, knives. Admission requirements: High school diploma or GED, application, essay, entrance exam, interview.

Location: Historic lower downtown Denver, two blocks from Coors Field, an area known for restaurants and night life.

Contact: Morey Hecox, Manager, Cook Street School of Advanced Culinary Arts, 123 Cook St., #201, Denver, CO 80206 US; 303-377-9886, Fax 303-377-7499, URL http://www.cookstreet.com.

COOKING SCHOOL OF THE ROCKIES

Boulder/Year-round *(See also page 197) (See display ad above)*

This private school offers a 6-month Professional Culinary Arts diploma program that consists of 5 months of training in Boulder, the sixth month in Carpentas, Provence, France. Also a 10-month evening program. Established in 1996. Accredited by Dept. of Higher Education of the State of Colorado. Calendar: semester. Curriculum: culinary. Admission dates: January, July. Total enrollment: 24. 12 each admission period. Student to teacher ratio 4:1. 90% of graduates obtain employment within six months. Facilities: Modern, fully-equipped professional kitchen.

Courses: Emphasis on classic, modern, and regional French and Italian cuisine. Prepare meals/cook daily. Tasting/palette development, including wine knowledge. Exploration of culinary careers. Part-time externships available in local and area restaurants and catering establishments.

Faculty: Chef James Moore, executive chef of the American Lamb Council; Chef Andrew Floyd, former chef de partie at the Occidental Grill in Washington, DC; Chef Diana Hoguet, formerly pastry chef at the Quilted Giraffe and 21 Club in NYC.

Costs: $18,500 tuition for day program (includes airfare and room and board in France), $14,000 for evening program (does not include France). Admission requirements: High school diploma or GED, application, essay, entrance exam, interview.

Location: Boulder, 25 miles northwest of Denver; Carpentas, 20 miles west of Avignon in the South of France.

Contact: Joan Brett, Director, Cooking School of the Rockies, Professional Culinary Arts Program, 637 S. Broadway, Ste. H, Boulder, CO 80303 US; 303-494-7988, Fax 303-494-7999, E-mail csrockies@aol.com, URL http://www.cookingschoolrockies.com.

CULINARY INSTITUTE OF COLORADO SPRINGS

Colorado Springs/Year-round

This division of Pikes Peak Community College offers a certificate in Culinary Arts, 2-year AAS degree in Culinary Arts/Food Management. Established in 1986. Accredited by NCA. Calendar: semester. Curriculum: core. Admission dates: August, January, June. Total enrollment: 35. 10-15 each admission period. 100% of applicants accepted. 90% receive financial aid. 0% enrolled part-time. Student to teacher ratio 10:1. 100% of graduates obtain employment within six months. Facilities: Include kitchen, student-run cafeteria and catering service/classroom.

Courses: Food preparation, restaurant management, wine and spirits, food and beverage management, and sanitation.Other required courses: computer, English, math, accounting. Continuing education: language classes available.

Faculty: 3 full-time, 1 part-time. Qualifications: ACFEI certification. Includes George J. Bissonnette, CCE, CEC, Dept. Chair; Robert Hudson, CSC.

Costs: Annual tuition in-state $4,000, out-of-state $4,800. Average off-campus housing cost $250 and up. Admission requirements: High school diploma or GED and admission test. Scholarships: yes. Loans: yes.

Location: The 15,000-student campus is in an urban setting 2 miles from Colorado Springs.

Contact: George Bissonnette CCE, CEC, Dept. Chair, Culinary Institute of Colorado Springs, TISO Div., PPCC, 5675 S. Academy Blvd., Colorado Springs, CO 80906 US; 719-540-7371, Fax 719-540-7453, E-mail Bissogj@ppcc.cccoes.edu, URL http://www.ppcc.cccoes.edu.

JOHNSON & WALES UNIVERSITY AT VAIL

Vail/Year-round

This private nonprofit career institution offers a 1-year accelerated AAS degree in Culinary Arts for those with a bachelor's degree or higher. Established in 1993. Accredited by NEASC, ACICS-CCA. Curriculum: culinary. Admission dates: June. Total enrollment: 46. 40 each admission period. 94% of applicants accepted. 90% receive financial aid. Student to teacher ratio 20:1. 98% of graduates obtain employment within six months. Facilities: 5,000-square-foot of kitchen and restaurant space, 2 additional restaurants.

Courses: Laboratory courses, professional development, menu planning, nutrition, sanitation, cost control, garde manger, advanced patisserie/dessert, advanced dining room procedures, international cuisine and classical French cuisine. Paid co-ops at Vail Valley restaurants.

Faculty: 8 full/part-time. Includes Carl Calvert, BS, Tim Cameron, CEC, MA, Karl Guggenmos, BS, GMC, David Hendrickson, AOS, CWC, CCE, Michael Koons, CEC, CCE, Todd Rymer, MS, Paul Ferzacca, AOS.

Costs: Tuition $19,354, $477 general fee. Admission requirements: Bachelor's degree from an accredited institution. Request for waiver may be directed to the Director of Culinary Admissions in Norfolk, VA. Scholarships: yes. Loans: yes.

Contact: Todd Rymer, Director, Johnson & Wales University at Vail, 616 W. Lionshead Circle, Vail, CO 81657 US; 970-476-2993, Fax 970-476-2994, E-mail admissions@jwu.edu, URL http://www.jwu.edu. Additional contact: Director of Admissions, 2428 Almeda Ave., #316, Norfolk, VA 23513; 800-277-CHEF.

PUEBLO COMMUNITY COLLEGE

Pueblo/Year-round

This two-year college offers 25 credit-hour certificate and 67 credit-hour AAS degree programs in Culinary Arts. AAS degree students choose either management or production track. Calendar: semester. Curriculum: core. 95% of graduates obtain employment within six months. Facilities: Kitchen and dining areas of the college.

Courses: Food preparation, cost controls, purchasing, customer service, merchandising, production techniques, nutrition.

Faculty: Department Chair is Carol Himes, M.Ed.

Costs: Tuition is $73 per credit-hour in-state, $263 per credit-hour out-of-state.

Contact: Carol Himes, M.Ed., Dept. Chair, Pueblo Community College, 900 West Orman Ave., Pueblo, CO 81004-1499 US; 719-549-3200, Fax 719-543-7566, URL http://www.pcc.cccoes.edu.

SCHOOL OF NATURAL COOKERY

Boulder/Year-round *(See also page 199)*

This private trade school specializing in vegetarian cuisine offers Personal Chef Training: 13 weeks includes 5 weeks fundamentals of cooking, baking and pastry, repertoire and training in home kitchens; Teacher training: 2 weeks plus practice teaching, requirements of 5-week fundamentals. Established in 1991. Calendar: semester. Curriculum: culinary. Total enrollment: 32 per year. 4-16 each admission period. 95% of applicants accepted. Student to teacher ratio 4:1. 98% of graduates obtain employment within six months. Facilities: Teaching kitchen.

Courses: Fundamentals covers technique, grains, food energetics, intuitive cooking, meal composition; Teacher Training includes practice teaching; Personal Chef Training covers marketing and management; Baking and Pastry covers sugar, wheat, and dairy alternatives. Foodservice externship for graduates, some animal food cooking. Continuing education: 1-day and 1-week seminars for professional chefs.

Faculty: 3-4 instructors.

Costs: Personal Chef Training $6,980 (includes fundamentals $3,500, baking $1,430, repertoire $1,250, client work $800). Teacher Training $4,840 (includes fundamentals $3,500, teacher training $1,340). Deposit $500. Admission requirements: High school grad or equivalent or written essay & personal interview. Students must qualify for personal chef & teacher training.

Location: 35 miles from Denver.

Contact: Joanne Saltzman, Director, School of Natural Cookery, PO Box 19466, Boulder, CO 80308 US; 303-444-8068, Fax On request, E-mail snc@sprynet.com.

WARREN OCCUPATIONAL TECHNICAL CENTER

Golden/August-May

This public institution offers a 1-semester (options for 2nd and 3rd semesters) certificate in Restaurant Arts. Program started 1974. Accredited by State, NCA. Admission dates: August, January. Total enrollment: 60. 45 each admission period. 98% of applicants accepted. 2% receive financial aid. 25% enrolled part-time. Student to teacher ratio 20:1. 95% of graduates obtain employment within six months. Facilities: Kitchen, 60-student classroom, 2 dining rooms and a restaurant.

Courses: Production, nutrition, baking, safety and sanitation. Externship provided.

Faculty: 3 full-time with master's degree in vocational education.

Costs: Tuition in-state $1,600 per semester, out-of-state $2,464 per semester. Parking fee $50, materials $10. Off-campus housing cost is $350 per month. Admission requirements: High school diploma or GED.

Location: The 1,200-student campus is in a suburban setting.

Contact: Sharron K. Pizzuto, C.F.E., Warren Occupational Technical Center, 13300 W. 2nd Pl., Lakewood, CO 80228-1256 US; 303-982-8555, Fax 303-982-8547, URL http://jeffco.k12.co.us.

CONNECTICUT

CONNECTICUT CULINARY INSTITUTE

Farmington/Year-round

This private occupational school offers a 52-credit Advanced Culinary Arts program (12 months/1,390 hours), 18-credit Professional Training program (3 months/360 hours), 10 credit Professional Pastry and Baking program (2-1/2 months/250 hours). Part-time and evenings also available. Established in 1988. Accredited by ACCSCT, State Dept. of Higher Education. Calendar: semester. Curriculum: culinary. Admission dates: Continuous enrollment. Total enrollment: 200+. 11 each admission period. 42% enrolled part-time. Student to teacher ratio 11:1. 92% of graduates obtain employment within six months. Facilities: Multiple kitchens, all newly renovated. Restaurant labs open-to-the-public, fine dining venue.

Courses: Hands-on classes and small class size. Students practice techniques in classroom kitchens, then in school-run restaurants, then on a 4-month paid externship. Advanced Culinary Arts Program includes a 4-month, 500-hour paid externship. Continuing education courses include sanitary food handling and low-cholesterol cooking.

Faculty: Includes 16 full-time chefs, 25 adjunct. Average industry experience 10+ years with awards and honors.

Costs: Tuition: Advanced Culinary Arts Program $13,900, Professional Training Program $5,900, Professional Pastry and Baking Program $4,100. Application/registration fee $125, materials/supplies $498-$675. Admission requirements: High school diploma or equivalent, personal interviews. Loans: yes.

Location: Farmington.

Contact: Admissions Department, Connecticut Culinary Institute, Loehmann's Plaza, 230 Farmington Ave., Farmington, CT 06032 US; 860-677-7869, Fax 860-676-0679. E-mail ct.culinary.inst@snet.net, URL http://www.ctculinary.com.

GATEWAY COMMUNITY-TECHNICAL COLLEGE

New Haven/Year-round

This college offers a 1-year certificate in Culinary Arts, 2-year degree in Food Service Management. Program started 1987. Accredited by State, NEAS. Calendar: semester. Curriculum: culinary. Admission dates: September, January. Total enrollment: 140. 20-30 each admission period. 100% of applicants accepted. 40% receive financial aid. 60% enrolled part-time. Student to teacher ratio 15:1. 100% of graduates obtain employment within six months. Facilities: Include 1 lab, many classrooms, restaurant.

Courses: Externship provided.

Faculty: 2 full-time, 4 part-time.

Costs: Annual tuition in-state $1,690, out-of-state $5,106. Average off-campus housing $500 per month. Admission requirements: High school diploma or equivalent and admission test. Loans: yes.

Contact: Stephen Fries, Director, Gateway Community-Technical College, Hospitality Management, 60 Sargent Dr., New Haven, CT 06511 US; 203-789-7067, Fax 203-777-8637.

INSTITUTE OF GASTRONOMY & CULINARY ARTS

(See display ad above)

West Haven/Year-round

The University of New Haven offers a 15-week (one day per week) certificate program plus 6 college credits. Program started 1997. Accredited by NEASC and state. Calendar: semester. Curriculum: culinary. Admission dates: January, May, September. Total enrollment: 24. 24 each admission period. 100% of applicants accepted. 100% enrolled part-time. Student to teacher ratio 8:1. 100% of graduates obtain employment within six months. Facilities: Cooking lab plus dining room in School of Hotel/Restaurant and Dietetics.

Courses: French cooking techniques with emphasis on health and sanitation principles.

Faculty: Director is Patrick Boisjot, chef and past director of the French Institute.

Costs: $3,850. Admission requirements: Commitment to fine cooking.

Location: Southeast Connecticut shore, accessible from Interstate 95.

Contact: Patrick Boisjot, Director, University of New Haven, 300 Orange Ave., West Haven, CT 06516 US; 800-DIA-LUNH, Fax 203-932-7083, E-mail bubbles@charger.newhaven.edu.

MANCHESTER COMMUNITY TECHNICAL COLLEGE

Manchester/September-May

This college offers a 1-year certificate. Program started 1977. Accredited by ACFEI. Admission dates: September, January. Total enrollment: 54. 20 each admission period. 25% receive financial aid. 50% enrolled part-time. Student to teacher ratio 18:1. 95% of graduates obtain employment within six months. Facilities: Include classrooms and 2 kitchens.

Courses: Include 4 culinary, 2 baking, sanitation, nutrition, and co-op ed. Externship provided. Continuing education in decorative work, wines and spirits, hospitality industry, cost control.

Faculty: 7 full-time with master's degree or equivalent.

Costs: Tuition in-state $907 per semester (full time, 17 credits). Application fee $10. Admission requirements: High school diploma or equivalent.

Location: 160-acre campus in a suburban setting 10 miles from Hartford.

Contact: G. S. Lemaire, Program Coordinator, Manchester Community Technical College, Culinary Arts Dept., 60 Bidwell St., Manchester, CT 06040 US; 860-647-6140, Fax 860-647-6238, E-mail ma_lemaire@commnet.edu, URL http://www.mctc.commnet.edu.

NAUGATUCK VALLEY COMMUNITY-TECHNICAL COLLEGE

Waterbury/September-May

This 2-year college offers an AS degree (4 semesters full-time) in Foodservice Management. Established 1982. Accredited by NEASC. Calendar: semester. Curriculum: culinary, core. Total enrollment: 100. 100% of applicants accepted. 20% receive financial aid. 40% enrolled part-time. Student to teacher ratio 12:1. 100% of graduates obtain employment within six months. Facilities: Commercial kitchen lab, formal dining room, modern computer labs and classrooms.

Courses: Professional food preparation and service, management and supervision of foodservice operations. Cooperative education component.

Faculty: 7 full- and part-time faculty with advanced degrees and industry experience.

Costs: $907/semester in-state full-time (12 credits or more). Scholarships: yes. Loans: yes.

Location: An hour from New York City, 30 minutes from New Haven, CT.

Contact: Todd Jones, Program Coordinator, Naugatuck Valley Community-Technical College, E-512, 750 Chase Pkwy., Waterbury, CT 06708 US; 860-575-8175, E-mail MT_MARKOS_17@apollo.commnet.edu, URL http://www.nvctc.commnet.edu.

NORWALK COMMUNITY-TECHNICAL COLLEGE

Norwalk/August-May

This two-year community college offers a 32-week/10-month/1-year (30 credit) certificate in Culinary Arts, associate degree in Restaurant/Foodservice Management. Program started 1992. Calendar: semester. Curriculum: culinary, core. Admission dates: August, January. Total enrollment: 40. 20 each admission period. 100% of applicants accepted. 50% receive financial aid. 90% enrolled part-time. Student to teacher ratio 16:1. 100% of graduates obtain employment within six months. Facilities: High tech, fully equipped kitchen.

Courses: 2 baking courses, 4 food prep courses, 1 sanitation, safety and maintenance course, 1 nutrition course. 400-hour paid co-op work experience. Continuing education available.

Faculty: 2 full- and 6 part-time instructors.

Costs: $1,814 per year tuition and fees, $500 other expenses. Admission requirements: High school diploma & placement test. Scholarships: yes. Loans: yes.

Contact: Tom Connolly, Coordinator, Hospitality Management & Culinary Arts, Norwalk Community-Technical College, Culinary Arts, 188 Richards Ave., Norwalk, CT 06854-1655 US; 203-857-7355, Fax 203-857-3327, E-mail nk_connolly@commnet.edu.

DELAWARE

DELAWARE TECHNICAL & COMMUNITY COLLEGE

Newark/September-May

This college offers a 2-year AAS degree in Culinary Arts and a 1-year diploma and 2-year AAS degree in Foodservice Management. Started 1994. Accredited by MSA. Calendar: semester. Curriculum: culinary. Admission dates: August. Total enrollment: 45. 24 per admission period. 80% of applicants accepted. 50% receive financial aid. 80% part-time. Student to teacher ratio 12:1. 100% of graduates obtain jobs within six months. Facilities: Training kitchen and dining room.

Courses: Current trends.

Faculty: 3 full-time (2 are CEC), 4 part-time (1 is FMP).

Location: Five miles from Wilmington.

Contact: David Nolker, CEC, Dept. Chair, Delaware Technical & Community College, 400 Stanton-Christiana Rd., Newark, DE 19713 US; 302-453-3757, Fax 302-368-6620, E-mail dnolker@hopi.dtcc.edu, URL http://www.dtcc.edu.

FLORIDA

THE ACADEMY/AMERICAN CULINARY ARTS PROGRAM

Lakeland/Year-round

This private school offers a specialized associate degree in Culinary Arts. Program started 1995. Accredited by ACCSCT. Curriculum: culinary. Admission dates: Every 7 weeks. Total enrollment: 90. 20 each admission period. 95% of applicants accepted. 95% receive financial aid. 0% enrolled part-time. Student to teacher ratio 12:1. 95% of graduates obtain employment within six months. Facilities: Modern, well-equipped facility including classroom and kitchen space.

Courses: American and international cuisines, pastry, cost and inventory control, 9 National Restaurant Certifications. 10-week paid externship, 70% at major Fla. resorts and properties. Continuing ed classes available.

Faculty: 7 full-time chefs with 125+ years experience. Qualifications on staff include CCE, CEC, CFBM, CC de Cuisine.

Costs: $13,978 for program, includes tuition, books, culinary tools, uniforms.

Admission requirements: High school diploma or equivalent. Scholarships: yes. Loans: yes.

Location: Suburban area midway between Tampa and Orlando.

Contact: Anne E. McSoley, V.P., The Academy/American Culinary Arts Program, 3131 Flightline Dr., Lakeland, FL 33811 US; 800-532-3210, Fax 941-648-2204, URL http://www.theacademy.net.

ART INSTITUTE OF FT. LAUDERDALE – SCHOOL OF CULINARY ARTS

(See display ad page 8)

Ft. Lauderdale/Year-round

This proprietary school offers a 12-month diploma in The Art of Cooking, 21-month AS degree program in Culinary Arts. Program started 1991. Accredited by ACCSCT. Calendar: quarter. Admission dates: January, April, July, October. Total enrollment: 250. 15-40 each admission period. 90% of applicants accepted. 80% receive financial aid. Student to teacher ratio 19:1. 98% of graduates obtain employment within six months. Facilities: 3 kitchens and classrooms and a student-run restaurant.

Courses: General education; 1st year includes basic cooking, product identification, baking and pastry, knife skills, and nutrition; 2nd year includes garde manger, art history, internat'l cuisine, menu planning, kitchen layout, wine appreciation and restaurant service.

Faculty: 7 full-time ACF-certified instructors.

Costs: Annual tuition $10,980. Application fee $50. On-campus housing $1,290 per quarter. Average off-campus housing $600 per month. Admission requirements: High school diploma or equivalent. Scholarships: yes. Loans: yes.

Contact: Eileen Northrop, V.P. Director of Admissions, Art Institute of Ft. Lauderdale-School of Culinary Arts, 1799 S.E. 17th St., Ft. Lauderdale, FL 33316 US; 954-463-3000/800-275-7603, Fax 954-728-8637, URL http://www.artinstitute.edu.

ATLANTIC VOCATIONAL TECHNICAL CENTER

Coconut Creek/Year-round

This public institution offers a 1,080-hour certificate in Culinary Arts. Program started 1976. Accredited by SACS, ACFEI. Calendar: quarter. Curriculum: core. Admission dates: Open. Total enrollment: 100. 100 each admission period. 90% of applicants accepted. 30% receive financial aid. 1% enrolled part-time. Student to teacher ratio 15:1. 85% of graduates obtain employment within six months. Facilities: Include student-run restaurant.

Courses: Hot foods, cold foods, bakery, nutrition, sanitation, supervision, dining room, and management. Industry Cooperative Education.

Faculty: 8 full-time.

Costs: Annual tuition $700. Textbook and workbook $35, uniforms $60. Admission requirements: Admission test and basic academic skills. Scholarships: yes. Loans: yes.

Location: The campus is in a suburban setting.

Contact: Moses Ball, CCE, CEC, AAC, Atlantic Vocational Technical Center, Culinary Arts, 4700 N.W. Coconut Creek Pkwy., Coconut Creek, FL 33066 US; 305-977-2066, Fax 305-977-2019.

CHARLOTTE COUNTY VOCATIONAL TECHNICAL CENTER

Port Charlotte/Year-Round

This vocational-technical post-secondary school offers a 5-quarter certificate of completion in Commercial Foods & Culinary Arts. Established 1980. Accredited by COE. Calendar: quarter. Curriculum: culinary. Admission dates: July, October, January, April. Total enrollment: 30. 99% of applicants accepted. 60% receive financial aid. Student to teacher ratio 15:1. 100% obtain jobs within six months. Facilities: Kitchen, lab, classroom.

Faculty: 2 full-time faculty, 1 part-time aide.

Costs: $1.25 per hour.

Contact: Dick Santello, Charlotte Voc-Tech Center, 18300 Toledo Blade Blvd., Pt. Charlotte, FL 33948; 941-255-7500, Fax 941-255-7509, E-mail cvtc@cyberstreet.com.

DAYTONA BEACH COMMUNITY COLLEGE

Daytona Beach/Year-round

This college offers a 3-year Certificate of Apprenticeship or AS degree in Culinary Management. Program started 1980. Accredited by ACFEI Apprenticeship Program. Calendar: trimester. Curriculum: culinary. Admission dates: Open. Total enrollment: 50. 80% of applicants accepted. Student to teacher ratio 25:1. 100% of graduates obtain employment within six months. Facilities: Include kitchen and classroom.

Courses: Culinary apprenticeship, sanitation, supervision, nutrition. One-semester internship for AS program.

Faculty: 5 part-time.

Costs: State funded vocational program for apprenticeship. Standard community college tuition rates for AS program. Admission requirements: High school diploma or equivalent and admission test. Loans: yes.

Contact: Jeff Conklin, CEC, Program Manager, Daytona Beach Community College, PO Box 2811; Bldg #39, Room #149, Daytona Beach, FL 32120-2811 US; 904-255-8131 x3735, Fax 904-254-3063, E-mail conklij@dbcc.cc.fl.us, URL http://www.dbcc.cc.fl.us.

FLORIDA CULINARY INSTITUTE

(See display ad above)
West Palm Beach/Year-round

This proprietary institution, a division of New England Institute of Technology at Palm Beach, offers an 18-month Specialized Associate degree in Culinary Arts, International Baking and Pastry, and Food and Beverage Management. Established in 1987. Accredited by ACFEI, COE. Calendar: quarter. Curriculum: culinary. Admission dates: January, April, July, October. Total enrollment: 560. 140 each admission period. 95% of applicants accepted. 75% receive financial aid. Student to teacher ratio 18:1. Facilities: Include 8 kitchens and 8 classrooms.

Courses: 6 quarters including food preparation, facilities planning, nutrition, purchasing, baking, classical American and international cuisine, and management. Day, afternoon, or evening classes meet 5 hours daily, Monday-Thursday. Internships are served in practicum facility with Cafe Protege, the Institute's gourmet restaurant.

Faculty: 20 full-time. Includes Department Chair David Pantone, CEPC, Michael Barber, CCE, Dan Birney, CCE, William Boetcher, Jack Marshall, CMB, Manfred Schmidtke, CMB.

Costs: $9,800 per academic year. Average off-campus housing cost is $350-$500 per month. Admission requirements: High school diploma or equivalent. Advanced credit awarded through a testing program. Scholarships: yes. Loans: yes.

Location: 5-acre campus on Florida's southeast coast, 5 miles from Palm Beach Intl. Airport.

Contact: Alex F. Guerino, Admissions Coordinator, Florida Culinary Institute, 1126 53rd Court, West Palm Beach, FL 33407-9985 US; 800-826-9986/561-842-8324, Fax 561-842-9503, E-mail alex38@bellsouth.com, URL http://www.floridaculinary.com.

GULF COAST COMMUNITY COLLEGE

Panama City/Year-round

This college offers a 2-year AS degree in Culinary Management. Program started 1988. Accredited by SACS, ACFEI. Calendar: semester. Curriculum: core. Admission dates: Fall, spring. Total enrollment: 60. 20 each admission period. 100% of applicants accepted. Student to teacher ratio 16:1. Facilities: Include a student-run restaurant.

Faculty: 2 full-time, 1 part-time. Includes Travis Herr, CEC, CCE, John Holley, CCC, Jon Bullard, CCE, CEC.

Costs: Tuition in-state $30 per credit, out-of-state $110 per credit. Lab fees $8-$9. Admission requirements: High school diploma or equivalent. Scholarships: yes.

Location: The campus is in a suburban setting in Florida's Panhandle.

Contact: Travis Herr, Chef/Coordinator, Gulf Coast Community College, Culinary Management, 5230 W. U.S. Hwy 98, Panama City, FL 32401 US; 850-872-3850, Fax 850-872-3836, URL http://www.gc.cc.fl.us.

INSTITUTE OF THE SOUTH FOR HOSPITALITY & CULINARY ARTS

Jacksonville/Year-round

This two-year college offers a 2-year AS degree. Established in 1990. Accredited by SACS, ACFEI. Calendar: semester. Curriculum: core. Admission dates: August, January. Total enrollment: 120. 40 each admission period. 100% of applicants accepted. 30% receive financial aid. 80% enrolled part-time. Student to teacher ratio 20:1. 95% of graduates obtain employment within six months. Facilities: 3 kitchens, 4 classrooms, 2 restaurants.

Courses: Culinary. 2 externships at 150 hours, throughout the community at major properties.

Faculty: 4 full-time, 6 part-time. Includes Chefs Rick Grigsby and Joe Harrold. Qualifications: ACF certified. Professors Melanie Thompson and Margaret Wolson.

Costs: Annual tuition in-state $832, out-of-state $3,328. Admission requirements: High school diploma or equivalent. Scholarships: yes. Loans: yes.

Contact: Program Manager, Florida Community College at Jacksonville, Institute of the South for Hospitality & Culinary Arts, 4501 Capper Rd., Jacksonville, FL 32218 US; 904-766-6652, Fax 904-766-6654, E-mail scooper@fccj.cc.fl.us.

JOHNSON & WALES UNIVERSITY AT NORTH MIAMI

North Miami/Year-round

This private nonprofit career institution offers a 2-year AS in Culinary Arts and Baking and Pastry Arts, 4-year BS program in Culinary Arts. Established in 1992. Accredited by NEASC, ACICS. Calendar: quarter. Curriculum: culinary, core. Admission dates: Rolling. Total enrollment: 782. 76% of applicants accepted. 82% receive financial aid. 14% enrolled part-time. Student to teacher ratio 20:1. 98% of graduates obtain employment within six months. Facilities: Laboratory kitchens, academic classrooms, library, computer laboratory, conference center.

Courses: Culinary fundamentals, advanced culinary technologies, culinary principles. Other required courses: professional studies and academic courses. Term-long internships scheduled for all Baking and Pastry, Culinary, Hotel-Restaurant majors. Continuing education: The Culinary Arts Weekend program.

Faculty: 18 full-time.

Costs: College of Culinary Arts annual tuition $15,378. Other costs: $477 general fee, $125 orientation fee, on-campus housing (250 spaces) $3,930. Admission requirements: High school diploma or equivalent. Scholarships: yes. Loans: yes.

Location: The 8-acre campus is in South Florida.

Contact: Jeffrey Greenip, Director of Admissions, Johnson & Wales University at North Miami, Admissions Office, 1701 N.E. 127th St., North Miami, FL 33181 US; 800-232-2433 x7600, Fax 305-892-7020, E-mail admissions@jwu.edu, URL http://www.jwu.edu.

McFATTER SCHOOL OF CULINARY ARTS

Davie/Year-round *(See display ad page 33)*

This public, post-secondary occupational educational facility offers a 1-year certificate in Culinary Arts. Established in 1996. Accredited by COE. Calendar: quarter. Curriculum: culinary, core. Admission dates: Each 9-week term. Total enrollment: 50. 10-20 each admission period. 100% of applicants accepted. 50-75% receive financial aid. 10% enrolled part-time. Student to teacher ratio 20:1. 100% of graduates obtain employment within six months. Facilities: Cafeteria, cafe, dining room.

Courses: Baking and pastries, food production, garde manger, meat fabrication. Other required courses: 30 hours each of sanitation, nutrition, supervisory management.

Faculty: 2 full-time, 2 part-time. Includes Program Coordinator/Dept. Head V. Paul Citrullo, Jr., CEC, Chef Kay Bolm, Chef Ken Carver CEC.

Costs: Tuition is $245 per term full time. Other costs include books, uniforms, supply fee. Admission requirements: Basic skills testing. Scholarships: yes.

Location: 5 miles from Ft. Lauderdale.

Contact: V. Paul Citrullo, Jr., CEC, Exec. Chef/Director of Culinary Arts, McFatter Vocational Tech Center, 6500 Nova Dr., Davie, FL 33317 US; 954-424-4161, Fax 954-370-1647, E-mail chef-paul2@yahoo.com, URL http://www.gate.net/~mcfatter.

MID-FLORIDA TECHNICAL INSTITUTE

Orlando

This institution offers an 1,800-hour certificate. Program started 1970. Accredited by SACS. Admission dates: Open. Total enrollment: 80. 100% of applicants accepted. Student to teacher ratio 15-20:1. 100% of graduates obtain employment within six months.

Faculty: 2 full-time, 6 part-time.

Costs: $630 per semester full time, $315 per semester part time.

Contact: Dale Pennington, Chef Instructor, Mid-Florida Technical Institute, Commercial Cooking-Culinary Arts, 2900 W. Oakridge Rd., Orlando, FL 32809 US; 407-855-5880, Fax 407-855-5880.

NORTH TECHNICAL EDUCATION CENTER

Riviera Beach/Year-round

This public institution offers an 1,800-hour certificate in Commercial Foods and Culinary Arts. Program started 1970. Accredited by COE. Calendar: semester. Curriculum: culinary. Admission dates: August, October, January, March, June. Total enrollment: 40. 30 day and 30 evening each admission period. 100% of applicants accepted. 50% receive financial aid. 50% enrolled part-time. Student to teacher ratio 15-20:1. 90% of graduates obtain employment within six months. Facilities: Kitchen and classroom.

Faculty: 2 full-time, 1 part-time.

Costs: Annual tuition in-state $400-$800. Textbook, uniform and shoes approximately $200.

Contact: R. Robertson, Instructor, North Technical Education Center, Commercial Foods & Culinary Art, 7071 Garden Rd., Riviera Beach, FL 33404 US; 561-881-4600, Fax 561-881-4668.

PENSACOLA JR. COLLEGE CULINARY MANAGEMENT PROGRAM

Pensacola/Year-round

This college offers a 2-year (64-credit hours) AS degree, 420-hour vocational certificate. Program started 1995. Calendar: semester. Curriculum: culinary, core. Admission dates: August, January, May. Total enrollment: 75. 20 each admission period. 90% of applicants accepted. 90% receive financial aid. 50% enrolled part-time. Student to teacher ratio 10:1. 100% of graduates obtain employment within six months. Facilities: New $250,000 kitchen, public dining room.

Courses: Technique, presentation, public service. 150 hours required at good service establishment.

Faculty: 1 full-time, 4 part-time; all ACF-certified.

Costs: Tuition $41 per hour in-state, $125 per hour out-of-state; lab fee $50 per class. Admission requirements: High school grad or GED. Scholarships: yes. Loans: yes.

Contact: Howard Aller, CEC, CCE, Director of Culinary Mgt., Pensacola Jr. College, 1000 College Blvd., Pensacola, FL 32504 US; 904-484-1422, Fax 904-484-1543, E-mail haller@pjc.cc.fl.us.

PINELLAS TECHNICAL EDUCATIONAL CENTER

N. Clearwater/Year-round

This college offers a 15-month diploma in Culinary Arts. Program started 1965. Accredited by ACFEI, COE. Calendar: trimester. Curriculum: culinary. Admission dates: Monthly. Total enrollment: 60. 6-10 each admission period. 100% of applicants accepted. 75% receive financial aid. Student to teacher ratio 15:1. Facilities: 2 kitchens, 2 classrooms and a student-run restaurant.

Faculty: 4 full-time.

Costs: Tuition in-state $1.25 per student contact hour, ~$375 per trimester.

Contact: Vincent Calandra, Dept. Chair, Pinellas Technical Educational Center, Culinary Arts Dept., 6100 154th Ave., N. Clearwater, FL 33760 US; 813-538-7167 x1140, Fax 813-538-7203, E-mail vcalandra@ptecclw.pinellas.k12.fl.us, URL http://www.ptecclw.pinellas.k12.fl.us.

PINELLAS TECHNICAL EDUCATIONAL CENTER

St. Petersburg/Year-round

This trade school offers an 1,800-hour diploma in Culinary Arts. Accredited by SACS, ACFEI. Admission dates: Open. Total enrollment: 52. 10 each admission period. 100% of applicants accepted. 60% receive financial aid. Student to teacher ratio 17:1. 95% of graduates obtain employment within six months. Facilities: Include kitchen, baking lab and classroom.

Faculty: 2 full-time, 1 part-time. Dr. Warren Laux, Dr. Tom Maas, Alvin Miller, Fred Lemiesz.

Costs: Annual tuition $495 resident. Books $42. Admission requirements: Admission test.

Location: The 1,950-student campus is in a suburban setting 21 miles from Tampa.

Contact: Alvin Miller, Pinellas Technical Educational Center, Culinary Arts Dept., 901 34th St. South, St. Petersburg, FL 33711 US; 813-893-2500 x1104.

ROBERT MORGAN VOCATIONAL TECHNICAL INSTITUTE

Miami/Year-round

This Dade County Public School offers 720-hour courses (two 15-week trimesters) in Commercial Cooking and Commercial Baking, 212-hour course in Cake Decoration. Established 1979. Accredited by SACS, COE. Calendar: trimester. Curriculum: culinary. Admission dates: Open. Total enrollment: 220-250. 100-120 each admission period. 100% of applicants accepted. 30% receive financial aid. 30% enrolled part-time. Student to teacher ratio 15:1. 95% of graduates obtain jobs.

Courses: Food preparation and presentation, safety, sanitation, use of leftovers, storage, customer relations, employability, personal hygiene, menu planning, principles of entrepreneurship.

Faculty: 3 full- and 2 part-time Florida certified instructors.

Costs: $0.67 per hr in-state, $2.91 per hour out-of-state, $4.73 per hour part-time. Admission requirements: 10th grade. Scholarships: yes. Loans: yes.

Contact: Giorgio Moro, Food Services Coordinator, Robert Morgan Voc-Tech Institute, Culinary Arts Program, 18180 SW 122nd Ave., Miami, FL 33177 US; 305-253-9920, Fax 305-253-3023.

SARASOTA COUNTY TECHNICAL INSTITUTE

Sarasota

This institution offers a 2-year/1,485-hour certificate. Program started 1967. Accredited by SACS. Admission dates: Open. 90% of applicants accepted.

Faculty: 1 full-time.

Costs: $2,005 per year in-state, $10,080 per year out-of-state.

Contact: Timothy Carroll, Sarasota County Technical Institute, Culinary Arts, 4748 Beneva Rd., Sarasota, FL 34233 US; 941-924-1365, Fax 941-924-1365, URL http://www.careerscape.org.

SOUTH FLORIDA COMMUNITY COLLEGE

Avon Park/Year-round

This college offers a vocational certificate in Food Management, Production, and Services. AS degree in Hospitality Management. Established in 1965. Accredited by SACS. Calendar: semester.

Curriculum: culinary. Admission dates: August, December. Total enrollment: 15-30. Varies each admission period. 75-80% of applicants accepted. 50% enrolled part-time. Student to teacher ratio 12:1. 80-90% of graduates obtain employment within six months. Facilities: College-owned historic Hotel Jacaranda, in downtown Avon Park.

Courses: Food production, general hospitality and education.

Faculty: 1 full- and 2 part-time.

Location: 60-70 miles from Orlando, 70 miles from Tampa.

Contact: K. Viviano, Professor of Hospitality Management, South Florida Community College, 600 W. College Dr., Avon Park, FL 33825 US; 941-453-6661 x1337, Fax 941-453-8023, E-mail viviano_k@popmail.firn.edu.

THE SOUTHEAST INSTITUTE OF CULINARY ARTS

St. Augustine/Year-round

This school offers a 1-year (900-hour) certificate and 2-year (900-hour) diploma programs in Commercial Foods and Culinary Arts. Established in 1970. Accredited by ACFEI, COE. Calendar: quinmester. Curriculum: culinary. Admission dates: Every 9 weeks. Total enrollment: 600+. Student to teacher ratio 15-20:1. 99% of graduates obtain employment within six months.

Courses: Rotation through basic stations, participation in intensives that include epicurean service, advanced bakeshop, buffet catering, purchasing and receiving, and a la carte. Specialized certificates awarded for portions of course. 270-hour internship for diploma. Hospitality management programs at area colleges offer credit toward AS degree.

Faculty: 15 instructors, all ACF-certified or pending, 5 support staff.

Costs: Total cost is $2,100 in-state; out-of-state residents should call for fees. $15 registration fee, $179 per quinmester fee, $50 refundable book deposit, $500 for supplies. Off-campus lodging is $400 per month. Admission requirements: 16 years or older, high school diploma or equivalent on completion. Scholarships: yes.

Location: St. Augustine, the nation's oldest city and home of the American Culinary Federation, on Florida's northeast coast.

Contact: Chef David S. Bearl, CCC, CCE, Program Coordinator/Division Head, The Southeast Institute of Culinary Arts, 2980 Collins Ave., St. Augustine, FL 32094-9970 US; 904-829-1060/904-829-1061, Fax 904-824-6750.

SOUTHEASTERN ACADEMY'S CULINARY TRAINING CENTER

Kissimmee/Year-round

This proprietary institution offers a 30-week diploma in Culinary Arts. Established in 1990. Accredited by ACCSCT, approved for veterans' training. Calendar: quarter. Curriculum: culinary. Admission dates: Every 5 weeks. Total enrollment: 120. 30 each admission period. 95% of applicants accepted. 90% receive financial aid. Student to teacher ratio 10:1. 85% of graduates obtain employment within six months. Facilities: Include 12 cooking labs and classrooms, student-run restaurant, bakery.

Courses: Culinary arts, baking/pastry arts, sanitation, nutrition, professional development, restaurant management. Externship provided.

Faculty: 10 full-time, 2 part-time. Includes David S. Nina, CEC, AAC, Baking and Pastry Chef Instructor Glen Rhoades, Chef de Cuisine David Weir.

Costs: Annual tuition $8,030, fees and equipment $665. Dorm (30 weeks) $3,600; off-campus $450-$600 per month. Admission requirements: High school diploma or equivalent. Loans: yes.

Location: Suburban Orlando.

Contact: David L. Peoples, President, Southeastern Academy, 233 Academy Dr., Box 421768, Kissimmee, FL 34742-1768 US; 407-847-4444, Fax 407-847-8793, E-mail peoples@gdi.net, URL peoples@gdi.net.

GEORGIA

ART INSTITUTE OF ATLANTA – SCHOOL OF CULINARY ARTS

Atlanta/Year-round *(See display ad page 8)*

This private institution offers an 18-month AA degree in Culinary Arts. Established in 1991. Accredited by ACFEI, SACS. Calendar: quarter. Curriculum: culinary, core. Admission dates: January, April, July, October. Total enrollment: 400. 50-75 each admission period. 95% of applicants accepted. 75% receive financial aid. 10% enrolled part-time. Student to teacher ratio 20-25:1. 96% of graduates obtain employment within six months. Facilities: 10,000-square-foot facility with 4 teaching kitchens, 3 classrooms, and a full-service teaching restaurant.

Courses: Culinary skills, food production, baking and pastry, garde manger. Faculty-conducted weekend workshops, certificate of completion in Baking and Pastry.

Faculty: 17 full- and part-time instructors, all with industry experience.

Costs: Quarterly tuition $3,712. Application fee $50, quarterly lab fee $250. On-campus housing $450 per month. Admission requirements: High school diploma or equivalent and writing sample. Scholarships: yes. Loans: yes.

Location: The 1,300-student campus is in Atlanta's uptown Buckhead area.

Contact: Todd Knutson, Director of Admissions, The Art Institute of Atlanta-The School of Culinary Arts, 3376 Peachtree Rd. NE, Atlanta, GA 30326 US; 800-275-4242, Fax 404-848-9551, E-mail knutsont@aii.edu, URL http://www.aii.edu.

ATLANTA AREA TECHNICAL SCHOOL

Atlanta

This career institution offers an 18-month diploma. Program started 1967. Accredited by SACS. Admission dates: Quarterly. Student to teacher ratio 12:1. 92% of graduates obtain employment within six months.

Faculty: 6+ full-time.

Costs: Tuition $296 per quarter full-time, $23 per credit part-time. Admission requirements: High school diploma or equivalent and admission test.

Contact: Barbara Boyd, Culinary Chair, Atlanta Area Technical School, Culinary Arts Program, 1560 Stewart Ave. S.W., Atlanta, GA 30310 US; 404-756-3700 x3727, Fax 404-756-0932.

AUGUSTA TECHNICAL INSTITUTE

Augusta/Year-round

This state trade school offers a 4-quarter certificate program in Food Service, 6-quarter diploma in Culinary Arts. Program started 1985. Accredited by SACS. Calendar: quarter. Curriculum: culinary. Admission dates: September, March. Total enrollment: 36. 15 each admission period. 90% of applicants accepted. 100% receive financial aid. 3% enrolled part-time. Student to teacher ratio 18:1. 100% of graduates obtain employment within six months. Facilities: Include 1 kitchen, 2 classrooms, and local restaurants.

Courses: Basic food preparation, introduction to baking, garde manger, nutrition and menu management, safety and sanitation, and consumer education. Other required courses: office accounting, computer literacy. 150 hour salaried internship. Diploma from Educational Foundation of the NRA on completion of 6 quarters. Continuing education: catering, cake decoration, sanitation.

Faculty: 2 full-time.

Costs: Quarterly tuition $274 in-state, $548 out-of-state. Application fee $15. Average off-campus housing cost $500 per month. Admission requirements: High school diploma or GED.

Location: The 50-acre, 5,000-student suburban campus is 7 miles from Augusta.

Contact: Willie Mae Crittenden, Department Head, Augusta Technical Institute, Culinary Arts,

3116 Deans Bridge Rd., Augusta, GA 30906 US; 706-771-4000, Fax 706-771-4016, E-mail WCritten@augusta.tec.ga.us.

EAST CENTRAL TECHNICAL INSTITUTE

Fitzgerald/Year-round

This trade-technical school offers Basic Culinary Arts, 12- and 18-month programs in Culinary Arts. Established in 1967. Accredited by COEI. Calendar: quarter. Curriculum: culinary. Admission dates: January, April, July, October. Total enrollment: 10. 10 each admission period. 100% of applicants accepted. 90% receive financial aid. 0% enrolled part-time. Student to teacher ratio 10:1. 100% of graduates obtain employment within six months. Facilities: Full facilities, catering kitchen, 400-seat banquet facility.

Faculty: Larry Roberson, CCE, CCC, has a BS in education.

Location: South central Georgia.

Contact: Larry Roberson, Instructor, East Central Technical Institute, PO Box 1069, Fitzgerald, GA 31750 US; 912-468-7487, Fax 912-468-5550, E-mail LRoberso@eastcentral.tec.ga.us, URL http://www.eastcentral.tec.ga.us.

SAVANNAH TECHNICAL INSTITUTE

Savannah/Year-round

This public institution offers 6- and 4-quarter diploma programs in Culinary Arts. Program started 1981. Accredited by SACS, ACFEI. Calendar: quarter. Curriculum: culinary, core. Admission dates: Biannually (spring and fall). Total enrollment: 24. 10 each admission period. 100% of applicants accepted. 54% receive financial aid. 10% enrolled part-time. Student to teacher ratio 12:1. 100% of graduates obtain employment within six months. Facilities: Include kitchen, classroom, student-run restaurant.

Courses: Culinary arts, sanitation and equipment, food preparation, baking, garde manger, nutrition, and management. Other required courses: English, math, psychology. Externship: 150 hours, restaurant. Continuing education: sanitation, cake decorating, restaurant management.

Faculty: 1 full-time, 1 part-time. Includes Marvis T. Hinson, CFBE, M.Ed and Executive Chef John E. Thompson.

Costs: Tuition in-state $276 per quarter, out-of-state $528 per quarter. Application fee $15. 2 uniforms $50, shoes $40, knives $100, books $100 per quarter. Admission requirements: High school diploma or equivalent and admission test. Scholarships: yes. Loans: yes.

Contact: Marvis Hinson, Department Head, Savannah Technical Institute, Culinary Arts, 5717 White Bluff Rd., Savannah, GA 31405-5594 US; 912-351-4553, Fax 912-352-4362.

HAWAII

KAPIOLANI COMMUNITY COLLEGE

Honolulu/August-May

This college offers a 2-year AS degree in Culinary Arts, Patisserie, School Food Service, and Food Service-Health care; 18-month and 1-semester certificate program in Culinary Arts and Patisserie. Program started 1947. Accredited by ACFEI, WASC. Calendar: semester. Curriculum: core. Admission dates: Fall, spring. Total enrollment: 450. 100% of applicants accepted. 25% receive financial aid. 50% enrolled part-time. Student to teacher ratio 20:1. 98% of graduates obtain employment within six months. Facilities: 9 kitchens, 8 classrooms, 5 restaurants.

Courses: Asian/Pacific and international cuisine, garde manger, confisserie. Externship: available at local and neighbor island hotels.

Faculty: 14 full-time, 6 part-time. Qualifications: Industry experience.

Costs: Annual tuition in-state $504, out-of-state $3,096. $10 fee per semester. Average off-campus housing cost $500 per month. Admission requirements: Age 18, or age 17 with high school diploma or GED.

Location: The 7,500-student, 47-acre suburban campus is 2 miles from Waikiki.

Contact: Frank Leake, Special Assistant to the Provost, Kapiolani Community College, Food Service & Hospitality Education, 4303 Diamond Head Rd., Honolulu, HI 96816 US; 808-734-9204, Fax 808-734-9208, E-mail yonemori@kccada.kcc.hawaii.edu, URL http://www.hawaii.edu.

KAUAI COMMUNITY COLLEGE

Lihue

This two-year college offers a 1-year certificate and 2-year AAS degree in Culinary Arts. Calendar: semester. Curriculum: core.

Costs: $41.50 per credit in-state, $240.50 per credit out-of-state.

Contact: Al Spencer, Kauai Community College, 3-1901 Kaumualii Hwy., Lihue, HI 96766 US; 808-245-8311 x242, Fax 808-245-8297.

LEEWARD COMMUNITY COLLEGE

Pearl City/Year-round

This two-year college offers a 2-year AAS degree in Food Service, 1.5-year certificate in Food Service, 1-semester certificate in Food Preparation, 2-semester certificates in Baking and Dining Room. Program started 1974. Accredited by WASC, ACFEI. Calendar: semester. Curriculum: core. Admission dates: August, January. Total enrollment: 80-100. 25-30 each admission period. 100% of applicants accepted. 40% receive financial aid. 20% enrolled part-time. Student to teacher ratio 15-20:1. 90% of graduates obtain employment within six months. Facilities: 3 kitchens, 5 classrooms, 1 restaurant, 2 cafe/dining rooms.

Courses: Provides students with technical knowledge and basic skills training for a professional food service career. Minimum 75 hours in a commercial kitchen required second semester, paid or non-paid stations.

Faculty: 5 full-time with 15-20 years industry experience, 3 part-time. Includes Tommylynn Benavente, Stanley Ikei, Fern Tomisato, Mary Schaefer, Steven Jones.

Costs: Semester tuition in-state $468, out-of-state $2,856. Admission requirements: Age 18 or 17 with high school diploma. Scholarships: yes. Loans: yes.

Location: The 6,000-student, 49-acre campus is on the Pearl City peninsula between Pearl City and Waipahu.

Contact: Tommylynn Benavente, Program Coordinator, Leeward Community College, Food Service, 96-045 Ala Ike, Pearl City, HI 96782 US; 808-455-0298/808-455-0300, Fax 808-455-0471, E-mail tlbenave@hawaii.edu.

MAUI COMMUNITY COLLEGE/CULINARY ARTS PROGRAM

Kahului/August-June

This two-year college offers a 1-year certificate, 2-year AAS degree-Culinary, 2-year AAS degree-Baking. Program started 1969. Accredited by ACFEI. Calendar: semester. Curriculum: culinary, core. Admission dates: August, January. Total enrollment: 140. 30 each admission period. 100% of applicants accepted. 25% receive financial aid. 20% enrolled part-time. Student to teacher ratio 10:1. 98% of graduates obtain employment within six months. Facilities: Include 2 kitchens, 2 classrooms, 1 restaurant, 1 cafe/dining room.

Courses: Culinary Arts or Baking Specialties. Apprenticeship, internships, externships, co-op ed available. Continuing Education: Specialty courses available.

Faculty: 5 full-time, 5 part-time.

Costs: Annual tuition in-state $477, out-of-state $2,890 per semester. 2 bedroom cost is $997 per semester, 3 bedroom cost is $857 per semester. Scholarships: yes. Loans: yes.

Contact: Karen Tanaka, Coordinator, Maui Community College, Culinary Arts, 310 Kaahumanu Avenue, Kahului, HI 96732 US; 808-984-3225, Fax 808-984-3314, URL http://www.hawaii.edu/welcome/mcc.html.

IDAHO

BOISE STATE UNIVERSITY

Boise/Year-round

This university offers a 1-year certificate, 2-year AAS degree. Program started 1969. Accredited by ACFEI. Calendar: semester. Admission dates: August, January. Total enrollment: 35-45. Student to teacher ratio 10:1. 98% of graduates obtain employment within six months.

Faculty: 3 full-time, 1 part-time.

Costs: Tuition $1,236 per semester in-state, $2,940 per semester out-of-state. Admission requirements: High school diploma or equivalent and admission test. Scholarships: yes.

Contact: Vern Hickman, CCC, CCE, Boise State University, Culinary Arts, 1910 University Dr., Boise, ID 83725 US; 208-385-4199, Fax 208-385-1948.

COLLEGE OF SOUTHERN IDAHO

Twin Falls/September-May

This community college offers a 2-year AAS, 1-year technical certificate. Program started 1986. Calendar: semester. Curriculum: core. Admission dates: Open. Total enrollment: 20. 12 each admission period. 32% of applicants accepted. 90% receive financial aid. Student to teacher ratio 10:1. 90% of graduates obtain employment within six months.

Courses: Culinary arts, Food production. Internships provided in summer. Cooperative education during school year.

Faculty: 2 full-time, 6 part-time.

Costs: Tuition in-state $615 per semester, out-of-state $1,615 per semester. On-campus room and board is $1,625; average off-campus housing cost is $350 per month. Admission requirements: High school diploma or equivalent. Scholarships: yes.

Location: South Central Idaho, 120 miles east of Boise.

Contact: Larry Motzner, Program Coordinator, College of Southern Idaho, Hotel-Restaurant Mgmt., 315 Falls Ave., Box 1238, Twin Falls, ID 83303-1238 US; 208-733-9554 x2407, Fax 208-736-2136, E-mail LMotzner@evergreen2.csi.cc.id.us, URL http://www.csi.cc.id.us.

IDAHO STATE UNIVERSITY SCHOOL OF APPLIED TECHNOLOGY

Pocatello/Year-round

This university offers a 2-semester certificate program in Culinary Arts Technology, AAS degree in Business Technology. Calendar: semester. Curriculum: core. Admission dates: August, October, January, March, May.

Courses: Food preparation and service, nutrition, purchasing, human relations, job search.

Costs: Tuition for the certificate program is $992 per semester (resident). Supplies $556.

Contact: Philip Gibson, Dept. Chair, Idaho State University School of Applied Technology, Campus Box 8380, Pocatello, ID 83209-8380 US; 208-236-2507, Fax 208-236-4641, E-mail gibsphil@isu.edu, URL http://www.isu.edu.

NORTH IDAHO COLLEGE

Coeur d'Alene/Year-round

This college offers a 36.5 credit-hour certificate in Culinary Arts.

Courses: Breakfast cooking and catering, prep, pantry, stock, soup, sauce preparation, line and grill cooking, baking.

Costs: Annual tuition is $1,128-$1,687 per credit in-state, $3,884-$4,443 per credit out-of-state. Room and board $3,510 per year. Books and supplies $225-$2,300 per year.

Contact: Richard Schultz, North Idaho College, 1000 W. Garden Ave., Coeur d'Alene, ID 83814 US; 208-769-3458, E-mail rick_schultz@nidc.edu, URL http://www.nidc.edu.

ILLINOIS

THE ART OF STYLING FOOD

Elmwood Park/April-May, September-October

Food stylist Donna Lafferty offers 4-day certificate courses (limit 6 students) in styling food for still and motion photography. Established in 1993. 6 students each admission period. Student to teacher ratio 1:1. Facilities: 250-square-foot professional kitchen with work stations and overhead mirror.

Courses: 4-day course offers 32 hours of instruction and a shopping trip to specialty shops. Other activities: students work with a food photographer for their own food set-up and photo.

Faculty: Donna Lafferty has 19 years of experience styling for still and motion advertising. Her clients include Edy's Grand Ice Cream, Kraft Foods, The Oprah Winfrey Show, and edible ornaments for the White House Christmas Trees as requested by the First Lady.

Costs: $1,250 for 4 days. Motels/hotels recommended. Admission requirements: Students should know how to cook.

Location: U.S. classes: a Chicago suburb.

Contact: Donna Lafferty, International Food Stylist, The Art of Styling Food, 2733 N. 75th Ct., Elmwood Park, IL 60707-1433 US; 708-456-8415.

BELLEVILLE AREA COLLEGE

Granite City/Year-round

This college offers 1-year certificate programs in Culinary Arts (15 credit-hours), Foodservice (16 credit-hours), and Hospitality Food Service (31 credit-hours); 2-year AAS degree in Hospitality Food Service Management (71 credit-hours). Calendar: semester.

Costs: $43.50 per credit-hour in-district, $107 out-of-district, $150 for uniforms and supplies.

Contact: J. Michael Hayes, Program Coordinator, Belleville Area College, 4950 Maryville Rd., Granite City, IL 62040-2699 US; 618-931-0600 x689, URL http://www.bacnet.edu.

COLLEGE OF DUPAGE

Glen Ellyn/Year-round

This college offers a 1-year certificate and 2-year AAS degree in Food Service Administration and Culinary Arts, 1-year certificate in Pastry Arts. Program started 1966. Accredited by NCA, RBA, NRA Educational Foundation, ACFEI. Calendar: quarter. Curriculum: culinary, core. Admission dates: September, January, March, June. Total enrollment: 200. 50-125 each admission period. 100% of applicants accepted. 25% receive financial aid. 30% enrolled part-time. Student to teacher ratio 15:1. 100% of graduates obtain employment within six months. Facilities: Include kitchen, pastry shop, classrooms, restaurant, dining room.

Courses: Food preparation, classical cuisine, baking and pastry, merchandising, cake decorating, garde manger, wines and international cuisine. Externship provided.

Faculty: 4 full-time, 12 part-time. Includes George C. Macht, CHA, CFE, FMP, Chris Thieman, CWC, Rolfe Sick, CMP, Jim Zielinski, FMP.

Costs: In-district $32 per credit hour. Admission fee $10. Texts, uniforms, tools and fees vary. Average off-campus cost is $500 per month. Admission requirements: High school diploma or equivalent. Scholarships: yes. Loans: yes.

Location: The 33,000-student campus is in a suburban setting 25 miles from Chicago.

Contact: Catherine Leveille, Program Assistant, College of DuPage, Culinary Arts/Pastry Arts, 425 22nd St., Glen Ellyn, IL 60137 US; 630-942-3663, Fax 630-858-9399, E-mail machtg@cdnet.cod.edu.

COLLEGE OF LAKE COUNTY

Grayslake/August-May

This college offers a 1-year certificate in Culinary Arts or Food Service Management, AAS in Food

Service Management. Program started 1987. Accredited by NCA. Calendar: semester. Curriculum: core. Admission dates: August, January, June. Total enrollment: 40. 10 each admission period. 95% of applicants accepted. 10% receive financial aid. 50% enrolled part-time. Student to teacher ratio 10:1. 98% of graduates obtain employment within six months. Facilities: Include 2 kitchens and 3 classrooms.

Courses: Cooking, baking, nutrition and menu planning. Externship: 16-week part-time.

Faculty: 5 full-time. Includes C. Wener, M. Eskenazy, J. Lempke, J. Bress. Qualifications: minimum 2-year culinary school or formal internship.

Costs: Tuition $51 per credit hour. Other fees: lab, equipment, uniforms.

Location: Lake County, IL.

Contact: Cliff Wener, Coordinator, College of Lake County, Food Service Program, 19351 W. Washington St., Grayslake, IL 60030-1198 US; 847-543-2823, Fax 847-223-7248, E-mail crwener-fsm@clc.cc.il.us, URL http://www.clc.cc.il.us.

COOKING ACADEMY OF CHICAGO

Chicago/Year-round

This private culinary school offers a 6-month (17-month part-time) Culinary Career certificate, 6-month (17-month part-time) Baking and Pastry certificate. Established in 1992. Accredited by State Board of Education. Curriculum: culinary. Admission dates: January, May, September. Total enrollment: 50. 15 each admission period. 100% of applicants accepted. 50% enrolled part-time. Student to teacher ratio 15:2. 100% of graduates obtain employment within six months. Facilities: 2 full kitchens, 2 classrooms.

Courses: Knife skills, soups and stocks, vegetables and starches, meats, fish, poultry, sauces, baking and pastry.

Faculty: 2 full-time, 2 part-time. Qualifications: minimum AAS and 5 years experience.

Costs: Tuition culinary $4,000, baking and pastry $4,000. Fees: $100. Admission requirements: High school or GED.

Location: 20 blocks west of Wrigley Field.

Contact: Nora Christensen, Director, Cooking Academy of Chicago, 2500 W. Bradley Pl., Chicago, IL 60618 US; 773-478-9840, Fax 773-478-3146, URL http://www.cookingacad.com.

COOKING AND HOSPITALITY INSTITUTE OF CHICAGO

(See also page 210) (See display ad page 42) **Chicago/Year-round**

This private institution offers a 2-year (69-credit-hour) associate degree in Culinary Arts, 30-week (27-credit-hour) certificate programs in Professional Cooking and Baking and Pastry, 17 credit-hour certificate program in Restaurant Management. Established in 1983. Accredited by ACCSCT, ACFEI. Calendar: semester. Curriculum: core. Admission dates: 6 times yearly. Total enrollment: 750. 100 each admission period. 80% of applicants accepted. 60% receive financial aid. 40% enrolled part-time. Student to teacher ratio 25:1. 95% of graduates obtain employment within six months. Facilities: 4 fully-equipped instructional kitchens, on-site restaurant.

Courses: Professional cooking covers qualitative and quantitative cooking, menu planning, recipe development, sanitation, and job search techniques. Baking and pastry covers production techniques, food as an art form, yeast breads, and decoration. Optional course offered at Etoile in Italy. Variety of continuing education and nonvocational classes.

Faculty: School founder Linda Calafiore is a past state coordinator of vocational training programs. Instructors are ACF certified and accomplished in their fields of expertise.

Costs: Tuition is $300 per credit hour. Registration fee is $100. Off-campus housing varies. Admission requirements: High school diploma or equivalent. Scholarships: yes. Loans: yes.

Contact: Jim Simpson, Director, The Cooking and Hospitality Institute of Chicago, 361 W. Chestnut, Chicago, IL 60610 US; 312-944-2725, Fax 312-944-8557, E-mail chic@chicnet.org, URL http://www.chicnet.org.

CULINARY SCHOOL OF KENDALL COLLEGE

Evanston/Year-round *(See display ad page 43)*

This nonprofit school, a division of Kendall College, offers a 21-month AAS degree in Culinary Arts, 2- and 4-year programs in Hospitality Management, certificates in cookery and pastry. Established in 1985. Accredited by ACFEI, NCA, UMCUS. Calendar: quarter. Curriculum: culinary. Admission dates: Continuous admissions. Total enrollment: 270. 40 each admission period. 80% of applicants accepted. 82% receive financial aid. 30% enrolled part-time. Student to teacher ratio 13:1. 90% of graduates obtain employment within six months. Facilities: Include demonstration, production, and display kitchens, 3 professional kitchens, pastry kitchen, dining room, banquet area, cafeteria.

Courses: Culinary skills, business management, nutrition, menu planning, classic and American cuisine. Beginning students work in cafeteria, bakery; advanced students work in restaurant. 10-week internship. Continuing education: culinary fundamental series.

Faculty: 25 chef/instructors and hospitality specialists as well as visiting lecturers and guest chefs. Include Michael Carmel, CEC, BA in CW, AOS, Mike Artlip, CEC, CCE, AAS, Hubert Martini, AAC, CEC, CCE, Lawrence Smith, Kader Temkkit.

Costs: Tuition $13,566 per year, including equipment and activities fee and 2 meals daily. Room and board is $764 double, $993 single per term. Admission requirements: High school diploma or equivalent and ACT/SAT scores or college experience.

Location: 35 minutes from Chicago's Loop by express train.

Contact: Jase Frederick, Public Relations, The Culinary School of Kendall College, 2408 Orrington Avenue, Evanston, IL 60201 US; 847-866-1297, Fax 847-866-1320. Additional contact: Mike Artlip, Director, Culinary Arts; 847-866-1362, Fax 847-866-1320.

ELGIN COMMUNITY COLLEGE
Elgin

This college offers a 2-year associate degree in Culinary Arts. Program started 1972. Accredited by Illinois Community College Board, ACFEI. Admission dates: August, January. Total enrollment: 140. 30-40 each admission period. 80% of applicants accepted. 30% enrolled part-time. Student to teacher ratio 12-30:1. 100% of graduates obtain employment within six months. Facilities: 4 kitchens, 3 classrooms and a culinary training center.

Faculty: 3 full-time, 8 part-time. Qualifications: ACF-certification.

Costs: $43.50 per credit hour in-district, $260.46 per credit hour out-of-district. Books and uniforms ~$200. Average off-campus housing cost $400-$600 per month. Admission requirements: High school diploma or equivalent.

Contact: Michael Zema, Director, Elgin Community College, Hospitality Dept., 1700 Spartan Dr., Elgin, IL 60123 US; 847-888-7461, Fax 847-888-7995, URL http://www.elgin.cc.il.us.

FRENCH PASTRY SCHOOL
Chicago/Year-round

Pastry Chefs Sebastien Canonne and Jacquy Pfeiffer offer 50+ 3- to 5-day hands-on dessert courses per year. 15-week pastry program beginning 1999. Program started 1997. Accredited by ACF. Calendar: semester. Curriculum: culinary. Admission dates: Year-round. 10 students per class. 100% of applicants accepted. Student to teacher ratio 8:1. 90% of graduates obtain employment within six months. Facilities: New teaching facility.

Courses: Pastry, sugar, ice cream and sorbet, candy and confectionery, chocolate, plated desserts, wedding cakes, competition skills.

Faculty: Award-winning pastry chefs Sebastien Canonne and Jacquy Pfeiffer, recipients of the

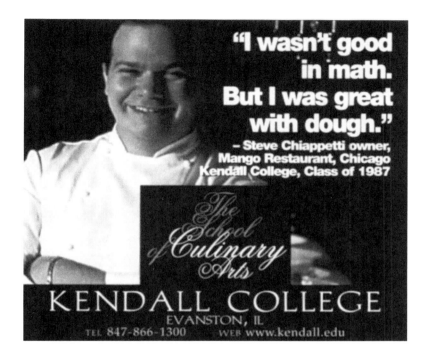

1997 Vice World Champion Coupe du Monde de la Patisserie, Lyon, France.

Costs: $525 per 3-day course, $780 per 5-day course.

Contact: Jacquy Pfeiffer, Co-owner, French Pastry School, Ecole de Patisserie Francaise, 1153 W. Grand Ave., Chicago, IL 60622 US; 312-243-3808, Fax 312-243-1430, E-mail Jpfeif0927@aol.com, URL http://www.frenchpastryschool.com.

JOLIET JUNIOR COLLEGE
Joliet/Year-round

This college offers a 2-year certificate/AAS degree in Culinary Arts. Program started 1970. Accredited by NCA, ACFEI. Calendar: semester. Curriculum: culinary. Admission dates: August, January, May, June. Total enrollment: 200. 100 each admission period. 98% of applicants accepted. 40% receive financial aid. Student to teacher ratio 20:1. 95% of graduates obtain employment within six months. Facilities: Include 3 kitchens, demonstration kitchen, 3 classrooms and a pastry shop.

Faculty: 9 full-time.

Costs: $49 per credit-hour in-state. Admission requirements: High school diploma or equivalent and admission test.

Location: The 11,000-student campus is in a small town setting.

Contact: Patrick F. Hegarty, CEC, CCE, Joliet Junior College, Culinary Arts/Hotel-Restaurant Management, 1215 Houbolt Ave., Joliet, IL 60431-8938 US; 815-729-9020 x2448, Fax 815-744-5507, URL http://www.jjc.cc.il.us.

LEXINGTON COLLEGE
Chicago/September-May

This independent institution offers a 2-year AAS degree in Hotel, Restaurant and Institutional Management. Program started 1977. Accredited by NCA. Calendar: semester. Curriculum: core. Admission dates: September, January. Total enrollment: 20. 20 each admission period. 50% of applicants accepted. 85% receive financial aid. 10% enrolled part-time. Student to teacher ratio 5:1. 95% of graduates obtain employment within six months. Facilities: Include a culinary lab, 3 classrooms, library, computer lab, bookstore, off-campus site for quantity foods.

Courses: Professional baking, pastries, food and beverage sales and service, garde manger, purchasing, quantity foods, basic food production, professional cooking, food service sanitation, and general education liberal arts and management courses. Other courses required. Externship provided. Continuing education: evenings and weekends.

Faculty: 4 part-time.

Costs: Tuition $3,250 per semester. Books and equipment $300-350 per semester, lab culinary fees $50-100 per semester. Average off-campus housing cost is $1,800 per semester. Admission requirements: High school diploma or equivalent, with 2.0 average or above. Scholarships: yes. Loans: yes.

Location: An urban setting, 20 miles from downtown Chicago.

Contact: Helen Straub, Director of Admissions, Lexington College, Admissions, 10840 S. Western Ave., Chicago, IL 60643-3294 US; 773-779-3800, Fax 773-779-7450.

LINCOLN LAND COMMUNITY COLLEGE
Springfield/Year-round

This community college offers a 2-year diploma, one-year certificate, specialty classes. Program started 1994. Accredited by NCA. Calendar: semester. Curriculum: culinary, core. Admission dates: Year-round. Total enrollment: 20. 20 each admission period. 100% of applicants accepted. 35% receive financial aid. 50% enrolled part-time. Student to teacher ratio 15:1. 100% of graduates obtain employment within six months. Facilities: Training kitchen.

Courses: Culinary with emphasis on nutrition. Students work in area hotels and restaurants and receive credit. Continuing education special topics classes.

Faculty: Eight instructors, three certified by the ACF, one instructor MSRD.

Costs: $39 per credit hour. Scholarships: yes. Loans: yes.

Location: Central Illinois.

Contact: Jay Kitterman, Director, Hospitality Management, Lincoln Land Community College, Hospitality Management, 5250 Shepherd Rd., Springfield, IL 62794 US; 217-786-2772, Fax 217-786-2495, E-mail jkitterm@cabin.llcc.cc.il.us, URL http://www.llcc.cc.il.us.

MORAINE VALLEY COMMUNITY COLLEGE

Palos Hills/Year-round

This community college offers certificate programs in Culinary Arts Management, Baking/Pastry Arts, Beverage Management, and Restaurant/Hotel Management; AAS degree programs in Culinary Arts Management and Restaurant/Hotel Management. Established 1967. Calendar: semester. Curriculum: culinary, core.

Costs: In-district (out-of-district, out-of-state) $44 ($179, $204) per credit-hour.

Location: 20 miles southwest of Chicago.

Contact: Anne Jachim, Moraine Valley Community College, 10900 S. 88th Ave., Palos Hills, IL 60465 US; 708-974-5320, E-mail jachim@moraine.cc.il.us, URL http://www.moraine.cc.il.us.

REND LAKE COLLEGE

Ina/Year-round

This college offers a 1-year/32 credit-hour occupational certificate in Culinary Arts Management. Calendar: semester. Curriculum: culinary.

Courses: Quantity food preparation, nutrition, menu planning, cost management, job strategy.

Costs: Tuition in-district $36 per credit-hour, out-of-district $124 per credit-hour, out-of-state $160 per credit-hour.

Contact: Tim March, Program Director and Instructor, Rend Lake College, 468 N. Ken Gray Pkwy., Ina, IL 62846 US; 618-437-5321/800-369-5321, Fax 618-437-5677, URL http://www.rlc.cc.il.us.

TRITON COLLEGE

River Grove/Year-round

This college offers a 2-year AAS degree in Culinary Management, AAS degree in Hotel Management. Program started 1970. Accredited by NCA, ACFEI. Calendar: semester. Curriculum: culinary, core. Admission dates: August, January. Total enrollment: 150. 30 each admission period. 90% of applicants accepted. 20% receive financial aid. 30% enrolled part-time. Student to teacher ratio 12:1. 97% of graduates obtain employment within six months. Facilities: 2 kitchens and classrooms, demonstration kitchen, ice carving facility, and student-run restaurant.

Courses: Garde manger, international cooking, ice carving, food production, food theory, menu planning, purchasing, cost control, nutrition, service, and baking. Externship provided. Continuing education: international cooking, sanitation, and nutrition.

Faculty: 3 full-time, 10 part-time.

Costs: $43 per credit hour in-district, $128.25 out-of-district. Application fee $25. Other fees approximately $100 per semester. Average off-campus housing cost $300 per month. Admission requirements: High school diploma or equivalent. Scholarships: yes. Loans: yes.

Contact: Jerome J. Drosos, Coordinator, Triton College, Hospitality Industry Administration, 2000 Fifth Ave., River Grove, IL 60171 US; 708-456-0300 x3624, URL http://www.triton.cc.il.us.

WASHBURNE TRADE SCHOOL

Chicago/Year-round

The city colleges of Chicago offers an 80-week certificate in Chef Training. Program started 1937. Accredited by City Colleges of Chicago. Curriculum: culinary. Admission dates: September, January, May. Total enrollment: 150. 25 e0ach admission period. 100% of applicants accepted.

75% receive financial aid. Student to teacher ratio 21:1. 98% of graduates obtain employment within six months. Facilities: 6 kitchens, 6 classrooms.

Courses: 85% hands-on training, 5 days per week. Continuing education: ice carving.

Faculty: 7 full-time.

Costs: Annual tuition of $4,600 includes books and uniforms. Application fee $200 if cash-paying student; none if financial aid. Cutlery set $272. Admission requirements: High school diploma or equivalent and admission test. Scholarships: yes.

Location: The 4-story campus occupies one block in Chicago.

Contact: Dean Jaramillo, Department Program Director, Washburne Trade Program, Chef Training Program, 3233 W. 31st St., Chicago, IL 60623 US; 312-579-6100/312-579-6109, Fax 312-376-5940.

WILLIAM RAINEY HARPER COLLEGE

Palatine/Year-round

This community college offers a 1-year certificate in Culinary Arts and Baking. Program started 1975. Accredited by NCA. Calendar: semester. Curriculum: culinary. Admission dates: Year-round. Total enrollment: 150. 25 each admission period. 95% of applicants accepted. 50% receive financial aid. 75% enrolled part-time. Student to teacher ratio 15:1. 100% of graduates obtain employment within six months. Facilities: Include 3 kitchens including production bakery, production kitchen, demo lab, and several classrooms.

Courses: Garde manger, classical cuisine, cake decorating, basic and advanced culinary, and basic and advanced baking. Other required courses: 200 management contact hours. A variety of long and short-term offerings.

Faculty: 3 full-time, 4 part-time.

Costs: Tuition is $50 per credit in-district, $210.18 per credit out-of-district. Other costs: application fee $20, lab fee $400, books $400. Average off-campus housing cost $500-$800 per month. Scholarships: yes. Loans: yes.

Location: The 27,000-student, 20-building campus is in a suburban setting, 35 miles from Chicago's loop.

Contact: Bruce Borher, Director of Admissions, William Rainey Harper College, 1200 W. Algonquin Rd., Palatine, IL 60067-7398 US; 847-925-6000, Fax 847-925-6031, E-mail bborher@harper.cc.il.us, pbeach@harper.cc.il.us, URL http://www.harper.cc.il.us.

WILTON SCHOOL OF CAKE DECORATING

Woodridge/February-November *(See also page 213)*

This private school offers career-oriented 4- to 10-day cake decoration and candy making diploma courses. Established in 1929. Admission dates: Open. 20 per class each admission period. Student to teacher ratio 15:1. Facilities: The 2,200-square-foot school includes a classroom, teaching kitchen, student lounge, and retail store.

Courses: The 10-day (70-hour) Master Course, which covers basic cake decorating and design techniques; 5-day courses cover chocolate, the Lambeth method, Australian techniques, and catering; 3- and 4-day courses cover gum paste and pulled sugar.

Faculty: Includes Sandra Folsom, Susan Matusiak, Nicholas Lodge, Wesley Wilton, and Elaine Gonzalez.

Costs: Range from $75 for 1 day to $650 for a 10-day course.

Location: A southwestern Chicago suburb, 25 miles from downtown.

Contact: School Secretary, Wilton School of Cake Decorating and Confectionery Art, 2240 W. 75th St., Woodridge, IL 60517 US; 630-963-7100 x211, Fax 630-963-7299.

INDIANA

BALL STATE UNIVERSITY
Muncie/Year-round

This university offers 2- and 4-year degree programs in Food Management. Established 1975. Admission dates: August, January, May. Total enrollment: 35. Student to teacher ratio 17:1. 100% obtain jobs within six months. Facilities: Classrooms, computer lab, 3 kitchen production labs with the latest equipment.

Faculty: All instructors have advanced college degrees and industry experience.

Costs: Annual tuition is $3,316 in-state, $8,872 out-of-state. Room and board $4,120. Admission requirements: Minimum score of 920 on SAT or 19 ACT, class rank in top 50%.

Contact: Lois Altman, Ed.D., Associate Professor, Ball State University, 2000 University Ave., Muncie, IN 47306; 765-285-5931, Fax 765-285-2314, E-mail 00laaltman@bsu.edu, URL http://www.bsu.edu.

IVY TECH STATE COLLEGE
East Chicago, Gary, Valparaiso/Year-round

This college offers a 2-year AAS degree and 1-year technical certificate in Hospitality Administration. Program started 1981. Accredited by NCA. Calendar: semester. Curriculum: core. Admission dates: January, May, August. Total enrollment: 100. 15 each admission period. 90% of applicants accepted. 50% receive financial aid. 30% enrolled part-time. Student to teacher ratio 12:1. 100% of graduates obtain employment within six months. Facilities: Include 4 kitchens at 3 campuses, catering facilities, restaurant, bakeshop.

Courses: Include portfolio program: wine, culinary/hotel specialty, catering, baking. Externship: 16 weeks, 144 hours. Continuing ed: French studies with Premier Sommelier and Chef of France.

Faculty: 2 full-time, 7 part-time. Program director Sharon Purdy.

Costs: $66.50 per credit-hour in-state, $123.50 out-of-state. Admission requirements: High school diploma or equivalent and admission test. Scholarships: yes. Loans: yes.

Contact: Sharon Purdy, Program Chair, Ivy Tech State College, Hotel & Restaurant Management/Culinary Arts, 410 E. Columbus Dr., East Chicago, IN 46312 US; 219-392-3600 x18/981-4400, Fax 219-981-4415. Additional address: 1440 E. 35th Ave., Gary, IN 46409. Additional phones: 219-981-1111/4400, URL http://www.ivy.tec.in.us.

IVY TECH STATE COLLEGE
Fort Wayne/Year-round

This college offers a 2-year AAS degree in Hospitality Administration with Culinary Arts or Pastry Arts Specialty. Program started 1981. Accredited by NCA, ACFEI. Calendar: semester. Curriculum: culinary. Admission dates: Year-round. Total enrollment: 140. 75 each admission period. 100% of applicants accepted. 33% receive financial aid. 42% enrolled part-time. Student to teacher ratio 10-12:1. 100% of graduates obtain employment within six months. Facilities: Include 5 kitchens and classrooms, pastry arts lab, large full service kitchen with top-of-the-line equipment.

Courses: Basic foods, soups, stocks and sauces, nutrition, meat cutting, special cuisines, classical cuisines, fish and seafood, pantry and breakfast, garde manger, catering, breads and pastries, cake decoration, chocolates and baking. Externship: 144-hour, salaried, in an ACF approved site.

Faculty: 2 full-time, 12 part-time. Includes Program Chair Alan Eyler, CCE, CFBE, and Chef Instructor Jerry Wilson with 35 years industry experience.

Costs: Annual tuition $1,835 in-state, $3,335 out-of-state. Other fees, books, uniforms, knife kits, specialty tools. Admission requirements: High school diploma or equivalent.

Location: The small campus is in an urban community about 200 miles from Chicago or Indianapolis.

Contact: Alan Eyler, CCE, CFBE, Program Chair, Ivy Tech State College, Hospitality Administration, 3800 N. Anthony Blvd., Fort Wayne, IN 46805 US; 219-480-4240, Fax 219-480-4171, E-mail aeyler@ivy.tec.in.us.

IVY TECH STATE COLLEGE

Indianapolis/Year-round

This college offers a 2-year AAS degree in Culinary Arts, Baking and Pastry Arts, and Hotel Restaurant Management. Program started 1986. Accredited by NCA, ACFEI, CAHM. Calendar: semester. Admission dates: August, January, May. Total enrollment: 200. 120 each admission period. 100% of applicants accepted. 75% receive financial aid. 50% enrolled part-time. Student to teacher ratio 10:1. 100% of graduates obtain employment within six months. Facilities: 3 kitchens, classrooms, 1 restaurant.

Courses: Include basic food theory and skills, sanitation, classical French techniques. Externship: 5 months. 2+2 programs set up with more than 28 4-year colleges and universities.

Faculty: 3 full-time, 12 part-time. Qualifications: Associate degree, 5 years experience, certifiable.

Costs: Annual tuition $3,000 in-state, $4,650 out-of-state. Average off-campus housing cost $300 per month. Admission requirements: High school diploma or equivalent and admission test.

Contact: Chef Vincent Kinkade, Chair, Ivy Tech State College, Hospitality Admin., One W. 26th St., Indianapolis, IN 46208 US; 317-921-4619, Fax 317-921-4753, URL http://www.ivy.tec.in.us/.

VINCENNES UNIVERSITY

Vincennes/August-May

This public institution offers a 2-year AS degree in Culinary Arts. Program started 1983. Accredited by NCA. Calendar: semester. Curriculum: culinary, core. Admission dates: Open. Total enrollment: 60. 30+ each admission period. 100% of applicants accepted. 90% receive financial aid. 5% enrolled part-time. Student to teacher ratio 12:1. 100% of graduates obtain employment within six months. Facilities: Include kitchen, 3 classrooms, hands-on lab and restaurant.

Courses: Quantity foods, pastry and bake shop, haute cuisine, food facility design, hospitality, sanitation, purchasing, supervision, core curriculum 22 to 23 hours. Externship provided. Continuing education: ACF regional chefs conduct hands-on classes.

Faculty: 2 full-, 1 part-time. Qualifications: combined 58 years in industry, AS and BA degrees.

Costs: Annual tuition $2,000 in-state, $5,200 out-of-state. Application fee $20. Student activities fee $18. On-campus housing: 3,000 spaces. Admission requirements: High school diploma or equivalent. Scholarships: yes. Loans: yes.

Contact: Robert H. Bird CCE, CFE, Assistant Professor, Indiana ACF State Rep, Vincennes University, Culinary Arts, Hoosier Hospitality Center, Governor's Hall, Vincennes, IN 47591 US; 812-888-5741, Fax 812-888-4586, E-mail rbird@vunet.vinu.edu, URL http://www.vinu.edu.

IOWA

DES MOINES AREA COMMUNITY COLLEGE

Ankeny/Year-round

This college offers a 2-year AAS degree in Culinary Arts. Program started 1975. Accredited by NCA, ACFEI. Admission dates: Fall, spring. Total enrollment: 50. 20 each admission period. 100% of applicants accepted. 60% receive financial aid. 25% enrolled part-time. Student to teacher ratio 15:1. 90% of graduates obtain employment within six months. Facilities: Include 2 kitchens, demonstration lab, several classrooms and restaurant.

Courses: Externship provided.

Faculty: 2 full-time, ACF-certified.

Costs: $59.40 per credit hour in-state, $112.40 out-of-state. Application fee $10. Average off-campus housing cost $300 per month. Admission requirements: High school diploma or equivalent and admission test.

Location: The 12,000-student suburban campus is 10 miles from Des Moines.

Contact: Robert Anderson, Des Moines Area Community College, Culinary Arts, 2006 S. Ankeny Blvd., Ankeny, IA 50021 US; 515-964-6532, Fax 515-964-6486, URL http://www.dmacc.cc.ia.us.

INDIAN HILLS COMMUNITY COLLEGE

Ottumwa/Year-round

This college offers an 18-month AAS degree in Culinary Arts, 9-month diploma in Culinary Assistant. Program started 1969. Accredited by NCA. Calendar: semester. Curriculum: culinary. Admission dates: Fall, spring. Total enrollment: 35. 15-20 each admission period. 100% of applicants accepted. 85% receive financial aid. Student to teacher ratio 12:1. 97% of graduates obtain employment within six months. Facilities: Include 3 kitchens, 3 classrooms, student-run dining room.

Faculty: 4 full-time. Includes Mary Kivlahan, Tom Shepard, Lisa Larson.

Costs: $57 per credit-hour in-state, $82 per credit-hour out-of-state. On-campus housing: 472 spaces. Average off-campus housing cost is $350 per month.

Contact: Tom Shepard, Program Director, Indian Hills Community College, Culinary Arts, 525 Grandview, Ottumwa, IA 52501 US; 515-683-5195, Fax 515-683-5184, URL http://www.ihcc.cc.ia.us.

IOWA LAKES COMMUNITY COLLEGE

Emmetsburg/Year-round

This college offers a 2-year AAS degree in Culinary Arts and/or Hotel, Motel, Restaurant Management program. Program started 1974. Accredited by NCA, NRA, AHMA. Calendar: semester. Curriculum: core. Admission dates: September, January. Total enrollment: 30-40. 95% of applicants accepted. 70% receive financial aid. Student to teacher ratio 10-12:1. 95% of graduates obtain employment within six months. Facilities: 2 kitchens, 2 classrooms, 1 restaurant, 1 banquet facility.

Courses: Externship provided.

Costs: Annual tuition $2,500 in-state, $3,000 out-of-state. Dormitory spaces available on campus. Admission requirements: High school diploma or equivalent and admission test. Scholarships: yes. Loans: yes.

Contact: R. Halverson, Professor/Coordinator, Iowa Lakes Community College, So. Attendance Ctr., Culinary Arts, 3200 College Dr., Emmetsburg, IA 50536 US; 712-852-3554 x256, Fax 712-852-2152, E-mail rhalverson@ilcc.cc.ia.us.

IOWA WESTERN COMMUNITY COLLEGE

Council Bluffs/August-May

This college offers a 2-year AAS degree in Culinary Arts. Program started 1974. Accredited by NCA. Admission dates: Fall, spring. Total enrollment: 30-40. 10-20 each admission period. 95% of applicants accepted. 80% receive financial aid. 1% enrolled part-time. Student to teacher ratio 10-12:1. 95% of graduates obtain employment within six months. Facilities: Include kitchen and 2 classrooms.

Faculty: 2 full-time, 3 part-time. P. Swope, B. Gauke, B. Leeder, CCE, N. Johnson, L. Harrill.

Costs: Tuition in-state $58 per credit hour, out-of-state $81 per credit hour. Room and board $1,500-$2,000 per semester. Admission requirements: High school diploma or equivalent.

Location: The 4,000-student school is in a suburban area across the river from Omaha.

Contact: Robert Graunke, Coordinator, Iowa Western Community College, Food Service Management/Culinary Arts, 2700 College Rd., Box 4-C, Council Bluffs, IA 51502 US; 712-325-3378/712-325-3236, Fax 712-325-3424.

KIRKWOOD COMMUNITY COLLEGE

Cedar Rapids/Year-round

This state institution offers a bakery certificate and 2-year AAS degree in Culinary Arts and Restaurant Management, AAS Dietetic Technician. Program started 1972. Accredited by NCA, ACFEI. Calendar: semester. Curriculum: culinary, core. Admission dates: Fall, spring, summer. Total enrollment: 140. 50-60 each admission period. 100% of applicants accepted. 50% receive financial aid. 25% enrolled part-time. Student to teacher ratio 12:1. 97% of graduates obtain employment within six months. Facilities: Include 1 kitchen, 2 classrooms, restaurant and bakery.

Courses: Food production, culinary arts, garde manger, food and culture, artistic display, bakery, wines, purchasing, menu planning, nutrition, restaurant law, sanitation, personnel management, and general education courses. Other required courses: 14 credit hours. Hospitality internships offered in summer sessions.

Faculty: 3 full-time, 6 part-time. Qualifications: college degrees and industry experience. Includes Mary Jane German, MS, RD, Mary Rhiner, MA, RD, David Dettman, Charles Allen, David Dukes, Janelle Kamerling, Amy Wyss, Linda Rollie.

Costs: Tuition in-state $60 per credit hour, out-of-state $120 per credit hour. Average off-campus housing cost $375 per month. Admission requirements: High school diploma or equivalent and admission test. Scholarships: yes. Loans: yes.

Location: The 10,000-student campus has 9 off-campus sites and 13 other buildings.

Contact: Mary Jane German, Instructor/Coordinator, Kirkwood Community College, 6301 Kirkwood Blvd. S.W., PO Box 2068, Cedar Rapids, IA 52406 US; 319-398-4981, Fax 319-398-5667, E-mail mgerman@kirkwood.cc.ia.us. Additional contact: Mary Rhiner.

SCOTT COMMUNITY COLLEGE

Bettendorf/Year-round

This college offers a 3-year AAS degree and 6,000-hour apprenticeship in sanitation and cook certification from ACF. Program started 1991. Accredited by ACFEI. Admission dates: Fall. Total enrollment: 30. 10-15 each admission period. 60% of applicants accepted. 75% receive financial aid. 5% enrolled part-time. Student to teacher ratio 10:1. 100% of graduates obtain employment within six months.

Courses: Nutrition, sanitation, menu planning, management, beverages, garde manger, hot food, baking, purchasing, and general education courses. Externship: 6,000-hour, salaried, in restaurant or hotel setting.

Faculty: 1 full-time, 8 part-time. Qualifications; chef instructors certified by ACF, lecture instructors 4-year degrees, all have industry experience and ACF membership.

Costs: $58.50 per credit hour in-state. Other costs: application fee $25, books, uniform, knives, ACFEI registration $650 (one-time). Off-campus housing cost approximately $300-$400 per month. Admission requirements: Admission test.

Contact: Brad Scott, Scott Community College, Culinary Arts, 500 Belmont Rd., Bettendorf, IA 52722-6804 US; 319-359-7531 x278, Fax 319-344-0384, URL http://otis.ciccd.cc.ia.us.

KANSAS

AMERICAN INSTITUTE OF BAKING

Manhattan/September-June *(See also page 313)*

This nonprofit educational and research institution offers a 16-week Baking Science and Technology course, 10-week Bakery Maintenance Engineering program. Established in 1919. Accredited by NCA. Calendar: semester. Curriculum: culinary. Admission dates: August/September, February. 90% of applicants accepted. 25% receive financial aid. Student to teacher ratio 15:1. 95% of graduates obtain employment within six months. Facilities: Include a bread shop with 1,500 loaves per hour capacity oven, cake shop with carbon dioxide freezer, in-store bakery, cookie and cracker plant.

Courses: The Baking Science courses include cake and sweet goods production, bread and roll production, and food product safety. The Maintenance Engineering courses include refrigeration, basic electricity, and motor controls. Continuing ed and correspondence courses include the 50-lesson Science of Baking course.

Faculty: 7 full-time.

Costs: 16-week program is $3,000. Registration fee is $45. Admission requirements: High school diploma or equivalent and at least 2 years of bakery experience (or completion of Bakery Science correspondence course). Scholarships: yes.

Location: The 87,000-square-foot facility is on 13 acres overlooking the Kansas State University campus, 120 miles west of Kansas City.

Contact: American Institute of Baking, 1213 Bakers Way, Manhattan, KS 66502 US; 800-633-5137/913-537-4750, Fax 913-537-1493.

JOHNSON COUNTY COMMUNITY COLLEGE

Overland Park/Year-round

This college offers a 2- to 3-year AOS degree. Program started 1975. Accredited by NCA, ACFEI. Calendar: semester. Admission dates: July, November. Total enrollment: 500. 140 each admission period. 80% of applicants accepted. 100% enrolled part-time. Student to teacher ratio 20:1. 100% of graduates obtain employment within six months.

Courses: Paid externship provided.

Faculty: 10 full-time.

Costs: Annual tuition $1,700 in-state, $5,100 out-of-state. Average off-campus housing cost $500 per month. Admission requirements: High school diploma or equivalent and admission test. Scholarships: yes. Loans: yes.

Contact: Jerry Vincent, Academic Director, Johnson County Community College, Business and Technology, 12345 College at Quivira, Overland Park, KS 66210-1299 US; 913-469-8500, Fax 913-469-2560, E-mail j.vincent@johnco.cc.ks.us.

KANSAS CITY KANSAS AREA VOCATIONAL TECHNICAL SCHOOL

Kansas City/August-May

This public institution offers a 720-hour certificate in Professional Cooking, certificate in Cooking and Baking. Program started 1975. Accredited by State. Calendar: quarter. Curriculum: culinary. Admission dates: Open. Total enrollment: 20. 99% of applicants accepted. 60% receive financial aid. 50% enrolled part-time. Student to teacher ratio 5:1. 88% of graduates obtain employment within six months. Facilities: Working kitchen, classroom, cafeteria, child care center, banquet facilities.

Courses: Food preparation, cooking and presentation; safety and sanitation, baking, serving. Externship provided.

Faculty: 4 full-time. Includes Sharyn Gassmann, BS, MS, M. Mollentine, L. Benson, S. Cole.

Costs: Annual tuition $780. Application fee $25. Other fees: $50. Admission requirements: Admission test.

Location: 5 miles west of downtown Kansas City.

Contact: Sharyn Gassmann, Program Director, Kansas City Kansas Area Vocational Technical School, 2220 W. 59th St., Kansas City, KS 66104 US; 913-596-5500, Fax 913-596-5509.

NORTHEAST KANSAS AREA VOCATIONAL TECHNICAL SCHOOL

Atchison/August-May

This public school offers a 1-year diploma in Culinary Arts. Program started 1969. Accredited by State. Admission dates: Open. Total enrollment: 12. 90% receive financial aid. 30% enrolled part-time. Student to teacher ratio 15:1. 100% of graduates obtain employment within six months. Facilities: Include kitchen and student-run restaurant.

Courses: Professional cooking, baking, purchasing, sanitation. Externship provided.

Faculty: 1 full-time.

Costs: Tuition in-state $598 per semester full-time ($299 part-time), out-of-state $3,990 per semester full-time ($1,995 part-time).

Contact: Marianne Estes, Northeast Kansas Area Vocational Technical School, 1501 West Riley, Atchison, KS 66002 US; 913-367-6204, Fax 913-367-3107.

WICHITA AREA TECHNICAL COLLEGE

Wichita/August-May

This technical college offers a 9-month certificate. Program started 1975. Accredited by State. Calendar: semester. Curriculum: core. Admission dates: August. Total enrollment: 20. 95% of applicants accepted. Student to teacher ratio 6:1. 95% of graduates obtain employment within six months.

Courses: Culinary knowledge and skills.

Faculty: 6 full-time, 2 certified culinary educators on staff.

Costs: Annual tuition is approximately $1,600. Admission requirements: High school diploma or equivalent and admission test.

Contact: Colette Baptista, CEC, CCE, Coordinator, Wichita Area Technical College, Food Service, 324 N. Emporia, Wichita, KS 67202 US; 316-833-4360, Fax 316-833-4341, URL http://www.watc.tec.ks.us. Additional contact: Dan Hypse, CCE, CCC.

KENTUCKY

JEFFERSON COMMUNITY COLLEGE

Louisville/August-May

This college offers a 2-year AAS degree. Program started 1974. Accredited by ACFEI, SACS. Curriculum: culinary, core. Admission dates: August. Total enrollment: 22. 22 each admission period. 90% of applicants accepted. 40% enrolled part-time. Student to teacher ratio 11:1. 96% of graduates obtain employment within six months. Facilities: 2 kitchens and 2 classrooms.

Courses: Food preparation, American and European pastries, garde manger, menu planning, and catering. Other required courses: nutrition, sanitation, management, food cost and portion control. Externship provided.

Faculty: 2 full-time, 1 part-time.

Costs: Annual tuition $1,140 in-state, $3,260 out-of-state. Admission requirements: High school diploma or equivalent and admission test.

Contact: Gail Crawford, Program Coordinator, Jefferson Community College, Downtown Campus, 109 E. Broadway, Louisville, KY 40202 US; 502-584-0181 x2317, Fax 502-584-0181 x2414, E-mail gcrawfor@pop.jcc.uky.edu.

KENTUCKY TECH – DAVIESS COUNTY CAMPUS

Owensboro/August-June

This independent institution offers a 4- to 5-semester certificate/diploma in Culinary Arts. Program started 1971. Accredited by SACS. Calendar: semester. Curriculum: culinary. Admission dates: August. Total enrollment: 24. 2-6 each admission period. 100% of applicants accepted. 90% receive financial aid. 50% enrolled part-time. Student to teacher ratio 18:1. 85% of graduates obtain employment within six months. Facilities: Include kitchen and classroom.

Courses: Include culinary and general education courses. 150 hours of co-op and occupational training. Continuing education: cake decorating.

Faculty: One full-time, Dudley Mitchell. Qualifications: BS in Home Economics/Occupational Foodservice, meets state requirements.

Costs: $175 per quarter. Application fee $25. Admission requirements: TABE test, high school diploma or equivalent. Scholarships: yes. Loans: yes.

Location: Western Kentucky, 125 miles from Louisville.

Contact: Kaye Evans, Counselor, Kentucky Tech-Daviess County Campus, Student Services, 15th and Frederica St., Owensboro, KY 42301 US; 502-687-7260, Fax 502-687-7208.

KENTUCKY TECH ELIZABETHTOWN

Elizabethtown/August-June

This public institution offers a 17-month diploma in Food Service Technology. Program started 1966. Accredited by COE. Calendar: semester. Curriculum: core. Admission dates: Continuous; registration July, January. Total enrollment: 18. 18 each admission period. 100% of applicants accepted. 80% receive financial aid. Student to teacher ratio 18:1. 95% of graduates obtain employment within six months. Facilities: Kitchen, classroom and restaurant.

Courses: Food service, quantity food production, short order cooking, bakery, and cake decorat-

ing. Externship: 5 months (150 hours). Continuing education courses available.

Faculty: 2 full-time, Brenda Harrington and Maxine Terrill.

Costs: Annual tuition $640 in-state, $1,280 out-of-state. Admissions fee $20 and books. Off-campus housing cost is $2,464. Admission requirements: High school diploma or equivalent and admission test. Scholarships: yes.

Location: The 20-acre campus is in a small town 40 miles from Louisville.

Contact: Rene J. Emond, Registrar, Kentucky TECH Elizabethtown, Food Service Technology, 505 University Dr., Elizabethtown, KY 42701 US; 502-766-5133 x122/128, Fax 502-737-0505, E-mail kte@kvnet.org.

SULLIVAN COLLEGE'S NATL. CENTER FOR HOSPITALITY STUDIES

Louisville/Year-round *(See display ad page 53)*

This division of Sullivan College offers 18-month AS degree programs in Culinary Arts, Baking and Pastry Arts, Hotel/Restaurant Management, Professional Catering, and Travel and Tourism. Established in 1987. Accredited by SACS, ACFEI. Calendar: quarter. Curriculum: core. Admission dates: January, March, June, September. Total enrollment: 324. 100 each admission period. 95% of applicants accepted. 93% of students receive financial aid. Student to teacher ratio is 17:1. 100% of graduates obtain employment within six months. Facilities: A la carte kitchen, three bakery labs, international, garde manger, basic skills and computer labs, retail bakery, fine dining restaurant (Winston's).

Courses: Include theory and skills, regional and international cuisine and pastry, business management, nutrition and meal planning, and menu design and layout. Students also participate in a 400-hour practicum at Winston's Restaurant. Externship provided.

Faculty: 45-member resident faculty and a 38-member adjunct faculty. Includes Dean Newal Hunter, Jr.

Costs: Tuition is $19,890. Comprehensive supplies fee is $850 per lab. Nearby apartments are $310 per month. Admission requirements: High school diploma or equivalent. Scholarships: yes. Loans: yes.

Location: Watterson Expressway and Bardstown Road in suburban Jefferson County.

Contact: Greg Cawthon, Director of Admissions, Sullivan College's National Center for Hospitality Studies, Watterson Expwy at Bardstown Rd., Box 33-308, Louisville, KY 40232 US; 800-844-1354/502-456-6504, Fax 502-454-4880, E-mail admissions@sullivan.edu.

WEST KENTUCKY TECHNICAL COLLEGE

Paducah

This career institution offers an 18-month diploma/degree. Program started 1979. Accredited by SACS. Admission dates: July, October, January, March, June. Total enrollment: 36. Student to teacher ratio 18:1. 80% of graduates obtain employment within six months.

Faculty: 2 full-time.

Costs: Annual tuition in-state $320 per semester full-time, out-of-state $640. Admission requirements: High school diploma or equivalent and admission test.

Contact: Mary Sanderson, Food Service Technology Instructor, West Kentucky Technical College, Culinary Arts, 5200 Blandville Rd., Box 7408, Paducah, KY 42002-7408 US; 502-554-4991, Fax 502-554-9754, URL http://www.wkytech.com.

LOUISIANA

BOSSIER PARISH COMMUNITY COLLEGE

Bossier City

This college offers a 9-month certificate. Program started 1986. Accredited by ACFEI. Admission dates: August. Total enrollment: 25. 98% of applicants accepted. Student to teacher ratio 13:1. 100% of graduates obtain employment within six months.

Faculty: 2 full-time, 4 part-time.

Costs: Annual tuition $3,100. Admission requirements: High school diploma or equivalent and admission test. Scholarships: yes. Loans: yes.

Contact: Elizabeth Dickson, Chef/Coordinator, Bossier Parish Community College, Culinary Arts, 2719 Airline Drive North, Bossier City, LA 71111 US; 318-747-4567, Fax 318-742-8664.

CHEF JOHN FOLSE CULINARY INSTITUTE

(See also page 216) (See display ad above) **Thibodaux/Year-round**

This university offers a 2-year AS degree and a 4-year BS degree in Culinary Arts. Established in 1994. Accredited by SACS. Calendar: semester. Curriculum: culinary, core. Admission dates: Rolling. Total enrollment: 110. 76% of applicants accepted. 60% receive financial aid. 25% enrolled part-time. Student to teacher ratio 15:1. Facilities: 2 newly-equipped teaching kitchens, 2 demonstration classrooms.

Courses: Classic culinary knowledge and regional American cuisine, culinary operations, product development. Sophomore and senior externship, each 360 hours paid work experience. Externships throughout U.S. Continuing education courses year-round.

Faculty: The 5 full- and 2 part-time instructors are Jerald W. Chesser, Ed.D., CEC, CCE, Chef John Folse, CEC, AAC, Carol Gunter, BS, Louis Jesowshek, CEC, AAC, Cornelia Price, MS, Clifford Whithem, Ph.D.

Costs: Semester tuition (full-time) $1,008 in-state, $2,304 out-of-state. Additional fees and equipment $725-$750 + $50-$70 per lab course. Off campus housing rates vary. Admission requirements: Minimum GPA, ACT score, or top 25% of high school class. Scholarships: yes.

Location: South Louisiana, 45 minutes from New Orleans, 75 minutes from Baton Rouge.

Contact: Dr. Jerald Chesser, Dean, Chef John Folse Culinary Institute, Nicholls State University, PO 2099, Thibodaux, LA 70310 US; 504-449-7100, Fax 504-449-7089, E-mail jfci-jwc@nich-nsunet.nich.edu, URL http://server.nich.edu/~jfolse.

CULINARY ARTS INSTITUTE OF LOUISIANA

(See display ad page 56) **Baton Rouge/Year-round**

This private institution offers a 15-month (1,800-clock-hour) AOS degree in Nutrition, Sanitation, and Restaurant Management. Established in 1988. Accredited by ACCSCT. Calendar: quarter. Curriculum: core. Admission dates: Monthly. Total enrollment: 60. 10-20 each admission period. 99% of applicants accepted. 75% receive financial aid. Student to teacher ratio Very low. 100% of graduates obtain employment within six months. Facilities: Include a white linen restaurant and commercial kitchen with stations for garde manger, patisserie, baking, and sate grill.

Courses: Include culinary theory and technique, restaurant management, ice carving, nutrition, and general subjects. Continuing education courses: advanced classical French, Cajun/Creole, tableside service.

Faculty: Industry professionals.

Costs: Tuition is $17,339, which includes meals, uniforms, insurance, textbooks. A $50 nonre fundable fee must accompany application 30 days in advance. Off-campus lodging is $200-$400 per month. Admission requirements: High school graduate with some cooking experience. Scholarships: yes. Loans: yes.

Location: A leased hotel building overlooking the Mississippi River.

Contact: Maureen Harrington, Director, Culinary Arts Institute of Louisiana, 427 Lafayette St., Baton Rouge, LA 70802 US; 800-927-0839/504-343-6233, Fax 504-336-4880, URL http://www.explore-br.com-caila.

DELGADO COMMUNITY COLLEGE

New Orleans/Year-round

This two-year college offers a 3-year/6,000-hour apprenticeship AAS degree program. Accredited by ACFEI. Calendar: semester. Curriculum: core.

Costs: $675 per semester in-state, $2,000 out-of-state.

Contact: Iva Bergeron, Director of Culinary Arts, Delgado Community College-Culinary Arts Dept., 615 City Park Ave., New Orleans, LA 70119 US; 504-483-4208, Fax 504-483-4893, E-mail iberg@dcc.edu, URL http://www.dcc.edu.

LAFAYETTE REGIONAL TECHNICAL INSTITUTE

Lafayette

This career institute offers a program in culinary arts and occupations.

Contact: Rafael Galindo, Lafayette Regional Technical Institute, 1101 Bertrand Dr., Lafayette, LA 70502 US; 318-262-5962, Fax 318-262-5122.

LOUISIANA TECHNICAL COLLEGE – BATON ROUGE

Baton Rouge/Year-round

This public institution offers a 1-year diploma/certificate in Culinary Arts. Program started 1974. Calendar: quarter. Curriculum: culinary. Admission dates: Year-round. Total enrollment: 10-15. 6 each admission period. 95% of applicants accepted. 45% receive financial aid. 20% enrolled part-time. Student to teacher ratio 12:1. 95% of graduates obtain employment within six months. Facilities: Include 2 kitchens.

Courses: Sanitation, nutrition, food and beverage management, professional cooking.

Faculty: 1 full-time. Michael Travasos. Qualifications: bachelor's degree, industry experience.

Costs: Annual tuition $420. Application fee $9.50. Books, uniforms, equipment $215. Average off-campus housing cost is $300 per month.

Contact: Rose Fair, Louisiana Technical College-Baton Rouge Campus, Admissions, 3250 N. Acadian Throughway, Baton Rouge, LA 70805 US; 504-359-9202, Fax 504-359-9296, URL http://www.brti.tec.la.us.

LOUISIANA TECHNICAL COLLEGE – SOWELA

Lake Charles/Year-round

This two-year technical college offers an 18-month (1800-clock-hour/6-quarter) diploma program in Culinary Arts and Occupations. Accredited by COE. Calendar: quarter. Curriculum: culinary. Admission dates: August, November, March, May. Total enrollment: 9-35. Student to teacher ratio 10-15:1. Facilities: Industrial kitchen and bakery with the latest equipment.

Courses: Emphasis on planning, selecting, purchasing, preparing and serving quality food products. Includes nutrition, menu planning, food processing and production, quality cooking, equipment use, sanitation and safety.

Faculty: 2 full- and 2 part-time executive chefs and professors.

Costs: $105 per quarter. Admission requirements: 10th grade level score on entrance test.

Contact: Susan Simmons, Dept. Head, Student Services, Louisiana Technical College, Culinary Arts & Occupations, 3820 J. Bennett Johnston Ave., Lake Charles, LA 70615 US; 800-256-0483, 318-491-2688, Fax 318-491-2054, URL http://www.techrome.latech.edu.

NEW ORLEANS REGIONAL TECHNICAL INSTITUTE

New Orleans/Year-round

This career college offers a program in culinary arts and occupations. Program started 1982. Accredited by ACFEI, COE. Calendar: quarter. Curriculum: culinary. Admission dates: Year-round. Total enrollment: 35. Open each admission period. 95% of applicants accepted. 50% receive financial aid. 20% enrolled part-time. Student to teacher ratio 12:1. 75% of graduates obtain employment within six months. Facilities: Commercial kitchen.

Courses: On-the-job training agreements. Certification for Management/Sanitation/Nutrition.

Faculty: Christina Nicosia, CCE, CEPC, FMP; John Ryder.

Costs: $630 per 6 terms + equipment and books. Admission requirements: TABE test.

Contact: Christina Nicosia, CCE, CEPC, FMP, Program Director, New Orleans Regional Technical Institute, Culinary Arts, 980 Navarre Ave., New Orleans, LA 70124-2710 US; 504-483-4626, Fax 504-483-4643.

NUNEZ COMMUNITY COLLEGE

Chalmette/Year-round

This two-year college offers a 1-year/33-credit-hour certificate and 2-year/68-credit-hour AAS degree in Culinary Arts and Occupations. Calendar: semester. Curriculum: culinary, core.

Courses: Basic food preparation, baking, meat, poultry, seafood, soups, stocks, sauces, garde manger, patisserie, food and beverage purchasing, cost control.

Costs: For 12+ credit-hours: in-state $488, out-of-state $1,523 plus technology fee.

Contact: Nunez Community College-Culinary Arts, 3700 LaFontaine St., Chalmette, LA 70043 US; 504-278-7440, Fax 504-278-7463, URL http://www.nunez.cc.la.us.

SCLAFANI'S COOKING SCHOOL, INC.

(See display ad page 58)

Metairie/Year-round

This proprietary school offers a 4-week (120-hour) certificate of completion program in Commercial Cooking/Baking. Established in 1987. Licensed by State Board of Elementary and Secondary Education. Calendar: quarter. Curriculum: culinary, core. Admission dates: Monthly. Total enrollment: 95. 8 each admission period. 85% of applicants accepted. 60% receive financial aid. Student to teacher ratio 8:1. 98% of graduates obtain employment within six months. Facilities: Classroom/dining room, commercial kitchen preparation room, storage area.

Courses: Culinary arts, baking, food cost math, sanitation and safety, supervisory skills. Continuing education: 5 points towards ACFEI certification, 120 points towards re-certification.

Faculty: 2 full-time. Includes Administrative Instructor Frank P. Sclafani, Sr., CEC, and chef/instructor Sable Bell.

Costs: Tuition is $2,145. Average off-campus housing cost: $15 per day and up. Admission requirements: Age 18 or older; must pass 7th grade level reading and math.

Location: 3 miles from New Orleans.

Contact: Frank P. Sclafani, Sr., CEC, President, Sclafani's Cooking School, Inc., Culinary Arts, 107 Gennaro Pl., Metairie, LA 70001 US; 504-833-7861/800-583-1282, Fax 504-833-7872, E-mail sclafani@gnofn.org, URL http://www.sclafanicookingschool.com.

SIDNEY N. COLLIER VOCATIONAL TECHNICAL INSTITUTE

New Orleans/Year-round

This career institution offers an 18-month certificate. Program started 1957. Accredited by SACS. Admission dates: Open. Total enrollment: 20. 100% of applicants accepted. Student to teacher ratio 20:1. 80% of graduates obtain employment within six months.

Faculty: 30 full-time, 1 part-time.

Costs: $105 per quarter, $630 for program (6 quarters/18 months).

Contact: Edward James, Instructor, Sidney N. Collier Vocational Technical Institute, Culinary Arts, 3727 Louisa St., New Orleans, LA 70126 US; 504-942-8333, Fax 504-942-8337.

MAINE

SOUTHERN MAINE TECHNICAL COLLEGE

South Portland/September-May

This state-owned institution offers a 2-year diploma and associate degree in Culinary Arts. Program started 1958. Accredited by NEASC. Calendar: semester. Curriculum: culinary, core. Admission dates: Rolling. Total enrollment: 70. 80 each admission period. 75% of applicants accepted. 50% receive financial aid. 10% enrolled part-time. Student to teacher ratio 16:1. 90% of graduates obtain employment within six months. Facilities: Include 8 kitchens and classrooms, restaurant.

Courses: Baking, food development, buffet, classical cuisine, dining room management, and general education courses. Continuing education: bartending and cake decorating.

Faculty: 6 full-time with college degrees and/or ACF certification.

Costs: Annual tuition in-state $3,100, out-of-state $5,800. On-campus housing: 100 spaces; off-campus housing cost $50 per week. Admission requirements: High school diploma or equivalent and admission test. Scholarships: yes. Loans: yes.

Contact: Robert Weimont, Director of Admissions, Southern Maine Technical College, 2 Fort Rd., South Portland, ME 04106 US; 207-767-9520, Fax 207-767-9671.

MARYLAND

BALTIMORE INTERNATIONAL COLLEGE

Baltimore and Ireland/Year-round *(See also page 311)*

This private nonprofit college specializing in culinary arts, hospitality, business, and management offers associate degrees in Professional Cooking, Professional Baking and Pastry, Professional Cooking and Baking, Food and Beverage Management, and certificates in Professional Cooking, Professional Baking and Pastry and Culinary Arts. Established in 1972. Accredited by MSA.

Calendar: semester. Curriculum: culinary, core. Admission dates: January, May, July, September. Total enrollment: 807. 275-560 each admission period. 95% of applicants accepted. 90% receive financial aid. Student to teacher ratio 13:1. 85%-90% of graduates obtain employment within six months. Facilities: Include 30 kitchens and classrooms, hotels, restaurants, bakeshops, lecture theaters, and 100-acre Park Hotel-Deer Park Lodge estate in Ireland.

Courses: Each program builds from a foundation of theories and techniques to advanced techniques and special projects. Core courses include math, science, English, economics, history, and psychology. Degree programs include 2-1/2 to 5 weeks study in Ireland plus a 3-month externship. Epicurean diploma, demonstration courses, Elderhostel.

Faculty: 45-member faculty. Academic faculty members hold degrees through the doctorate level. European-trained chefs at the College's European campus hold credentials from the City and Guilds of London.

Costs: $3,360 per semester tuition, $572-$1,420 per semester fees. On-campus lodging for 200 students begins at $1,680 per semester. Admission requirements: High school diploma or equivalent and satisfactory admissions test score. Scholarships: yes. Loans: yes.

Location: Baltimore's Inner Harbor area and County Cavan, Ireland, 50 miles from Dublin.

Contact: Raymond L. Joll, Vice President of Student Services, Baltimore International College, 17 Commerce St., Baltimore, MD 21202 US; 800-624-9926/410-752-4710, Fax 410-752-3730, URL http://www.bic.baltimore.md.us.

CHESAPEAKE INSTITUTE OF CULINARY STUDIES

Piney Point/Year-round

This private institution for maritime training and education offers a 9-month program in Culinary Arts with emphasis on shipboard cookery. Program started 1997. Curriculum: culinary. Admission dates: Twice monthly. Facilities: 2 working kitchens, bakeshop, classrooms, demo theatre, culinary and computer labs, restaurant and banquet facilities; swimming pool, gym, museum, library.

Courses: Food preparation, regional and international cuisines, baking and pastry, nutrition, menu planning, sanitation. Specialized maritime courses include lifeboat, firefighting, first aid, CPR, shipboard familiarization. Work-study is possible. Continuing education: cake decorating, ice carving, wine appreciation.

Faculty: 15 full-time instructors. Certified professionals teach extracurricular courses.

Costs: Tuition is $7,800. Room and board varies. Students live in the Paul Hall Conference Center and can take college credit courses as well as sailing and arts & crafts. Admission requirements: High school diploma or equivalent.

Location: Paul Hall Center for Maritime Training and Education in southern MD, 1.5 hours SE of Washington, DC and 1 hr SW of Baltimore.

Contact: Admissions Dept., Chesapeake Institute of Culinary Studies, Lundeberg School of Seamanship, Rte. 249, Piney Point, MD 20674 US; 301-994-0010, Fax 301-994-2180.

INTERNATIONAL SCHOOL OF CONFECTIONERY ARTS, INC.

(See display ad page 60) **Gaithersburg/Year-round**

This proprietary school offers 1-day to 1-week certificate courses in confectionery arts. Established in 1982. Admission dates: Weekly. Total enrollment: 400 per year. 12-16 each admission period. 100% of applicants accepted. 3% receive financial aid. Student to teacher ratio 4:1. 100% of graduates obtain employment within six months. Facilities: 2,400-square-foot area has 16 individual work stations, overhead mirrors, marble tables, and decorating and candy-making equipment.

Courses: Sugar pulling, blowing, and casting, chocolate decoration, Swiss candy making, cake decoration, gum paste.

Faculty: Ewald Notter has won gold and Olympic medals in international competitions and taught in Japan, Hong Kong, Denmark, Germany, England, Spain, and Finland. He judges international and Olympic competitions and wrote/published The Text Book of Sugar Pulling and Blowing and That's Sugar.

Costs: $120-$720, which includes breakfast, lunch and materials. Nearby lodging averages $59 per night.

Location: 2 miles from Gaithersburg, 10 miles from Washington, D.C.

Contact: Elizabeth Raisch, International School of Confectionery Arts, Inc., Admissions, 9209 Gaither Rd., Gaithersburg, MD 20877 US; 301-963-9077, Fax 301-869-7669, E-mail esnotter@aol.com.

L'ACADEMIE DE CUISINE

Gaithersburg/Year-round *(See also page 219) (See display ad page 61)*

This proprietary vocational school offers a 1-year full-time Culinary Career Training program, Pastry Arts program, 6- and 9-month part-time certificate courses, continuing education and nonprofessional courses, culinary and cultural program in Gascony, France. Established in 1976. Accredited by ACCET and approved by the Maryland Higher Education Commission. Calendar: semester. Curriculum: culinary. Admission dates: January, March, July, September. Total enrollment: 70 culinary, 40 pastry. 35 culinary, 20 pastry each admission period. 85% of applicants accepted. 61% receive financial aid. 90% enrolled part-time. Student to teacher ratio 13:1. 95% of graduates obtain employment within six months. Facilities: 35-station practice and pastry kitchen and 35-seat demonstration classroom.

Courses: Curriculum is based on classic French technique. Program covers food preparation and presentation, pastries and desserts, wine selection, catering, menu planning, kitchen management. Placement in fine dining restaurants in metropolitan D.C. Advanced/continuing ed courses: marzipan, wedding cakes, sugar, chocolate, catering, sanitation.

Faculty: 4 full-, 3 part-time. School President Francois Dionot, graduate of L'Ecole Hoteliere de la Societe Suisse des Hoteliers, founder of the IACP, Pascal Dionot, Bonnie Moore, Brian Patterson, David Arnold, L'Academie graduates Somchet Chumpapo, Mark Ramsdell.

Costs: Tuition is $15,000 for the full-time Culinary Career program, $8,750 for the Pastry Arts program. $75 application fee, supplies $475. Certificate course tuition $2,450-$3,100. Payment plans available. Admission requirements: High school diploma or equivalent. Scholarships: yes. Loans: yes.

Location: The Gaithersburg branch is 20 minutes north of Washington, DC; the Bethesda branch is 3 miles northwest of D.C.

Contact: Wendy Sisson, Admissions, L'Academie de Cuisine, 16006 Industrial Dr., Gaithersburg, MD 20877-1414; 301-670-8670/800-664-CHEF, Fax 301-670-0450, E-mail LAcademe@erols.com, URL http://www.washingtonpost.com/yp/LAcademie. Addl. contact: Patrice Dionot.

MASSACHUSETTS

BERKSHIRE COMMUNITY COLLEGE

Pittsfield

This college offers a 1-year certificate, 2-year AAS degree. Program started 1977. Accredited by NEASC. Calendar: semester. Curriculum: culinary, core. Admission dates: Fall, spring. Total enrollment: 15 per semester. Facilities: Full kitchen lab, use of local hotel kitchen/dining room for quantity food dinners.

Courses: Includes baking, garde manger, quantity foods production, sanitation, purchasing and food service management. Externship provided.

Faculty: 2 full-time, 2 part-time.

Costs: $88 per credit-hour in-state, $105 out-of-state. Admission requirements: High school diploma or equivalent and learning skills assessment. Loans: yes.

Contact: Nancy Simonds-Ruderman, Professor of Hotel & Restaurant Management, Berkshire Community College, Culinary Arts, 1350 West St., Pittsfield, MA 01201-5786 US; 413-499-4660 x229, Fax 413-447-7840, URL http://cc.berkshire.org.

BOSTON UNIVERSITY CULINARY ARTS

(See also page 220)

Boston/Year-round

This university offers a 4-month certificate in the Culinary Arts (started 1988) and Master of Liberal Arts with concentration in Gastronomy (started 1994). Program started 1986. Calendar: semester. Curriculum: core. Admission dates: January, September. Total enrollment: 24. 12 each admission period. 60% of applicants accepted. 50% receive financial aid. Student to teacher ratio 6:1. 100% of graduates obtain employment within six months. Facilities: Demonstration room with overhead mirror, classroom, 8 restaurant stations in the laboratory kitchen.

Courses: Cover basic classic and modern techniques and theory, ethnic and regional cuisine, food history, dining room theory and practice, purchasing. Master program includes 4 core courses in history and anthropology of food, nutrition and diet, 4 electives. Continuing education courses are available.

Faculty: 1 full-time, 30 part-time. Guest chefs have included Julia Child, Albert Kumin, Jacques Pépin, Roger Fessaguet, and Jasper White.

Costs: Certificate course tuition $6,600. Application fee $35. Admission requirements: Applicants must have some foodservice experience. Loans: yes.

Location: The 28,000-student campus is in Kenmore Square.

Contact: Rebecca Alssid, Director of Special Programs, Boston University Culinary Arts, 808 Commonwealth Ave., Boston, MA 02215 US; 617-353-9852, Fax 617-353-4130, E-mail ralssid@bu.edu, URL http://www.bu.edu.

BRISTOL COMMUNITY COLLEGE

Fall River/September-May

This college offers a 1-year certificate. Program started 1985. Accredited by NEASC. Calendar: semester. Curriculum: culinary. Admission dates: September. Total enrollment: 22-24. 22-24 each admission period. 80% of applicants accepted. 70% receive financial aid. 20% enrolled part-time. Student to teacher ratio 6:1. 75% of graduates obtain employment within six months.

Courses: All courses are designed to enhance the cook's skill and knowledge for restaurant work.

Faculty: 4 full-time.

Costs: Tuition in-state $3,000, out-of-state $6,800. Loans: yes.

Contact: John Caressimo, CCE, Culinary Director, Bristol Community College, Culinary Arts, 777 Elsbree St., Fall River, MA 02720 US; 508-678-2811 x2111, Fax 508-678-2811.

BUNKER HILL COMMUNITY COLLEGE

Boston/Year-round

This two-year college offers a 1-year/29-credit certificate and 2-year/64-credit AS degree in Culinary Arts. Calendar: semester. Curriculum: culinary, core. Admission dates: September, January. Total enrollment: 180. 80-90 each admission period. 95% of applicants accepted. 40% receive financial aid. 10% enrolled part-time. Student to teacher ratio 10-12:1. 99% of graduates obtain employment within six months.

Courses: Food and beverage management, purchasing, baking, facilities planning, banquet, classical, buffet catering, accounting. 660 hours of employment required.

Faculty: 4 full- and 2 part-time.

Costs: $78 per credit in-state, $248 per credit out-of-state. Admission requirements: High school diploma. Scholarships: yes. Loans: yes.

Contact: Arthur Buccheri, Prof. Dept. Chair, Bunker Hill Community College, 250 New Rutherford Ave., Boston, MA 02129-2991 US; 617-228-BHCC, Fax 617-228-2082.

CAMBRIDGE SCHOOL OF CULINARY ARTS

Cambridge/Year-round *(See also page 220) (See display ad above)*

This proprietary school offers a 37-week (726-clock-hour) Professional Chef's diploma program, year-round continuing education program, and culinary excursions to Europe. Established in 1973. Accredited by ACCSCT and Licensed by the Commonwealth of Massachusetts Dept. of Education. Calendar: quarter. Curriculum: culinary. Admission dates: September, January. Total enrollment: 110-200. 75-100 each admission period. 80% of applicants accepted. Student to teacher ratio 32:1 in lectures. 90% of graduates obtain employment within six months. Facilities: 3 large newly renovated kitchens, 3 large demonstration classrooms; gas and electric commercial appliances.

Courses: Emphasis on chemistry and principles of fine cooking. Curriculum includes: savory food basics, baking and pastry arts, cuisines of Italy, France and Asia, food and business management, applied food service sanitation, cheese and wine. Day and evening classes available. Continuing education classes for novices and experienced cooks.

Faculty: Instructors are state-approved teachers. President-Founder Roberta Avallone Dowling earned diplomas from Julie Dannenbaum, Marcella Hazan, Madeleine Kamman, and Richard Olney.

Costs: Tuition $11,200, payable in 4 installments; application fee $45. $100 deposit on acceptance, refundable within 5 days. Nearby lodging ranges from $700-$1,600 per month. Admission requirements: Applicants must be at least age 18 and have a high school diploma or equivalent.

Location: 5 miles from downtown Boston, 1 mile from Harvard Square.

Contact: Director of Admissions, The Cambridge School of Culinary Arts, 2020 Massachusetts Ave., Cambridge, MA 02140 US; 617-354-2020, Fax 617-576-1963, E-mail rldcsca@tiac.net, URL http://www.cambridgeculinary.com.

ESSEX AGRICULTURAL AND TECHNICAL INSTITUTE

Hathorne/September-May

This career institution offers a 2-year AAS degree in Culinary Arts and Food Service. Program started 1968. Accredited by NEASC. Calendar: semester. Curriculum: culinary. Admission dates: September, January. Total enrollment: 65. 40 each admission period. 50% of applicants accepted. 20% receive financial aid. 12% enrolled part-time. Student to teacher ratio 15:1. 60% of graduates obtain employment within six months. Facilities: Include 6 kitchens and classrooms, bakery and restaurant.

Courses: Restaurant operation, baking, garde manger, international cuisine, buffet, specialty food production, cakes and pastries, American regional cuisine, and nutrition. Externship provided.

Faculty: 4 full-time. Includes Division Chair C. Naffah, P. Kelly, L. Bassett, J. Cristello.

Costs: Annual tuition $2,400. Acceptance fee $30. Other fees approximately $1,830. Admission requirements: High school diploma or equivalent. Scholarships: yes. Loans: yes.

Location: The 180-acre campus is in a suburban setting 20 miles from Boston.

Contact: John Maguire, Director of Admissions, Essex Agricultural and Technical Institute, Admissions, 562 Maple St., Box 362, Hathorne, MA 01937-0362 US; 978-774-0050 x216, Fax 978-774-6530, E-mail aggie@eati.usal.com, URL http://www.usal.com/~nsag.

HOLYOKE COMMUNITY COLLEGE

Holyoke/September-May

This college offers a 1-year certificate in Culinary Arts. Program started 1991. Accredited by NEASC. Calendar: semester. Curriculum: culinary. Admission dates: September, January. Total enrollment: 75. 50 each admission period. 100% of applicants accepted. 50% receive financial aid. 20% enrolled part-time. Student to teacher ratio 12:1 lab. 90% of graduates obtain employment within six months. Facilities: Include 2 kitchens, bakeshop.

Courses: Food production management, advanced food production, and nutrition. Optional externship. Continuing ed.: evening courses in advanced baking, food service supervision, nutrition.

Faculty: 4 full-time. Includes H. Robert, D. Walsh, W. Grinnan.

Costs: Annual tuition in-state $2,605, out-of-state $6,275. Application fee $10. Advance payment fee $30. Admission requirements: High school diploma or equivalent. Scholarships: yes. Loans: yes.

Location: Western Massachusetts.

Contact: Hugh Robert, Department Chair, Holyoke Community College, Hospitality Management, 303 Homestead Ave., Holyoke, MA 01040 US; 413-538-7000, Fax 413-534-8975, E-mail hrobert@hcc.mass.edu, URL http://www.hcc.mass.edu.

MASSASOIT COMMUNITY COLLEGE

Brockton

This college offers a 2-year degree. Program started 1982. Accredited by State. Admission dates: September. Total enrollment: 60. Student to teacher ratio 20-35:1. 93.5% of graduates obtain employment within six months.

Courses: Externship provided.

Faculty: 3 full-time, 2 part-time.

Costs: Tuition in-state $72 per credit, out-of-state $204 per credit. Admission requirements: High school diploma or equivalent.

Contact: Culinary Director, Massasoit Community College, Culinary Arts, 1 Massasoit Blvd., Brockton, MA 02402 US; 508-588-9100.

MINUTEMAN REGIONAL VOCATIONAL TECHNICAL SCHOOL

Lexington

This independent trade school offers a 3-year diploma, 2-year post-graduate course, 90-day retraining courses in Culinary, Baking, Hotel and Restaurant Management. Program started 1973. Total enrollment: 150. 25 each admission period. 95% of applicants accepted. Student to teacher ratio 10:1. 99% of graduates obtain employment within six months. Facilities: Include 6 kitchens and classrooms.

Courses: Sanitation, nutrition, management, purchasing, computer skills, applied math, and applied science. Externship: 20 weeks, hotel/restaurant.

Costs: Annual tuition in-state $6,200. Uniform fee: $100. Admission requirements: Admission test.

Contact: John Fitzpatrick, Director, Minuteman Tech, Foodservice/Hospitality Management, 758 Marrett Road, Lexington, MA 02173 US; 781-861-6500, Fax 781-863-1254.

NEWBURY COLLEGE

Brookline/September-May

This nonprofit institution offers a 2-year AAS degree in Culinary Arts and Food Service Management. Program started 1981. Accredited by NEASC. Calendar: semester. Curriculum: culinary, core. Admission dates: September, January. Total enrollment: 240. 140 each admission period. 60% of applicants accepted. 80% receive financial aid. 10% enrolled part-time. Student to teacher ratio 17:1. 99% of graduates obtain employment within six months. Facilities: Include 7 production kitchens and the college dining room.

Courses: Include preparation and presentation of international and regional cuisines, equipment, sanitation, and nutrition, menu planning, and general education subjects. A 12- to 18-week salaried externship is provided.

Faculty: The 10 full- and 6 part-time faculty members are active industry professionals.

Costs: Annual tuition is $12,500 and culinary program fee is $1,170. On-campus room and board are provided for 400 students at $7,000 (double occupancy). Admission requirements: High school diploma or GED and a 2.0 GPA. Scholarships: yes. Loans: yes.

Location: A suburban setting, 4 miles from downtown Boston.

Contact: Donald J. MacDonald, Dean of Admission, Newbury College, Admissions Center, 129 Fisher Ave., Brookline, MA 02445 US; 617-730-7007/800-NEWBURY, Fax 617-731-9618, E-mail info@newbury.edu, URL http://www.newbury.edu.

MICHIGAN

GRAND RAPIDS COMMUNITY COLLEGE

Grand Rapids/August-May *(See display ad page 65)*

This two-year college offers a 2-year AAAS degree in Culinary Arts and Culinary Management and a certificate in Baking and Pastry Arts. Program started 1980. Accredited by ACFEI, NCA. Calendar: semester. Curriculum: culinary, core. Admission dates: January, August. Total enrollment: 340. 80 each admission period. 95% of applicants accepted. 60% receive financial aid. 20% enrolled part-time. Student to teacher ratio 16:1. 99% of graduates obtain employment within six months. Facilities: Include 8 kitchens, many classrooms, bar, 3 bakeries, bistro, banquet rooms, library, auditorium, 2 storerooms, student-run restaurant and deli-bakery.

Courses: Basic and advanced culinary and baking skills, ice carving, banquets and catering, restaurant operations, international studies. Externship: 240-hour, summer semester, throughout US and abroad. Continuing ed.: ~50 non-credit seminars per year, clubs, international study tours.

Faculty: 12 full-time, 7 part-time. Qualifications: equivalent of a bachelor's degree and minimum 6 years industry experience in management. Most have master's degrees or are master chefs.

Costs: Annual tuition $6,059 in-state, $6,935 out-of-state. Application fee $20. Books, uniforms, cutlery kit: $1,100. Average off-campus housing cost $250 per month. Admission requirements: High school diploma or equivalent and admission test. Scholarships: yes.

Location: The 14,000-student, 11-building campus is in an urban setting.

Contact: Robert B. Garlough, Director, Hospitality Education Dept., Grand Rapids Community College, Hospitality Education, 151 Fountain, N.E., Grand Rapids, MI 49503-3263 US; 616-234-3690, Fax 616-234-3698, E-mail bgarloug@post.grcc.cc.mi.us, URL http://www.grcc.cc.mi.us/hed.

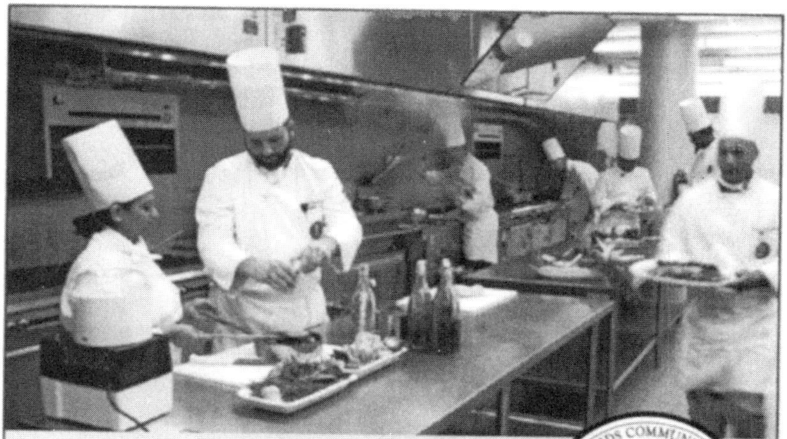

HENRY FORD COMMUNITY COLLEGE

Dearborn/Year-round

This two-year college offers an AS degree. Program started 1972. Accredited by ACF. Calendar: semester. Curriculum: culinary. Admission dates: Throughout the year. Total enrollment: 90. 30 each admission period. 100% of applicants accepted. 80% receive financial aid. 60% enrolled part-time. Student to teacher ratio 16:1. 90% of graduates obtain employment within six months. Facilities: 2 kitchens, student-run dining room.

Courses: 320-hour professional internship program. Year-round enrichment classes, continuing ed.

Faculty: 4 full-time, 6 part-time.

Location: A half mile from Detroit.

Contact: Dennis Konarski, CFE, CCE, Culinary Director, Henry Ford Community College, Culinary Arts, 5101 Evergreen Rd., Dearborn, MI 48128 US; 313-845-9651, Fax 313-845-9784, E-mail dennis@mail.henryford.cc.mi.us, URL http://www.henryford.cc.mi.us.

MACOMB COMMUNITY COLLEGE

Clinton Township/September-May

This two-year college offers a 2-year AAS degree in Culinary Arts, ACFEI-certified cook, NRA diploma, 1-year certificate in Supervision, 1-year certificate in Production. Program started 1972. Accredited by State, ACFEI. Calendar: semester. Curriculum: culinary. Admission dates: Fall, spring, summer. Total enrollment: 186. 100% of applicants accepted. 35% enrolled part-time. Student to teacher ratio 16-35:1. 99% of graduates obtain employment within six months. Facilities: 3 full-service kitchen labs, student/faculty restaurant.

Courses: ACFEI apprenticeship program.

Faculty: 3 full-time, 6 part-time.

Costs: Tuition in-county $53.50 per credit hour, out-of-county $81 per credit hour. Admission requirements: High school diploma or equivalent. Scholarships: yes. Loans: yes.

Contact: David Schneider, CEC, CCE, Dept. Coordinator, Macomb Community College, Culinary Arts, 44575 Garfield, Clinton Township, MI 48038 US; 810-286-2000, Fax 810-286-2250.

MONROE COUNTY COMMUNITY COLLEGE

Monroe/September-May

This independent 2-year college offers a 2-year AOS degree/certificate in Culinary Skills and Management. Program started 1981. Accredited by NCA, ACFEI. Calendar: semester. Curriculum: core. Admission dates: September. Total enrollment: 36. 20 each admission period. 80% of applicants accepted. 25% receive financial aid. 25% enrolled part-time. Student to teacher ratio 18:1. 87% of graduates obtain employment within six months. Facilities: Include 2 kitchens, classroom, restaurant.

Courses: Restaurant production, baking, buffet, institutional food, management, sanitation, nutrition, garde manger, ice carving, menu planning, purchasing and receiving, a la carte, dining room procedure and general education. Externship provided. Articulation agreement with Sienna Heights College.

Faculty: 2 full-time. Includes K. Thomas, CCE, CEC and one technician, V. LaValle.

Costs: Tuition is $47 per credit-hour in-county, $75 per credit-hour out-of-county, $83 per credit-hour out-of-state. Technology fee $3 per credit-hour. Application fee $21. Average off-campus housing cost is $400 per month. Admission requirements: High school diploma or equivalent and admission test. Scholarships: yes. Loans: yes.

Location: The 4,000-student campus is in a rural setting.

Contact: Kevin Thomas, Instructor of Culinary Skills, Monroe County Community College, Culinary Arts, 1555 S. Raisinville Rd., Monroe, MI 48161 US; 313-242-7300 x4104, Fax 313-242-9711, E-mail wguerriero@mail.monroe.cc.mi.us, URL http://www.monroe.lib.mi.us/mccc.

MOTT COMMUNITY COLLEGE

Flint/Year-round

This two-year college offers a 66 credit-hour/83 contact-hour AAS degree in Culinary Arts, 65 credit-hour/80 contact-hour AAS degree in Food Service Management.

Costs: Tuition is $56 per contact-hour in-state, $82 per contact-hour out-of-district, $109 per contact-hour out-of-state.

Contact: Grace Alexander, Instructor/coordinator, Mott Community College, 1401 E. Court St., Flint, MI 48503 US; 810-232-7845, URL http://www.mcc.edu.

NORTHERN MICHIGAN UNIVERSITY

Marquette/August-April

This university offers a 1-year certificate, 2-year AAS degree, 4-year BS degree, program in Culinary Arts, Restaurant and Institutional Management. Program started 1970. Accredited by NCA. Calendar: semester. Admission dates: September, January. Total enrollment: 74. 36 each admission period. 100% of applicants accepted. 80% receive financial aid. 25% enrolled part-time. Student to teacher ratio 18:1. 100% of graduates obtain employment within six months. Facilities: Include 1 kitchen and 4 classrooms, computer lab, restaurant and meat-cutting room.

Courses: Cooking, baking, garde manger, sanitation, purchasing, and general education.

Faculty: 4 full-time. Qualifications: bachelor's or master's degrees.

Costs: Tuition in state $1,493 per semester full-time, out-of-state $2,633 per semester full-time. Application fee $50. On-campus housing: 2,000 spaces at $2,300 per year. Average off-campus housing cost: $2,000 per year. Admission requirements: High school diploma or equivalent. Scholarships: yes. Loans: yes.

Location: The 9,000-student campus is in a small town setting.

Contact: David Sonderschafer, Professor, Northern Michigan University, Dept. of Consumer & Family Studies, College of Technology & Applied Sciences, Marquette, MI 49855 US; 906-227-2364, Fax 906-227-1549.

NORTHWESTERN MICHIGAN COLLEGE

Traverse City/Year-round

This college offers a 2-year AAS degree in Culinary Arts. Program started 1978. Accredited by ACFEI. Calendar: semester. Curriculum: culinary, core. Admission dates: Open. Total enrollment: 75. 30 each admission period. 100% of applicants accepted. 65% receive financial aid. 45% enrolled part-time. Student to teacher ratio 15:1. 98% of graduates obtain employment within six months. Facilities: Include 3 kitchens, many classrooms, bake shop.

Courses: Externship: 400-hour, salaried.

Faculty: 3 full- and 10 part-time.

Costs: Tuition in-county $51 per credit hour, out-of-county $84 per credit hour. Application fee $15. On-campus housing: 120 spaces; average cost $1,950 with meal plan. Admission requirements: High school diploma or equivalent. Scholarships: yes. Loans: yes.

Location: A small town campus, 150 miles from Grand Rapids.

Contact: Fred Laughlin, Dept. Chair, Northwestern Michigan College, Culinary Arts, 1701 E. Front St., Traverse City, MI 49686 US; 616-922-1197, Fax 616-922-1134, E-mail Flaughli@nmc.edu, URL http://www.nmc.edu.

OAKLAND COMMUNITY COLLEGE

Farmington Hills/September-June

This two-year college offers a 2-year AAS program in Culinary Arts, Food Service Management, Hotel Management, Chef Apprentice. Program started 1978. Accredited by ACFEI. Calendar: semester. Curriculum: core. Admission dates: September, January, May. Total enrollment: 175. 50 each admission period. 98% of applicants accepted. 35% receive financial aid. 50% enrolled part-time. Student to teacher ratio 15:1. 90% of graduates obtain employment within six months.

Facilities: Include 10 kitchens and classrooms and restaurant.

Courses: Cooking, baking, pastries, garde manger, front-of-house service, sanitation, purchasing, menu planning, cost control. Continuing education: garde manger, culinary competition, meat cutting, baking.

Faculty: 9 full-time, 6 part-time. Includes Chairperson Susan Baier, FMP; Kevin Enright, CEC, CCE; Roger Holder, CPC, CCE; Dan Rowlson, CCE; Robert Zemke, RD, MBA; Darlene Levinson, FMP.

Costs: Annual tuition in-county $46 per credit-hour. Lab fee $330 for production classes. Scholarships: yes. Loans: yes.

Contact: Susan Baier, Dept. Chair, Oakland Community College, Hospitality/Culinary Arts, 27055 Orchard Lake Rd., Farmington Hills, MI 48334 US; 248-471-7786, Fax 248-471-7553, E-mail smbaier@occ.cc.mi.us.

SCHOOLCRAFT COLLEGE

Livonia/August-April

This college offers a 2-year certificate/AAS degree in Culinary Arts and Culinary Management. Program started 1966. Accredited by NCA. Calendar: semester. Admission dates: January, August. Total enrollment: 156. 12-78 each admission period. 100% of applicants accepted. 40% enrolled part-time. Student to teacher ratio 14:1. 100% of graduates obtain employment within six months. Facilities: Include 8 kitchens and classrooms, pastry kitchen, butcher shop, bakery, and restaurant.

Courses: Baking, pastries, a la carte, international cuisine, butchery. Other required courses: sanitation, nutrition. Externship: 16-week, salaried, restaurants. Continuing education: vegetarian cuisine, art of cooking, low fat-high flavor, winter soups and stews.

Faculty: 5 full-time, 6 part-time. Qualifications: ACF certification.

Costs: Tuition $52 per credit-hour resident, $76 per credit-hour non-resident. Lab fees $35-$105. Admission requirements: High school diploma or equivalent and admission test, completion of CAP 090. Scholarships: yes. Loans: yes.

Location: This 13,900-student suburban campus is 20 miles from Detroit.

Contact: Kevin Gawronski, CMC, Culinary Manager, Schoolcraft College, Culinary Arts, 18600 Haggerty Rd., Livonia, MI 48152-2696 US; 313-462-4423, Fax 313-462-4581, E-mail kgawrons@schoolcraft.cc.mi.us, URL http://www.schoolcraft.cc.mi.us.

WASHTENAW COMMUNITY COLLEGE

Ann Arbor/September-June

This two-year college offers a 1-year certificate in Food Service Production Specialist and a 2-year AAS degree in Culinary Arts and Hotel and Restaurant Management. Program started 1975. Accredited by NCA. Calendar: semester. Curriculum: culinary. Total enrollment: 80-120. 50 each admission period. 90% of applicants accepted. 10% receive financial aid. 60% enrolled part-time. Student to teacher ratio 16-25:1. 95% of graduates obtain employment within six months. Facilities: Include kitchen, bake shop, and student-run dining room.

Courses: 50% culinary courses, 25% business courses, 25% general studies. Externship: 300-hour, salaried. Continuing education year-round.

Faculty: 4 full-time. Qualifications: bachelor's or master's degree and ACF certification.

Costs: $52 per cr-hour in-district, $77 per cr-hour out-of-district, $98 per cr-hour out-of-state. $23 registration fee per semester. Off-campus housing cost $400-$600 per month. Admission requirements: High school diploma or equivalent and admission test. Scholarships: yes.

Location: The suburban 11,000-student campus is 40 miles from Detroit.

Contact: David Placey, Director, Washtenaw Community College, Admissions, 4800 E. Huron River Dr., Ann Arbor, MI 48106-0978 US; 734-973-3525, Fax 734-677-5414, E-mail admissions@orchard.washtenaw.cc.mi.us, URL http://www.washtenaw.cc.mi.us. Additional contact: Don L. Garrett, Dept. Chair, Culinary and Hospitality Mgt. Program.

MINNESOTA

ART INSTITUTES INTERNATIONAL MINNESOTA

(See display ad page 8)

Minneapolis/Year-round

This private school offers a 6-quarter AAS degree in Culinary Arts. Program started 1998. Accredited by ACICS. Calendar: quarter. Curriculum: core.

Costs: $230-$245 per credit. Scholarships: yes. Loans: yes.

Contact: Audra Staebell, Asst. Director of Admissions, Art Institutes International Minnesota, 15 S. 9th St., Minneapolis, MN 55402 US; 612-332-3361, Fax 612-332-3934, URL http://www.aii.edu.

HENNEPIN TECHNICAL COLLEGE

Brooklyn Park, Eden Prairie/September-May

This career college offers a 50-credit diploma/certificate in Culinary Arts. Program started 1972. Accredited by NCA, ACFEI. Calendar: trimester. Curriculum: culinary. Admission dates: Fall, spring. Total enrollment: 60. 10-20 each admission period. 100% of applicants accepted. 40% receive financial aid. 10% enrolled part-time. Student to teacher ratio 15:1. 98% of graduates obtain employment within six months. Facilities: Include 3 kitchens, 3 classrooms and restaurant.

Courses: Externship: 6- to 8-week and post-graduate assistantships.

Faculty: 4 full-, 3 part-time. Qualifications: ACF certified or certifiable.

Costs: Annual tuition $2,228 in-state, $4,456 out-of-state. Approximately $550 for supplies. Scholarships: yes. Loans: yes.

Location: The campus is in a suburban setting, 10 miles from Minneapolis.

Contact: Carlo Castagneri, Lead Instructor, Hennepin Technical College-Brooklyn Park Campus, Culinary Arts, 9000 Brooklyn Blvd., Brooklyn Park, MN 55445 US; 612-425-3800 x2116, Fax 612-550-2119, E-mail Carlo.Castagneri@htc.mnscu.edu, URL http://www.htc.mnscu.edu.

HIBBING COMMUNITY COLLEGE

Hibbing/Year-round

This two-year college offers 103-credit diploma and 113-credit AAS degree programs in Food Service Management. Calendar: semester. Curriculum: culinary.

Costs: Tuition is $66 per credit in-state, $132 per credit out-of-state plus fees.

Contact: Dan Lidholm, Hibbing Community College, 2900 E. Beltline, Central Campus, Hibbing, MN 55746 US; 800-224-4422 x7228, Fax 218-262-7222, URL http://www.hibbing.tec.mn.us.

NATIONAL BAKING CENTER

Minneapolis/Year-round

This continuing education program offers 1-day to 3-week short courses and seminars geared to the professional and other serious bakers, including foodservice baking. Established in 1996. Calendar: trimester. Admission dates: Year-round. Student to teacher ratio 3-14:1. Facilities: One theater-style demonstration bakery, one hands-on bake shop, one classroom, test baking and research areas.

Courses: Artisan breads: includes French, Italian, sourdough. Pastries: international to basic American, includes decorating and chocolate. Basic and advanced levels. 3-week Opening a Bread Bakery course, includes baking, equipment layout, production and management. 26-week paid internships offered, 4 entry dates per year.

Faculty: 2 full-time with international experience and European professional certification.

Costs: Vary with courses and seminars.

Contact: Greg Tompkins, Director, National Baking Center, 818 Dunwoody Blvd., Minneapolis, MN 55403-1192 US; 612-374-3303, Fax 612-374-3332, E-mail nbc@dunwoody.tec.mn.us, URL http://www.dunwoody.tec.mn.us/nbc.htm.

NORTHWEST TECHNICAL COLLEGE

Moorhead/September-May

This career college offers a 2-year diploma in Chef Training. Program started 1966. Accredited by State, NCA. Calendar: semester. Curriculum: core. Total enrollment: 40-50. 25 each admission period. 80% receive financial aid. 10% enrolled part-time. Student to teacher ratio 20-25:1. 89% of graduates obtain employment within six months. Facilities: 2 kitchens, 1 classroom, 2 restaurants.

Courses: Quantity food preparation, food purchasing and cost controls, menu planning.

Faculty: 2 full-time. Includes Kim E. Brewster, CEC, CCE, Colleen Kraft.

Costs: Annual tuition in-state $1,996, out-of-state $3,840. Other fees: $60 per quarter meal fee, $40 per quarter uniform fee, $20 admission application fee. Off-campus housing cost is $250 per month. Admission requirements: High school diploma or equivalent. Loans: yes.

Contact: Kim Brewster, Dept. Chairperson, Northwest Technical College, Chef Training, 1900 28th Ave. S., Moorhead, MN 56560 US; 800-426-5603 x572, Fax 218-236-0342, E-mail bre372@nikola.ntc.mnscu.edu, URL http://www.ntc.mnscu.edu.

ST. CLOUD TECHNICAL COLLEGE – CULINARY ARTS

St. Cloud/August-May

This college offers a 37-week certificate of completion. Program started 1972. Accredited by NCA. Calendar: semester. Curriculum: core. Admission dates: Open. Total enrollment: 20. 90% receive financial aid. 100% enrolled part-time. Student to teacher ratio 20:1. 80% of graduates obtain employment within six months. Facilities: 2 kitchens.

Faculty: 1 full-time: Mike Costello, CCE, CEC, AAC; 1 part-time: Anita Zanardi.

Costs: $2,605.47 plus $635 for books and supplies. Scholarships: yes. Loans: yes.

Contact: Diane Wysoski, St. Cloud Technical College, Culinary Arts, 1540 Northway Dr., St. Cloud, MN 56303-1240 US; 320-654-5000, Fax 320-654-5981, E-mail mac@cloud.tec.mn.us, URL http://sctcweb.tec.mn.us.

ST. PAUL TECHNICAL COLLEGE

St. Paul/Year-round

This career college offers a 12-month diploma. Program started 1967. Accredited by NCA. Calendar: quarter. Curriculum: culinary. Admission dates: March, September, December. Total enrollment: 40-50. 25 each admission period. Student to teacher ratio 16:1. 95% of graduates obtain employment within six months.

Faculty: 3 full-time.

Costs: Tuition in-state $71 per credit hour, out-of-state $142 per credit hour. Admission requirements: High school diploma or equivalent and admission test.

Contact: Eberhard Werthmann, Culinary Director, St. Paul Technical College, Culinary Arts, 235 Marshall Ave., St. Paul, MN 55102 US; 612-221-1300, Fax 612-221-1416, E-mail ewerthma.@spt.edu.mn.us, URL http://www.sptc.tec.mn.us.

SOUTH CENTRAL TECHNICAL COLLEGE

North Mankato/September-July

This career college offers a 15-month 52-credit diploma and 2-year 72-credit AAS degree in Hotel, Restaurant and Institutional Cooking. Program started 1968. Accredited by NCA. Calendar: semester. Curriculum: culinary. Admission dates: August, January, June. Total enrollment: 25. 5-7 each admission period. 100% of applicants accepted. 90% receive financial aid. 10% enrolled part-time. Student to teacher ratio 17:1. 95% of graduates obtain employment within six months. Facilities: Include 2 kitchens, bakery and classroom.

Courses: Food preparation, inventory and cost control, management, menu design, job preparation, Other required courses: CPR, first aid, microcomputers, job search skills, interpersonal relations.

Faculty: 2 full-time.

Costs: Annual tuition in-state $74 per semester credit, out-of-state $148 per semester credit.

Books and uniforms $520. Off-campus housing cost $300 per month. Admission requirements: High school diploma or equivalent. Scholarships: yes. Loans: yes.

Contact: Jim Hanson, Instructor, South Central Technical College, Culinary Arts, 1920 Lee Blvd., PO Box 1920, North Mankato, MN 56003 US; 507-389-7229, Fax 507-388-9951, E-mail JimH@tc-mankato.scm.tec.mn.us, URL http://www.sctc.mnscu.edu.

MISSISSIPPI

HINDS COMMUNITY COLLEGE

Jackson/August-May

This two-year college offers 1-year certificate and 2-year AAS degree programs in Culinary Arts. Accredited by SACS. Calendar: semester. Curriculum: culinary, core. Admission dates: January, August. Total enrollment: 50. 20 each admission period. 100% of applicants accepted. 80% receive financial aid. 15% enrolled part-time. Student to teacher ratio 16:1. 100% of graduates obtain employment within six months. Facilities: Full-scale commercial kitchen with all major equipment.

Courses: Fundamentals of culinary arts, French techniques, international and American cuisine, baking, garde manger. 6-month paid internship at approved dining facility, at least 20 hours per week.

Faculty: 3 faculty: BS in Hospitality Management, certified chef/instructor with BBA, MS in Hospitality Management.

Costs: $515 for full-time students (12-19 hours), $25 registration fee. Deferred payment plan. Admission requirements: High school diploma or GED. Scholarships: yes. Loans: yes.

Contact: Kathleen Bruno, Hinds Community College, Culinary Arts Program, 3925 Sunset Dr., Jackson, MS 39213 US; 601-987-8130, Fax 601-982-5804, E-mail Gata1967@aol.com.

MISSISSIPPI UNIVERSITY FOR WOMEN

(See display ad above)

Columbus/Year-round

This public university offers a 4-year BS (52 semester-hours): required minor (18-21 semester-hours) in Entrepreneurship/Small Business Development, Food Journalism, Food Art, Nutrition/Wellness; minor (22 semester-hours) in Culinary Arts. Established in 1997. Accredited by SACS. Calendar: semester. Curriculum: core. Admission dates: August, January. Total enrollment: 25. Open each admission period. Student to teacher ratio 16:1. Facilities: 5 kitchens, classrooms in renovated building listed in National Register of Historic Places.

Courses: Classic cooking techniques, small quantity food preparation. Correlate minor required. 6 semester-hour internship required; international internship optional. Community education and professional development classes.

Faculty: Chef/Director, nutritionist.

Costs: $1,278 ($2,773) fees per semester in-state (out-of-state), $1,278.50 living expenses, $75 for food preparation classes. Admission requirements: College prep curriculum with 2.0 GPA/850 SAT, 2.5 GPA/760 SAT, or 3.2 SAT. Scholarships: yes. Loans: yes.

Contact: Sarah Labensky, Director, Mississippi University for Women, Interdisciplinary Studies,

Box W-1639, Columbus, MS 39701 US; 601-241-7472/7626, Fax 601-241-7627, E-mail cularts@muw.edu, URL http://www.muw.edu/interdisc.

MISSOURI

MISSOURI CULINARY INSTITUTE

Kansas City/Year-round *(See also page 226)*

This private school offers a 12-week certificate of completion. Established in 1995. State certificate, vocational rehabilitation approved. Calendar: quarter. Curriculum: culinary. Admission dates: Every 12 weeks starting in January. Total enrollment: Max. 48. 12 each admission period. Student to teacher ratio 12:1. Facilities: demo kitchen and lab kitchen with work stations.

Courses: Basic culinary and baking skills, safety, sanitation, nutrition, menu planning. Special classes on baking, candy making, wine tasting, and sauces.

Faculty: 2 full-time, 1 part-time. Terry Kopp, Chef de Cuisine/Instructor, owns several food service businesses and has appeared on NBC, Fox, and local PBS TV shows. Dorothy Kopp, with over 50 years restaurant experience, instructs front of the house operations.

Costs: Tuition $2,500. Application and registration fees are $150; other fees and supplies are $850. Average off-campus housing cost $150-$500. Admission requirements: High school graduation, GED, or school examination. Scholarships: yes. Loans: yes.

Location: A rural setting, 35 miles from Kansas City.

Contact: Terry Kopp, Chef de Cuisine/Instructor, Missouri Culinary Institute, Kansas City, MO 64067 US; 816-259-6464, E-mail tkopp@iland.net.

ST. LOUIS COMMUNITY COLLEGE – FOREST PARK

St. Louis/August-May

This two-year college offers a 2-year AAS degree, apprenticeship leading to ACF certification. Program started 1976. Accredited by NCA. Calendar: semester. Curriculum: culinary. Admission dates: August, January. Total enrollment: 150. 50 each admission period. 100% of applicants accepted. 40% enrolled part-time. Student to teacher ratio 20:1. 98% of graduates obtain employment within six months. Facilities: New 30,000-square-foot facility opens 1/99: 4 kitchens, student-operated restaurant, classrooms.

Courses: Meat analysis, garde manger, pastry, baking, nutrition, food specialties, catering, general education, involvement with Junior Chef Organization. Other required courses: 20 hours. Externship: 150-hour, salaried, each semester.

Faculty: 4 full-time, 25 part-time. Qualifications: masters degree and industry experience. Includes Dept. Chair Kathy Schiffman, Scott Vratarich, Reed Miller, Mike Downey, CCE, CCC.

Costs: Annual tuition: $1,200 in-state, $1,700 out-of-state. Parking $16. Off-campus housing cost $450 per month. Admission requirements: High school diploma or equivalent and admission test. Scholarships: yes. Loans: yes.

Contact: Kathy Schiffman, Dept. Chair, St. Louis Community College-Forest Park, Culinary Management, 5600 Oakland Ave., St. Louis, MO 63110 US; 314-644-9751, Fax 314-644-9992, E-mail falcon3083@aol.com.

MONTANA

UNIVERSITY OF MONTANA – COLLEGE OF TECHNOLOGY

Missoula/Year-round

This public career institution offers a 1-year certificate and 2-year AAS degree in Food Service Management. Program started 1973. Accredited by ACFEI, NASC. Calendar: trimester. Curriculum: culinary, core. Admission dates: August, January. Total enrollment: 62. 25 each admission period. 90% of applicants accepted. 30% receive financial aid. 5% enrolled part-time. Student to teacher ratio 15:1. 90% of graduates obtain employment within six months. Facilities: Include 4 kitchens and classrooms and 3 restaurants.

Courses: Cooking, baking, management, and general education.

Faculty: Qualifications: CEC, CPC, CC. Includes F. Sonnenberg, CEC, R. Lodahl, M.M. Barton, CPC, S. Bartos, CC.

Costs: Annual tuition $3,100 in-state, $5,600 out-of-state. Application fee $20. Other fees $400. Off-campus housing cost $300 per month. Admission requirements: High school or equivalent and admission test. Scholarships: yes. Loans: yes.

Contact: Frank Sonnenberg, CEC, Department Chair, College of Technology-University of Montana, Culinary Arts, 909 S. Ave. W., Missoula, MT 59801-7910 US; 406-243-7816, Fax 406-243-7899, URL http://www.umt.edu/mcot. Additional contact: Admissions Officer, 406-243-7882.

NEBRASKA

CENTRAL COMMUNITY COLLEGE

Hastings/September-June

This two-year college offers a 2-year certificate/AAS degree in Culinary Arts. Program started 1971. Accredited by NCA. Calendar: semester. Curriculum: core. Admission dates: Open. Total enrollment: 40. 6 each admission period. 100% of applicants accepted. 80% receive financial aid. Student to teacher ratio 20:1. 95% of graduates obtain employment within six months. Facilities: 1 kitchen, 4 classrooms, restaurant.

Courses: Bake shop, pantry, entrees, international cuisine, advanced sauces, pastries, garde manger.

Faculty: 2 full-time.

Costs: Annual tuition in-state $1,230, out-of-state $1,809. Uniform and supplies $40. On-campus dormitories available. Admission requirements: High school diploma or equivalent. Scholarships: yes. Loans: yes.

Location: The 600-acre campus is 100 miles from Lincoln.

Contact: Jaye Kieselhorst, Program Supervisor, Central Community College, Hotel, Motel, Restaurant Management, PO Box 1024, Hastings, NE 68902 US; 402-461-2458, Fax 402-461-2454.

METROPOLITAN COMMUNITY COLLEGE

Omaha/Year-round

This two-year college offers a 1- to 2-year AAS degree in Culinary Arts, Foodservice Management, Chef Apprentice, Research Chef, Bakery Arts. Program started 1976. Accredited by ACFEI, CAHM, OMA. Calendar: quarter. Curriculum: culinary. Admission dates: September, December, March, June. Total enrollment: 125. 25+ each admission period. 95% of applicants accepted. 68% receive financial aid. 48% enrolled part-time. Student to teacher ratio 12:1. 99% of graduates obtain employment within six months. Facilities: Include kitchens and classrooms, and restaurant.

Courses: Externship: 150-hour, restaurant setting.

Faculty: 3 full-time, 10 part-time.

Costs: Tuition in-state $28.50 per credit hour, out-of-state $32 per credit hour. Other fees: uniforms, tools. Admission requirements: High school diploma or equivalent. Scholarships: yes. Loans: yes.

Contact: Jim Trebbien, Division Representative, Metropolitan Community College, Food Arts & Management, Box 3777, Omaha, NE 68103-0777 US; 402-457-2510, Fax 402-457-2515, E-mail jtrebbien@metropo.mccneb.edu.

SOUTHEAST COMMUNITY COLLEGE

Lincoln/Year-round

This two-year college offers an 18-month AS degree in Culinary Arts. Program started 1988. Accredited by NCA. Calendar: quarter. Curriculum: core. Admission dates: September, March. Total enrollment: 20. 10 each admission period. 100% of applicants accepted. 30% enrolled part-time. Student to teacher ratio 15:1. 95-100% of graduates obtain employment within six months. Facilities: Include kitchen and 2 classrooms.

Courses: Advanced food, buffet, decorating and catering, professional baking. Externship: 220-hour, salaried.

Faculty: 1 full-time, 3 part-time. Qualifications: associate degree. Includes Gerrine Schreck Kirby, CCE, CWC; Jo Taylor, MA, RD; Lois Cockerham, BS; Erin Coudill, MS, RD; Keenan Cain, CWC.

Costs: Annual tuition in-state $28.50 per credit hour, out-of-state $33.50 per credit hour. Program reservation fee $25. Student activities fee $12.

Contact: Jo Taylor, Program Chair, Southeast Community College, Food Service, 8800 "O" St., Lincoln, NE 68520-9989 US; 402-437-2465, Fax 402-437-2404, E-mail BJTaylor@sccm.cc.ne.us, URL http://webster.sccm.cc.ne.us.

NEVADA

COMMUNITY COLLEGE OF SOUTHERN NEVADA – CHEYENNE

North Las Vegas

This two-year college offers a 1-year certificate and 2-year AAS degree in Hotel, Restaurant, and Casino Management with Culinary Arts emphasis. Program started 1990. Accredited by NASC, ACFEI. Calendar: semester. Curriculum: culinary, core. Admission dates: Open. Total enrollment: 400. 100% of applicants accepted. Student to teacher ratio 15:1. 100% of graduates obtain employment within six months.

Faculty: 4 full-time, 18 part-time.

Costs: Tuition is $39.50 per credit in-state. Out-of-state tuition is $57 per credit for 1-6 credits and $39.50 per credit plus $1,760 per semester for 7 credits or more.

Contact: Giovanni DelRosario, Culinary Director, Community College of Southern Nevada, Culinary Arts, 3200 E. Cheyenne Ave., Z1A, North Las Vegas, NV 89030 US; 702-651-4192, Fax 702-651-4558, URL http://www.ccsn.nevada.edu.

TRUCKEE MEADOWS COMMUNITY COLLEGE

Reno/August-May

This two-year college offers a 2-year AAS degree in Food Service Technology. Program started 1980. Accredited by State. Calendar: semester. Curriculum: core. Admission dates: January, September. Total enrollment: 72. 30-40 each admission period. 100% of applicants accepted. 30% receive financial aid. 70% enrolled part-time. Student to teacher ratio 16:1. 75% of graduates obtain employment within six months. Facilities: Include 8 kitchens and classrooms, and student-run restaurant.

Courses: Cooking, baking, pastry, garde manger, sauces, business, nutrition, sanitation, general ed.

Faculty: 1 full-time, 5 part-time. Includes George Skivofilakas, CEC, CCE, AAC.

Costs: Tuition in-state $34 per credit hour, out-of-state $1,100 + $31 per credit hour. Application fee $5. Off-campus housing cost is $400-500 per month. Admission requirements: High school diploma or equivalent. Scholarships: yes. Loans: yes.

Contact: Victor L. Bagan, Executive Chef/Instructor, Truckee Meadows Community College, Culinary Arts, 7000 Dandini Blvd., M-25, Reno, NV 89512 US; 702-673-7096, Fax 702-673-7108, E-mail bagan_victor@tmcc.edu, URL http://www.tmcc.edu.

NEW HAMPSHIRE

NEW HAMPSHIRE COLLEGE CULINARY INSTITUTE

Manchester/September-May *(See display ad page 75)*

This college offers a 2-year AAS degree in Culinary Arts, transferrable to a 4-year Hotel/Restaurant degree. Established in 1983. Accredited by ACFEI. Calendar: semester. Curriculum: culinary. Admission dates: Rolling. Total enrollment: 120. 75 each admission period. 80% of applicants accepted. 75% receive financial aid. 10% enrolled part-time. Student to teacher ratio 15:1. 100% of graduates obtain employment within six months. Facilities: New facility with the latest equipment, including 2 bakeshop labs, 4 production labs, computer center, and 5 classrooms.

Courses: Include culinary skills, bakeshop, food production, garde manger, nutritional cooking, and general education courses. Other required courses: dining room management and classical, regional, and international cuisine. Externship: 600-hour, salaried, with travel opportunities. 2+2 program toward BS in Restaurant Management.

Faculty: 5 full-time, 6 part-time.

Costs: Annual tuition $11,900. Knife set fee $200, books $200, uniforms $100. On-campus housing cost $5,980 annually. Admission requirements: High school diploma or equivalent. Scholarships: yes. Loans: yes.

Location: The 1,200-student campus is situated on 200- wooded acres in a rural setting.

Contact: Jennifer Kidwell, Asst. Director of Admissions/Culinary Coordinator, New Hampshire College Culinary Institute, Admissions, 2500 N. River Rd., Manchester, NH 03106 US; 800-642-4968, Fax 603-645-9693, E-mail admission@nhc.edu, URL http://www.nhc.edu/admissio/cul.htm.

NEW HAMPSHIRE COMMUNITY TECHNICAL COLLEGE

Berlin/Year-round

This career college offers a 2-year diploma/certificate/AAS in Culinary Arts. Program started 1966. Accredited by NEASC. Calendar: semester. Curriculum: core. Admission dates: Open. Total enrollment: 30. 20 each admission period. 76% of applicants accepted. 89% receive financial aid. 10% enrolled part-time. Student to teacher ratio 15:1. 100% of graduates obtain employment within six months. Facilities: 2 kitchens plus classrooms.

Courses: Soups, sauces, food production, meat, sanitation, baking, patisserie, desserts, garde manger, charcuterie, buffet, food sculpture, menu and restaurant design. Externship provided.

Faculty: 2 full-time. Includes R. Turgeon, CWC, CCE and K. Hohmeister, working Executive Chef.

Costs: Annual tuition $3,300 in-state, $4,950 New England regional, $7,590 out-of-state. Summer externship $110 per credit in-state, $165 per credit New England, out-of-state $253 per credit. Other fees: $90. Admission requirements: High school diploma or equivalent.

Contact: Kurt Hohmeister, Department Chair, New Hampshire Community Technical College, Culinary Arts, 2020 Riverside Dr., Berlin, NH 03570 US; 603-752-1113, Fax 603-752-6335.

NEW JERSEY

ATLANTIC COMMUNITY COLLEGE

Mays Landing/August-May *(See also page 230) (See display ad above)*

This two-year college offers a 2-year AAS degree in Culinary Arts and Food Service Management. Program started 1981. Accredited by MSA. Calendar: semester. Curriculum: culinary. Admission dates: September, January. Total enrollment: 400. 160 each admission period. 100% of applicants accepted. 54% receive financial aid. 15% enrolled part-time. Student to teacher ratio 20:1. 100% of graduates obtain employment within six months. Facilities: Include 8 kitchens, 4 classrooms, computer lab, a bake shop, retail store, and gourmet restaurant.

Courses: Include food purchasing, pastry, garde manger, hot food preparation, and nutrition. An internship is required. Also offers non-credit classes and ACF Apprenticeship program in entry level skills.

Faculty: 15 full-time international faculty with professional designations.

Costs: Approximately $3,000 per semester, in-state. Housing is available. Admission requirements: High school diploma or equivalent. Scholarships: yes. Loans: yes.

Location: Atlantic Community College, a 536-acre campus in New Jersey's Pinelands, is 17 miles west of Atlantic City's boardwalk and 45 miles from Philadelphia.

Contact: Bobby Royal, Executive Director of College Recruitment, Academy of Culinary Arts - Atlantic Community College, Admissions, 5100 Black Horse Pike, Mays Landing, NJ 08330-2699 US; 609-343-5000, Fax 609-343-4921, E-mail accadmit@atlantic.edu, URL http://www.atlantic.edu.

BERGEN COMMUNITY COLLEGE

Paramus/September-May

This college offers a 1-year (18-30 credit) certificate in Culinary Arts and 2-year (64-credit) degree in Hotel/Restaurant/Hospitality. Established in 1974. Accredited by MSA. Calendar: semester. Curriculum: culinary, core. Admission dates: September, January. Total enrollment: 125. 40-50 each admission period. 25% of applicants accepted. 10% receive financial aid. 20% enrolled part-time. Student to teacher ratio 25:1. 100% of graduates obtain employment within six months. Facilities: 2 cooking labs, 2 dining rooms, 1 computer lab.

Courses: Hands-on training integrated with management. Co-op training required.

Faculty: 4 full- and 2 part-time. All master's level and industry trained.

Location: 10 miles from New York City.

Contact: Prof. David Cohen, Prof. Don Delnero, Bergen Community College-Hotel/Restaurant/Hospitality, 400 Paramus Rd., Paramus, NJ 07652 US; 201-447-7192, Fax 201-612-5240.

CULINARY EDUCATION CENTER OF MONMOUTH COUNTY

Lincroft/Year-round

Brookdale Community College and Monmouth County Vocational School District offer 1-year (34.5-credit) certificate and 2-year (69.5-credit) AAS degree programs in Culinary Arts, 2-year AAS degree in Food Service Management. Program started 1998. Accredited by MSA. Calendar: semester. Curriculum: culinary, core. Admission dates: September, January. Total enrollment: 320. 50-85 each admission period. 100% of applicants accepted. 30% enrolled part-time. Student to teacher ratio 16:1. Facilities: Include 3 kitchens, 2 baking kitchens, 2 dining rooms, computer lab, classrooms.

Courses: Include recipe interpretation, cooking techniques and principles, sanitation and safety, storeroom, nutrition, garde manger, baking, patisserie, regional cuisine, service & management. 300 hours of externship required of all graduates in approved sites.

Faculty: 10 full- and part-time faculty with advanced degrees and certifications, as well as professional culinary work experience.

Costs: ~$3,500 per semester in-county. Total 2-year program cost ~$13,750 (includes tuition, fees, books, knife kit, uniforms). Admission requirements: High school diploma or GED. Scholarships: yes. Loans: yes

Location: Rural Monmouth County, 45 minutes from NYC, minutes from the New Jersey Shore. The Center is located in nearby Asbury Park.

Contact: Shirley Sesler, Enrollment Specialist, Brookdale Community College, 765 Newman Springs Rd., Lincroft, NJ 07738-1597 US; 732-224-2371, Fax 732-842-0203, E-mail cberg@brookdale.cc.nj.us, URL http://www.brookdale.cc.nj.us.

HUDSON COUNTY COMMUNITY COLLEGE

Jersey City/September-May

This two-year college offers a 2-year AAS degree, 1-year Culinary Certificate, specialized proficiency certificates, Sanitation Certificate (AHMA). Program started 1983. Accredited by MSA, State Commission on Higher Education. Calendar: semester. Curriculum: core. Admission dates: September, January. Total enrollment: 200. Student to teacher ratio 15:1. 95% of graduates obtain employment within six months. Facilities: 4 kitchens, 3 classrooms, 1 bar.

Courses: Include bakeshop, garde manger, buffet catering. 600-hour externship provided.

Faculty: 10 full-time.

Costs: Annual tuition $4,000 in-county, $6,200 out-of-county. Admission requirements: Open admission. High school diploma, GED, or over 18 yrs of age. Scholarships: yes.

Contact: Siroun Meguerditchian, Executive Director, Hudson County Community College, Culinary Arts Institute, 161 Newkirk St., Jersey City, NJ 07306 US; 201-714-2193, Fax 201-656-1522, URL http://www.hudson.cc.nj.us.

MIDDLESEX COUNTY COLLEGE

Edison/Year-round

This college offers a certificate in Culinary Arts, 2-year AAS degree in Hotel Restaurant, and Institution Management with Culinary Arts Management Option. Program started 1987. Accredited by MSA. Calendar: semester. Curriculum: culinary, core. Admission dates: January, May, September. Total enrollment: 32. 10 e0ach admission period. 100% of applicants accepted. 50% receive financial aid. 40% enrolled part-time. Student to teacher ratio 16:1. 100% of graduates obtain employment within six months. Facilities: Include 2 kitchens, 3 classrooms.

Courses: Food selection and preparation, baking, food production, garde manger, food and beverage cost controls and purchasing, sanitation, beverage management, and general education.

Externship: 180 hours.

Faculty: 6 full-time, 3 part-time.

Costs: Annual tuition in-state $2,300, out-of-state $4,600. Application fee $25. Admission requirements: High school diploma or equivalent. Loans: yes.

Location: The 200-acre suburban campus is 30 miles from New York City.

Contact: Marilyn Laskowski-Sachnoff, Department Chair, Middlesex County College, Hotel Restaurant & Institution Management, 2600 Woodbridge Ave., PO Box 3050, Edison, NJ 08818-3050 US; 732-906-2538, Fax 732-906-7745, E-mail mlsachnoff@aol.com, URL http://www.middlesex.cc.nj.us. Also: sachnoff@email.njin.net.

MORRIS COUNTY SCHOOL OF TECHNOLOGY

Denville/September-June

This trade-technical school offers a 1- or 2-year sequential program in Culinary Arts with certificates in Hospitality Management, Food Preparation/Production, and ServSafe Sanitation. Established in 1977. Accredited by Dept. of Vocational Education. Curriculum: culinary, core. Admission dates: September. Total enrollment: 110. 60-80 each admission period. 100% of applicants accepted. 10% receive financial aid. 90% enrolled part-time. Student to teacher ratio 15:1. 95% of graduates obtain employment within six months. Facilities: 2 commercial kitchens with the latest equipment, including a bakeshop in one.

Courses: Includes Food ID, Pantry, Cost Control, Soups, Stocks and Sauces, Purchasing, Intro and Advanced Baking, Menu Development, Intro and Advanced Garde Manger, Meat, Poultry, and Seafood Cookery, Customer Skills, American Regional Cuisine. Guest speakers, field trips, job shadowing, internships, coop work, mentorships. Cook-Apprentice 3-year (9-course) certificate program for FMP and ACF certification credits.

Faculty: 4 instructors with culinary degrees, teaching degrees, and industry experience.

Costs: $1,050 per year part-time, $2,100 per year full-time. Admission requirements: Junior/senior level in high school, high school graduate. Scholarships: yes. Loans: yes.

Location: Northern NJ, 30 miles from NYC, 30 miles from PA, 20 miles from upstate NY.

Contact: Kim Letourneau, Work-based Learning Coordinator, Morris County School of Technology, 400 E. Main St., Denville, NJ 07834 US; 973-627-4600 x269, Fax 973-627-6979, E-mail letourneau@mcvts.org, URL http://www.mcvts.org.

PASSAIC COUNTY TECHNICAL INSTITUTE

Wayne/September-June

This vocational high school offers a high school diploma in Culinary Arts. Program started 1970. Accredited by NJ State Dept. of Education. Calendar: semester. Curriculum: core. Admission dates: September. Total enrollment: 120+. 120+ each admission period. 0% enrolled part-time. Student to teacher ratio 10:1. Facilities: 3 kitchens (special needs, faculty, student), 3 dining rooms, Chez Technique restaurant, bakeshop, food lab.

Courses: Baking, cafeteria, production, pantry, table service. Students prepare 1,850 student meals and 400 faculty/staff meals per day.

Faculty: 8 culinary arts teachers: M. Adams, E. Cannataro, R.M. Halas, J. Nuzzo, M.E. Norbe, M. Majette, L. Walden, C. Zaleski.

Location: 30 minutes west of NYC.

Contact: Michael W. Adams, Passaic County Technical Institute, 45 Reinhardt Rd., Wayne, NJ 07470 US; 201-389-4296, E-mail madams@pcti.tec.nj.us, URL http://www.pcti.tec.nj.us/culinary3.html.

SALEM COUNTY VOCATIONAL TECHNICAL SCHOOLS

Woodstown

This career institution offers a 2-year certificate in Culinary Arts. Program started 1976. Accredited by MSA. Admission dates: September, January. Total enrollment: 30. 75% of applicants accepted.

Student to teacher ratio 15:1. 85% of graduates obtain employment within six months.

Faculty: 20 full-time.

Costs: Annual tuition $3,000.

Contact: Eva Hoffman, Culinary Arts Instructor, Salem County Vocational Technical Schools, Culinary Arts, Box 350, Woodstown, NJ 08098 US; 609-769-0101, Fax 609-769-4214, URL http://www.scvts.org.

NEW MEXICO

ALBUQUERQUE TECHNICAL VOCATIONAL INSTITUTE

Albuquerque/Year-round

This community career institution offers a Baking certificate, Quantity Foods certificate, Food Service Management certificate, AAS degree. Program started 1965. Accredited by ACFEI, RBA. Calendar: trimester. Curriculum: culinary, core. Admission dates: December-January, April-May, August-September. Total enrollment: 200. 150 each admission period. 100% of applicants accepted. 65% receive financial aid. 33% enrolled part-time. Student to teacher ratio 12:1. 100% of graduates obtain employment within six months. Facilities: Baking lab, 2 quantity foods labs, computer lab.

Courses: Baking, quantity foods, food service management. Co-operative education.

Faculty: 11 full- and part-time. Includes Carmine Russo, CCC, CCE, Joyce Jones, CWC, Darcy Buland, Martin Samudio, CCC, Rudy Garcia, Masters Management, Kevin Zink, Michael Williams, Elizabeth McGeehan.

Costs: $22.50 plus equipment and uniforms; tuition-free for New Mexico residents. Off-campus housing available. Admission requirements: Age 18+, High school diploma, GED or concurrent HS. Scholarships: yes. Loans: yes.

Contact: Carmine J. Russo, CCC, CCE, Program Director, Albuquerque Technical Vocational Institute-TVI Community College, Culinary Arts, 525 Buena Vista SE, Albuquerque, NM 87106 US; 505-224-3755, Fax 505-224-3781, E-mail crusso@tvi.cc.nm.us, URL http://www.tvi.cc.nm.us.

SANTA FE COMMUNITY COLLEGE

Santa Fe/Year-round

This two-year college offers a 2-year certificate/AAS degree in Culinary Arts, 1-year certificate. Program started 1985. Accredited by NCA. Calendar: quarter. Curriculum: culinary, core. Admission dates: August 20, January 10, June 1. Total enrollment: 190. 30 each admission period. 100% of applicants accepted. 20% receive financial aid. 70% enrolled part-time. Student to teacher ratio 14:1. 100% of graduates obtain employment within six months. Facilities: Include 2 kitchens and classrooms, culinary lab and restaurant.

Courses: Culinary courses, apprenticeship program, specialty topics, Southwestern cuisine. Other required courses: nutrition, sanitation, food and beverage management, general studies. Externship provided, 2 credit-hours or 200 working-hours. Continuing ed: 12 courses per semester.

Faculty: 1 full-time, 8 part-time. Qualifications: working executive chef.

Costs: Annual tuition in-county $17 per credit hour, out-of-county $24 per credit hour, out-of-state $45 per credit hour. Lab fees. Admission requirements: High school diploma or equivalent. Scholarships: yes. Loans: yes.

Location: The new 9,000-student campus is in a country setting.

Contact: Bill Weiland, CEC, Director of Culinary Arts, Santa Fe Community College, Culinary Arts/Hospitality, 6401 Richards Ave., Santa Fe, NM 87502 US; 505-466-0315, E-mail bweiland@prism.santafe.cc.nm.us.

NEW YORK

ADIRONDACK COMMUNITY COLLEGE

Queensbury/September-May

This proprietary college offers 1-year certificate and 2-year AAS degree programs in Food Service. Program started 1969. Accredited by MSA. Admission dates: Fall, spring. Total enrollment: 25. 70-80% receive financial aid. 75% enrolled part-time. Student to teacher ratio 10:1. 90% of graduates obtain employment within six months. Facilities: Include 3 kitchens, classroom and restaurant.

Courses: 1 year of food preparation, 1/2 year of spa cuisine, 1/2 year of American regional cuisine. Externship: 1,000-1,200 hours. Continuing education: baking, wines and other topics.

Faculty: 1 full-time, 3 part-time.

Costs: Tuition $1,075 per semester in-state, $2,150 per semester out-of-state plus activity fees. Admission requirements: High school diploma or equivalent.

Contact: William Steele, Program Coordinator, Adirondack Community College, Commercial Cooking, Bay Rd., Queensbury, NY 12804 US; 518-743-2200 x374, Fax 518-743-2317, E-mail info@acc.sunyacc.edu, URL http://www.suny.edu.

ALFRED STATE COLLEGE – CULINARY ARTS

Alfred/August-May

This two-year college offers a 2-year AOS degree program. Established in 1966. Accredited by MSA. Calendar: semester. Curriculum: culinary. Admission dates: January, August. Total enrollment: 55. 33 each admission period. 85% of applicants accepted. 80% receive financial aid. 0% enrolled part-time. Student to teacher ratio 8:1. Facilities: Comparative with the food industry.

Courses: Emphasis on the principles of the food service industry to prepare students for supervisory trainee positions, food production, or food service positions. A minimum of 320 hours employment in the food service industry is required during the summer.

Costs: Tuition $3,200 in-state, $5,000 out-of-state; books, uniforms, tools ~$300 per semester; room and board $4,856. Admission requirements: High school average of C or better. Scholarships: yes. Loans: yes.

Contact: Deborah Goodrich, Director of Admissions, Alfred State College, Upper College Dr., Alfred, NY 14802 US; 800-4AL-FRED, Fax 607-587-4299, E-mail admissions@alfredtech.edu, URL http://www.alfredtech.edu.

THE CULINARY INSTITUTE OF AMERICA

Hyde Park/Year-round *(See also pages 13 and 234) (See display ad page 81)*

This independent, not-for-profit educational institution offers 21-month associate degree programs in Culinary Arts and Baking and Pastry Arts, 38-month bachelor's degree programs in Culinary Arts Management and Baking and Pastry Arts Management, continuing education and nonvocational courses. Established in 1946. Accredited by ACCSCT; curricula registered with NY State Educational Dept. Curriculum: culinary. Admission dates: Associate degree 16 (8) times per year for culinary (baking and pastry) arts; bachelor's degree upper-level courses 3 times per year. Total enrollment: 2,065. 72 (50) assoc. (bachelor's) degree each admission period. 72% of applicants accepted. 85% receive financial aid. Student to teacher ratio 18:1. Facilities: 36 kitchens and bakeshops, 50,800-volume library, 2,800 instructional videos, learning resource and nutrition centers, bookstore, 4 student-run restaurants.

Courses: Associate degree programs give students comprehensive, hands-on experience in the theory and techniques of foodservice. Bachelor's degree programs provide additional managerial and conceptual skills. All degree programs include 18-week paid externships off campus. Continuing Ed. Dept. offers 1- to 30-week courses in cooking, baking, and management.

Faculty: Over 125 chefs and instructors from more than 20 countries, including the highest concentration of ACF-Certified Master Chefs.

Costs: Estimated cost: $32,955 associate degree programs, $26,960 bachelor's degree programs (junior and senior years), $695-$15,400 cont ed courses. On-campus dorms house 1,079 students for $1,375-$2,055 per semester. Admission requirements: High school diploma or equivalent and at least 6 months foodservice experience, including work in a professional kitchen. Scholarships: yes. Loans: yes.

Location: The 150-acre campus overlooks the Hudson River on U.S. Route 9 in Hyde Park, about 90 minutes north of New York City.

Contact: Susan Weatherly, Dir. of Enrollment Services, The Culinary Institute of America, 433 Albany Post Rd., Hyde Park, NY 12538-1499 US; 800-CUL-INAR, E-mail admissions@culinary.edu, URL http://www.ciachef.edu.

DELHI COLLEGE OF TECHNOLOGY

Delhi/Year-round

This two-year college offers an AAS degree in Culinary Arts. Program started 1994. Calendar: semester. Curriculum: culinary, core. Admission dates: September. Total enrollment: 40. 20 each admission period. Student to teacher ratio 10:1. Facilities: Hospitality center with catering and restaurant kitchens, catering facility, beverage lounge.

Courses: Summer internship required.

Faculty: 2: certified chef, bachelor's degree.

Costs: Tuition is $3,200 per year in-state, $5,200 per year out-of-state. Admission requirements: High school graduate with 1 yr general biology. Scholarships: yes. Loans: yes.

Contact: Rosalie Higgins, Dept. Chair, Delhi College of Technology, Main St., Delhi, NY 13753 US; 607-746-4400, Fax 607-746-4769, E-mail higginrl@delhi.edu, URL http://www.delhi.edu.

ERIE COMMUNITY COLLEGE, CITY CAMPUS

Buffalo/September-May

This two-year college offers a 2-year AOS degree in Hotel Technology/Culinary Arts. Program started 1985. Accredited by MSA. Calendar: semester. Curriculum: culinary. Admission dates: Fall and spring. Total enrollment: 80. 80 each admission period. 60% of applicants accepted. 70% receive financial aid. 10% part-time. Student to teacher ratio 28:1. 75% of graduates obtain jobs within six months. Facilities: Include 5 kitchens and classrooms, computer lab and restaurant.

Faculty: 5 full-time, 3 part-time.

Costs: Annual tuition: in-state $1,980, out-of-state $3,960. Other fees $100.

Contact: Paul J. Cannamela, CCE, AAC, Assistant Professor, Erie Community College, Hotel Management/Culinary Arts, 121 Ellicott St., Buffalo, NY 14203 US; 716-851-1034, Fax 716-851-1129, URL http://davey.sunyerie.edu.

THE FRENCH CULINARY INSTITUTE

New York/Year-round *(See also page 235) (See display ads pages 83, 85, 87)*

This proprietary institution offers 600-hour 6-month full-time (5-days per week) and 9-month part-time (3 evenings per week) Grande Diplôme programs in Classic Culinary Arts and in Classic Pastry Arts; 6-week full-time 210-hour Diplôme du Boulanger program in the Art of International Bread Baking. Established in 1984. Accredited by ACCSCT. Calendar: quarter. Curriculum: culinary. Admission dates: Every 6 weeks. Total enrollment: 450. 22 each admission period. 90% of applicants accepted. 75% receive financial aid. 60% enrolled part-time. Student to teacher ratio 11:1. 95% of graduates obtain employment within six months. Facilities: 15,000-square-feet of kitchens, newly-equipped pastry and bread kitchens, demonstration amphitheater, L'École open-to-the-public restaurant.

Courses: Culinary: Classic Culinary Arts with emphasis on classic French Techniques. Pastry Arts: Classic French pastry technique, traditional dessert composition. Bread Baking: Classic artisanal breads of France, Italy, Germany, restaurant breads, viennoiseries. Cooking experience at Restaurant L'École. Placement assistance provided after completion of Level I.

Faculty: 40 full-time, 10 assistant chef-instructors. Includes Senior Dean of Studies Alain Sailhac, Dean of Special Programs Jacques Pépin, Dean of Pastry Arts Jacques Torres, Master Chef/Senior Lecturer André Soltner, Dean of Bread Baking Daniel Leader.

Costs: 1999 tuition is $22,760 each for the Classic Culinary Arts and Classic Pastry Arts programs, $4,900 for the Bread Baking program, which includes uniform, equipment, registration fee. Application fee $100, deposit $400. Housing assistance is available. Admission requirements: High school diploma or equivalent, 150-word essay, personal interview. Loans: yes.

Location: New York's historic SoHo district, adjacent to Chinatown and Little Italy.

Contact: Cynthia C. Marchese, Director of Admissions, The French Culinary Institute, 462 Broadway, New York, NY 10013-2618 US; 888-FCI-CHEF/212-219-8890, Fax 212-431-3054, E-mail cmarchese@frenchculinary.com, URL http://www.frenchculinary.com.

JEFFERSON COMMUNITY COLLEGE

Watertown

This two-year college offers a 2-year certificate/AAS degree. Program started 1975. Accredited by MSA. Admission dates: August, January. Total enrollment: 120. 100% of applicants accepted. Student to teacher ratio 25:1. 95% of graduates obtain employment within six months.

Courses: Externship provided.

Faculty: 4 full-time.

Costs: Annual tuition in-state $1,250 full-time, $84 per credit part-time; out-of-state $2,318 per semester, $168 per credit. Admission requirements: High school diploma or equivalent.

Contact: Deborah Running, Dept. Chair, Jefferson Community College, Hospitality & Tourism, Outer Coffeen St., Watertown, NY 13601 US; 315-786-2200, Fax 315-786-2459, URL http://www.sunyjefferson.edu.

MOHAWK VALLEY COMMUNITY COLLEGE

Rome/August-May *(See display ad above)*

This college offers a 1-year Chef Training certificate, 2-year AOS degree in Food Service, 2-year AAS degrees in Restaurant Management and Hotel Technology:Meeting Services Management, and a 2-year AS degree in Nutrition and Dietetics. Program started 1978. Accredited by MSA. Calendar: semester. Curriculum: culinary. Admission dates: August, January. Total enrollment: 120. 30 each admission period. 90% of applicants accepted. 80% receive financial aid. 40% enrolled part-time. Student to teacher ratio 15-25:1. 98% of graduates obtain employment within six months. Facilities: Include 6 kitchens and classrooms, student-run restaurant, cafe.

Courses: Include sanitation, food preparation, food merchandising, and baking. A 225-hour externship is provided.

Faculty: 3 full-time, 6 part-time. Includes Director Dennis R. Baumeyer, BS, MS, HRM.

Costs: Annual tuition is $2,600 in-state, $5,200 out-of-state. Application fee is $30. Other fees: student activity and insurance fee is $66. Admission requirements: High school diploma or equivalent and admission test. Scholarships: yes. Loans: yes.

Location: The 2,500-student college campus is 12 miles from Utica.

Contact: Dennis R. Baumeyer, Director, Mohawk Valley Community College, Hospitality Programs, 1101 Floyd Ave., Rome, NY 13440 US; 315-334-7710/7702, Fax 315-334-7762, E-mail dbaumeyer@mvcc.edu, URL http://www.mvcc.edu.

MONROE COMMUNITY COLLEGE

Rochester/September-May

This two-year college offers a 2-year certificate/AAS degree. Program started 1967. Accredited by MSA. Calendar: semester. Curriculum: core. Admission dates: Fall, spring. Total enrollment: 175. 60 each

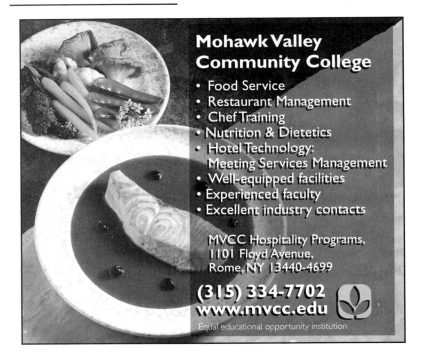

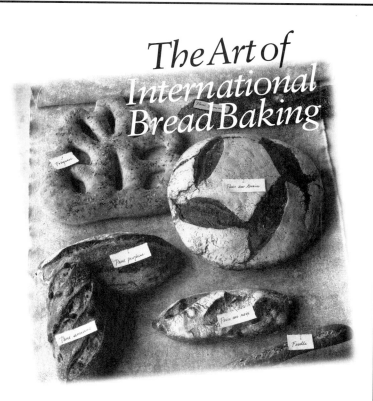

admission period. 80% of applicants accepted. 60% receive financial aid. 20% part-time. Student to teacher ratio 18:1. 96% of graduates obtain jobs within six months. Facilities: 4 kitchens, dining room.

Faculty: 8-10 full- and part-time. Includes Chair E.F. Callens, CEC, CCE, CFE.

Costs: Tuition $1,205 per semester in-state full-time, $105 per credit-hour part-time. Out-of-state tuition is double. Admission requirements: High school diploma or equivalent.

Contact: Eddy Callens, Chairperson, Monroe Community College, Food, Hotel, & Tourism Management, 1000 E. Henrietta Rd., Rochester, NY 14623 US; 716-292-2000 x2586, Fax 716-427-2749, URL http://www.monroe.cc.edu.

THE NATURAL GOURMET COOKERY SCHOOL

New York/Year-round *(See also page 238) (See display ad above)*

This proprietary school devoted to health supportive, natural foods cooking and theory offers 600+-hour 4-month full-time/9-month part-time/new 15-month Sunday Chef's Training Program (CTP), individual nonvocational courses for the public. Established in 1977. Chef's Training Program curriculum licensed by NY State Dept. of Education. Curriculum: culinary. Admission dates: CTP begins 9 times per year. 16 each admission period. 20% receive financial aid. 33% enrolled part-time. Student to teacher ratio 16:1. 75% of graduates obtain employment within six months. Facilities: Include 2 kitchens, classroom, and bookstore.

Courses: Vegetarian focus, emphasis on natural foods, health supportive cooking, contemporary presentation, preparation, techniques. Curriculum includes knife skills, limited poultry and fish, baking and desserts, career opportunities, theoretical approaches to health. A 95-hour externship in an outside food establishment is required. 30 classes, open to the public.

Faculty: 15 full/part-time faculty, includes founder Annemarie Colbin, MA, Certified Health Education Specialist, author of Food and Healing, The Book of Whole Meals, The Natural Gourmet; Co-Presidents Diane Carlson and Jenny Matthau, graduates of the school.

Costs: $11,450 tuition includes knife kit, garde manger kit, books, other materials. $100 application fee. Nearby lodging begins at $500 per month. Admission requirements: High school diploma or equivalent. Scholarships: yes. Loans: yes.

Contact: Merle Brown, Director of Admissions, The Natural Gourmet Cookery School, 48 W. 21st St., 2nd Floor, New York, NY 10010 US; 212-645-5170/212-627-COOK, Fax 212-989-1493, URL http://www.naturalgourmetschool.com.

NEW SCHOOL CULINARY ARTS

New York/Year-round *(See also page 238)*

The New School for Social Research offers master class certificate courses in Cooking, Baking, Professional Catering, Italian Cooking, Restaurant Management; other programs for professionals and nonprofessionals. Established 1919. Calendar: trimester. Total enrollment: 12 per Master class. 100% enrolled part-time. Student to teacher ratio 12:1. Facilities: Restored landmark townhouse with indoor and outdoor dining areas and a fully-equipped instructional kitchen.

Courses: Master Class: basic skills, cuisine and pastry preparation, presentation, recipe develop-

ment; Baking: pastries, breads and doughs, cake decoration, chocolate; Catering: food preparation and instruction in the business of catering. Graduates are eligible for apprenticeships. Include restaurant management, cake decorating, opening a coffee bar, creating and selling a new food.

Faculty: 50+ faculty headed by Gary Goldberg, co-founder Martin Johner, includes Miriam Brickman, Richard Glavin, Arlyn Hackett, Micheal Krondl, Harriet Lembeck, Lisa Montenegro, Robert Posch, Dan Rosati, Stephen Schmidt, Marie Simmons, Karen Snyder, Carole Walter.

Costs: Master Class $2,375 (+$450 materials fee) for Cooking, $1,420 (+$200) for Baking, $940 (+$195) for Catering. Most other professional courses $70 per session (+$5-$14).

Contact: Gary A. Goldberg, Executive Director, New School Culinary Arts, 100 Greenwich Ave., New York, NY 10011 US; 212-255-4141/800-544-1978, Fax 212-229-5648, E-mail NSCulArts@aol.com, URL http://www.newschool.edu.

NEW YORK CITY TECHNICAL COLLEGE

Brooklyn/September-May

This career college offers a 2-year AAS degree and 4-year BS degree in Hospitality Management. Program started 1947. Accredited by MSA, state, ACPHA. Calendar: semester. Curriculum: core. Admission dates: September, February. Total enrollment: 800. 125 each admission period. 50% receive financial aid. 50% enrolled part-time. Student to teacher ratio 15:1. 90% of graduates obtain employment within six months. Facilities: Include 5 kitchens, dining room, 3 classrooms, restaurant.

Courses: Food and beverage cost control, culinary arts, baking and pastry arts, wines, beverage management. Externship: 8 weeks, in Europe, if qualified. Continuing education courses.

Faculty: 13 full-time, 20-40 part-time. Includes Patricia S. Bartholomew, Chair.

Costs: Annual tuition in-state $3,200, out-of-state $6,400. CUNY fee is $35. Textbooks $1,000, materials fee $10, uniforms $120. Average off-campus housing cost is $600 per month. Admission requirements: High school diploma or equivalent and admission test. Scholarships: yes. Loans: yes.

Contact: Dr. Patricia Bartholomew, Chair, New York City Technical College, Hospitality Management, 300 Jay St., #N220, Brooklyn, NY 11201 US; 718-260-5630, Fax 718-260-5997, E-mail psbny@cunyvm.cuny.edu.

NEW YORK FOOD AND HOTEL MANAGEMENT SCHOOL

New York/Year-round

This proprietary institution offers a 9-month certificate in Commercial Cooking and Catering. Established in 1935. Accredited by ACCST. Curriculum: culinary. Admission dates: Every 4 to 6 weeks. Total enrollment: 90. 20 each admission period. 80% of applicants accepted. Student to teacher ratio 16:1. Facilities: Include 4 kitchens, 5 classrooms and restaurant.

Courses: Skills development, quantity food production, food preparation, catering, restaurant operation, food purchasing, sanitation, baking and pastry production. Externship: 3 months.

Faculty: 10 full-time and part-time.

Costs: Annual tuition $7,515. Registration fee $100. Books $235, food lab fee $1,490, kits and uniforms $155. Admission requirements: High school diploma or equivalent or admission test.

Contact: Harold Kaplan, Vice-President, New York Food and Hotel Management School, Admissions, 154 W. 14th Street, New York, NY 10011 US; 212-675-6655, Fax 212-463-9194.

NEW YORK INSTITUTE OF TECHNOLOGY – CULINARY ARTS CENTER

Central Islip/Year-round

This independent institution offers a 19-month AOS degree in Culinary Arts. Program started 1987. Accredited by MSA, ACFEI. Calendar: semester. Curriculum: culinary, core. Admission dates: Fall, spring. Total enrollment: 170. 95/20 each admission period. 90% of applicants accepted. 95% receive financial aid. 65% part-time. Student to teacher ratio 16:1. 99% of graduates obtain jobs within six months. Facilities: 4 kitchens, bakery, 2-10 classrooms, 2 restaurants.

Courses: Externship: 3-month, in public restaurants and food service establishments.

Faculty: 9 full-time, 3 part-time.

Costs: Annual tuition $13,000. On-campus housing: 150 spaces; average cost: $2,500 per semester.

Contact: Prof. Susan Sykes Hendee, Dept. Chair, New York Institute of Technology-Culinary Arts Center, 300 Carleton Ave., #66-101, PO Box 9029, Central Islip, NY 11722-9029 US; 516-348-3290, Fax 516-348-3247, E-mail shendee@iris.nyit.edu, URL http://www.nyit.edu/culinary.

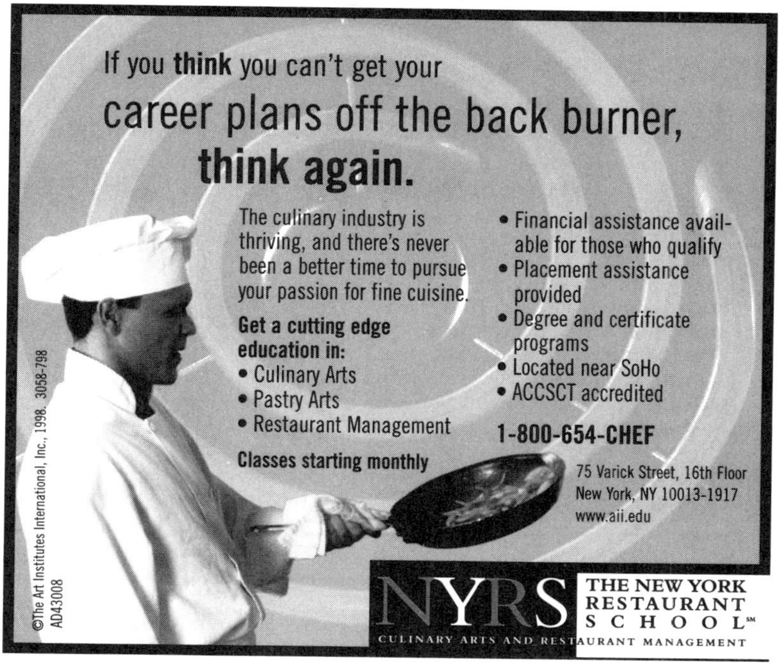

NEW YORK RESTAURANT SCHOOL

(See display ad above)

New York/Year-round

This private college offers a 17-month AOS degree program in Culinary Arts and a 6- to 9-month credit-bearing certificate program in Pastry Arts and Culinary Skills. Established in 1980. Accredited by ACCSCT, NYS Bd. of Regents. Curriculum: culinary. Admission dates: Rolling admission. Total enrollment: 1000. 125 each admission period. 80% of applicants accepted. 90% receive financial aid. 25% enrolled part-time. Student to teacher ratio 21:1 lab. 93% of graduates obtain employment within six months. Facilities: The 36,000-square-foot recently renovated facility includes 7 newly-equipped kitchens, 5 classrooms, and resource center.

Courses: Include basic food preparation and knife skills, food handling, meat, vegetable, fish, and seafood preparation, international cuisine, and garde manger. Programs conclude with a 3-month local externship.

Faculty: The 28 faculty members all have a minimum of 5 years experience and many have 4-year degrees and prior teaching background.

Costs: $16,075 for Culinary Arts, $9,977 for Pastry Arts, $10,129 for Culinary Skills, $26,583 for AOS degree program. $50 application fee required. Admission requirements: Vary. Scholarships: yes. Loans: yes.

Contact: Sarah Cirrincione, Director of Admissions, New York Restaurant School, Admissions, 75 Varick St., New York, NY 10013 US; 212-226-5500/800-654-CHEF, Fax 212-226-5664.

NEW YORK UNIVERSITY

New York/Year-round *(See also page 238) (See display ad above)*

The Dept. of Nutrition and Food Studies, School of Education offers bachelor's, master's, and doctorate degree programs in Food Studies, Food and Restaurant Management, and Nutrition. Established in 1986. Accredited by MSA. Calendar: trimester. Curriculum: core. Total enrollment: 60 in food program, many part-time. 100% of graduates obtain employment within six months. Facilities: New teaching kitchen and library, computer, academic resources.

Courses: Over 50 in food science and management, food culture and history, food writing, nutrition. Internship required for all students. Food studies and food management.

Faculty: 9 full-time academic, 45 part-time academic and professional.

Costs: ~$650 per credit. Some housing available. Scholarships: yes. Loans: yes.

Location: Washington Square, Greenwich Village.

Contact: Dr. Marion Nestle, Dept. Chair, New York University, Nutrition & Food Studies, 35 W. 4th St., 10th Flr., New York, NY 10012-1172 US; 212-998-5588/212-998-5580, Fax 212-995-4194, E-mail marion.nestle@nyu.edu, URL http://www.nyu.edu/education/nutrition.

NIAGARA COUNTY COMMUNITY COLLEGE

Sanborn/Year-round

This two-year college offers a 2-year/66 credit-hour AOS degree in Food Service with Professional Chef emphasis. Calendar: semester. Curriculum: core.

Costs: $88 per credit-hour in-state.

Contact: Sam Sheusi, Coordinator/Instructor, Niagara County Community College, 3111 Saunders Settlement Rd., Sanborn, NY 14132 US; 716-731-3271 x248, Fax 716-731-4053, URL http://www.sunyniagara.cc.ny.us.

ONONDAGA COMMUNITY COLLEGE

Syracuse

This two-year college offers a 1-year certificate, 2-year AAS degree. Program started 1979. Accredited by State. Admission dates: Fall, spring. Total enrollment: 90-100. Student to teacher ratio 16:1.

Courses: Externship provided.

Costs: Tuition in-state $1,225 per semester, out-of-state $3,675 per semester.

Contact: Jim Drake, Culinary Director, Onondaga Community College, Culinary Arts, Onondaga Hill, Syracuse, NY 13215 US; 315-469-7741 x2231.

PAUL SMITH'S COLLEGE

Paul Smiths/Year-round *(See display ad page 91)*

This college offers a 2-year AAS degree in Culinary Arts and a 1-year Baking certificate. Program started 1980. Accredited by MSA. Admission dates: September, January. Total enrollment: 150. 85% of applicants accepted. 80% receive financial aid. Student to teacher ratio 14:1. 99% of graduates obtain employment within six months. Facilities: Include 4 campus Foods Laboratories, an a la carte kitchen, and a 60-seat dining room.

Courses: Baking certificate curriculum covers journeyman skills, including advertising, merchandising, and management. Students produce goods for an on-campus bakery. During year two, 1 semester is spent in the College's Hotel Saranac, the other is an externship.

Faculty: 66 full-time, 12 part-time. Includes Robert Brown, CM, and Paul Sorgule, CCE, 1988 Culinary Olympics gold medalist.

Costs: Annual tuition $18,000. Culinary Arts program comprehensive fee is $545 per semester. $25 application fee. Housing averages $2,300 per year; board is $2,480 per year. Admission requirements: High school transcript or GED. The Baking Certificate program is open to those who have industry experience or have completed a culinary arts program.

Location: On the shore of the Lower St. Regis Lake, surrounded by 13,100 acres of college-owned forests and lakes in the Adirondack Mountains.

Contact: Enrico A. Miller, VP for Enrollment Management, Paul Smith's College, Admissions, PO Box 265, Paul Smiths, NY 12970 US; 800-421-2605, Fax 518-327-6161, URL http://www.paulsmiths.edu.

PETER KUMP'S NEW YORK COOKING SCHOOL

New York/Year-round *(See also page 239) (See display ads pages 93, 239, 240)*

This private school offers 20- and 26-week diploma programs in Culinary Arts and Pastry and Baking, 38-week weekend diploma program in Culinary Arts, 10- to 24-week certificate programs in cooking and baking, continuing education and business courses, programs for nonprofessionals. Established in 1974. Career program curricula licensed by NY State Dept. of Education. Curriculum: culinary. Admission dates: 35 times per year. Total enrollment: 400. Student to teacher ratio 13:1. Facilities: 27,000-square-foot facility (opens 12/98) includes 9 kitchens, wine studies center, confectionery lab, rooftop herb garden.

Courses: In addition to theory and hands-on training in the preparation and presentation of classic cuisines, culinary arts courses cover Italian and Asian cuisine, kitchen management, butchering, pastry and baking, and wine. 6-week apprenticeship at a restaurant/pastry shop in New York City or France follows diploma courses. Over 800 non-credit classes offered year-round.

Faculty: 12 full- and 30+ part-time and guest instructors. Founder Peter Kump was founding president of The James Beard Foundation and president of the IACP. Director of Pastry and Baking Nick Malgieri was one of Pastry Art & Design Magazine's top 10 pastry chefs in 1998.

Costs: Tuition for the 20-week day (26-week evening) diploma programs is $12,500 ($9,950) for Culinary Arts and $11,900 ($9,950) for Pastry & Baking. Tuition for the 38-week weekend diploma in Culinary Arts is $9,950. Admission requirements: Proof of 2 years college or 4 years profes-

sional work experience (any profession). Scholarships: yes. Loans: yes.

Location: East 92nd Street and West 23rd Street in New York City.

Contact: Steve Tave, Director, Peter Kump's New York Cooking School, 50 W. 23rd St., New York, NY 10011 US; 800-522-4610/212-242-2882, Fax 212-242-0127, E-mail user700766@aol.com, URL http://www.pkcookschool.com.

SUNY COLLEGE OF AGRICULTURE & TECHNOLOGY

Cobleskill/August-May

This college offers a 2-year AOS degree in Culinary Arts. Program started 1971. Accredited by MSA, ACFEI. Calendar: semester. Curriculum: culinary. Admission dates: Fall, spring. Total enrollment: 120. 60 each admission period. 85% of applicants accepted. 95% receive financial aid. 5% enrolled part-time. Student to teacher ratio 15:1. 99% of graduates obtain employment within six months. Facilities: Include 5 kitchens.

Courses: Externship provided. Fellowship available.

Faculty: 10 full-time.

Costs: Annual tuition $3,300 in-state, $5,000 out-of-state. Admission requirements: High school diploma or equivalent. Scholarships: yes.

Location: The 750-acre campus is 30 miles from Albany, Schenectady, and Troy.

Contact: Alan Roer, Department Chair, SUNY College of Agriculture & Technology, Culinary Arts, Hospitality & Tourism, Cobleskill, NY 12043 US; 518-234-5011, Fax 518-234-5333, E-mail RoerAH@cobleskill.edu.

SCHENECTADY COUNTY COMMUNITY COLLEGE

Schenectady/Year-round

This two-year college offers a 2-year degree, 1-year certificate. Program started 1980. Accredited by MSA, ACFEI. Calendar: semester. Curriculum: culinary. Admission dates: September, January, June. Total enrollment: 340. 100 each admission period. 86% of applicants accepted. 22% enrolled part-time. Student to teacher ratio 20:1. 88% of graduates obtain employment within six months. Facilities: 7 kitchens, restaurant, 2 dining rooms, banquet room.

Courses: Other required courses: 600 hours work experience. Externship available (Walt Disney World college program, ARAMARK at Kentucky Derby).

Faculty: 13 full-time, 22 part-time. Includes David Brough, CEC, Paul Hiatt, CCE, Wayne Maibe, CCE, Jim Rhodes, CEC, American Academy of Chefs, Toby Strianese, ACFEI Accrediting Team.

Costs: Annual tuition $2,340 in-state, $4,680 out-of-state. Admission requirements: High school diploma or equivalent and placement testing. Scholarships: yes. Loans: yes.

Contact: Toby Strianese, Chair and Professor, Schenectady County Community College, Hotel, Culinary Arts, 78 Washington Ave., Schenectady, NY 12305 US; 518-381-1391, Fax 518-346-0379, E-mail strianaj@gw.sunysccc.edu.

SULLIVAN COUNTY COMMUNITY COLLEGE

Loch Sheldrake/September-April

This two-year college offers a 2-year AAS degree in Professional Chef and Hotel Technology. Program started 1965. Accredited by MSA. Calendar: semester. Curriculum: culinary. Admission dates: September, January. Total enrollment: 126. 75 each admission period. 90% receive financial aid. 22% enrolled part-time. Student to teacher ratio 14:1. 90% of graduates obtain employment within six months. Facilities: Include 7 kitchens and classrooms, restaurant.

Faculty: 8 full-time. Qualifications: bachelor's or master's degree.

Costs: Annual tuition for 1997-98 is estimated to be $2,456 in-state, $4,556 out-of-state. Application fee $25. On-campus housing: 300 spaces. Average off-campus housing cost: $4,600. Admission requirements: High school diploma or equivalent and admission test. Scholarships: yes. Loans: yes.

Location: The 405-acre campus is 35 miles from Middletown.

Contact: Mark Sanok, Chairperson, Sullivan County Community College, Hospitality, 1000 LeRoy Rd., Box 4002, Loch Sheldrake, NY 12759-4002 US; 914-434-5750, Fax 914-434-4806, URL http://www.sullivan.suny.edu.

WESTCHESTER COMMUNITY COLLEGE
Valhalla/September-May

This two-year college offers a 2-year AAS degree in Food Service Administration. Program started 1971. Accredited by MSA. Calendar: semester. Curriculum: culinary. Admission dates: All year. Total enrollment: 100. 50 each admission period. 20% receive financial aid. 25% enrolled part-time. Student to teacher ratio 15:1. 100% of graduates obtain employment within six months. Facilities: Include lab/demo kitchen, baking kitchen, production kitchen, bar/beverage management lab, instructional dining room.

Courses: Food preparation, quantity food production, buffet catering, advanced foods, garde manger, bar/beverage management, and menu planning. Continuing education provided.

Faculty: 4 full-time. Qualifications: MS required. Includes Curriculum Chair D. Nosek, D. Salvestrini, J. Snyder, T. Cousins.

Costs: Annual tuition in-state $1,075 per semester. Lab fees $15. Admission requirements: High school diploma or equivalent. Scholarships: yes.

Location: 30 miles north of New York City.

Contact: Daryl Nosek, Curriculum Chair, Westchester Community College, Restaurant Management, 75 Grasslands Rd., Valhalla, NY 10595-1698 US; 914-785-6600, Fax 914-785-6765.

NORTH CAROLINA

ASHEVILLE BUNCOMBE TECHNICAL COMMUNITY COLLEGE
Asheville/Year-round

This two-year career college offers a 2-year AAS degree in Culinary Technology. Program started 1968. Accredited by SACS. Calendar: semester. Curriculum: culinary. Admission dates: Begins September of year previous to official enrollment. Total enrollment: 70. 35 each admission period. 100% of applicants accepted. 16% enrolled part-time. Student to teacher ratio 11:1. 100% of graduates obtain employment within six months. Facilities: Include 2 kitchens, 4 classrooms, restaurant 1 day per week.

Courses: Food prep, baking, garde manger, classical lab, palate development, butchering, dining room personnel, sanitation, and international and American regional cuisine. Internship provided.

Faculty: 3 full-time, 5 part-time. Sheila Tillman, BS, MA, Brian McDonald, CEC, CCE, Scott Gerken, CWPC, Lance Etheridge, John Hofland, CEC, Jodee Sellers, Jeff Frank, CEC, Karen Spradley.

Costs: Annual tuition $763 in-state, $4,966 out-of-state. Activity fee $7 quarterly. Off-campus housing cost $300 per month. Admission requirements: High school diploma or equivalent and admission test. Scholarships: yes.

Location: The 4,000-student, 127-acre campus is in a suburban setting.

Contact: Sheila Tillman, Chairperson, Asheville Buncombe Technical Community College, Hospitality Education, 340 Victoria Rd., Asheville, NC 28801 US; 704-254-1921 x232, Fax 704-251-6355, E-mail stillman@asheville.cc.nc.us, URL http://www.asheville.cc.nc.us.

CENTRAL PIEDMONT COMMUNITY COLLEGE
Charlotte/Year-round

This two-year college offers a 2-year AAS degree in Culinary Arts. Certificate programs in baking, culinary, garde manger, hot foods. Program started 1974. Accredited by SACS. Calendar: semester. Curriculum: culinary, core. Admission dates: Fall, winter, spring, summer. Total enrollment: 500. 100-150 each admission period. 75% of applicants accepted. 10% receive financial aid. 25% enrolled part-time. Student to teacher ratio 15:1. 98% of graduates obtain employment within six months. Facilities: Include 4 kitchens, 3 classrooms, baking lab, small quantities lab, computer lab, restaurant.

Courses: Short-term (8-week) courses. Cooperative education/work experiences readily available. Short-term courses.

Faculty: 3 full-time C.I.A. graduates, 5 part-time.

Costs: Annual tuition in-state $800, out-of-state $4,800. Off-campus housing cost $400-$500 per month. Admission requirements: High school diploma or equivalent and admission test.

Location: Near downtown Charlotte.

Contact: Robert G. Boll, FMP, CFE, Department Head, Central Piedmont Community College, Culinary Arts, PO Box 35009, Charlotte, NC 28235 US; 704-330-6721, Fax 704-330-6581.

GUILFORD TECHNICAL COMMUNITY COLLEGE

Jamestown/Year-round

This two-year community college offers 1- and 2-year programs in Culinary Technology, Hotel/Restaurant Management, Travel and Tourism. Program started 1989. Calendar: semester. Curriculum: core. Admission dates: Year-round. Total enrollment: 102. 35 each admission period. 90% of applicants accepted. 50% receive financial aid. 50% enrolled part-time. Student to teacher ratio 7:1. 98% of graduates obtain employment within six months.

Courses: Garde manger, baking and pastry, dining room management, nutritional cuisine, customer service. Co-op training.

Faculty: 2 full-time, 4 part-time. Qualifications: ACFEI certified and minimum 5 years experience. Includes Ronald Wolf, CCC, CCE and Keith Gardiner, CEC.

Costs: Tuition in-state $20 per quarter-hour, out-of-state $163 per quarter-hour. Admission requirements: Placement exam. Scholarships: yes.

Contact: Keith Gardiner, Department Chair, Guilford Technical Community College, Culinary Technology, Box 309, Jamestown, NC 27282 US; 336-334-4822 x2347, Fax 336-841-4350, E-mail gardinerk@gtcc.cc.us.nc, URL http://technet.gtcc.nc.us.

SANDHILLS COMMUNITY COLLEGE

Pinehurst/Year-round

This two-year college offers a 2-year/70 semester-hour degree in Culinary Technology. Calendar: semester. Curriculum: core. 95% of graduates obtain employment within six months.

Courses: Basic and advanced culinary skills, food science, baking, garde manger, international and American regional cuisine, pastry and confections, purchasing, food and beverage service.

Costs: Full-time students $280 per semester in-state, $2,282 per semester out-of-state. Scholarships: yes.

Contact: Ted Oelfke, Sandhills Community College, 2200 Airport Rd., Pinehurst, NC 28374 US; 910-695-3756, Fax 910-695-1823, E-mail oelfketd@sandpiper.sandhills.cc.nc.us, URL http://www.sandhills.cc.nc.us.

WAKE TECHNICAL COMMUNITY COLLEGE

Raleigh/Year-round

This two-year community college offers a 2-year associate degree in Culinary Arts. Program started 1985. Accredited by SACS. Calendar: semester. Curriculum: core. Admission dates: Year-round. Total enrollment: 70. 60 each admission period. 75% of applicants accepted. 10% enrolled part-time. Student to teacher ratio 10:1. 98% of graduates obtain employment within six months. Facilities: Include kitchen and restaurant.

Courses: Foods, nutrition, sanitation, cost control, wine, inventory control, general education. Externship provided.

Faculty: 5 full-time, 1 part-time. Qualifications: BS, HRM, certified chefs. Includes Richard Roberts, Fredi Morf, Carolyn House, Jane Broden, James Hallett.

Costs: Annual tuition in-state $2,300. Off-campus housing cost is $500+ per month. Admission requirements: High school diploma or equivalent. Scholarships: yes. Loans: yes.

Location: Suburban.

Contact: Richard Roberts, Dept. Head, Wake Technical Community College, Culinary Technology, 9101 Fayetteville Rd., Raleigh, NC 27603 US; 919-662-3417, Fax 919-779-3360.

NORTH DAKOTA

NORTH DAKOTA STATE COLLEGE OF SCIENCE

Wahpeton/August-May

This college offers an 18-month diploma/AAS degree in Culinary Arts. Program started 1903. Accredited by NCA. Calendar: semester. Admission dates: August, January. Total enrollment: 30-35. 18 each admission period. 100% of applicants accepted. 80% receive financial aid. Student to teacher ratio 18:1. 100% of graduates obtain employment within six months. Facilities: Include 2 kitchens and classrooms.

Courses: Food preparation, baking, catering, gourmet, short order. Other required courses: 23 semester hours. Co-op training between 1st and 2nd year.

Faculty: 2 full-time.

Costs: Annual tuition $1,761 in-state, $4,353 out-of-state. On-campus housing: 1,700 spaces, average cost $1,000. Admission requirements: High school diploma or equivalent. Scholarships: yes. Loans: yes.

Location: The 2,492-student campus is on 125 acres in a small town, 45 miles from Fargo.

Contact: Neil Rittenour, Program Director, North Dakota State College of Science, Culinary Arts, 800 N. 6th St., Wahpeton, ND 58076 US; 800-342-4325, Fax 701-671-2126, E-mail rittenou@plains.nodak.edu, URL http://www.ndscs.nodak.edu.

OHIO

ASHLAND COUNTY – WEST HOLMES CAREER CENTER

Ashland/September-May

This joint vocational school-adult education center offers a 9-month diploma in Culinary Arts. Program started 1995. Curriculum: culinary. Admission dates: June, July, August. Total enrollment: 42. 42 each admission period. 95% of applicants accepted. 90% receive financial aid. 95% of graduates obtain employment within six months. Facilities: Professional kitchen.

Courses: The program runs a small restaurant Friday nights the last 6 months of training.

Costs: $3,650. Scholarships: yes.

Location: 5 miles from Ashland, in central Ohio.

Contact: Allen Wright, Financial Aid Officer/Recruiter, Ashland County-West Holmes Career Center, Culinary Chef Institute-1783 State Rte. 60, Ashland, OH 44805 US; 419-289-3313, Fax 419-289-3729, E-mail AWHJ_WRIGHT@tccsa.ohio.gov.

CINCINNATI STATE TECHNICAL & COMMUNITY COLLEGE

Cincinnati/Year-round

This two-year career college offers a 2-year ASOB degree in Chef Technology, 36-hour certificate program. Program started 1980. Accredited by ACFEI, NCA. Calendar: quinmester. Curriculum: core. Admission dates: Open. Total enrollment: 130. 30 each admission period. 100% of applicants accepted. 40% receive financial aid. 20% enrolled part-time. Student to teacher ratio 15:1. 100% of graduates obtain employment within six months. Facilities: Include commercial kitchen.

Courses: 7 culinary courses. Other required courses: 23 other courses. Co-op externship provided.

Faculty: 4 full-time.

Costs: Annual tuition $3,025 in-state, $5,000 out-of-state. Off-campus housing cost $350 per month. Admission requirements: High school diploma or equivalent and admission test. Scholarships: yes.

Contact: Richard Hendrix, Dept. Chair, Cincinnati State Technical & Community College,

Business Division, 3520 Central Pkwy., Cincinnati, OH 45223 US; 513-569-1500, Fax 513-569-1467, E-mail hendrixr@cinstatecc.oh.us.

THE CLEVELAND RESTAURANT COOKING SCHOOL

(See also page 243) **Cleveland/Year-round**

This restaurant offers a 10-week professional program, year-round classes for hobbyists, 1-week seminars. Established in 1986. Curriculum: culinary. Admission dates: 1-2 times per year. Total enrollment: 6. 6 each admission period. 95% of applicants accepted. Student to teacher ratio 6:1. 100% of graduates obtain employment within six months. Facilities: Include a teaching kitchen, demonstration area, a restaurant kitchen.

Courses: Cover restaurant planning, food costing and importing, catering management, food journalism,organic farming. Two months of daily classes are followed by a 2-week externship.

Faculty: Chef Parker Bosley, owner of Parker's Restaurant and Catering, and the restaurant staff.

Costs: $3,500 for the program. Admission requirements: Personal interview.

Location: 5 minutes from downtown Cleveland.

Contact: The Cleveland Restaurant Cooking School, 2801 Bridge Ave., Cleveland, OH 44113 US; 216-771-7130, Fax 216-771-8130.

COLUMBUS STATE COMMUNITY COLLEGE

(See also page 243) (See display ad page 244) **Columbus/Year-round**

This two-year college offers a 3-year AAS degree in Culinary Apprenticeship. Program started 1978. Accredited by ACFEI, NCA. Calendar: quarter. Admission dates: September, March. Total enrollment: 75. 35-40 each admission period. 100% of applicants accepted. 100% enrolled part-time. Student to teacher ratio 15:1. 100% of graduates obtain employment within six months. Facilities: Classrooms, labs, off-campus professional kitchens.

Courses: General education, business, foodservice management, culinary courses. 40 hours per week on apprenticeship. Columbus State Culinary Academy offers 3-hour cooking classes for general public.

Faculty: 7 full-time. Includes Chair C. Kizer, M. Steiskal, D. Cobler, T. Atkinson, J. Taylor.

Costs: Annual tuition $2,124 in-state (if full-time). Admission requirements: High school graduate, letters of reference, interview, math and English placement test. Scholarships: yes. Loans: yes.

Contact: Carol Kizer, Chairperson, Columbus State Community College, Hospitality Management, 550 E. Spring St., Columbus, OH 43215 US; 614-227-2579, Fax 614-227-5973, E-mail ckizer@cscc.edu, URL http://www.cscc.edu.

CUYAHOGA COMMUNITY COLLEGE

Cleveland/Year-round

This two-year college offers a 2-year AAB degree in Culinary Arts, Restaurant Food Service Management, and Hotel/Motel Management. Pro management courses and certification. Program started 1969. Accredited by NCA. Calendar: semester. Curriculum: core. Admission dates: August, January. Total enrollment: 175. 60-80 each admission period. 95% of applicants accepted. 60% receive financial aid. 60% enrolled part-time. Student to teacher ratio 10:1. 95% of graduates obtain employment within six months. Facilities: Include 3 kitchens, 2 classrooms, computer lab, restaurant.

Courses: Management/food prep, haute cuisine, garde manger. Other required courses: purchasing, accounting, menu planning. Continuing education: safety and sanitation, nutrition, food prep.

Faculty: 5 full-time, 9 part-time. Qualifications: degree and industry experience, CEC, CCE.

Costs: Annual tuition in-state (out-of-county, out-of-state) $61.50 ($76.85, $158.20) per semester-hour. Application fee $10, lab fees $300. Off-campus housing cost is $500 per month. Admission requirements: Testing in English/math. Scholarships: yes. Loans: yes.

Location: Metro area.

Contact: Jan DeLucia, Program Manager, Cuyahoga Community College, Hospitality Management, 2900 Community College Ave., Cleveland, OH 44115 US; 216-987-4082, Fax 216-987-4086, E-mail jan.delucia@tri-c.cc.oh.us, URL http://www.tri-c.cc.oh.us.

HOCKING COLLEGE

Nelsonville/Year-round *(See display ad above)*

This career college offers a 2-year certificate/AAS degree. Program started 1979. Accredited by NCA, ACFEI. Calendar: quarter. Curriculum: culinary. Admission dates: September, January, March, June. Total enrollment: 150. 100% of applicants accepted. 65% receive financial aid. Student to teacher ratio 15:1. 95% of graduates obtain employment within six months.

Courses: Training in for-profit Quality Inn. Internship available.

Faculty: 5 full-time.

Costs: Annual tuition in-state $2,157, out-of-state $4,665. Admission requirements: High school graduate, GED, or ability to benefit. Scholarships: yes. Loans: yes.

Contact: Dr. John Pierce, Dean, School of Business & Hospitality, Hocking College, 3301 Hocking Pkwy., Nelsonville, OH 45764 US; 740-753-3591, Fax 740-753-9018, E-mail admissions@hocking.edu, URL http://www.hocking.edu.

THE LORETTA PAGANINI SCHOOL OF COOKING

Chesterland/Year-round *(See also page 245)*

This private school in conjunction with Lakeland Community College offers a Professional Chef Training program, Baking and Pastry Arts certificate program. Established in 1989. Accredited by ACF. Calendar: semester. Curriculum: culinary. Total enrollment: 36. 12 each admission period. 80% of applicants accepted. 0% receive financial aid. 80% enrolled part-time. Student to teacher ratio 12:1. 80% of graduates obtain employment within six months. Facilities: Fully-equipped 600-square-foot professional kitchen and overhead mirror.

Courses: Classical European with emphasis on technique and hands-on training. Opportunities to apprentice at local restaurant or assist in classes. Continuing Chef's Training available.

Faculty: 25 full- and part-time.

Costs: Total of $1,690, includes tuition of $995 for each program plus core classes. Admission requirements: Successful completion of core classes: Basic and Advanced Techniques of Cooking and Baking.

Location: 25 miles east of Cleveland.

Contact: Loretta Paganini, Director, Lakeland Community College, Gingerbread, 8613 Mayfield Rd., Chesterland, OH 44026 US; 440-729-COOK, Fax 440-729-6459, E-mail LPSCInc@msn.com.

OWENS COMMUNITY COLLEGE

Toledo

This two-year college offers a 2-year AAB degree in Food Service Management. Program started 1968. Accredited by NCA. Calendar: semester. Curriculum: core. Admission dates: August, January, June. Total enrollment: 38. 100% of applicants accepted. 55% enrolled part-time. Student to teacher ratio 18:1. 95% of graduates obtain employment within six months. Facilities: Production kitchen and dining room.

Courses: Food service management. 300 hours of cooperative work experience in summer.

Faculty: 1 full-time, 4 part-time.

Costs: $79 per cr-hour in-state, $148 per cr-hour out-of-state. Admission requirements: Assessment of reading, writing, and math skills. Scholarships: yes.

Contact: Janell Lang, Dept. Chair, Owens Community College, HRI, PO Box 10,000 - Oregon Rd., Toledo, OH 43699-1947 US; 419-661-7214, Fax 419-661-7251, E-mail janlang@owens.cc.oh.us, URL http://www.owens.cc.oh.us.

SINCLAIR COMMUNITY COLLEGE

Dayton/Year-round

This two-year college offers a 2-year associate degree in Hospitality Management/Culinary Arts Option. Program started 1993. Accredited by ACFEI. Calendar: quarter. Curriculum: culinary, core. Total enrollment: 200. 100% of applicants accepted. 30% receive financial aid. 40% enrolled part-time. Student to teacher ratio 20:1. 100% of graduates obtain employment within six months. Facilities: 3 kitchens, 150-seat dining room, classrooms.

Courses: Food preparation, garde manger, butchery and fish management, pastry and confectionery, classical foods, baking.

Faculty: 3 full-time, 8 part-time. Qualifications: certified by ACF and NRA. Steven Cornelius, CCE, FMP, Assoc. Professor; Frank Leibold, CEC, CWPC, Asst. Professor; Derek Allen, CHE, Instructor.

Costs: Tuition in-county $31, in-state $49. Admission requirements: High school diploma or GED. Scholarships: yes. Loans: yes.

Contact: Steven Cornelius, Dept. Chair, Sinclair Community College, Hospitality Management, 444 W. Third St., Dayton, OH 45402-1460 US; 937-512-5197, Fax 937-512-4530, E-mail scorneli@sinclair.edu, URL http://www.sinclair.edu.

UNIVERSITY OF AKRON

Akron/September-April

This university offers a 1-year certificate and 2-year AAS degree in Culinary Arts. Program started 1968. Accredited by State. Calendar: semester. Curriculum: culinary, core. Admission dates: Fall, spring. Total enrollment: 100. 30+ per admission period. 100% of applicants accepted. 50% receive financial aid. 50% part-time. Student to teacher ratio 15:1. 95% of graduates obtain jobs.

Courses: Hospitality management courses in addition to culinary arts. Externship provided.

Faculty: 4 full-time.

Costs: Tuition $140 ($338) per credit-hour in-state (out-of-state). Loans: yes.

Contact: Jan Eley, Coordinator, University of Akron, Hospitality Management, Gallucci Hall #104, Akron, OH 44325-7907 US; 330-972-7026/800-221-8308, Fax 330-972-5101, URL http://www.commtech.uakron.edu/commtech/bus_tech/hosp.htm.

OKLAHOMA

MERIDIAN TECHNOLOGY CENTER

Stillwater/August-May

This career institution offers a 1,050-hour certificate. Program started 1975. Accredited by State. Calendar: semester. Curriculum: core. Admission dates: August. Total enrollment: 36. Student to teacher ratio 18:1. Facilities: 1 kitchen, 1 classroom, 2 restaurants.

Faculty: 3 full-time.

Costs: Annual tuition in-district $1,400, out-of-district $2,800. Admission requirements: Assessment, interview.

Contact: Doug Major, Meridian Technology Center, Culinary Arts, 1312 S. Sangre Rd., Stillwater, OK 74074 US; 405-377-3333, Fax 405-377-9604, E-mail DougM@meridian-technology.com, URL http://www.meridian-technology.com.

OKLAHOMA STATE UNIVERSITY

Okmulgee

This university offers 24-month AAS degree and 20-month diploma programs in Food Service Management. Program started 1946. Accredited by NCA. Admission dates: August, January, April. Total enrollment: 120. 35 each admission period. 90% of applicants accepted. 65% receive financial aid. 10% enrolled part-time. Student to teacher ratio 16:1. 90% of graduates obtain employment within six months. Facilities: Include 4 kitchens and 4 classrooms.

Courses: Pastry production, food preparation, garde manger, hot food production, meat identification, dining room management, nutrition, and general education courses. Externship provided. Continuing education: advanced cooking and sauces, beginning cake decoration, meat identification.

Faculty: 4 full-time. Qualifications: 5 years experience, college level culinary arts, ACF certified.

Costs: Tuition in-state $47 per credit hour, out-of-state $119 per credit-hour. On-campus housing: 1,000 spaces, average cost $200-$250 per month. Admission requirements: Admission test.

Location: The 2,000-student campus is in a small town, 45 miles from Tulsa.

Contact: Dean Daniel, Department Head, Oklahoma State University, Hospitality Services Technology, 1801 E. 4th St., Okmulgee, OK 74447 US; 918-756-6211, Fax 918-756-1315.

PIONEER TECHNICAL CENTER

Ponca City

This career institution offers a 1-year certificate. Program started 1972. Accredited by NCA. Total enrollment: 36. Student to teacher ratio 6:1. 100% of graduates obtain employment within six months.

Courses: Externship provided.

Faculty: 3 full-time.

Costs: Annual tuition $962 for residents, $1,924 for non-residents. Admission requirements: High school diploma or equivalent.

Contact: Steve Ellenwood, Instructor, Pioneer Technical Center, Commercial Foods, 2101 N. Ash, Ponca City, OK 74601 US; 580-762-8336, Fax 580-765-5101, E-mail SteveE@pioneertech.org, URL http://www.pioneertech.org.

SOUTHERN OKLAHOMA AREA VOC-TECH SCHOOL

Ardmore/August-May

This vocational-technical career institution offers a 2-year (1,050-hour) certificate in Culinary Arts. Program started 1966. Accredited by State. Calendar: quarter. Curriculum: culinary. Admission dates: August. Total enrollment: 40. 35-40 each admission period. 80% of applicants accepted. 40% receive financial aid. 20% enrolled part-time. Student to teacher ratio 10:2. 15% of graduates obtain employment within six months. Facilities: 1 kitchen with baking, and pantry hot foods; salads; fast food/catering.

Courses: Catering. OSU.

Faculty: 1 full-time, 2 part-time.

Costs: Annual tuition in-district $0, out-of-district $1,600. Scholarships: yes. Loans: yes.

Contact: J.W. Reese, CEC, Chef/Culinary arts Instructor, Southern Oklahoma Area Vocational Technical School, Culinary Arts, 2610 San Noble Pkwy., Ardmore, OK 73401 US; 405-223-2070 x210/276-3977, Fax 405-226-9389, E-mail jreese@sotc.org.

OREGON

CASCADE CULINARY INSTITUTE

Bend/September-June

Central Oregon Community College offers a certificate of completion in Culinary Arts. Program started 1993. Calendar: quarter. Curriculum: culinary. Admission dates: December, March, September. Total enrollment: 24. 24 each admission period. 100% of applicants accepted. 90%

receive financial aid. Student to teacher ratio 16:1. 100% of graduates obtain employment within six months. Facilities: Kitchen, dining room, deli operation.

Courses: Hot Food Production, Baking, Garde Manger, Dining Room Service and Management, Sanitation, Food Service Nutrition. Internship required.

Faculty: 1 full- and 1 part-time CEC.

Location: East side of Cascade Mountains.

Contact: Timothy H. Hill, Cascade Culinary Institute, 2600 NW College Way, Bend, OR 97701 US; 541-383-7713, Fax 541-383-7508, E-mail thill@cocc.edu, URL http://www.cocc.edu.

INTERNATIONAL SCHOOL OF BAKING

(See display ad above) **Bend/Year-round**

This private school offers 1- to 20-day customized courses that focus on European breads and pastries. Established in 1986. Curriculum: culinary. Total enrollment: 1 or 2 per course. Facilities: Modern baking facility.

Courses: Cover ingredient function, bakery start-up, troubleshooting, all types of European artisan breads and pastries. Students can select own curriculum. Schedule: 8-hour daily hands-on sessions.

Faculty: Director Marda Stoliar has taught European bread making since 1965, owned a French bakery, and is a baking consultant in China, Hong Kong, and Macau for U.S. Wheat Associates.

Costs: From $350 per day per student. A list of nearby lodging is provided on request.

Location: In Oregon's Cascade Mountains, a 3-minute drive from downtown Bend, 15 minutes from the Redmond Oregon Airport.

Contact: Marda Stoliar, Director, International School of Baking, 1971 NW Juniper Ave., Bend, OR 97701 US; 541-389-8553, Fax 541-389-3736, E-mail domocorp@empnet.com, URL http://www.empnet.com/domocorp/.

LANE COMMUNITY COLLEGE

Eugene/September-June

This independent college offers a 1-year certificate and 2-year AAS degree in Culinary Arts and Culinary option. Program started 1976. Calendar: quarter. Curriculum: core. Admission dates: Open. Total enrollment: 85-105. 25-50 each admission period. 100% of applicants accepted. 60-70% receive financial aid. 3% enrolled part-time. Student to teacher ratio 18:1. 95% of graduates obtain employment within six months. Facilities: Include 2 kitchens, 1 dining room, deli/bake shop, and 4 to 6 classrooms.

Courses: Introduction to foods, restaurant lab, buffet, baking, sanitation, safety, menu planning, and general education courses. Externship: 325-450 hour, in commercial, institutional, or single proprietor settings. Post-graduate.

Faculty: 3 full-time, 3 part-time. Qualifications: ACF certifications with BS, AAS degrees, and training from European cooking schools. Includes Willie Kealoha, Guy Plaa, Don Savoie, Wendy McDaniel, Peter Lohr, Duane Partain.

Costs: Annual tuition $34 per credit-hour in-state, $116 per credit-hour out-of-state. Lab fees, student fees. Average off-campus housing cost $350-$800 per month. Admission requirements: Limited enrollment.

Contact: Willie Kealoha, Lane Community College, Culinary Food Service, 4000 E. 30th Ave., Eugene, OR 97405-0640 US; 541-747-4501, Fax 541-744-4159.

LINN-BENTON COMMUNITY COLLEGE

Albany/September-June

This two-year college offers a 2-year AAS degree in Culinary Arts/Hospitality Services with Chef Training Option. Program started 1973. Accredited by NASC. Calendar: quarter. Curriculum: core. Admission dates: September, January, March. Total enrollment: 30. 15 each admission period. 100% of applicants accepted. 50% receive financial aid. 5% enrolled part-time. Student to teacher ratio 5:1. 100% of graduates obtain employment within six months. Facilities: Include bakery production facility, restaurant.

Courses: Chef training. Fellowships awarded.

Faculty: 6 full-time. Includes S. Anselm, M. Whitehead, M. Young.

Costs: Annual tuition in-state $36 per credit, out-of-state $123 per credit. Application fee $20. Approximately $300 for tools and uniforms. Off-campus housing cost is ~$250 per month. Admission requirements: High school diploma or equivalent.

Location: The 16,000-student campus in a small town setting is 24 miles from Salem and 42 miles from Eugene.

Contact: Scott Anselm, Linn-Benton Community College, Culinary Arts, 6500 SW Pacific Blvd., Albany, OR 97321 US; 541-917-4999, Fax 541-917-4395, E-mail anselm@gw.lbcc.cc.or.us.

ROBERT REYNOLDS, NORTHWEST FORUM

Oregon/Year-round *(See also page 247)*

This private school offers an 8-week course with professional focus for those seeking a foundation in cooking, 1-week master class for those seeking advanced culinary study. Program started 1997. Curriculum: culinary. 6 each admission period. Student to teacher ratio 6:1. Facilities: Restaurant kitchen in Portland, winery kitchen in the countryside, cooking school classroom in Seattle, restaurant kitchen in Provence.

Courses: Theory, methods, practice, techniques, physics, chemistry, history, culture: advanced study leading to cook's independence.

Faculty: Robert Reynolds operated Le Trou Restaurant in San Francisco for 15 years; operated aux Gastronomes cooking school in France; created and directed professional program in Colorado; co-directs Provence 3D cooking program in France.

Costs: $7,500 for 8-week professional course (France), $1,250 ($2,450) for master class in Portland and Seattle (Niort, France) includes meals and local transport. Master class in France also

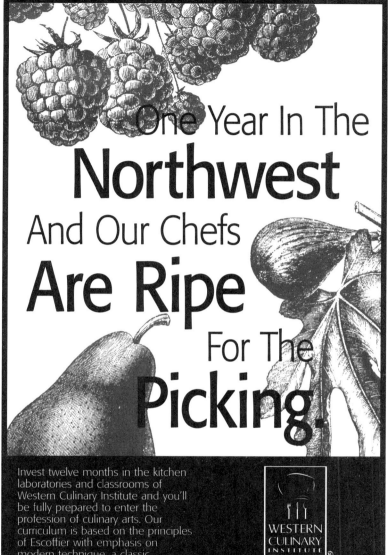

includes 1st class hotel lodging.

Location: OR: Downtown Portland and wine country, 45 minutes S of Portland. WA: Bellevue, Seattle. France: St Remy de Provence, 30 minutes from Avignon.

Contact: Robert Reynolds, Northwest Forum for Advanced Culinary Arts, 222 SE 18th Ave., Portland, OR 97214 US; 888-733-3391, Fax 503-233-1934, E-mail rowbear@ibm.net, URL http://www.RobertReynoldsCooks.com.

WESTERN CULINARY INSTITUTE

Portland/Year-round *(See display ad page 105)*

This private school offers a 12-month accelerated diploma program in Culinary Arts. Established in 1983. Accredited by ACFEI, ACCSCT. Curriculum: culinary. Admission dates: Every 6 weeks. Total enrollment: 500. 72 each admission period. 85% of applicants accepted. 80% receive financial aid. Student to teacher ratio 15-35:1. 97% of graduates obtain employment within six months. Facilities: Include up-to-date, well-equipped kitchens and an open-to-the-public restaurant.

Courses: Curriculum is 80% participation, based on the principles of Escoffier with emphasis on modern techniques and trends. The 18 courses include culinary fundamentals, purchasing and cost control, international cuisines, nutrition, baking and pastry, and wines. 6-week internship in an approved foodservice operation.

Faculty: The 18-member faculty is made up of individuals with international experience and training, many of whom have won culinary awards.

Costs: Tuition $14,770, which includes cutlery, uniforms, lab fees. A $25 nonrefundable application fee and $100 enrollment fee required. Admissions representatives assist in finding suitable lodging. Admission requirements: High school diploma or equivalent. Scholarships: yes. Loans: yes.

Location: Near Portland State University in downtown Portland.

Contact: Mary Harris, Director of Admissions, Western Culinary Institute, Admissions, 1201 SW 12th Ave., #100, Portland, OR 97205 US; 800-666-0312/503-223-2245, Fax 503-223-5554, E-mail mary@westernculinary.com, URL http://www.westernculinary.com.

PENNSYLVANIA

ART INSTITUTE OF PHILADELPHIA – SCHOOL OF CULINARY ARTS

Philadelphia/Year-round *(See display ad page 8)*

This proprietary school offers a 6-quarter/18-month AS degree program in Culinary Arts. Established in 1997. Accredited by ACCSCT. Calendar: quarter. Curriculum: culinary. Admission dates: October, January, April, July. Total enrollment: 150. 50 each admission period. Facilities: New 20,000-square-foot facility. Three kitchens: baking and pastry, a la carte, skills.

Courses: Hands-on instruction in culinary techniques, baking and pastry. Day and evening options.

Faculty: Chef director is Michael Baskette, CEC, CCE, AAC.

Costs: $3,420 per quarter. Admission requirements: High school diploma or GED, interview.

Contact: Jim Palermo, Director of Admissions, The Art Institute of Philadelphia-School of Culinary Arts, 1622 Chestnut St., Philadelphia, PA 19103 US; 800-275-2474/215-567-7080, Fax 215-246-3358, E-mail palermoj@aii.edu, URL http://www.aii.edu.

BUCKS COUNTY COMMUNITY COLLEGE

Newtown/Year-round

This two-year college offers a 2-year AA degree, 3-year degree/apprenticeship program. Program started 1968. Accredited by MSA. Calendar: semester. Curriculum: culinary, core. Admission dates: Year-round. Total enrollment: 240 in HRIM and Chef Program. 40 each admission period. 80% enrolled part-time. Student to teacher ratio 15-18:1. 90% of graduates obtain employment within six months. Facilities: Include kitchen, dining room, lab, computer labs.

Courses: Hospitality Management/Restaurant/Institutional Foodservice and Chef Apprenticeship. HRIM: 200 hours, co-op + 600 hour summer internship. Chef apprenticeship:

6,000 hours paid OJT. Continuing education available.

Faculty: 2 full-time, 4 part-time. Qualifications: ACF-certification and degree preferred.

Costs: Tuition $71 per credit-hour in-county, $142 per credit-hour out-of-county. Admission requirements: High school diploma or GED, college placement tests. Scholarships: yes. Loans: yes.

Location: 20 miles north of Philadelphia.

Contact: Earl R. Arrowood, Jr., HRIM Programs Coordinator-Professor, Bucks County Community College, Business Dept., Swamp Rd., Newtown, PA 18940 US; 215-968-8241, Fax 215-504-8509, E-mail arrowood@storm.bucks.edu, URL http://www.bucks.edu.

COMMUNITY COLLEGE OF ALLEGHENY COUNTY

Monroeville

This two-year college offers a 2-year certificate/AAS degree. Program started 1967. Accredited by MSA. Admission dates: Open. Total enrollment: 175. Student to teacher ratio 15:1. 100% of graduates obtain employment within six months.

Courses: Externship provided.

Faculty: 2 full-time, 5 part-time.

Costs: Annual tuition in-state $68 per credit, out-of-county $136 per credit, out-of-state $204 per credit. Admission requirements: High school diploma or equivalent.

Contact: Linda Sullivan, Community College of Allegheny County, Hospitality Management, 595 Beatty Rd., Monroeville, PA 15146 US; 412-371-8651, URL http://www.ccac.edu.

COMMUNITY COLLEGE OF ALLEGHENY COUNTY

Pittsburgh

This two-year college offers a 2-year AAS degree in Culinary Arts. Program started 1974. Accredited by MSA. Admission dates: Fall, spring. Total enrollment: 60. 20-15 each admission period. 25% of applicants accepted. 50% receive financial aid. 50% enrolled part-time. Student to teacher ratio 2:1. 90% of graduates obtain employment within six months.

Courses: Include basic foods, culinary artistry, nutrition, baking, costing. Externship: 240 hours.

Faculty: 12 faculty members with bachelor's degree or CEC.

Costs: Tuition is $68 per credit in-state, $136 per credit out-of-county, $204 per credit out-of-state.

Contact: Willie Stinson, CEC, AAC, Community College of Allegheny County, Culinary Arts, 808 Ridge Ave., Jones Hall, Rm. 012, Pittsburgh, PA 15212 US; 412-237-2511, Fax 412-237-4678, URL http://www.ccac.edu.

COMMUNITY COLLEGE OF BEAVER COUNTY

Monaca/Year-round

This two-year college offers a 2-semester/34 credit-hour certificate in Culinary Arts, 3-semester/34 credit-hour certificate in Mastery of Culinary Arts, 2 year/64 credit-hour AAS degree in Culinary Arts. Calendar: semester. Curriculum: culinary, core.

Courses: Commercial foods, moist and dry heat preparation principles, stocks/soups/sauces/starches/eggs, line cooking/advanced buffet, management, food preperation and services.

Costs: $1,728 per year.

Contact: Dr. David Blumer, Director of Business and Technologies, Community College of Beaver County, 1 Campus Dr., Monaca, PA 15061-2588 US; 800-335-0222/412-775-8561 x215, Fax 412-775-4055, URL http://www.ontv.com/college/ccbc.htm.

COMMUNITY COLLEGE OF PHILADELPHIA

Philadelphia/Year-round

This two-year college offers a 63 credit-hour AAS degree in Culinary Arts-Chef, ACF-approved chef-apprenticeship program. Calendar: semester. Curriculum: core.

Costs: Tuition is $75 per credit-hour in-state, $225 per credit-hour out-of-state.

Contact: Mark Kushner, Community College of Philadelphia, 1700 Spring Garden St., Philadelphia, PA 19130 US; 215-751-8000/8797, URL http://www.ccp.cc.pa.us.

DREXEL UNIVERSITY

Philadelphia/Year-round

This university offers a 2-year full-time, 3-year part-time degree in Culinary Arts. Established in 1894. Accredited by MSA. Calendar: trimester. Curriculum: core. Admission dates: Rolling. 65% of applicants accepted. 80% receive financial aid. 20% enrolled part-time. Student to teacher ratio 10:1. 99% of graduates obtain employment within six months. Facilities: 10,000 square feet of facilities, including 4 kitchens and a restaurant.

Costs: $16,842 per year plus fees. Scholarships: yes. Loans: yes.

Contact: John Canterino, Director, Drexel University-Hotel & Restaurant Management-Culinary Arts, 1314 Chestnut St., Nesbitt College, Philadelphia, PA 19104 US; 215-895-4923, Fax 215-895-5939, E-mail canterji@dunyl.ocs.drexel.edu, URL http://www.drexel.edu.

HARRISBURG AREA COMMUNITY COLLEGE

Harrisburg/Year-round

This two-year college offers a 2-year certificate/AA program in Culinary Arts. Program started 1965. Accredited by MSA, ACBSP. Calendar: semester. Curriculum: culinary, core. Admission dates: August, January, May. Total enrollment: 250. 48 each admission period. 76% of applicants accepted. 50% receive financial aid. 50% enrolled part-time. Student to teacher ratio 15-20:1. 100% of graduates obtain employment within six months. Facilities: Include production kitchen, demonstration kitchen, culinary classroom, weekly luncheons.

Courses: Culinary arts, quantity foods, and 20 other courses. Externship: 3-month, salaried.

Faculty: 3 full-time, 3 part-time. Qualifications: bachelor's degree, master's preferred, certifiable by ACF.

Costs: Annual tuition in-state $130.50 per credit hour, out-of-state $197.25 per credit hour. $25 to enroll. Equipment and uniforms about $300. Admission requirements: Admission test and portfolio. Scholarships: yes. Loans: yes.

Location: The 3 campuses (Harrisburg, Lebanon, and Lancaster) with 11,000 students are 100 miles from Philadelphia.

Contact: Marcia W. Shore, M.S.Ed., CCE, Assistant Professor, Coordinator Culinary Arts Prog, Harrisburg Area Community College, One HACC Dr., Harrisburg, PA 17110-2999 US; 717-780-2674, Fax 717-231-7670.

HIRAM G. ANDREWS CENTER

Johnstown/Year-round

This proprietary school offers a 4-month diploma in Kitchen Helper and AST Culinary. Program started 1975. Accredited by ACCSCT. Calendar: trimester. Curriculum: core. Admission dates: Every 4 months. Total enrollment: 30. 12-15 each admission period. 40% receive financial aid. 0% enrolled part-time. Student to teacher ratio 15:1. 100% of graduates obtain employment within six months. Facilities: Include 3 kitchens and classrooms and a part-time restaurant.

Courses: Baking, sanitation, nutrition, food preparation, cooking methods and techniques, menu writing, and table service. Externship: 2 months.

Faculty: 3 full-time.

Costs: Annual tuition $11,936. On-campus housing: 400 spaces. Admission requirements: High school diploma or equivalent.

Location: The 66-acre campus is in a suburban setting 80 miles from Pittsburgh.

Contact: Jack B. Demuth, Voc. Supv., Hiram G. Andrews Center, Culinary Arts, 727 Goucher St., Johnstown, PA 15905-3092 US; 814-255-8288, Fax 814-255-3406.

INDIANA UNIVERSITY OF PENNSYLVANIA

(See display ad above) **Indiana/Year-round**

This university offers a 16-month certificate in Culinary Arts, 42 credits guaranteed transfer to an IUP BS degree in Hospitality Management. Established in 1989. Accredited by MSA, ACFEI. Calendar: semester. Curriculum: culinary. Admission dates: September. 100 each admission period. 75% of applicants accepted. 90% receive financial aid. Student to teacher ratio 13:1. 99% of graduates obtain employment within six months. Facilities: Include 5 production and 2 demonstration kitchens, computer lab, classroom.

Courses: Include cuisine and pastry preparation, purchasing, nutrition, wine appreciation, international cuisine, and menu and facility design. A 450-hour salaried externship is required.

Faculty: 8 full-time, 1 part-time.

Costs: Tuition $4,700 per semester. Application fee is $30. A $115 activity fee is required each semester. On campus lodging is $931 ($1,424) per semester for a double (single) room. Admission requirements: High school diploma or equivalent. Scholarships: yes. Loans: yes.

Location: 25 miles north of Indiana in rural Punxsutawney.

Contact: Kelly Barry, Admissions Coordinator, IUP Academy of Culinary Arts, Reschini Building, IUP, Indiana, PA 15705 US; 800-727-0997, Fax 412-357-6200, E-mail culinary-arts@grove.iup.edu, URL http://www.iup.edu/cularts.

INTERNATIONAL CULINARY ACADEMY

(See display ad page 111) **Pittsburgh/Year-round**

This division of Computer Tech career institution offers 2-year (1,800-clock-hour) AST degrees in Culinary and Pastry Arts. Established in 1989. Accredited by ACICS, ACFEI. Calendar: semester. Admission dates: March, August, October. Total enrollment: 350. 75 each admission period. 90% of applicants accepted. 80% receive financial aid. Student to teacher ratio 25:1. 92% of graduates obtain employment within six months. Facilities: Include professional-size commercial kitchens, classrooms, resource center, executive dining room, student lounge.

Courses: Culinary arts covers kitchen and storeroom operations, hands-on preparation and presentation of classic international cuisines. Pastry arts covers baking principles, international and specialty breads, pies and cakes. French pastry, and advanced decoration and design. Salaried externship.

Faculty: 8 ACF-Certified Culinary Educators.

Costs: Tuition for each program is $16,550, which includes uniforms, books, equipment and all food costs. Student housing is offered. Admission requirements: High school diploma or equivalent and aptitude test.

Location: In Pittsburgh's Golden Triangle, overlooking the Allegheny River.

Contact: Debbie Love, Director of Admissions, International Culinary Academy, A Division of Computer Tech, 107 Sixth St., Fulton Bldg., Pittsburgh, PA 15222 US; 412-471-9330, Fax 412-391-4224, E-mail info@intlculinary.com, URL http://www.intlculinary.com.

JNA INSTITUTE OF CULINARY ARTS

Philadelphia/Year-round *(See display ad above)*

This private culinary institute offers a 30-week diploma in Food Service Training (Professional Cooking), 20-week diploma in Specialized Food Service Management, 60-week associate degree in Culinary Arts/Restaurant Management. Established in 1988. Accredited by ACCSCT. Curriculum: culinary. Admission dates: Rolling. Total enrollment: 60. 15 each admission period. 70% of applicants accepted. 90% receive financial aid. 10% enrolled part-time. Student to teacher ratio 10:1. 98% of graduates obtain employment within six months. Facilities: 2 kitchens, 1 demonstration area, classrooms.

Courses: Courses are taught through a combination of hands-on labs, demos, and projects. Lectures are a part of training. Externship, most paid. NRA Diploma Partners, ProMgmt Program, FMP Certification, Applied Food Service Certification.

Faculty: 9 full- and 4 part-time, all with formal training and/or experience. Includes Joseph DiGironimo, CFE, FMP, John Knorr, Bernadette Aichhorn, Chefs Donald Smith, CFE, Valerie Jamison, Al Passalacqua, James Panetta, Cyrus Knower.

Costs: Program based: $3,350-$12,000. $75 registration fee. Admission requirements: High school diploma or equivalent. Scholarships: yes. Loans: yes.

Contact: Diane Goldstein, Admissions, JNA Institute of Culinary Arts, 1212 S. Broad St., Philadelphia, PA 19146 US; 215-468-8800, Fax 215-468-8838, E-mail admissions@culinaryarts.com, URL http://www.culinaryarts.com.

KEYSTONE COLLEGE

La Plume/September-May

This private college offers a 2-year AAS degree in Culinary Arts. Program started 1996. Calendar: semester. Curriculum: culinary, core. Admission dates: August. Total enrollment: 20. 20 each admission period. 100% of applicants accepted. 90% receive financial aid. 20% enrolled part-time. Student to teacher ratio 6-12:1.

Courses: 500-hour externship.

Faculty: 1 full-, 2 part-time.

Costs: Tuition is $9,250 per year, room and board is $5,640 per year. Scholarships: yes. Loans: yes.

Contact: Mayra Zehnal, Chef/Instructor, Keystone College, Harris Hall, La Plume, PA 18440 US; 717-945-5141, Fax 717-945-7916, URL http://199.224.64.217.

MERCYHURST COLLEGE – THE CULINARY AND WINE INSTITUTE

North East/Year-round

This four-year college offers a 2-year AS degree in Culinary Arts. Established in 1995. Accredited by MSA. Calendar: trimester. Curriculum: core. Admission dates: September, November, March. Total enrollment: 70. 30-35 each admission period. 57% of applicants accepted. 100% receive financial aid. 18% enrolled part-time. Student to teacher ratio 15:1. 100% of graduates obtain employment within six months. Facilities: 3 professional kitchens, including bake shop, 30-seat

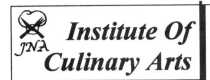

dining room, receiving and storage area.

Courses: Specialized courses in wines and wine-making; emphasis on management and thinking skills along with traditional culinary courses. 420-hour paid externship after 3 terms in quality dining facility.

Faculty: 5 full-time, 5 part-time, with industry experience and educational background. Includes Stephen C. Fernald, CCC, John Harrison, Deborah Hilbert, MS, RD, William Kunz, Douglas Moorhead.

Costs: Tuition is $7,264 per year. Fees are $1,500, books $450. On-campus dormitories available at $4,176 room and board per year. Admission requirements: High school graduate or equivalent, math and English placement test. Scholarships: yes. Loans: yes.

Location: 15 miles from Erie.

Contact: James Theeuwes, Director of Admissions, Mercyhurst College-North East, The Culinary & Wine Institute, 501 E. 38th St., Erie, PA 16546 US; 800-825-1926 x2238, Fax 814-824-2179.

NORTHAMPTON COMMUNITY COLLEGE

Bethlehem/Year-round

This two-year college offers a 45-week specialized diploma in Culinary Arts. Program started 1993. Accredited by MSA. Calendar: trimester. Curriculum: culinary. Admission dates: March, September. Total enrollment: 46. 23 each admission period. 100% of applicants accepted. 30% receive financial aid. Student to teacher ratio 20:1. 98% of graduates obtain employment within six months.

Courses: Baking, pastry, nutrition, pantry, skill development, garde manger, restaurant operations.

Faculty: 4 full-time. Qualifications: culinary degree and 10 years professional experience. Includes Duncan Howden and Scott Kalamar.

Costs: Annual tuition in-state $1,770, out-of-state $5,500. Other fees $565. On-campus housing:

145 spaces; average cost: $2,400 per year. Average off-campus housing cost: $350 per month. Admission requirements: High school diploma or equivalent. Scholarships: yes. Loans: yes.

Location: 65 miles southeast of Philadelphia.

Contact: Duncan Howden, Assoc. Professor, Northampton Community College, Culinary Arts, 3835 Green Pond Rd., Bethlehem, PA 18017 US; 610-861-5593, Fax 610-861-5093, E-mail dch@mail.nrhm.cc.pa.us, URL http://www.nrhm.cc.pa.us.

ORLEANS TECHNICAL INSTITUTE

Philadelphia

This career institution offers a 30-week specialized diploma in Food Preparation. Program started 1978. Admission dates: Open. 85% of graduates obtain employment within six months.

Faculty: 1 to 2 full-time.

Costs: Tuition is $2,550 (Food Service) for the 480-hour class. Admission requirements: Admission test.

Contact: Chandra Davis, Orleans Technical Institute, Culinary Arts, 1330 Rhawn St., Philadelphia, PA 19111 US; 215-728-4488.

PENNSYLVANIA COLLEGE OF TECHNOLOGY

Williamsport/Year-round *(See display ad above)*

This career college offers a 2-year AAS degree in Food/Hospitality Management, Baking/Pastry Arts, Culinary Arts, Dietary Manager Technology. Program started 1981. Accredited by MSA, ACFEI. Calendar: semester. Curriculum: core. Admission dates: Fall, spring. Total enrollment: 98. 72 each admission period. 95% of applicants accepted. 86% receive financial aid. 15% enrolled part-time. Student to teacher ratio 12:1. 100% of graduates obtain employment within six

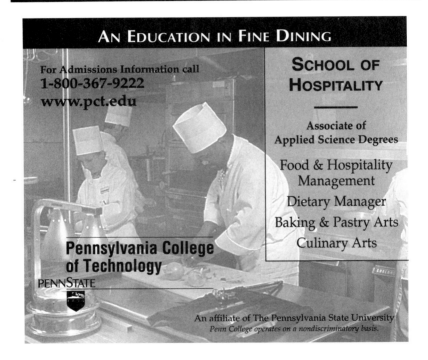

months. Facilities: Include 10 kitchens and classrooms, retail restaurant, catering and meeting facilities, theatre lounge, bed & breakfast, conference center, retreat center.

Courses: Cooking, baking, service, sanitation, supervision, management skills, and nutrition. Semester-long internships provided.

Faculty: 8 full-time, 4 part-time. Qualifications: college degree and ACF certification. 70% ACFEI certified, 100% certifiable.

Costs: Annual tuition in-state $218 per credit hour, out-of-state $238 per credit hour. Application fee $20. Lab fee $13. Off-campus housing cost $250-$450 per month. Admission requirements: High school diploma or equivalent and admission test. Scholarships: yes. Loans: yes.

Location: The 53-acre, 4,300-student campus is in a small town 200 miles from Pittsburgh and Philadelphia.

Contact: Chet Schuman, Director of Admissions, Pennsylvania College of Technology, Hospitality Division, One College Ave., Williamsport, PA 17701-5799 US; 717-326-3761 x4761, Fax 717-327-4503, E-mail cschuman@pct.edu, URL http://www.pct.edu.

PENNSYLVANIA CULINARY

Pittsburgh/Year-round *(See display ad page 113)*

This private career institution offers 16-month associate degrees (12 months instruction, 4-month paid internship) in Specialized Technology in Culinary Arts (77.75 credits) and Specialized Business in Hotel and Restaurant Management (83 credits). Established in 1986. Accredited by ACCSCT, ACFEI. Calendar: semester. Curriculum: culinary. Admission dates: January, March, May, June, September, October. Total enrollment: 1,300+. 135-330 each admission period. 85% of applicants accepted. 93% receive financial aid. Student to teacher ratio 22-24:1. 98% of graduates obtain employment within six months. Facilities: Include 9 kitchens with the latest equipment, 15 classrooms, full-service dining room lab and mixology lab, library resource center and computer lab.

Courses: Include food preparation and skill development, advanced classical and international cuisine, nutrition, wines and spirits, menu planning, dining room management. Both programs include a 16-week paid externship, nationwide.

Faculty: 41 faculty members: 23 certified chef-instructors, 18 related studies/management instructors.

Costs: Tuition $4,487 per semester. Application fee $50. On-campus housing provided for 1,000 students. Off-campus lodging ranges from $200-$400 per month. Admission requirements: High school diploma or equivalent required, foodservice experience desirable, letters of recommendation encouraged. Scholarships: yes. Loans: yes.

Location: Pittsburgh's cultural district.

Contact: Cindy Chalovich, Director of Admissions, Pennsylvania Institute of Culinary Arts, 717 Liberty Ave., Admissions Dept., Pittsburgh, PA 15222 US; 800-432-2433, Fax 412-566-2434, URL http://www.paculinary.com.

THE RESTAURANT SCHOOL

Philadelphia/Year-round *(See also page 251) (See display ad page 115)*

This proprietary career institution offers 15-month programs leading to a specialized associate degree in Chef Training (79 credits), Pastry Chef Training (71 credits), Restaurant Management (98 credits), Hotel Management (93 credits). Established in 1974. Accredited by ACCSCT. Calendar: quarter. Curriculum: core. Admission dates: March, June, October, December. Total enrollment: 500. 250 each admission period. 95% of applicants accepted. 95% receive financial aid. 20% enrolled part-time. Student to teacher ratio 18:1. 98% of graduates obtain employment within six months. Facilities: Include 4 classroom kitchens, two 85-seat demonstration kitchens, pastry shop, and 4 student-run restaurants.

Courses: Chef Training combines classroom instruction with apprenticeship; includes business management, dining room service, wines, and 7 certification courses. Pastry Chef Training covers culinary and baking skills, baking science, chocolate, candies, and 6 certification courses.

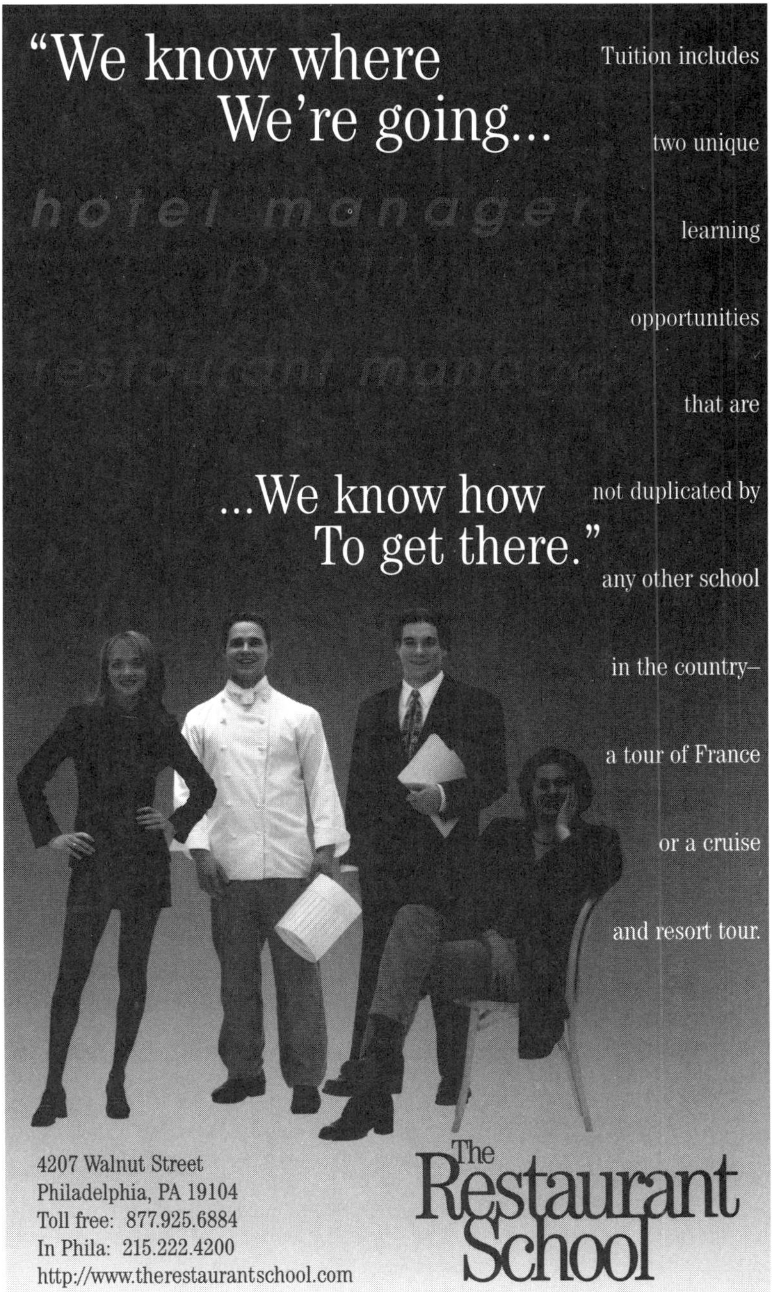
"We know where
We're going...
hotel manager
...We know how
To get there."
Tuition includes
two unique
learning
opportunities
that are
not duplicated by
any other school
in the country–
a tour of France
or a cruise
and resort tour.
4207 Walnut Street
Philadelphia, PA 19104
Toll free: 877.925.6884
In Phila: 215.222.4200
http://www.therestaurantschool.com
The Restaurant School

Externships in area restaurants, hotels, and abroad. Both programs include an 8-day tour of France.

Faculty: The 18-member professional faculty have a minimum of 12 years experience in the restaurant, foodservice, and hotel industry. 4 ACF-certified chefs, 1 master pastry chef.

Costs: The $18,550 cost of each program includes trip to France. Other fees $1,000. On-campus dorm and apartments are available. Admission requirements: High school diploma or equivalent, reference letters, basic achievement test. Scholarships: yes. Loans: yes.

Location: Restored mansion in University City.

Contact: Karl D. Becker, Director of Admissions, The Restaurant School, 4207 Walnut St., Philadelphia, PA 19104 US; toll-free 877-925-6884, 215-222-4200 x3011, Fax 215-222-4219, URL http://www.therestaurantschool.com.

WESTMORELAND COUNTY COMMUNITY COLLEGE

Youngwood/Year-round

This two-year college offers a 2- and 3-year AAS degree in Culinary Arts with apprenticeship option, 1-semester Culinary Arts certificate, 2-year Baking and Pastry degree, 1-semester Baking/Pastry/Deli certificate, 1-semester Dining Room Management certificate. Program started 1981. Accredited by ACFEI. Calendar: semester. Curriculum: core. Admission dates: August, January. Total enrollment: 132. 132 each admission period. 100% of applicants accepted. 37% receive financial aid. 54.5% enrolled part-time. Student to teacher ratio 15-20:1. 100% of graduates obtain employment within six months. Facilities: Include 4 specially equipped kitchens and classroom projects simulate student-run restaurant.

Courses: Garde manger, quantity foods, purchasing and storage, baking, food specialties, and hospitality marketing, as well as ACF Laurel Highlands Chapter Membership. 1 semester externships and 3-year apprenticeships provided.

Faculty: 4 full-time, 18 part-time. Qualifications: ACF certification, experience in field, academic requirements. Includes Mary B. Zappone, CCE, Cheryl Shipley, RD, Carl Dunkel, CWC, CCE.

Costs: Annual tuition in-county $1,440, out-of-county $2,880. Application fee is $10, lab fee $20 per culinary course. Admission requirements: High school diploma or equivalent and admission test. Scholarships: yes.

Location: Southwestern Pennsylvania; 30 miles from Pittsburgh.

Contact: Mary Zappone, Professor, Westmoreland County Community College, Hospitality Dept., Armbrust Rd., Youngwood, PA 15697 US; 724-925-4016, Fax 724-925-4293, E-mail zapponm@astro.westmoreland.cc.pa.us, URL http://www.westmoreland.cc.pa.us. Additional contact: Chef Carl Dunkel.

WINNER INSTITUTE OF ARTS & SCIENCES

Transfer/Year-round

This private school offers a 38-week diploma program in Culinary Arts. Program started in 1997. School is registered with the Ohio Board of Proprietary Schools. Calendar: trimester. Curriculum: culinary. Admission dates: January, March, July, September. Total enrollment: 36. Up to 30 each admission period. 98% of applicants accepted. 90% receive financial aid. 75% enrolled part-time. Student to teacher ratio 7:1. 100% of graduates obtain employment within six months. Facilities: Kitchen with the latest equipment, banquet room, 4 classrooms, fully-equipped computer lab and library.

Courses: Hands-on training preparing students for careers as foodservice professionals. 200-hour paid externship often leading to full-time employment. Evening cooking classes for the public, professional seminars.

Faculty: 3 full-time ACF-certified chef instructors, 1 part-time chef instructor, 1 part-time academic instructor.

Costs: Tuition $11,470 ($3,990 per term), fees $1,250. Admission requirements: High school diploma or GED & passing score on entrance exam. Loans: yes.

Location: Rural campus 90 minutes from Pittsburgh, Cleveland, and Erie.

Contact: John Matsis, Executive Director, Winner Institute of Arts & Sciences, One Winner Place, Transfer, PA 16154 US; 888-414-CHEF/724-646-CHEF, Fax 724-646-0218, E-mail wias@infonline.net, URL http://www.by1.com/pa/winnerchefs.

PUERTO RICO

INSTITUTO DE EDUCACION UNIVERSITY

Hato Rey/Year-round

This school offers a certificate program. Program started 1991. Accredited by ACCSCT. Calendar: trimester. Admission dates: February, September. Total enrollment: 164 (culinary), 60 (baking). 50-60 each admission period. 95% of applicants accepted. 98% receive financial aid. Student to teacher ratio 20:1. Facilities: 2 kitchens and classrooms.

Courses: Include safety and hygiene, cooking methods, basic sauces, menu planning; bread formulas, cost production. Externship: 120 hours, nearby hotels, restaurants, and bakeries.

Faculty: 6 full-time culinary, 2 full-time baking. Includes Chef Phillipe Chapuis.

Contact: Willie Lucca, Coordinator of Culinary Arts, Instituto de Educacion University, Culinary Arts, Barbosa Ave. #404, Hato Rey, PR 00930 US; 787-766-2443/787-767-2000, Fax 787-767-4755.

INSTITUTO DEL ARTE MODERNO, INC.

Hato Rey

This career institution offers a 1,000-hour certificate. Program started 1987. Accredited by ACCSCT. Admission dates: January, August. Total enrollment: 150. Student to teacher ratio 25:1.

Faculty: 5 full-time.

Costs: Annual tuition about $2,900. Admission requirements: High school dilploma or equivalent.

Contact: Myriam Aporte, Director, Instituto del Arte Moderno, Inc., Culinary Arts, Ave. Monserrate FR-5, Villa Fontana, Carolina, PR 00983-3912 US; 787-768-2532.

RHODE ISLAND

JOHNSON & WALES UNIVERSITY

Providence/Year-round

This private nonprofit career institution offers 2-year AAS and 4-year BS degree programs in Culinary Arts and Baking & Pastry Arts. Other degrees offered at different campus locations. Established in 1914. Accredited by NEASC, ACICS. Calendar: quarter. Curriculum: culinary, core. Admission dates: March, September, December. Total enrollment: 2,119 full-, 194 part-time. 84% of applicants accepted. 82% receive financial aid. 8% enrolled part-time. Student to teacher ratio 30:1. 98% of graduates obtain employment within six months. Facilities: Modern teaching facilities, including 5 student-run restaurants.

Courses: Culinary arts includes basic cooking and baking, classic and international cuisines, food preparation, nutrition, communication, menu design; baking and pastry includes basic ingredients, production techniques, French pastries, desserts, chocolate and sugar artistry. Sophomore year internship at Radisson Airport Hotel and the University's foodservice training facilities.

Faculty: 70 full-, 3 part-time.

Costs: Annual tuition $15,378, general fee $477, orientation fee $125, room and board $5,550 per year plus $672 for optional weekend meal plan. Admission requirements: High school diploma or equivalent. Scholarships: yes. Loans: yes.

Location: Near Providence's cultural and recreational facilities. Other campuses are located in South Carolina, Virginia, Florida and Colorado.

Contact: Licia Dwyer, Director, Culinary Admissions, Johnson & Wales University, College of Culinary Arts, 8 Abbott Park Place, Providence, RI 02903 US; 800-342-5598, Fax 401-598-2948, E-mail admissions@jwu.edu, URL http://www.jwu.edu.

SOUTH CAROLINA

GREENVILLE TECHNICAL COLLEGE

Greenville/Year-round

This career college offers 1-year certificate and 2-year degree programs in Food Service Management. Program started 1977. Accredited by ACFEI, SACS, ACBSP. Admission dates: Semesters (3). Total enrollment: 115. 45 each admission period. 50% of applicants accepted. 25% receive financial aid. 5.5% enrolled part-time. Student to teacher ratio 15:1. 100% of graduates obtain employment within six months. Facilities: Include kitchen and 3 classrooms.

Courses: A la carte, competition, buffet, nutrition, food production.

Faculty: 4 full-time, 2 part-time.

Costs: Tuition $525 per semester, application fee $25. Admission requirements: High school diploma or equivalent and admission test. Scholarships: yes.

Location: 150 miles from Atlanta and 170 miles from Charlotte.

Contact: Alan Scheidhauer, CEC, Dept. Head, Greenville Technical College, Hospitality Education Dept., PO Box 5616, Station B, Greenville, SC 29606-5616 US; 803-250-8303, Fax 803-250-8455, E-mail scheidajs@gvltec.edu, URL http://www.gvltec.edu.

HORRY-GEORGETOWN TECHNICAL COLLEGE

Conway/Year-round

This career college offers a 2-year degree in Culinary Arts Technology, certificate in Food Service. Program started 1985. Accredited by SACS, ACFEI. Calendar: semester. Curriculum: core. Admission dates: August, December, April. Total enrollment: 105. 40 each admission period. 100% of applicants accepted. 60% receive financial aid. 10% part-time. Student to teacher ratio 10:1. 100% of graduates obtain jobs within six months. Facilities: 4 kitchens, 3 dining rooms, 3 restaurants.

Courses: Food production, sanitation, nutrition, a la carte, buffet, menu planning. Externship provided. Continuing education courses available.

Faculty: 12 full- and part-time. Dept. Head C. Catino, K. Gerba, C. LaMarre, S. DePalma.

Costs: Tuition in-state $540 per semester, out-of-state $1,080 per semester. Application fee $15. Average off-campus housing cost $250-$350 per month. Admission requirements: High school diploma or equivalent and admission test. Scholarships: yes. Loans: yes.

Contact: Carmen Catino, Dept. Head, Horry-Georgetown Technical College, Culinary Arts, PO Box 1966, 2050 Hwy. 501 East, Conway, SC 29526 US; 803-347-3186, Fax 803-347-4207, E-mail catino@hor.tec.sc.us, URL http://www.hor.tec.sc.us.

JOHNSON & WALES UNIVERSITY AT CHARLESTON

Charleston/Year-round

This private nonprofit career institution offers AAS degrees in Culinary Arts and Baking and Pastry Arts. Established in 1984. Accredited by ACICS, NEASC. Calendar: quarter. Curriculum: culinary, core. Admission dates: Rolling. Total enrollment: 925. 86% of applicants accepted. 80% receive financial aid. 2% enrolled part-time. Student to teacher ratio 20:1. 99.1% of graduates obtain employment within six months. Facilities: 10 kitchens, 12 classrooms.

Courses: Culinary Arts. Other required courses: cooperative education practicum. Externship: 11 weeks, resorts/hotels/restaurants.

Faculty: 20 full-, 4 part-time.

Costs: Annual tuition $13,212 for College of Culinary Arts, $11,139 for Hospitality College. Other costs: $477 general fee, $125 orientation fee, on-campus housing (124 apartment units) $3,738. Admission requirements: High school diploma or equivalent. Scholarships: yes. Loans: yes.

Contact: Mary Hovis, Director of Admissions, Johnson & Wales University at Charleston, Admissions Office, 701 E. Bay St., PPC Box 1409, Charleston, SC 29403 US; 800-868-1522, Fax 803-763-0318, E-mail admissions@jwu.edu, URL http://www.jwu.edu.

TRIDENT TECHNICAL COLLEGE
Charleston/Year-round

This career college offers a 4-semester diploma and associate degree in Culinary Arts. Program started 1986. Accredited by SACS, ACFEI. Calendar: semester. Curriculum: culinary. Admission dates: Open. Total enrollment: 85. 75-85 each admission period. 93% of applicants accepted. 40% receive financial aid. 5% enrolled part-time. Student to teacher ratio 15:1. 100% of graduates obtain employment within six months. Facilities: 10 kitchens and classrooms and student-run restaurant.

Courses: Required internships. Assistance with internship available. Continuing education: 6-10 courses per semester.

Faculty: 3 full-time.

Costs: Annual tuition $1,572 in-county, $1,836 out-of-county, $2,743 out-of-state. Application fee $20. Off-campus housing $400 per month. Admission requirements: High school diploma or equivalent and admission test. Scholarships: yes. Loans: yes.

Location: The 9,700-student campus is in downtown Charleston.

Contact: Betty Howe, Dean, Trident Technical College, Division of Hospitality & Tourism, PO Box 118067, HT-P, Charleston, SC 29423-8067 US; 803-722-5541, Fax 803-720-5614, E-mail zphowe@trident.tec.sc.us, URL http://www.charleston.net/trident.tec.

MITCHELL TECHNICAL INSTITUTE
Mitchell/September-May

This career school offers an 18-month diploma, certificate, or AAS degree in Culinary Arts. Program started 1968. Accredited by NCA. Calendar: semester. Curriculum: culinary. Admission dates: February. Total enrollment: 40. 18-20 each admission period. 100% of graduates obtain employment within six months. Facilities: Include 3 kitchens, 3 classrooms, a 54-seat restaurant.

Courses: Program is being re-structured.

Faculty: 2 full-time.

Costs: Tuition $50 per credit-hour in-state, $82 per credit-hour out-of-state. Other fees $80. Off-campus housing cost $4,200. Admission requirements: High school diploma or equivalent and admission test.

Contact: John Weber, Mitchell Technical Institute, Cook/Chef, 821 N. Capitol, Mitchell, SD 57301 US; 605-995-3030, E-mail questions@mti.tec.sd.us, URL http://mti.tec.sd.us.

TENNESSEE

MEMPHIS CULINARY ACADEMY
Memphis/Year-round

This private trade school offers a 40-week diploma program that consists of 10-week basic, 5-week intermediate, and 15-week advanced courses based on classic French and European cuisines. Established in 1984. Accredited by TN Higher Education Commission. Calendar: quarter. Curriculum: culinary. Admission dates: January, April, June, September. Total enrollment: 40. 10 each admission period. Student to teacher ratio 6:1. 95% of graduates obtain employment within six months.

Courses: Include culinary skills, baking, nutrition, and kitchen rotation. Seminars in pastry, garde manger, kitchen management.

Costs: Tuition is approximately $3,000 for the basic course, $300 for intermediate and advanced courses, $3,600 for all three. Admission requirements: High school diploma. Scholarships: yes.

Contact: Joseph Carey, Memphis Culinary Academy, 1252 Peabody Avenue, Memphis, TN 38104 US; 901-722-8892.

OPRYLAND HOTEL CULINARY INSTITUTE
Nashville/Year-round

Opryland Hotel offers a 3-year certificate of Apprenticeship plus AAS degree through Volunteer

State Community College. Established in 1987. Accredited by ACFEI. Curriculum: culinary, core. Admission dates: August. Total enrollment: 60. ~20-25 each admission period. 20-25% of applicants accepted. 100% receive financial aid. 0% enrolled part-time. Student to teacher ratio 8-25:1. 100% of graduates obtain employment within six months. Facilities: Opryland Hotel.

Faculty: Richard Gerst, CEC, Executive Chef; Dina Starks, M.S., R.D., Apprenticeship Coordinator.

Costs: Tuition, fees, books, uniforms provided by Opryland; temporary housing available through Opryland Hotel. Admission requirements: High school diploma or GED; some college & work experience preferred.

Contact: Dina Starks, M.S., R.D., Culinary Apprenticeship Coordinator, Opryland Hotel Culinary Institute, 2800 Opryland Dr., Nashville, TN 37214 US; 615-871-7765, Fax 615-871-6942.

TEXAS

ART INSTITUTE OF HOUSTON – SCHOOL OF CULINARY ARTS

Houston/Year-round *(See display ad page 8)*

This proprietary school offers an 18-month AAS degree in Culinary Arts. Program started 1992. Accredited by ACCSCT, ACFEI, candidate with SACS. Calendar: quarter. Curriculum: core. Admission dates: January, April, July, September. Total enrollment: 275. 40-80 each admission period. 65% of applicants accepted. 75% receive financial aid. 10% enrolled part-time. Student to teacher ratio 20:1. 90% of graduates obtain employment within six months. Facilities: Include 3 teaching kitchens, bakery, deli, open-to-the-public restaurant.

Courses: Basic cooking, food production, garde manger, a la carte, baking, nutrition, sanitation, purchasing and cost controls, dining room management, management by menu. Other required courses: 24 hours of general education. 2 internships/externships are required, in fifth and sixth quarters.

Faculty: 12 full- and part-time. Includes Michael F. Nenes, CEC, CCE, Larry Matson, CWC, Peter Lehr, CEC, Charles Prince, CEC, Charles Boley, CMC.

Costs: Tuition $3,285 per quarter. Application fee $50, general fee $125, lab fee $250 per quarter, supply kit $625. Off-campus housing, currently 30 units, cost is $1,130 per quarter. Admission requirements: High school diploma or equivalent and interview, essay, ASSET test. Scholarships: yes. Loans: yes.

Location: The 1,100-student, 70,000-square-foot facility is in the primary business area of Houston, the Galleria.

Contact: Rick Simmons, Director of Admissions, Art Institute of Houston-School of Culinary Arts, 1900 Yorktown, Houston, TX 77056 US; 800-275-4244/713-623-2040, Fax 713-966-2797, E-mail aihadm@aih.aii.edu, URL http://www.aih.aii.edu. Additional contact: Steve Shroyer.

CREATIVE CUISINE & CATERING

Austin/Year-round *(See also page 255)*

This private career school offers five certificate programs in Culinary Arts, Pastry Arts, and Catering Management. These range from 6 weeks (150 contact hours) to one year (1,316 contact hours). Established 1998. Accredited by Texas Workforce Commission. Calendar: quarter. Curriculum: culinary. Admission dates: October, January, April, July; rolling. Total enrollment: 25. 10-15 each admission period. 70% of applicants accepted. 50-75% receive financial aid. 20% enrolled part-time. Student to teacher ratio 8:1. 100% of graduates obtain employment within six months. Facilities: Well-equipped commercial kitchen, bakeshop, catering operation, 2 classrooms, dining room, demonstration kitchen, resource room/computer lab.

Courses: Sequential curriculum covers food production, bakeshop, regional and international cuisine, culinary culture and history, nutrition, sanitation, purchasing analysis and cost controls, menu planning, brigade system, and business fundamentals. Externships provided. Internships with restaurants and bakeries. Continuing education courses, specialized classes, and workshops for culinary professionals.

Faculty: 2 full-time CECs, 3 part-time. Two advisory boards. Culinary professionals are guest lecturers.

Costs: $1,800-$11,750. Includes meals. Admission requirements: 18 years old, high school diploma or equivalent, interview. Scholarships: yes. Loans: yes.

Location: Two-story building in Central Austin, shared with a private, for-profit catering company, Private Affairs Catering. Near Texas hill country.

Contact: Glenn Mack, Director of Education, Creative Cuisine & Catering, 2823 Hancock Dr., Austin, TX 78731 US; 512-451-5743, Fax 512-467-9120, E-mail chefs@texas.net, URL http://chefs.home.texas.net.

CULINARY INSTITUTE ALAIN & MARIE LENÔTRE

(See also page 256) (See display ads above and page 95) **Houston/Year-round**

This private French export corporation offers two 20-week diploma courses: cooking (300 recipes) and baking/pastry (400 recipes). 30 hours per week, 90% hands-on. Each course offered as 5-week intermediate (II), and 5-week advanced (III). French language optional (6 hours per week, 10 weeks minimum). Program started 1998. Curriculum: culinary. Admission dates: All year. Total enrollment: 296. 37 each admission period. 25% receive financial aid. Student to teacher ratio 10-15:1. 90% of graduates obtain employment within six months. Facilities: 4 classrooms, meeting room, student cafeteria in a 28,000-square-foot free-standing modern new building with the newest equipment.

Courses: Restaurant cooking, catering, professional baking/pastry, breads, ice cream, chocolate, sugar, decor, international topics, guest chefs. Enrollment by the week available for all courses (space permitting), special topic 1-week programs.

Faculty: Technical Director Patrick LeNôtre, 4 full-time (day courses) and 4 part-time (evening

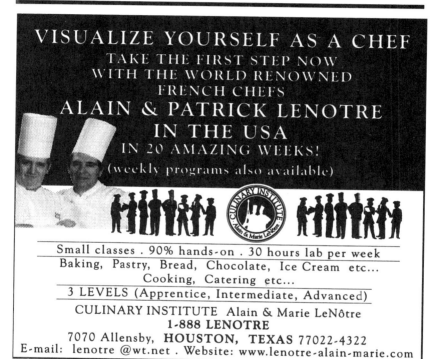

courses) French and European chefs rotating every 3 years.

Costs: 1 week $980 (II) and $1,100 (III), 5 weeks $4,250 Level II, 5 weeks $4,750 Level III, 10 weeks $7,500 Level I, $16,500 per 20 weeks (6 months), $29,700 per 40 weeks (1 year). Admission requirements: 1 year experience for intermediate, 2 years for advanced.

Location: Central Houston, 30 minutes from NASA Space Center, 1 hour from Galveston Island on the Gulf of Mexico.

Contact: Alain LeNôtre, President, Culinary Institute Alain & Marie LeNôtre, 7070 Allensby, Houston, TX 77022 US; 888-LeNotre/713-692-0077, Fax 713-692-7399, E-mail lenotre@wt.net, URL http://www.lenotre-alain-marie.com.

DEL MAR COLLEGE

Corpus Christi/Year-round

This state-supported institution offers a 1-year certificate, 2-year AAS degree, 4-year BS degree in Restaurant Management and Culinary Arts. Program started 1963. Accredited by SACS. Calendar: semester. Admission dates: June, September, January. Total enrollment: 175. 60 each admission period. 100% of applicants accepted. 40% receive financial aid. 30% enrolled part-time. Student to teacher ratio 15:1. 100% of graduates obtain employment within six months. Facilities: Include 3 restaurants, 4 classrooms and laboratory.

Courses: Saucier, garde manger, elementary baking, advanced pastry, and restaurant management. 3 semesters paid internships.

Faculty: 4 full-time, 11 part-time.

Costs: Annual tuition in-state $1,000, out-of-state $1,800. Average off-campus housing cost $250-$350. Admission requirements: High school diploma or equivalent and admission test. Scholarships: yes.

Contact: D.W. Haven, Professor and Chair, Del Mar College, Dept. of Hospitality Management, Baldwin at Ayers, Corpus Christi, TX 78404 US; 512-698-1734, Fax 512-698-1829, E-mail DHaven@camino.delmar.edu, URL http://www.delmar.edu.

EL CENTRO COLLEGE

Dallas/Year-round

This college offers a 2-year AAS degree in Food and Hospitality Services. Program started 1971. Accredited by SACS. Calendar: semester. Curriculum: culinary, core. Admission dates: January, August, May-June. Total enrollment: 400. 350-400 each admission period. 100% of applicants accepted. 40% receive financial aid. 65% enrolled part-time. Student to teacher ratio 20-35:1. 90% of graduates obtain employment within six months. Facilities: Include 3 kitchens, 4 classrooms, pastry/bakery labs.

Courses: Hands-on classes in food preparation and baking. Apprenticeship-ACF available.

Faculty: 4 full-time, 12 part-time.

Costs: Annual tuition in-county $500, out-of-county $900. Admission requirements: High school diploma or equivalent and admission test. Scholarships: yes. Loans: yes.

Location: Downtown/Center City.

Contact: C. Gus Katsigris, Director Food & Hospitality Services Institute, El Centro College, Food & Hospitality Services Institute, Main at Lamar St., Dallas, TX 75202 US; 214-860-2202, Fax 214-860-2335, E-mail cxk5531@dcccd.edu.

EL PASO COMMUNITY COLLEGE

El Paso/Year-round

This two-year public college offers 1-year certificate and 2-year AAS degree programs in Food Service, Culinary Arts. Program started 1992. Accredited by SACS. Calendar: semester. Curriculum: culinary, core. Admission dates: July/August, December/January, April/May. Total enrollment: 40. 15 each admission period. 90% of applicants accepted. 70% receive financial aid. 50% enrolled part-time. Student to teacher ratio 18:1. 100% of graduates obtain employment within six months.

Facilities: Currently: small kitchen. Planned: full training kitchen and baking kitchen.

Courses: Cooking, service, management. Co-op working experiences. Concurrent continuing ed courses.

Faculty: 1 full-time (bachelor's degree, 19 years experience), 2 part-time instructors (1 AAS and approaching CCF, 1 master's degree, registered dietitian).

Costs: $75 ($200-$250) per 1 credit hour + $20 ($50) ea additional + $135 fees (12 credits) for residents (nonresidents). Admission requirements: High school diploma, GED, TASP exam. Scholarships: yes. Loans: yes.

Contact: M.J. Linney, Program Coordinator, El Paso Community College, 919 Hunter Dr., El Paso, TX 79915 US; 915-831-2217, Fax 915-831-2155, E-mail mjlinney@aol.com.

GALVESTON COLLEGE – CULINARY ARTS ACADEMY

Galveston/Year-round

This college offers 1-year certificates in Food Preparation and Hospitality Management, 2-year AAS degree in Culinary Arts/Hospitality Management. Established in 1987. Accredited by SACS. Calendar: semester. Curriculum: core. Admission dates: January, June, July, August. Total enrollment: 40-50. 30-40 each admission period. 75% of applicants accepted. 100% receive financial aid. 10% enrolled part-time. Student to teacher ratio 15:1. 80% of graduates obtain employment within six months. Facilities: Include kitchen, bakeshop and classroom.

Courses: Externship provided.

Faculty: 3 full-time. Leslie Bartosh, CCC, FMP, Joseph M. Tabaracci, Charles Collins, CCC, CWPC.

Costs: Annual tuition $400 in-state, $600 out-of-state. Admission requirements: High school diploma or equivalent and admission test. Scholarships: yes. Loans: yes.

Contact: Leslie Bartosh, CCC, FMP, Director of Culinary Arts, Galveston College-Culinary Arts Academy, 4015 Ave. Q, Galveston, TX 77550 US; 409-763-6551 x304, Fax 409-762-9367, E-mail chef@gc.edu, URL http://www.gc.edu/FACULTY/LBARTOSH/Cover.htm.

HOUSTON COMMUNITY COLLEGE SYSTEM

Houston/Year-round

This college offers a certificate in Culinary Arts. Established in 1972. Accredited by SACS. Calendar: semester. Curriculum: core. Total enrollment: 200. 60% receive financial aid. 10% enrolled part-time. Student to teacher ratio 15-25:1.

Faculty: 5 full-time, 4 part-time.

Costs: Annual tuition $1,176 in-district, $1,974 out-of-district, $4,284 out-of-state, includes fees.

Contact: William H. Pile, Dept. Chair, Houston Community College System, Culinary Services, Houston, TX 77004 US; 713-718-6010, Fax 713-718-6054, URL http://www.hccs.cc.tx.us.

LE CHEF COLLEGE OF HOSPITALITY CAREERS

Austin/Year-round

This independent nonprofit institution offers a 17-month diploma program in Culinary Arts, 2-year AAS degree program in Culinary Arts and Food and Beverage Management, Advanced Pastry certificate program. Established in 1985. Accredited by COE. Curriculum: culinary, core. Admission dates: Continuous. Total enrollment: 200. Varies per term each admission period. 66% of applicants accepted. 90% receive financial aid. Student to teacher ratio 10:1. 94% of graduates obtain employment within six months. Facilities: 36-station culinary lab, 6 lecture rooms, learning center, computer lab, conference room, 2 student lounges.

Courses: Diploma and degree programs cover cuisine and pastry preparation, pantry production and garde manger, production, control, planning and presentation. Degree program also includes food and beverage management and general education core courses. The diploma program includes a 720-hour paid externship. Short courses designed for industry professionals.

Faculty: Founder/President Ronald Boston, CDM, CFBE, Chefs Andre Touboulle, CMC, Matt Collins, CWC, Nick Bogert, CEC, and others; 5-member general ed faculty with advanced degrees.

Costs: $12,882 for Culinary Arts diploma, $20,229 for AAS degree. A $75 nonrefundable registration fee is required. Admission requirements: High school diploma or GED or (Culinary Arts program only) pass test to show Ability to Benefit. Scholarships: yes. Loans: yes.

Location: North central Austin, less than 30 minutes from Texas Hill Country.

Contact: Shawn Fortner, Executive Assistant, Le Chef College of Hospitality Careers, 6020 Dillard Circle, Austin, TX 78752 US; 888-5LeChef/512-323-2511, Fax 512-323-2126, E-mail LeChef@onr.com, URL http://www.lechef.org.

THE NATURAL EPICUREAN ACADEMY OF CULINARY ARTS

Austin/Year-round *(See also page 258)*

This academy of natural whole foods cooking with vegetarian/vegan emphasis offers 104 hours course work, 75-hour assistantship, 160-hour internship. Certificate granted upon completion of all course requirements. Program started 1994. Curriculum: culinary. Admission dates: Rolling. Total enrollment: 36. 20 each admission period. 100% of applicants accepted. 10% receive financial aid. Student to teacher ratio 20:1. 100% of graduates obtain employment within six months. Facilities: Professionally equipped homestyle kitchen.

Courses: Health-supporting whole foods, vegetarian/vegan cooking techniques, natural remedies. 160-hour internship in-state, out-of-state, international. Continuing ed specialty classes, Oriental diagnosis program.

Faculty: 3 full- and 3 part-time. Teachers must have completed a qualifying chef program.

Costs: $2,060 tuition, $215 fees. Scholarships: yes. Loans: yes.

Location: Central Austin.

Contact: Elizabeth Foster, CEO, The Natural Epicurean Academy of Culinary Arts, 902 Norwalk Ln., Austin, TX 78703 US; 512-476-2276, Fax 512-476-2298.

ODESSA COLLEGE

Odessa/Year-round

This college offers a 2-year certificate/AAS degree. Program started 1990. Accredited by SACS. Calendar: semester. Curriculum: culinary, core. Admission dates: Open. Total enrollment: 35-50. 35-50 each admission period. 75% of applicants accepted. 65% receive financial aid. 10-15% enrolled part-time. Student to teacher ratio 10-15:1. 100% of graduates obtain employment within six months. Facilities: Training kitchen/laboratory, dining room.

Courses: Externships provided.

Faculty: 2 full-time.

Costs: Tuition $207 for first three credit-hours, $19 for each additional credit-hour. Admission requirements: High school diploma or equivalent and admission test. Scholarships: yes.

Contact: Peter Lewis, Dept. Chair, Odessa College, Culinary Arts, 201 W. University, Odessa, TX 79764 US; 915-335-6320, Fax 915-335-6860.

ST. PHILIP'S COLLEGE

San Antonio/Year-round

This college offers a 2-year AAS degree, 3-year apprenticeship in Hospitality Management. Program started 1979. Accredited by SACS. Admission dates: August, January, June. Total enrollment: 250. 250 each admission period. 80% receive financial aid. 50% enrolled part-time. Student to teacher ratio 50:1. 85% of graduates obtain employment within six months. Facilities: Include 7 kitchens and classrooms and a restaurant.

Courses: Garde manger, international food preparation, and baking principles. Externship: 8 weeks. Continuing education courses available.

Faculty: 5 full-time.

Costs: Annual tuition in-district $504, out-of-district $966, out-of-state $1,932. Admission requirements: High school diploma or equivalent and admission test.

Contact: William Thornton, Associate Professor, St. Philip's College, Hospitality Operations, 2111 Nevada, San Antonio, TX 78203 US; 210-531-3315, E-mail WThornton@accd.edu, URL http://www.accd.edu.

SAN JACINTO COLLEGE NORTH – CULINARY ARTS

Houston/September-July

This college offers a 2- and 3-year AAS degree. Program started 1986. Accredited by SACS. Calendar: semester. Curriculum: culinary. Admission dates: September, January. Total enrollment: 20-30. 25 each admission period. 90% of applicants accepted. 50% receive financial aid. Student to teacher ratio 12:1. 70% of graduates obtain employment within six months.

Faculty: 3 full-time. ACF chef, CEC/CCE.

Costs: Tuition $262 per semester in-district, $430 per semester out-of-district. Admission requirements: H.S. diploma-GED. Scholarships: yes. Loans: yes.

Contact: George J. Messinger, CEC, CCE, Dept. Chairman/Executive Chef, San Jacinto College North, Culinary Arts, 5800 Uvalde, Houston, TX 77049 US; 281-459-7150, Fax 281-459-7132, E-mail dpayne@sjcd.cc.tx.us, URL http://www.sjcd.cc.tx.us. Additional contact: Dr. Donna Payne, (713) 459-7174.

TEXAS STATE TECHNICAL COLLEGE

Waco/Year-round

This two-year technical-vocational college offers 1-year certificate and 2-year AAS degree programs in Food Service/Culinary Arts. Established in 1970. Calendar: quinmester. Curriculum: core. Admission dates: September, December, March, June. Total enrollment: 68. 25 each admission period. 100% of applicants accepted. 50% receive financial aid. 10% enrolled part-time. Student to teacher ratio 15:1. 100% of graduates obtain employment within six months. Facilities: Former Air Force base Officers' Club.

Courses: Job entry training, basic skills. 3-month co-op, not mandatory. DMA certificate.

Faculty: 3 faculty: one with an AAS in Vocational Education, one with a BS and RD, one with an MS in Meat Science and Food Technology.

Costs: Tuition is $25.50 per credit-hour in-state, $80 per credit-hour out-of-state. Admission requirements: CPT. Scholarships: yes. Loans: yes.

Contact: Homer Jones, Dept. Chair, Texas State Technical College, 3801 Campus Dr., Waco, TX 76705 US; 254-867-4868, Fax 254-867-3663, URL http://www.tstc.edu/waco.html.

UTAH

BRIDGERLAND APPLIED TECHNOLOGY CENTER

Logan/Year-round

This state applied technology center offers a 750-hour program consisting of basic food preparation courses (food production, sanitation, garde manger, baking, catering). Program started 1998. Calendar: quarter. Curriculum: culinary. Admission dates: Open admission. 100% of applicants accepted. 80% receive financial aid. 90% enrolled part-time. Student to teacher ratio 7-10:1. 100% of graduates obtain employment within six months. Facilities: 2 classrooms, large production lab, on-site cafeteria.

Courses: Job skills related to employment.

Faculty: 2 full time.

Costs: $330 per quarter. Loans: yes.

Location: 85 miles north of Salt Lake City.

Contact: Anne Parish, Bridgerland Applied Technology Center, 1301 N. 600 West, Logan, UT 84321 US; 435-750-3021, Fax 435-752-2016, E-mail aparish@m.batc.tec.ut.us, URL http://www.batc.tec.ut.us.

SALT LAKE COMMUNITY COLLEGE

Salt Lake/Year-round

This two-year college offers a 2-year full-time and 3-year part-time Apprentice Chef program. Program started 1984. Accredited by NASC, ACFEI. Calendar: quarter. Curriculum: culinary, core. Admission dates: Rolling. Total enrollment: 110. 40-50 each admission period. 90% of applicants accepted. 40-50% enrolled part-time. Student to teacher ratio 14:1. 100% of graduates obtain employment within six months. Facilities: Include kitchen, 8 classrooms, video and reference library.

Courses: Food preparation, sanitation, baking, menu design, and nutrition. Other required courses: AAS degree requires 24 credits in general education. Continuing education: specialized classes and workshops available for culinary professionals.

Faculty: 8 full-time and part-time.

Costs: Full-time tuition $771 per quarter in-state, $2,427 out-of-state. $20 application fee. Average off-campus housing cost is $300 per month. Admission requirements: High school diploma or equivalent and admission test.

Contact: Joe Mulvey, Apprenticeship Director, Salt Lake Community College, PO Box 30808, Salt Lake City, UT 84130-0808 US; 801-957-4066, Fax 801-957-4612, URL http://www.slcc.edu.

UTAH VALLEY STATE COLLEGE

Orem/August-April

This college offers a 2-year AAS degree in Culinary Arts. Program started 1992. Accredited by NASC. Admission dates: Open. Total enrollment: 35. 15 each admission period. 90% of applicants accepted. 80% receive financial aid. 10% enrolled part-time. Student to teacher ratio 12:1. 100% of graduates obtain employment within six months. Facilities: Include 3 kitchens, 3 classrooms, restaurant and food service operation.

Courses: Food production, nutrition, sanitation, garde manger, and buffet. Externship: 5-week, salaried, in hotels or restaurants.

Faculty: 3 full-time. Qualifications: certified chef, work experience.

Costs: Tuition in-state $760 per semester, out-of-state $2,387 per semester. Other fees: $100 class fee, $300 supplies fee. Average off-campus housing cost $200 per month. Admission requirements: High school diploma.

Contact: Greg Forte, Utah Valley State College, Business Dept., 800 W. 1200 South, Orem, UT 84058 US; 801-222-8000, Fax 801-226-5207/225-1229, URL http://csc.uvsc.edu.

VERMONT

NEW ENGLAND CULINARY INSTITUTE

Montpelier and Essex/Year-round *(See also page 260) (See display ad page 127)*

This private career institution offers a 2-year AOS degree program in Culinary Arts, upper level one-and-a-half-year bachelors degree in Food and Beverage Management, 40-week certificate program in Basic Cooking. Established in 1979. Accredited by State of Vermont, ACCSCT. Calendar: semester. Curriculum: culinary. Admission dates: September, December, March, June. Total enrollment: 700. 84 each admission period. 90% of applicants accepted. 80% receive financial aid. Student to teacher ratio 7:1. 100% of graduates obtain employment within six months. Facilities: 11 kitchens, 14 classrooms. Montpelier and Essex: 2 restaurants each, bakery, catering, banquet dept. Burlington: restaurant and bakery.

Courses: 75% of class time is spent preparing food for the public. Remaining class time covers cooking theory, food and wine history, wine and beverage management, tableservice, service management and purchasing. At least 45 hours of a structured physical fitness plan. Each year consists of a 24-week on-campus residency, followed by an 18- to 20-week internship. Short courses for the professional.

WHY WE CUT OUR CLASS SIZE TO THE BONE.

Less is more.
The fewer students in a class, the more time the teacher can spend with each student. Our student/teacher ratio is 7 to 1.
Our unique, 2 year program also offers:

- Hands-on training
- Paid, personalized internships
- Financial aid for qualified students
- Advanced Placement available
- A.O.S degree in Culinary Arts
- Bachelor's degree program in Service and Management

Call or write for our four-color catalog.

NEW ENGLAND CULINARY INSTITUTE

http://www.neculinary.com

250 Main St., Dept. SH
Montpelier, VT 05602-9720
(802) 223-6324

Accredited Member ACCSCT

Faculty: The 57-member faculty are chosen on the basis of experience and teaching ability. The Institute has a 19-member administrative staff and 3 advisory boards. Includes Michel LeBorgne, David Miles, and Robert Barral.

Costs: AOS: $20,950 per year includes room, board, uniforms; nonrefundable $25 application fee, $550 books and equipment. BA: $32,000 per 2 academic years includes room, board, uniforms. Dormitory lodging for 160 students, other lodging nearby. Admission requirements: High school diploma or equivalent and 3 reference letters; advanced placement second year students must pass an exam. Scholarships: yes. Loans: yes.

Location: The rural Montpelier campus is 3 hours from Boston. The Essex Junction campus is at The Inn at Essex country hotel, in a Bulington suburb.

Contact: Alison Czekelius, Associate Director of Admissions, New England Culinary Institute, Admissions Dept., 250 Main St., Dept. S, Montpelier, VT 05602 US; 802-223-6324, Fax 802-223-0634, URL http://www.neculinary.com.

VIRGINIA

ATI CAREER INSTITUTE – SCHOOL OF CULINARY ARTS

Falls Church/Year-round

This private career school offers a 12-month diploma in Culinary Arts. Program started 1990. Accredited by COE. Curriculum: core. Admission dates: Every 6 weeks. Total enrollment: 250. 48 each admission period. 80% of applicants accepted. 80% receive financial aid. 0% enrolled part-time. Student to teacher ratio 18:1. 95% of graduates obtain employment within six months. Facilities: Include 3 kitchens, 5 classrooms.

Courses: Culinary theory, nutrition, sanitation, sauces and entrees, baking, garde manger, hospitality management, and accounting. Externship: 12 weeks required.

Faculty: 12 full-time ACF-certified chefs. Guest chefs are frequent lecturers.

Costs: Tuition $14,783, including books and equipment. Nonrefundable application fee $65. Average off-campus housing cost $600 per month. Admission requirements: High school diploma or equivalent. Scholarships: yes. Loans: yes.

Location: The 20,000 square-foot campus, located on the Beltway, is 15 miles from Washington, D.C.

Contact: John W. Martin, CEC, CCE, CFE, Director of Culinary Arts, ATI Career Institute, School of Culinary Arts, 7777 Leesburg Pike, Ste. 100 South, Falls Church, VA 22043 US; 703-821-8570, Fax 703-556-9289, E-mail jmarti7031@aol.com.

J. SARGEANT REYNOLDS COMMUNITY COLLEGE

Richmond/Year-round

J. Sargeant Reynolds Community College offers a 68 credit-hour AAS degree in Culinary Arts. 2 years for non-apprentices, 3 years for ACF-registered apprentices. Established in 1973. Accredited by SACS. Calendar: semester. Curriculum: culinary, core. Admission dates: Ongoing. Total enrollment: 60. 100% of applicants accepted. 40% receive financial aid. 80% enrolled part-time. Student to teacher ratio 18:1. 100% of graduates obtain employment within six months. Facilities: Multipurpose facility consisting of classroom laboratory, food service, and conference space.

Courses: Competency-based technical and managerial education and training, ACF cook apprenticeship. To accommodate apprentices who work full-time, classes meet Mondays from 8am-7:40pm. 360-hour unpaid coordinated internship or 6,000-hour paid ACF-registered apprenticeship. ACF re-certification courses, advanced theory and technique for professionals and managers.

Faculty: 12 instructors. Credentials include Certified Hotel Administrator (CHA), Registered Dietitian (RD), and Ph.D.

Costs: Full degree program $3,335 in-state, $10,771 out-of-state. No deposit. Full refund first week of semester. No lodging on campus. Nearby apartments average $400 per month. Admission requirements: High school graduate or GED. Scholarships: yes. Loans: yes.

Contact: David J. Barrish, CHA, Program Head, J. Sargeant Reynolds Community College, PO Box 85622-Center for Hospitality Development, Richmond, VA 23285-5622 US; 804-786-2069, Fax 804-786-5465, E-mail dbarrish@jsr.cc.va.us, URL http://www.jsr.cc.va.us/DtcBusDiv/Hospitality.

JOHNSON & WALES UNIVERSITY AT NORFOLK

Norfolk/Year-round

This private nonprofit career institution offers a 12- and 18-month AAS degree in Culinary Arts, certificate in Culinary Arts. Established in 1986. Accredited by NEASC, ACICS. Curriculum: core. Admission dates: Rolling. Total enrollment: 546. 82% of applicants accepted. 19% enrolled part-time. Student to teacher ratio 14.5:1. 98% of graduates obtain employment within six months. Facilities: Newly-equipped kitchens, mixology and dining room labs, classrooms, computer lab.

Courses: Emphasis is on learning by doing.

Faculty: 13 full-, 10 part-time.

Costs: Annual tuition $12,867, general fee $477, orientation fee $125. Admission requirements: High school diploma or equivalent. Scholarships: yes. Loans: yes.

Contact: Torri Butler, Director of Student Affairs, Johnson & Wales University, 2428 Almeda Ave., Norfolk, VA 23513 US; 800-277-CHEF, Fax 757-857-4869, E-mail admissions@jwu.edu, URL http://www.jwu.edu.

NORTHERN VIRGINIA COMMUNITY COLLEGE

Annandale/Year-round

This two-year college offers a 1-year certificate program in Culinary Arts. Program started 1997. Accredited by SACS. Calendar: semester. Curriculum: culinary. Admission dates: Fall, summer, spring semesters. Total enrollment: 30. 30 each admission period. 100% of applicants accepted. 30% receive financial aid. 80% enrolled part-time. Student to teacher ratio 20:1. 90% of graduates obtain employment within six months. Facilities: Completed in 1997, facilities include 2 classrooms, computer lab, fully-equipped commercial kitchen and dining room.

Courses: Skills for culinary positions. Apprenticeship program available.

Faculty: 6 full-time, 1 part-time.

Costs: Tuition is $48 per credit-hour in-state, $162 per credit-hour out-of-state. Admission requirements: Open enrollment. Scholarships: yes. Loans: yes.

Location: 10 miles from downtown Washington, DC, near neighboring Maryland suburbs.

Contact: Benita Wong, CCC, CCE, Culinary Arts Instructor, Northern Virginia Community College, 8333 Little River Tpk., Annandale, VA 22003-3796 US; 703-323-3457, Fax 703-323-3509, E-mail nvwongb@nv.cc.va.us, URL http://www.nv.cc.va.us.

WASHINGTON

ART INSTITUTE OF SEATTLE – SCHOOL OF CULINARY ARTS

(See display ad page 131)

Seattle/Year-round

This two-year college offers a 7-quarter AAA degree in Culinary Arts. Program started 1996. Accredited by ACCSCT. Calendar: quarter. Curriculum: core. Admission dates: Rolling. Total enrollment: 300. 60 each admission period. 65% of applicants accepted. 75% receive financial aid. Student to teacher ratio 17:1. Facilities: Newly constructed kitchens and classroom space, dining room overlooking Puget Sound.

Courses: Basic skills, baking and pastry, desserts, American regional, classical, international, and Mediterranean cuisines, health-related cooking, charcuterie, cost management, menu and facility planning, dining room operations, show platter design and competition, general ed. Internship program with restaurants and resorts in the Western U.S.

Faculty: 18 culinary instructors with industry experience.

Costs: Total tuition $23,661. Student housing available. Admission requirements: High school diploma and admissions interview. Scholarships: yes. Loans: yes.

Location: On the waterfront overlooking Puget Sound.

Contact: Doug Worsley, VP, Dir. of Admissions, Art Institute of Seattle, Admissions Dept., 2323 Elliott Ave., Seattle, WA 98121 US; 800-275-2471/206-448-6600, Fax 206-448-2501, E-mail adm@ais.edu, URL http://www.ais.edu.

BELLINGHAM TECHNICAL COLLEGE

Bellingham/September-July

This two-year technical college offers certificates of completion in Culinary Arts and Baking, Pastry, and Confections. Program started 1957. Accredited by ACFEI, NACS. Calendar: quarter. Curriculum: culinary, core. Admission dates: Quarterly and other times with instructor's permission. Total enrollment: 40. 40 each admission period. 90% of applicants accepted. 30% receive financial aid. 20% enrolled part-time. Student to teacher ratio 20:1. 94% of graduates obtain employment within six months. Facilities: Instructional space, industrial kitchen/bakeshop, fine dining restaurant, deli/baking.

Courses: Culinary students operate an on-campus full-service restaurant. Baking students operate an on-campus bakeshop and prepare desserts for the restaurant.

Faculty: Michael Baldwin, CEC, and William Pifer, Master Baker, CMB.

Costs: Quarterly tuition and fees are $647 for Culinary Arts, $525 for Baking, Pastry and Confections. No on-campus housing. Admission requirements: 18 years or older, high school graduate, placement test. Scholarships: yes. Loans: yes.

Location: Northwestern Washington, 86 miles from Seattle, 80 miles from Vancouver, BC.

Contact: Michael Baldwin, Culinary Arts instructor, Bellingham Technical College, Culinary Arts, 3028 Lindberg Ave., Bellingham, WA 98225 US; 360-715-8350 x400, Fax 360-676-2798, E-mail mbaldwin@belltc.ctc.edu, URL http://www.ctc.edu. Additional contact: bpifer@belltc.ctc.edu.

CLARK COLLEGE CULINARY ARTS PROGRAM

Vancouver/Year-round

This community college offers a 1-year certificate and 2-year AAS degree programs in cooking, baking, and bakery and restaurant management. Established in 1958. Accredited by NWACC. Calendar: quarter. Curriculum: culinary, core. Admission dates: January, March, June, September. Total enrollment: 80. 15 (cooking), 10 (baking) each admission period. 80% of applicants accepted. 50% receive financial aid. Student to teacher ratio 5:1. 95% of graduates obtain employment within six months. Facilities: Modernized facility operates like a hotel kitchen. Students make all foods sold on-campus. Baking students operate the campus' retail bakery.

Courses: Cooking includes food preparation, advanced meat cutting, ice carving, wine appreciation, cake decoration and pastillage. Baking includes fundamentals the 1st year and specialized courses the 2nd year; includes theory, merchandising, bake shop management. 5-week internships available. Additional afternoon and evening specialized classes.

Faculty: 12-member faculty. Includes cooking instructors Larry Mains, CEC, CCE, AAC, George Akau, CCE, AAC, and Glenn Lakin and baking instructors Per Zeeberg and Jean Williams.

Costs: Cost for each 2-year program is $4,000 in-state, $10,092 out-of-state. Program can be 9 or 18 months. Off-campus housing provided close to campus. Admission requirements: High school diploma or equivalent. Scholarships: yes. Loans: yes.

Location: Minutes from Portland, Oregon.

Contact: Larry Mains, Director, Culinary Arts, Clark College, Culinary Arts, 1800 E. McLoughlin Blvd., Vancouver, WA 98663-3598 US; 360-992-2143, Fax 360-992-2839.

EDMONDS COMMUNITY COLLEGE

Lynwood/October-July

This community college offers a 6-quarter ATA. Program started 1988. Accredited by State. Calendar: quarter. Curriculum: core. Admission dates: Fall, winter, spring. Total enrollment: 45. 15 each admission period. 90% of applicants accepted. 25% receive financial aid. Student to teacher

ratio 20:1. 100% of graduates obtain employment within six months. Facilities: 1 kitchen, 1 classroom, 1 restaurant.

Courses: Contemporary Northwest cuisine, fine dining service, restaurant/food service management. Other required courses: service and management. Externship provided.

Faculty: 2 full-time, 2 part-time. Qualifications: CWC, CCE. Includes Walter Bronowitz, CWC, CCE; John Casey.

Costs: Annual tuition in-state $505 per quarter, out-of-state $1,987 per quarter. Admission requirements: High school diploma or equivalent. Scholarships: yes. Loans: yes.

Location: The 12,000-student suburban campus is 20 minutes from Seattle.

Contact: Walter N. Bronowitz, CCE, AAC, Chef/Instructor, Edmonds Community College, Culinary Arts, 20000 - 68th Ave. West, Lynnwood, WA 98036 US; 425-640-1329, Fax 425-771-3366, URL http://www.edcc.edu.

NORTH SEATTLE COMMUNITY COLLEGE

Seattle/September-June

This college offers a 1-year certificate and 2-year AAS degree in Culinary Arts, Hospitality and Restaurant Cooking. Program started 1970. Accredited by NASC. Calendar: quarter. Curriculum: core. Admission dates: Quarterly. Total enrollment: 80. 25 each admission period. 90% of applicants accepted. 25% receive financial aid. Student to teacher ratio 18:1. 90% of graduates obtain employment within six months. Facilities: Include 2 kitchens and classrooms, restaurant, bakery.

Courses: Restaurant cooking and commercial cooking.

Faculty: 4 full-time.

Costs: Annual tuition $1,750 in-state, $6,000 out-of-state. Uniform, supplies $750. Average off-campus housing cost $750 per month. Admission requirements: High school diploma or equivalent and admission test.

Contact: Darrell Mihara, Associate Dean, Culinary Arts & Hospitality, North Seattle Community College, Culinary Arts, 9600 College Way North, Seattle, WA 98103-3599 US; 206-528-4402, Fax 206-527-3635, E-mail DMihara@seaccd.sccd.ctc.edu, URL http://www.nsccux.sccd.ctc.edu.

OLYMPIC COLLEGE

Bremerton/September-May

This college offers a 3-quarter certificate, 2-year ATA degree. Program started 1978. Accredited by State. Calendar: trimester. Curriculum: culinary. Admission dates: Continuous enrollment. Total enrollment: 38. 28-32 each admission period. 85% of applicants accepted. 60% receive financial aid. 15% enrolled part-time. Student to teacher ratio 16:1. 90% of graduates obtain employment within six months. Facilities: Central kitchen, one classroom, two restaurants.

Courses: Classical cooking, restaurant baking, dining room service, restaurant management. Other required courses: math, English, computers, business. Continuing education: English composition, computers, business management.

Faculty: 2 full-time, 1 part-time.

Costs: Annual tuition: in-state $1,488, out-of-state $5,763. $50 lunch fee quarterly. Average off-campus housing cost $275 per month. Admission requirements: High school diploma or equivalent. Loans: yes.

Location: Puget Sound (west), 60 miles from Seattle or 1-hour ferry ride.

Contact: Steve Lammers, Chef Instructor, Olympic College, Commercial Cooking/Food Service, 16th and Chester, Bremerton, WA 98310-1688 US; 360-478-4576, Fax 360-478-4650, E-mail njgiovan@olympic.ctc.edu, URL http://www.olympic.ctc.edu.

RENTON TECHNICAL COLLEGE

Renton/Year-round

This career college offers a 1,620-hour certificate/AAS degree in Culinary Arts/Chef and Culinary Arts/Baker. Program started 1968. Accredited by ACFEI, NASC. Admission dates: Open. Total enrollment: 30. 10-30 each admission period. 100% of applicants accepted. 30% receive financial aid. Student to teacher ratio 12:1. 100% of graduates obtain employment within six months. Facilities: Include kitchen, bakery, demonstration classroom and 3 restaurants.

Courses: Externship provided.

Faculty: 2 instructors and 5 assistants full-time. Includes Chef Instructor David Pisegna, CEC, CCE, and Baking Instructor Erhard Volcke, CMB.

Costs: Tuition $640 per quarter. Other costs for books, uniforms. Admission requirements: High school diploma or equivalent and admission test.

Location: 20 minutes south of Seattle.

Contact: David Pisegna, Executive Chef Instructor, Renton Technical College, Culinary Arts, 3000 N.E. Fourth St., Renton, WA 98056 US; 425-235-2352, Fax 425-235-7832, E-mail DPisegna@ctc.edu, URL http://www.ctc.edu.

ROBERT REYNOLDS, NORTHWEST FORUM (See page 104)

SEATTLE CENTRAL COMMUNITY COLLEGE

Seattle/Year-round

This two-year college offers a 6-quarter Culinary Arts certificate, 2-year Culinary AAS degree, a 4-quarter Specialty Desserts and Breads certificate, 7-quarter Hospitality Management AA degree. Program started 1942. Accredited by ACFEI. Calendar: quarter. Curriculum: culinary. Admission dates: Quarterly. Total enrollment: 100-125. 30 each admission period. 85% of applicants accepted. 30% receive financial aid. 2% enrolled part-time. Student to teacher ratio 20:1. 97% of graduates obtain jobs within six months. Facilities: 3 kitchens, 8 classrooms, cafeteria, cafe, gourmet restaurant.

Courses: Professional cooking, restaurant cooking, baking, specialty desserts and breads, nutrition, buffet catering, costing, computerized menu planning, management, ice carving. Last quarter internship in local restaurant, catering business, or hotel. Occasional nutrition courses for culinary professionals; summer classes in breads and chocolate.

Faculty: 7 full-time, 2 part-time. Includes Keijiro Miyata, CEC, Linda Hierholzer, CCE, Diana Dillard, CIA graduate, David Madayag, CEC, John Balmores, Cynthia Wilson, Deb Hermansen, Regis Bernard, Don Reed.

Costs: In-state tuition $460 per quarter, out-of-state $1,830 per quarter. Off-campus housing cost is $300-$500 per month. Admission requirements: Admissions test or college transcripts for English & Math skills. Scholarships: yes. Loans: yes.

Location: Campus is located in Capitol Hill district of Seattle, a short walk to downtown.

Contact: Joy Gulmon-Huri, Program Manager, Seattle Central Community College, Hospitality & Culinary Arts, 1701 Broadway, Mailstop 2BE2120, Seattle, WA 98122 US; 206-587-5424, Fax 206-344-4323, E-mail jgulmo@sccd.ctc.edu, URL http://seaccd.sccd.ctc.edu/~cculhosp/index.html.

SKAGIT VALLEY COLLEGE

Mt. Vernon/September-May

This college offers a 1-year certificate and 2-year ATA degree in Culinary Arts/Hospitality Management. Program started 1979. Accredited by State, ACF. Calendar: quarter. Curriculum: core. Admission dates: Open. Total enrollment: 60. 6 each admission period. 100% of applicants accepted. 60% receive financial aid. Student to teacher ratio 15:1. 100% of graduates obtain employment within six months. Facilities: Include kitchen, classrooms, restaurant.

Courses: Externship provided.

Faculty: 3 full-time.

Costs: Annual tuition in-state $1,125, out-of-state $5,500. Admission requirements: High school diploma or equivalent.

Location: 60 miles north of Seattle.

Contact: Lyle Hildahl, Director, Skagit Valley College, Culinary Arts/Hospitality Management, 2405 College Way, Mt. Vernon, WA 98273 US; 360-416-7618, Fax 360-416-7890.

SOUTH PUGET SOUND COMMUNITY COLLEGE

Olympia/September-June

This two-year college offers a 2-year ATA degree, Food Service Tech. and Food Service Management. Program started 1989. Accredited by State. Calendar: quarter. Curriculum: culinary. Admission dates: September, January, April. Total enrollment: 40. 12 each admission period. 85% of applicants accepted. 60% receive financial aid. 10% enrolled part-time. Student to teacher ratio 12-15:1. 95% of graduates obtain employment within six months. Facilities: Bake shop, institutional foods, gourmet cooking, table-side cooking.

Courses: Lab classes are a reflection of the industry. W.S.U.

Faculty: 2 full-time, 4 part-time.

Costs: Annual tuition in-state $48 per credit hour, out-of-state $192 per credit hour. Admission requirements: High school diploma or equivalent and admission test. Scholarships: yes. Loans: yes.

Contact: Fred Durinski, Food Service Director, South Puget Sound Community College, Food Service Technology, 2011 Mottman Rd., SW, Olympia, WA 98512 US; 360-754-7711 x347, Fax 360-664-0780, E-mail twoodnut@spscc.ctc.edu, URL http://www.spscc.ctc.edu. Additional contact: Bill Wiklend.

SOUTH SEATTLE COMMUNITY COLLEGE

Seattle/Year-round

This two-year college offers an 18-month certificate/AAS degrees in Culinary Arts/Food Service Production and Pastry/Specialty Baking. Program started 1975. Accredited by ACFEI, NASC. Calendar: quarter. Curriculum: culinary, core. Admission dates: September, January, March, June. Total enrollment: 130-160. 30-40 each admission period. 100% of applicants accepted. 25% receive financial aid. Student to teacher ratio 15:1. 98% of graduates obtain employment within six months. Facilities: Include 4 kitchens, 6 classrooms and 2 waited-service dining rooms.

Courses: Quantity, fine dining and casual food production; professional dining room service; hospitality supervision and management; pastry and specialty baking.

Faculty: 7 full-time, 8 part-time. Qualifications: extensive industry experience.

Costs: Annual tuition in-state $3,800. Off-campus housing cost $400-$500 monthly. Scholarships: yes. Loans: yes.

Location: The 35-acre campus is in a suburban setting.

Contact: Daniel Cassidy, Associate Dean, South Seattle Community College, Hospitality & Food Science Div., 6000 16th Ave. S.W., Seattle, WA 98106-1499 US; 206-764-5344, Fax 206-768-6728, E-mail dcassidy@ssccmail.sccd.ctc.edu, URL http://www.chefschool.com.

SPOKANE COMMUNITY COLLEGE

Spokane/September-June

This two-year college offers a 2-year AAS degree in Culinary Arts, Pastry, and Baking. Program started 1962. Accredited by NASC, ACFEI. Calendar: quarter. Curriculum: culinary, core. Admission dates: September, January, March. Total enrollment: 75-100. 30 each admission period. 95% of applicants accepted. 40% receive financial aid. Student to teacher ratio 20:1. 90% of graduates obtain employment within six months. Facilities: Include 2 kitchens, bakeshop, pastry shop, 6 classrooms, restaurant.

Courses: Externship: 3-6 months, in area hotels and restaurants.

Faculty: 4 full-time.

Costs: Annual tuition approximately $1,400. Admission requirements: High school diploma or equivalent and admission test. Scholarships: yes. Loans: yes.

Contact: Doug Fisher, Program Coordinator, Spokane Community College, Culinary Arts, 1810 N. Greene St., Spokane, WA 99207 US; 509-533-7284/509-533-7372, Fax 509-533-8059, E-mail dfisher@ctc.edu, URL http://www.scc.spokane.cc.wa.us.

WEST VIRGINIA

SHEPHERD COMMUNITY COLLEGE

Shepherdstown/Year-round

This two-year college offers a 2-year AAS degree in Culinary Arts. Calendar: semester. Curriculum: core.

Costs: Annual tuition is $2,228 in-state, $5,348 out-of-state. Room and board is $4,139 per year.

Contact: Paul Saab, Shepherd Community College, Shepherdstown, WV 25443 US; 304-876-5212/5203, URL http://www.shepherd.wvnet.edu/ctcweb.

SYMPOSIUM FOR PROF. FOOD WRITERS AT THE GREENBRIER

White Sulphur Springs/March

This resort offers a 3-day conference for professional food writers that consists of lectures, seminars, informal discussions, receptions. 90 each admission period. Facilities: The Mobil 5-star, AAA 5-diamond Greenbrier resort.

Courses: Topics include food writing for newspapers and magazines, recipe development and writing for cookbooks, culinary history, and food writing for film.

Faculty: Over a dozen noted professional food writers, publishers and editors.

Location: The Allegheny Mountains, 15 minutes from the airport in Lewisburg, 75 miles from Roanoke, Va. Amtrak service available.

Contact: Townley Aide, Symposium Coordinator, Symposium for Professional Food Writers, The Greenbrier, 300 W. Main St., White Sulphur Springs, WV 24986 US; 800-624-6070/304-536-7892, Fax 304-536-7893.

WEST VIRGINIA NORTHERN COMMUNITY COLLEGE

Wheeling

This two-year college offers a 2-year certificate/AAS degree. Program started 1975. Accredited by NCA. Admission dates: Open. Total enrollment: 26. 100% of applicants accepted. Student to teacher ratio 9:1. 85% of graduates obtain employment within six months.

Faculty: 3 full-time.

Costs: Tuition in-state $1,486 per year, out-of-state $2,039 per year plus books. Admission requirements: High school diploma or equivalent and admission test.

Contact: James Panacci, West Virginia Northern Community College, Culinary Arts, College Square, Wheeling, WV 26003 US; 304-233-5900, URL http://www.northern.wvnet.edu.

WISCONSIN

BLACKHAWK TECHNICAL COLLEGE

Janesville/Year-round

This college offers a 1-year/34 credit-hour certificate and 2-year/68 credit-hour AS degree in Culinary Arts. Established in 1972. Accredited by ACFEI. Calendar: semester. Total enrollment: 32. 16 each admission period. 25% enrolled part-time. Student to teacher ratio 8:1. 98% of graduates obtain employment within six months. Facilities: Modern, well-equipped facility, student-run gourmet restaurant.

Courses: Training by area professionals. Externships are provided.

Faculty: 1 full-time, 4 part-time.

Costs: $4,450 per year in-state plus books and uniforms. Scholarships: yes. Loans: yes.

Location: South-central Wisconsin.

Contact: Joe Wollinger, CEC, CCE, Program Coordinator, Blackhawk Technical College, 6004 Prairie Rd., PO Box 5009, Janesville, WI 53547 US; 608-757-7730, Fax 608-757-9407, URL http://www.blackhawk/tec.wi.us.

FOX VALLEY TECHNICAL COLLEGE

Appleton/September-May

This independent career college offers a 2-year diploma/degree in Culinary Arts and Food Service Production. Program started 1972. Accredited by NCA, ACF. Calendar: semester. Curriculum: core. Admission dates: Fall, winter. Total enrollment: 70. 40 each admission period. 90% of applicants accepted. 60-80% receive financial aid. 50% enrolled part-time. Student to teacher ratio 6:1. 100% of graduates obtain jobs within six months. Facilities: 5 kitchens and classrooms including full quantity production kitchen, full bakery, full restaurant kitchen, and student-run restaurant.

Courses: Quantity production, catering, restaurant cooking, baking, deli operations, and general education courses. Externship provided.

Faculty: 5 full-time. Includes R. Kimball, A. Exenberger, M. Lang, C. Hribal, J. Igel.

Costs: Annual tuition in-state $2,100, out-of-state $10,000. Application fee $20. On-campus housing average cost: $250 per month. Average off-campus housing (shared apartments) cost $150 per month. Admission requirements: High school diploma or equivalent and admission test. Scholarships: yes.

Contact: Chef Jeff, Culinary Arts Manager, Fox Valley Technical College, Culinary Arts, 1825 N. Bluemound Dr., PO Box 2277, Appleton, WI 54913 US; 414-735-5643, Fax 414-831-5410, E-mail igel@foxvalley.tec.wi.us, URL http://www.foxvalley.tec.wi.us.

MADISON AREA TECHNICAL COLLEGE

Madison/August-May

This career college offers a 2-year AAS degree in Culinary Arts. Program started 1950. Accredited by ACFEI. Calendar: semester. Curriculum: core. Admission dates: August, January. Total enrollment: 60. 36 each admission period. 75% of applicants accepted. 50% receive financial aid. 20% enrolled part-time. Student to teacher ratio 15:1. 100% of graduates obtain employment within six months. Facilities: Include 3 large labs and classrooms.

Courses: Baking, sanitation, nutrition, gourmet foods, decorative foods, food costs and purchasing analysis, and general education. Limited courses available.

Faculty: Qualifications: certified by state and ACFEI. Includes D. McNicol, M. Egan, P. Short.

Costs: Tuition in-state $54.20 per credit hour. Advanced registration fee is $50. Application fee $25. Average off-campus housing cost is $400-$870. Admission requirements: High school diploma or equivalent and assessment test. Scholarships: yes. Loans: yes.

Contact: Mary G. Hill, Associate Dean, Madison Area Technical College, Culinary Trades Dept., 3550 Anderson St., Madison, WI 53704 US; 608-243-4455, Fax 608-246-6316, E-mail mhill@madison.tec.wi.us, URL http://ted.ele.madison.tec.wi.us.

MILWAUKEE AREA TECHNICAL COLLEGE

Milwaukee

This career college offers a 2-year AAS degree. Program started 1955. Accredited by NCA, ACFEI. Calendar: semester. Curriculum: culinary, core. Admission dates: August, January. Total enrollment: 150. 80 each admission period. Student to teacher ratio 20:1. 98% of graduates obtain employment within six months.

Faculty: 11 full-time.

Costs: Annual tuition in-state $2,260, out-of-state $8,244. Admission requirements: High school diploma or equivalent and admission test.

Contact: Barbara Cannell, Culinary Director, Milwaukee Area Technical College, Culinary Arts, 700 W. State St., Milwaukee, WI 53233 US; 414-278-6836, Fax 414-297-7733.

MORAINE PARK TECHNICAL COLLEGE

Fond du Lac/August-May

This career college offers a 2-year AA degree in Culinary Arts, a 1-year technical diploma in Food Service Production, Culinary Basics certificate, Deli/bakery certificate, Food Production certificate,

School Food Service certificate. Program started 1980. Accredited by NCA. Calendar: semester. Curriculum: core. Admission dates: July-August, November-December, flexible. Total enrollment: 36. 16 each admission period. 90% of applicants accepted. 40% receive financial aid. 30% enrolled part-time. Student to teacher ratio 10:1. 95% of graduates obtain employment within six months. Facilities: Include 3 kitchens, 2 classrooms.

Courses: Food production, sanitation, meat analysis, restaurant management, catering. Other required courses: various general education (270 hours). Continuing education: school food service, deli-bakery, IDDA certification.

Faculty: 3 full-time, 2 part-time. Includes Ron Speich, David Weber.

Costs: Tuition for Culinary Arts degree is $3,057 per year including books, $751 for certificate program, $3,001 for 1-year diploma. Off-campus housing cost $300 per month. Admission requirements: High school diploma, placement test, and interview. Scholarships: yes. Loans: yes.

Location: East central Wisconsin.

Contact: Donna Leet, Dean, Human Services & Hospitality, Moraine Park Technical College, Human Services & Hospitality, 235 N. National Ave., PO Box 1940, Fond du Lac, WI 54936-1940 US; 414-924-3289, Fax 414-929-2478, E-mail dleet@moraine.tec.wi.us, URL http://www.job.careernet.org.

NICOLET AREA TECHNICAL COLLEGE

Rhinelander/August-May

This career college offers a 1-year diploma in Food Service Production, 2-year associate degree in Culinary Arts, certificate in Baking, Catering, Kitchen Assistant, Food Service Management, School Food Service Assistant. Accredited by NCA. Calendar: semester. Curriculum: core. Admission dates: Fall. Total enrollment: 15 per program. 15 each admission period. 85% receive financial aid. 10% enrolled part-time. Student to teacher ratio 10:1. 90% of graduates obtain employment within six months.

Courses: Culinary fundamentals for restaurant and institutional cooking. Internships encouraged in summer between 1st and 2nd year.

Faculty: Linda Arndt, BS, MS, trained in culinary arts at La Varenne, Kyle M. Gruening, BS, MS.

Costs: Off-campus housing is available, cost varies. Admission requirements: High school diploma or equivalent, basic competency scores on ASSET. Scholarships: yes. Loans: yes.

Contact: Linda Arndt, Culinary Instructor, Nicolet Area Technical College, Culinary Arts, PO Box 518, Rhinelander, WI 54501 US; 715-365-4446, Fax 715-365-4596, E-mail larndt@nicolet.tec.wi.us, URL http://www.nicolet.tec.wi.us/.

POSTILION SCHOOL OF CULINARY ART

Fond du Lac/Year-round

This private school offers a diploma course consisting of four 100-hour participation sessions that emphasize classic French technique, economy, building a chef's larder. Established in 1951. Accredited by State of Wisconsin. Student to teacher ratio 8 students. Facilities: The professionally constructed teaching kitchen of a Victorian home.

Courses: 2 weeks (100 hours minimum) each of basic, advanced, menu planning, and cost accounting. A catering course is optional. Other Courses: professional pastry, butchering, sausage making, ice cream.

Faculty: Owner/instructor Mme. Kuony was educated in Belgium, France, and Switzerland.

Costs: Each 2-week segment is $1,600, which includes meals. Nonrefundable $200 registration fee required. Inexpensive lodging is available at nearby motels. Admission requirements: Students must begin with the basic class, regardless of experience.

Location: On the south side of Fond du Lac at the southern tip of Lake Winnebago, about an hour from Milwaukee and 120 miles from Chicago.

Contact: Mme. Liane Kuony, Owner, The Postilion School of Culinary Art, 220 Old Pioneer Rd., Fond du Lac, WI 54935 US; 414-922-4170.

WAUKESHA COUNTY TECHNICAL COLLEGE

Pewaukee/August-May

This career college offers a 1-year diploma and 2-year associate degree in Culinary Management., 3-year ACF apprenticeship in Culinary Arts. Program started 1971. Accredited by NCA, ACFEI. Calendar: quarter. Admission dates: August, January. Total enrollment: 65. 30 each admission period. 90% of applicants accepted. 40% receive financial aid. 35% enrolled part-time. Student to teacher ratio 12:1. 95% of graduates obtain employment within six months. Facilities: Include 3 kitchens, 4 classrooms, beverage lab, restaurant lab, computer lab.

Courses: Technical culinary arts training and principles of business management. Internship: semester long, working under certified ACF chef.

Faculty: 4 full-time, 2 part-time. Qualifications: All have college degrees and industry experience. Includes James Holden, CEC, CCE, Timothy Graham, CFBE, FMP, Keith Owsiany, Michael Leitzke, CEC, Jack Kaestner, CEC.

Costs: Annual tuition: in-state $54.20 per credit hour, out-of-state $80 per credit hour. Cutlery $150, uniforms $100. Scholarships: yes. Loans: yes.

Contact: William R. Griesemer, Associate Dean, Waukesha County Technical College, Center for Culinary Arts Studies, 800 Main St., Pewaukee, WI 53072 US; 414-691-5254, Fax 414-691-5155, E-mail WGriesemer@waukesha.tec.wi.us, URL http://www.waukesha.tec.wi.us.

AUSTRALIA

CANBERRA INSTITUTE OF TECHNOLOGY

Canberra City/Year-round

This career institute offers a 6-month certificate, 3-year diploma and 3-year part-time trade certificate. Program started 1992. Calendar: semester. Curriculum: culinary, core. Admission dates: February, July. Total enrollment: 450. 75 each admission period. 50% of applicants accepted. 50% enrolled part-time. Student to teacher ratio 15:1. 100% of graduates obtain employment within six months. Facilities: 6 kitchens, 4 restaurants, computer lab, butchery, bakery, bars.

Courses: Practical hands-on approach addressing industry needs.

Faculty: 30 full-time, 50 part-time. Qualifications: industry and educational.

Costs: Available on request.

Contact: John Wardrop, Head, Culinary Skills, Canberra Institute of Technology, School of Tourism & Hospitality, PO Box 826, Canberra City, 2601 Australia; (61) (0)6-2073184, Fax (61) (0)6-2073209, E-mail gordon.mcdonald@cit.act.edu.au, URL http://www.cit.act.edu.au/faculty/t&h/school/main.htm. Additional contact: Gordon McDonald.

CASEY INSTITUTE OF TECHNICAL AND FURTHER EDUCATION

Dandenong/Year-round

This institute offers an 8-week certificate one, 20-week certificate two, 3-year certificate three. Program started 1986. Accredited by National. Calendar: semester. Curriculum: culinary, core. Admission dates: February, April, July, October. Total enrollment: 250. 80% of applicants accepted. 75% enrolled part-time. Student to teacher ratio 15-20:1. 100% of graduates obtain employment within six months. Facilities: 4 kitchens, including one fully-equipped commercial kitchen.

Courses: Cookery, management and short courses.

Faculty: 25 full-time, 15 part-time.

Costs: A$500 in-state, A$7,000 out-of-state.

Location: 30 minutes from CBD of Melbourne.

Contact: Geoffrey Loosmore, Culinary Program Manager, Casey Institute of Technical and Further Education, School of Hospitality & Tourism, PO Box 684, Dandenong, Victoria, 3175 Australia; (61) (0)3-9212-5417, Fax (61) (0)3-9212-5459, E-mail gloosmore@casey.vic.edu.au, URL http://www.casey.vic.edu.au.

CROW'S NEST COLLEGE OF TAFE

Sydney/February-November

The New South Wales government offers a 1-year certificate in Asian cooking and Western cooking. Program started 1989. Accredited with vocational training board and TAFE Commission. Calendar: semester. Curriculum: culinary. Admission dates: January, July. Total enrollment: 150. 50 each admission period. 90% of applicants accepted. 5% receive financial aid. 60% enrolled part-time. Student to teacher ratio 15:1. 90% of graduates obtain employment within six months. Facilities: Include 2 kitchens, 4 classrooms and commercial bar/cellar, 2 dining rooms, coffee shop, food science lab, computer labs, library, learning resource ctr.

Courses: Professionally equipped Chinese/Asian kitchen. Industry-specific courses, fee-for-service customized programs.

Faculty: 6 full-time, 12 part-time; trade qualifications and education degrees.

Costs: A$350 per year. Admission requirements: Education certificate.

Contact: Geoff Tyrrell, Senior Head Teacher, Crow's Nest College of Tafe, Tourism & Hospitality, 149 West St., Crows Nest, Sydney, NSW, 2065 Australia; (61) (0)2-99654433, Fax (61) (0)2-99654408, E-mail geoff.tyrrell@tafensw.edu.au, URL http://www.tafensw.edu.au.

LE CORDON BLEU – SYDNEY

(See also page 154) (See display ad page 149) **Sydney/Year-round**

This private school opened at Northern Sydney Institute of TAFE, one of 5 locations worldwide, offers Cuisine and Pastry Certificates: 10-week courses in basic and intermediate levels (Superior levels only at Paris, London, and Tokyo schools), nonvocational courses. Established in 1996. Accredited by Through international hospitality management schools. Calendar: trimester. Curriculum: culinary. Facilities: Professionally equipped kitchens; individual workspaces with refrigerated marble tops; demonstration rooms with video.

Courses: Two levels of Cuisine, two levels of Pastry, taken consecutively or together. Courses cover basic-complex technique; classic, regional, ethnic, contemporary cuisines; wine and food pairing; presentation; decoration and execution.

Faculty: School Director is Ted Davis. Culinary staff consists of French and Australian Master Chefs from Michelin-starred restaurants and fine hotels.

Costs: From $100 for a full-day course to $5,250 for 10 weeks.

Contact: Ted Davis, Director, Le Cordon Bleu, Ryde College of TAFE, 250 Blaxland Rd., Ryde, NSW, 2112 Australia; (61) (0)2-808-8307, Fax (61) (0)2-809-3346, URL http://cordonbleu.net. US toll-free phone: 800-457-CHEF.

WILLIAM ANGLISS INSTITUTE

(See display ad page 140) **Melbourne/Year-round**

This career institute specializing in the hospitality, travel and food industries offers certificate programs in Commercial Cookery and Advanced Culinary Skills, diploma and advanced diploma in Hospitality. Students can link to degree courses at Victorian universities. Program started 1940. Accredited by national, state, and local agencies. Calendar: semester. Curriculum: core. Admission dates: Every 1 or 2 months (commercial cookery), 3 times a year (cookery), February and July (advanced course). Total enrollment: 4,000, varies each admission period. Student to teacher ratio 15:1. 91% of graduates obtain employment within six months. Facilities: $25 million teaching facility with 4 well-equipped bakeries, 6 kitchens, 3 restaurants, bars, computer rooms, butchery and confectionery centers.

Courses: Apprenticeship: $3\frac{1}{2}$ to 4 years in breadmaking and baking, pastry, cookery and butchering. Continuing education: evening courses in cooking and wine, accreditation available to degree courses.

Faculty: More than 100 full-time. Includes Director Dr. Christine French.

Costs: Courses vary in cost depending on length. Admission requirements: Cookery open to all applicants over 18. Advanced course open to those who have completed an apprenticeship or secondary education.

Contact: Chris Coates, Associate Director, William Angliss Institute, 555 La Trobe St., PO Box 4052, Melbourne, 3000 Australia; (61) (0)3-96062111, Fax (61) (0)3-96701330, URL http://www.angliss.vic.edu.au.

CANADA

ALGONQUIN COLLEGE
Nepean/Year-round

This college offers a 2-year diploma in Culinary Management, 1-year certificate in Cook Training, 40-week certificate in Baking Techniques. Program started 1960. Calendar: semester. Curriculum: core. Admission dates: September, January. Total enrollment: 140. 90 and 50 each admission period. 50% of applicants accepted. Student to teacher ratio 15-20:1. 90% of graduates obtain employment within six months. Facilities: Include 2 production kitchens, 3 demonstration labs.

Courses: Baking, menu planning, food demonstration and applications, institutional cooking, food and beverage control, management and computer applications. Continuing education: cake decorating, bread baking, Italian regional cooking.

Faculty: 5 full-time, 5 part-time. Includes Philippe Dubout, Mike Durrer, Serge Desforges, Alain Peyrun-Berron, Roger Souffez, Alan Fleming, Mario Ramsay.

Costs: Annual tuition in-state C$637.50 per semester, out-of-state C$4,607.50 per semester. Books, supplies, uniforms: C$850. Admission requirements: Secondary school diploma or 19 years of age.

Contact: Mike Durrer, Coordinator, Cook/Culinary Programs, Algonquin College, Admissions Office, 1385 Woodroffe Ave., Nepean, Ottawa, ON, K2G 1V8 Canada; 613-727-0002, Fax 613-727-7632, E-mail DurrerM@algonquinc.on.ca, URL http://www.algonquinc.on.ca.

CANADORE COLLEGE OF APPLIED ARTS & TECHNOLOGY
North Bay/September-April *(See also page 275)*

This college offers a 2-year diploma in Culinary Management, 3-year diploma in Culinary Administration, 1-year certificate in Chef Training. Program started 1984. Accredited by Canadian Federation of Chefs de Cuisine. Calendar: semester. Curriculum: culinary. Admission dates: September. Total enrollment: 50. 50 first year, 20 second year each admission period. 10% of applicants accepted. 75% receive financial aid. 10% enrolled part-time. Student to teacher ratio 20:1. 90-100% of graduates obtain employment within six months. Facilities: Include kitchen with specialized equipment, restaurant.

Courses: Food preparation, baking, sanitation, food and beverage management, nutrition, wines, contemporary cuisine, cost control, menu planning, quantity cooking, garde manger, international cuisine. Externship provided.

Faculty: 8 full-time.

Costs: Annual tuition in-state C$1,700, foreign C$8,900. Admission requirements: High school diploma or equivalent. Scholarships: yes.

Contact: Daniel Esposito, Professor/Advisor, Canadore College, School of Hospitality &

Tourism, 100 College Dr., PO Box 5001, North Bay, ON, P1B 8K9 Canada; 705-474-7600, Fax 705-494-7462, E-mail espositd@canadorec.on.ca, URL http://www.canadorec.on.ca.

CULINARY INSTITUTE OF CANADA

Charlottetown/Year-round

This two-year career school, a division of Atlantic Tourism & Hospitality Institute, offers an 80-week diploma program. Established in 1983. Calendar: trimester. Curriculum: culinary, core. Admission dates: September, April. Total enrollment: 190. 80 September, 30 April each admission period. 60% of applicants accepted. 90% receive financial aid. 5% enrolled part-time. Student to teacher ratio 16:1. 95% of graduates obtain employment within six months. Facilities: Include 6 training kitchens, 14 classrooms, 6 labs, 2 restaurants.

Courses: 75% practical. A 16-week externship is included in the program. Continuing education: short courses available.

Faculty: 40.

Costs: Annual tuition C$8,955 ($9,900 U.S.). Average off-campus housing cost is C$500 per month. Admission requirements: High school diploma. Scholarships: yes.

Contact: Richard MacDonald, Executive Director, Culinary Institute of Canada, 4 Sydney St., Charlottetown, PEI, C1A 1E9 Canada; 902-894-6899, Fax 902-894-6801, E-mail dmacdonald@athi.pe.ca, URL http://www.athi.pe.ca. Additional contact: David Harding.

DUBRULLE INTERNATIONAL CULINARY & HOTEL INSTITUTE

(See also page 277) **Vancouver/Year-round**

This private school offers a 17-week (34-week) Professional Culinary Training (evening) program, 17-week Professional Pastry and Desserts program, Advanced Culinary Training program, 1-year Hospitality Management diploma program, 2-year Culinary and Business Management diploma program. Established in 1982. Accredited by Canadian Education and Training Accreditation Commission. Calendar: trimester. Curriculum: culinary. Admission dates: January, April/May, July (Breadmaking), September/October. Total enrollment: 36. 72 each admission period. 90% of applicants accepted. Student to teacher ratio 12:1. 75%-85% of graduates obtain employment within six months. Facilities: The 6,000-square-foot facility has classrooms, 3 teaching kitchens with fully-equipped working stations, student dining areas.

Courses: Emphasis is on classic French methods and techniques; 80% practical, 20% theory. Accreditation may be given towards the B.C. Apprenticeship program.

Faculty: 5 classically-trained chef-instructors.

Costs: Culinary and Pastry programs C$7,050 each. Accommodation assistance is available. Admission requirements: Grade 10, 19 years of age. Scholarships: yes. Loans: yes.

Location: A block from Broadway and Granville, 3 hours from Seattle.

Contact: Robert Sung, Director of Admissions, Dubrulle International Culinary & Hotel Institute of Canada, 1522 W. 8th Ave., Vancouver, BC, V6J 4R8 Canada; 604-738-3155/800-667-7288, Fax 604-738-3205, E-mail cooking@dubrulle.com, URL http://www.dubrulle.com.

GEORGE BROWN COLLEGE OF APPLIED ARTS & TECHNOLOGY

Toronto/Year-round

This college offers 20 1- and 2-year full-time certificate and diploma programs, including 2-year Culinary Management, 1-year Chef Training, 1-year Culinary Arts, 1-year Baking and Pastry Arts, Prof. Sommelier, Chinese Cooking, Italian (post-diploma). 30 part-time prog. Program started 1965. Accredited by Ontario College Standards and Accreditation. Calendar: semester. Curriculum: culinary, core. Admission dates: September, January annually. Some programs may begin at other times as announced. Total enrollment: 2,500 full-time, 3,500 part-time. Varies each admission period. Percentage of applicants accepted varies. 50% receive financial aid. 10% enrolled part-time. Student to teacher ratio 24:1. 85% of graduates obtain employment within six months. Facilities: Include 6 culinary and 3 bake laboratories, each with 24 individual work stations; demonstration kitchens with overhead mirrors.

Courses: Culinary Management: basic, classical food preparation and presentation with theory, demonstrations, preparations for the student-run restaurant. Baking and Pastry Arts: breads, pastries, cakes, decorating. 3 weeks of scheduled industry experience in hotels and restaurants. 3,500 students in 40 different courses.

Faculty: More than 35 full-time internationally-trained chef and pastry professors as well as 30 full-time former hotel general managers and food and beverage professionals. Includes Canadian wine expert Jacques Marie.

Costs: Resident (nonresident) tuition for most diploma programs is approximately C$2,400 (C$10,000) for 32 weeks; for certificate programs it's C$544 to C$1,694 per program. Furnished room averages C$200-C$250 per month. Admission requirements: Minimum requirement for admission to a diploma program is an Ontario Secondary School Diploma or an equivalent from within North America. Scholarships: yes.

Location: The 4-story facility, completed in 1987, is in downtown Toronto at 300 Adelaide St. E. at the college's St. James campus.

Contact: Ron Thompson, Dean, George Brown College, Hospitality/Tourism Centre, 300 Adelaide St. E., Toronto, ON, M5A 1N1 Canada; 800-263-8995/416-415-2230, Fax 416-415-2501, URL http://www.gbrownc.on.ca. Additional contact: Dan Borrowec (same address).

GEORGIAN COLLEGE OF APPLIED ARTS AND TECHNOLOGY

Barrie/September-April

This college offers a 2-year diploma in Culinary Management. Program started 1988. Admission dates: August. Total enrollment: 85. 50 each admission period. 50% of applicants accepted. Student to teacher ratio 24:1. 100% of graduates obtain employment within six months. Facilities: Include 1 large-quantity kitchen, 2 small-quantity kitchens, bake lab, classrooms, student-run restaurant.

Courses: Bake theory/lab, menu planning, food/beverage control, creative cuisine.

Faculty: 7 full-time, 1 to 2 part-time.

Costs: Annual tuition in-country C$2,200, out-of-country C$10,000. Application fee C$50. On-campus housing: 252 spaces; average cost: C$405 per month. Average off-campus housing cost: C$400-800. Admission requirements: High school diploma or equivalent.

Location: The E-building, 3,500-student campus is in an urban/rural area.

Contact: Chris Cutler, Coordinator, Georgian College, Department of Hospitality & Tourism, One Georgian Dr., Barrie, ON, L4M 3X9 Canada; 705-728-1968 x1280, Fax 705-722-5135, E-mail ccutler@central.georcoll.on.ca, URL http://www.georcoll.on.ca.

HUMBER COLLEGE OF APPLIED ARTS & TECHNOLOGY

Etobicoke

This career college offers a 1-year certificate, 2-year diploma, 3-year AS degree. Program started 1975. Admission dates: September, January. Total enrollment: 150. Student to teacher ratio 20:1. 95% of graduates obtain employment within six months.

Courses: Externship provided.

Faculty: 14 full-time, 5 part-time.

Costs: Annual tuition in-state C$1,530, foreign C$9,600 plus C$292 fees. Admission requirements: High school diploma or equivalent and admission test.

Contact: John Walker, Chairman, Humber College, School of Hospitality, Tourism & Leisure, 205 Humber College Blvd., Etobicoke, ON, M9W 5L7 Canada; 416-675-3111, Fax 416-675-9730, URL http://www.humber.on.ca.

LAMBTON COLLEGE

Sarnia/September-April

This college offers a 2-year program in Culinary Management, cook apprenticeship. Established in 1967. Accredited by Ontario Ministry of Education and Training. Calendar: semester. Curriculum:

culinary. Admission dates: March. Total enrollment: 60. 35 each admission period. 50% of applicants accepted. 65% receive financial aid. 15% enrolled part-time. Student to teacher ratio 12:1. 100% of graduates obtain employment within six months. Facilities: Teaching kitchen, 45-seat restaurant, labs, classrooms.

Courses: Focus on basics with emphasis on contemporary cuisine.

Faculty: 1 full- and 5 part-time instructors, all certified chefs.

Costs: Tuition C$1,800, books C$200, uniforms and knives C$500. Admission requirements: High school graduate or mature student. Scholarships: yes. Loans: yes.

Contact: Bob Henry, Dean, Lambton College Culinary Programs, 1457 London Rd., Sarnia, ON, N7T 6K4 Canada; 519-542-7751, Fax 519-542-6696, E-mail BHenry@lambton.on.ca, URL http://www.lambton.on.ca.

LE CORDON BLEU OTTAWA CULINARY ACADEMY

(See also page 154) (See display ad page 149) **Ottawa/Year-round**

This private school offers 10-week certificate courses in French cuisine and pastry (4 of the 6 Classic Cycle courses), 1-day to 1-month intensives; specialized pastry programs, catering courses, evening classes. Established in 1988. Calendar: trimester. Curriculum: culinary. Admission dates: March, June, October. Total enrollment: 150-200. 50 each admission period. 90% of applicants accepted. Student to teacher ratio 12:1. 75% of graduates obtain jobs within six months. Facilities: demonstration room, fully-equipped kitchen with individual work spaces, specialized equipment.

Courses: Basic and Intermediate Cuisine, Basic and Intermediate Pastry. Students who wish to receive the Cuisine/Pastry Diplomas or Le Grand Diplome can complete the Superior Cuisine and/or Patisserie courses at Le Cordon Bleu in Paris or London. Specialized pastry workshops.

Faculty: All French chefs trained in France. Instructors are professional master chefs Philippe Guiet, Jean-Claude Petibon, pastry chefs Jean Michel Poncet, former member of the Canadian National Culinary Olympic Team, and Michel Denis.

Costs: Tuition is C$4,850 for Basic Cuisine, C$5,100 for Intermediate Cuisine, C$4,350 for Basic Pastry, C$4,850 for Advanced Pastry. Lodging ranges from C$600-C$1,200. Admission requirements: High school diploma, minimum age 18 years old. Scholarships: yes.

Location: Chateau Royale Professional Building, about 20 minutes from downtown Ottawa.

Contact: Sandra MacInnis, Administrative Manager, Le Cordon Bleu Ottawa Culinary Academy, 1390 Prince of Wales Dr., #400, Ottawa, ON, K2C 3N6 Canada; 613-224-8603/800-457-CHEF, Fax 613-224-9966, E-mail rhanna@cordonbleu.net, URL http://cordonbleu.net.

MALASPINA UNIVERSITY-COLLEGE

Nanaimo/Year-round

This university college offes a 12-month certificate program in Cook Training. Established in 1968. Curriculum: culinary. Admission dates: January, March, August, October. Total enrollment: 110. 110 each admission period. 95% of applicants accepted. 9% receive financial aid. 0% enrolled part-time. Student to teacher ratio 18:1. 100% of graduates obtain employment within six months.

Courses: Eight weeks each on: diet and nutrition, breakfast and meat cutting, meat/poultry/seafood, garde manger/vegetables/starch, pastry/desserts, a la carte. One hour theory/6 hours kitchen lab daily. 2-week live-in practicum for B-average student at Pan Pacific Hotel, Chateau Whistler Resort. Apprenticeship training.

Faculty: Chefs with experience in Canadian and European restaurants who have been awarded gold medals in national and international competitions.

Costs: C$1,572 tuition, C$300 nonrefundable application fee, C$122 student fee, C$406 supplies. Admission requirements: Completion of 10th grade, age 17 minimum, interview, assessment test (most students have grade 12 or cooking experience). Scholarships: yes. Loans: yes.

Location: Central Vancouver Island.

Contact: Alex Rennie, Coordinator, Malaspina University-College, 900 Fifth St., Nanaimo, BC, V9R 5S5 Canada; 250-753-3245 x2296, Fax 250-741-2683, URL http://www.mala.bc.ca.

McCALL'S SCHOOL OF CAKE DECORATION, INC.

Etobicoke/September-May *(See also page 279)*

This trade school offers full-time certificate courses in baking, commercial cake decorating (10 days each), Swiss chocolate techniques (5 days); programs for all levels. Established in 1976. Student to teacher ratio 10 per class. Facilities: 1,000 sq ft of teaching space with overhead mirrors and two 20-seat classrooms.

Faculty: Includes school director Nick McCall, and Kay Wong.

Costs: Professional courses range from C$480-C$750.

Location: A western subdivision of Toronto.

Contact: Nick McCall, President, McCall's School of Cake Decoration, Inc., 3810 Bloor St. West, Etobicoke, ON, M9B 6C2 Canada; 416-231-8040, Fax 416-231-9956, E-mail decorate@mccalls-cakes, URL http://www.mccalls-cakes.com.

NIAGARA COLLEGE HOSPITALITY AND TOURISM CENTRE

Niagara Falls/September-June

The College of Applied Arts and Technology offers 3-year cook and baker apprenticeship training, 2-year diploma in Culinary Skills. Program started 1989. Admission dates: September, January. Total enrollment: 120. 72 and 20 each admission period. 50% of applicants accepted. 60% receive financial aid. Student to teacher ratio 24:1. 100% of graduates obtain employment within six months. Facilities: Production kitchen, baking lab, mixology lab, computer lab, 3 food labs, lecture theatre, learning resource center, 90-seat student-run restaurant.

Courses: The Culinary Skills program covers food theory and preparation, kitchen management, nutrition, sanitation, and general education courses. The 6,000-hour apprenticeships are administered by the Ministry of Education and Training. A variety of continuing education courses are available.

Faculty: 6 full-time, 4 part-time. Chef professors are all Certified Chefs de Cuisine.

Costs: Annual tuition is C$1,312 for Canadian residents, C$7,750 for non-residents. A C$50 application fee and C$735 equipment fee are required. Off-campus lodging ranges from C$90-C$100 per week. Admission requirements: Secondary school diploma or equivalent.

Location: Niagara Falls City.

Contact: David Taylor, Director, Niagara College, Hospitality and Tourism Division, 5881 Dunn St., Niagara Falls, ON, L2G 2N9 Canada; 905-735-2211 x3600.

PACIFIC INSTITUTE OF CULINARY ARTS

Vancouver/Year-round

This private school offers five full-time programs: Culinary Arts and Baking and Pastry Arts (each 6 months), Breadmaking and Baking Arts and A La Carte Desserts and Pastry Arts - Advanced Level (each 3 months), Basic Techniques (1 month). Established in 1996. Calendar: trimester. Curriculum: culinary. Admission dates: January, April, July, September. Total enrollment: 132 maximum. 33 maximum each admission period. 0% enrolled part-time. Student to teacher ratio 9-12:1. 83% of graduates obtain employment within six months. Facilities: 4 commercial training kitchens, on-site white linen teaching restaurant and bakeshop.

Courses: French and international cuisines, baking, breads, pastries. First 3 months of 6-month programs cover basics, last 3 months consist of training in the Institute's restaurant. One-week practicum.

Faculty: 7 full-time chef instructors, all with international experience.

Costs: C$8,950 for each 6-month program; C$4,750 for each 3-month program, C$1,600 for the 1-month program. Lodging assistance provided. Ranges from C$350-C$1,000 per month. Admission requirements: High school diploma or equivalent. Scholarships: yes. Loans: yes.

Location: City of Vancouver, at entrance to Granville Island Market.

Contact: Sue Singer, Director of Admissions, Pacific Institute of Culinary Arts, 1505 W. 2nd Ave.,

Vancouver, BC, V6H 3Y4 Canada; 604-734-4488/800-416-4040, Fax 604-734-4408, E-mail admissions@picularts.bc.ca, URL http://www.picularts.bc.ca.

RED RIVER COMMUNITY COLLEGE

Winnipeg/Year-round

This two-year college offers 1-year certificate and 2-year diploma programs in Culinary Arts consisting of seven 3-month terms: 5 on campus and 2 off-campus work experience in Manitoba's hotels, restaurants, or private clubs. Calendar: trimester. Curriculum: culinary. Admission dates: September, March. 80% of graduates obtain employment within six months. Facilities: Prairie Lights Restaurant, a full-service open-to-the-public restaurant serving lunch and dinner.

Courses: Nutrition, food preparation, garde manger, patisserie, meat cutting, menu design, dining room service, business management, advanced cooking, communication, microcomputer productivity. Coop education: two 3-month terms.

Faculty: Dept. Chair David Rew and 9 instructors.

Costs: C$3,440 (C$1,070) first year tuition (books/supplies), C$2,638 (C$850) second year tuition (books/supplies). $35 application fee. Students have use of laptop computer during program. Admission requirements: Manitoba Senior 2 or equivalent secondary school prep or adult 10. Loans: yes.

Location: A 160-acre site in Winnipeg near the International Airport.

Contact: David Rew, Dept. Chair, Red River Community College, 2055 Notre Dame Ave., Winnipeg, MB, R3H 0J9 Canada; 204-632-2309/2285, Fax 204-632-9661, E-mail drew@rrcc.mb.ca, URL http://www.rrcc.mb.ca/~hospital/culhome.htm.

ST. CLAIR COLLEGE

Windsor/September-April

This college offers a 4-semester diploma in Culinary Arts. Program started 1993. Accredited by Province of Ontario. Calendar: semester. Curriculum: culinary, core. Admission dates: September. Total enrollment: 80. 40 each admission period. 75% of applicants accepted. 50% receive financial aid. 10% enrolled part-time. Student to teacher ratio 22-25:1. 98% of graduates obtain employment within six months. Facilities: Include 4 kitchens and classrooms, 140-seat restaurant.

Courses: Culinary arts, food preparation, culinary practice, hospitality marketing, nutrition and menu writing, garde manger, management techniques. Externship: 16-week, 7 hours per week, in hotels and restaurants. Continuing education: bartending.

Faculty: 3 full-time.

Costs: Annual tuition in-state C$1,200, out-of-state C$8,000. Scholarships: yes.

Location: An urban setting, 2 hours from London, Ontario.

Contact: Marg Jeffrey, Chairperson, St. Clair College, Business Hospitality Dept., 2000 Talbot Rd. W., Windsor, ON, N9A 6S4 Canada; 519-972-2727, Fax 519-972-0801, E-mail MJeffrey@stclairc.on.ca, URL http://www.stclairc.on.ca.

SOUTHERN ALBERTA INSTITUTE OF TECHNOLOGY

Calgary/Year-round

This nonprofit institution offers a 1-year diploma in Professional Cooking and 3- to 8-week Apprentice Cooking sessions. Program started 1949. Calendar: semester. Curriculum: culinary. Admission dates: September, January, May (November and March for Apprentice). Total enrollment: 195 professional, 216 apprentice. 65 (professional), 48 (apprentice) each admission period. 50% of applicants accepted. 50% receive financial aid. 20% enrolled part-time. Student to teacher ratio 15:1. 96% of graduates obtain employment within six months. Facilities: 1 test and 4 commercial kitchens, 2 commercial bakeries, 2 labs, 7 demo and 2 lecture classrooms, dining room, computer lab, ice plant.

Courses: Includes garde manger, patisserie, kitchen management, fat and ice sculpting. Other required courses include breakfast and short order cooking, technical writing, meat portioning, preparation of food in front of customers. Continuing education: includes bar mixology, bed and breakfast, fusion cooking.

Faculty: 31 full-time, 14 part-time. Qualifications: Journeymans and Red Seal in Cooking, Chef de Cuisine certification, Master Baker.

Costs: Semester tuition is C$1,200; other fees are C$400 per semester. On-campus housing: 204 spaces; average cost: C$15-C$20 per day. Average off-campus housing cost: C$400-C$500 per month. Admission requirements: Transcript, resume, statement of career goals. Scholarships: yes. Loans: yes.

Contact: Reg Hendrickson, Dean, Southern Alberta Institute of Technology, Hospitality Careers, 1301 16th Ave. N.W., Calgary, AB, T2M 0L4 Canada; 403-284-8612, Fax 403-284-7034, E-mail reg.hendrickson@sait.ab.ca, URL http://www.sait.ab.ca.

STRATFORD CHEFS SCHOOL

Stratford/November-March

This nonprofit training school offers a 2-semester full-time diploma. Established in 1983. Province of Ontario accreditation available upon passing exam. Calendar: semester. Curriculum: culinary. Admission dates: November. Total enrollment: 70. 35-40 each admission period. 25% of applicants accepted. Student to teacher ratio 12:1. 100% of graduates obtain employment within six months.

Courses: Gastronomy, nutrition, food styling, wine appreciation, kitchen management, menu preparation, food costing. Second-year students research, prepare and serve theme menus in a restaurant setting. Externship provided.

Faculty: 15 full-time. Founders/directors are restaurateurs Eleanor Kane and James Morris.

Costs: Annual tuition C$3,950 for Ontario residents, C$10,500 out-of country. Average off-campus housing cost is $300-400 per month.

Location: A small town, 90 minutes from Toronto.

Contact: Elisabeth Lorimer, Program Administrator, Stratford Chefs School, 68 Nile St., Stratford, ON, N5A 4C5 Canada; 519-271-1414, Fax 519-271-5679, E-mail stratfordchef@cyg.net, URL http://www.cyg.net/~stratfordchef. Additional contact: Gary Kaiser.

UNIVERSITY COLLEGE OF THE CARIBOO

Kamloops/September-May

This university college offers a 12-month certificate in Culinary Arts. Program started 1972. Curriculum: culinary. Admission dates: September, November, March. Total enrollment: 50. 18 each admission period. 80% of applicants accepted. 40% receive financial aid. Student to teacher ratio 12:1. 75% of graduates obtain employment within six months. Facilities: Include 4 kitchens and classrooms and dining room.

Faculty: 6 full-time.

Costs: Annual tuition C$1,980. Application fee C$15, student fees C$428. Average on-campus housing cost C$300 per month. Average off-campus housing cost C$300 per month. Admission requirements: High school diploma or equivalent. Scholarships: yes. Loans: yes.

Contact: Kurt Zwingli, Department Chair, University College of the Cariboo, Food Training/Tourism, Box 3010, 900 McGill Rd., Kamloops, BC, V2C 5N3 Canada; 250-828-5352, Fax 250-828-5086, E-mail zwingli@cariboo.bc.ca, URL http://www.cariboo.bc.ca/psd/tourism/pcook.htm. Additional contact: Peter Nielsen.

CHINA

CHOPSTICKS COOKING CENTRE

Kowloon/March-June, September-December *(See also page 281)*

This private trade school offers a 1-week Intensive Course for individuals and groups, a 4-week Intensive Course for individuals, 1/2- and 1-day courses, classes, programs tailored to individual needs. Established in 1971. Calendar: quarter. Curriculum: culinary. Admission dates: Year-round. Total enrollment: Maximum 10 per class. Flexible each admission period. 99% of applicants accepted. 60% enrolled part-time. Student to teacher ratio 1-10:1. 90% of graduates obtain employment within six months. Facilities: Professional kitchen with facilities for practical sessions.

Courses: Chinese regional dishes, dim sums, Chinese roasts, cakes and pastries, breads, Chinese festive cuisine and snacks.

Faculty: School principal Cecilia J. Au-Yang, domestic science graduate and author of a series of 40 cookbooks; Director Caroline Au-yeung, a graduate of HCIMA; professional chefs from various hotels and restaurants.

Costs: 1-week course $1,500 per person; 4-week course $3,000 basic, $4,000 intermediate. Short courses and classes range from $60-$900. Registration fee is $50. Local lodging averages $650 per week.

Contact: Caroline Au-yeung, Director, Chopsticks Cooking Centre, 108 Boundary St., G/Fl., Kowloon, Hong Kong, China; (852) 2336-8433, Fax (852) 2338-1462, E-mail cauyeung@netvigator.com.

ENGLAND

BUTLERS WHARF CHEF SCHOOL

(See also page 283)

London/Year-round

This private vocational school offers full-time 6-week Chefs and Restaurant certificate programs, 4-month Chefs and Restaurant diploma programs, 6-month Restaurant Advanced diploma and apprenticeship programs. Established in 1995. Accredited by National Vocational Qualifications. Curriculum: culinary. Admission dates: Flexible. Total enrollment: 250. 250 each admission period. 90% of applicants accepted. 60% receive financial aid. 40% enrolled part-time. Student to teacher ratio 8:1. 100% of graduates obtain employment within six months. Facilities: Training/production kitchen, specialized demonstration theatre, restaurant, study areas.

Courses: Realistic training in culinary and front-of-house skills.

Faculty: The 11-member faculty includes Director John Roberts, and chefs Gary Witchalls, Nicky Hopkins, Denzil Newton, Perry Reeves. Guest faculty includes chefs, food and wine experts, and restaurateurs.

Costs: Full-time 6-week programs £1,600 each, 4-month Chefs (Restaurant) diploma program £5,000 (£2,500), 6-month Restaurant Advanced diploma and apprenticeship programs £3,500. Part-time Chefs, Restaurant and Restaurant Advanced diploma programs £1,200 each. Admission requirements: Enthusiasm and commitment. Scholarships: yes.

Location: Central London.

Contact: John Roberts, Director, Butlers Wharf Chef School, Cardamom Bldg., 31 Shad Thames, London, SE1 2YR England; (44) (0)171-357-8842, Fax (44) (0)171-403-2638.

COOKERY AT THE GRANGE

(See also page 284)

Frome/Year-round

This private school offers a 4-week program, The Essential Cookery Course. Established in 1981. Curriculum: culinary. Admission dates: Rolling. Total enrollment: 140 per year. 14-20 each admission period. Student to teacher ratio 7:1. 90% of graduates obtain employment within six months. Facilities: Main kitchen, cold kitchen, herb garden.

Courses: Methods and principles of cookery with emphasis on classic techniques using fresh, natural ingredients and styles from around the world.

Faculty: Jane and William Averill (Grange trained) and teaching staff.

Costs: Tuition, including meals and shared housing, is £1,990 inclusive of VAT. The Grange has double and single bedrooms. Single room supp. £40 per week.

Location: The Grange, situated in converted farm buildings surrounded by gardens in rural England, is 90 minutes from London by train.

Contact: Jane and William Averill, Cookery at The Grange, Whatley, Frome, Somerset, BA11 3JU England; (44) (0)1373-836579, Fax (44) (0)1373-836579, E-mail cookery-grange@clara.net, URL http://www.hi-media.co.uk/grange-cookery.

THE CORDON VERT COOKERY SCHOOL

Altrincham/Year-round *(See also page 284)*

The Vegetarian Society UK, a registered charity and membership organization, offers four 1-week (or 2 weekend) Foundation courses leading to the Cordon Vert diploma, the 4-day Cordon Vert Vegetarian Catering certificate course, a variety of weekend and day courses. Established in 1984. Curriculum: culinary.

Courses: Basic and advanced techniques of vegetarian cookery, international cuisines, catering.

Faculty: Sarah Brown began the courses in 1982, based on her BBC-TV series, Vegetarian Kitchen. Tutors include Rachel Markham, Lyn Weller, Deborah Clark, Chico Francisco, John Williams.

Costs: Resident (non-resident) tuition is £340 (£280) for the 1-week Foundation courses, £450 + VAT for the Catering course. Resident tuition includes full board and lodging. Lodging is twin-bedded.

Location: Ten miles south of Manchester.

Contact: Lyn Weller, Cordon Vert Cookery School, The Vegetarian Society, Parkdale, Dunham Road, Altrincham, Cheshire, WA14 4QG England; (44) (0)161-928-0793, Fax (44) (0)161-926-9182, E-mail vegsoc@vegsoc.demon.co.uk, URL http://www.veg.org/veg/orgs/vegsocuk/.

LE CORDON BLEU – LONDON

London/Year-round *(See also page 154) (See display ad page 149)*

This private school acquired by Le Cordon Bleu-Paris in 1990 offers 3-, 5- and 10-week Classic Cycle Certificate courses in French Cuisine and Patisserie leading to the Grand Diplome, 5-week catering program. Optional theory classes in all certificate courses. Established in 1933. Calendar: quarter. Curriculum: culinary. Admission dates: January, February, April, June, August, September, October. Total enrollment: 120 per quarter. 100-120 each admission period. 100% of applicants accepted. 15% receive financial aid. 30% enrolled part-time. Student to teacher ratio 9:1. 95% of graduates obtain employment within six months. Facilities: Professionally equipped kitchens; individual workspaces with refrigerated marble tops; demonstration rooms with video.

Courses: Classic Cycle consists of 3 cuisine and 3 pastry courses, taken consecutively. Covers basic to complex technique, classic, ethnic and contemporary cuisines, planning, presentation, decoration and execution. Optional theory classes available. Students assist Master Chefs and are assisted in finding an internship. Continuing education: Master Chef Catering Course with necessary prerequisites.

Faculty: All staff full time. Chefs all professionally qualified with experience in Michelin-starred and fine quality culinary establishments. Includes President Andre Cointreau and School Directors Leslie Grey and Susan Eckstein.

Costs: Tuition ranges from approximately £250 for a week-long or evening course to £2,300-£4,000 for an intensive 10-week Classic Cycle course. Students are assisted in finding housing. Scholarships: yes.

Location: London's West End, close to Oxford and Bond Streets.

Contact: Anne-Laure Trehorel, Enrollment Officer, Le Cordon Bleu, Enrollment Office, 114 Marylebone Lane, London, W1M 6HH England; (44) (0)171-935-3503, Fax (44) (0)171-935-7621, E-mail information@cordonbleu.co.uk, URL http://cordonbleu.net. US toll-free phone: 800-457-CHEF.

LEITH'S SCHOOL OF FOOD AND WINE

London/Year-round *(See also page 287)*

This private school offers a 1-year or 2-term diploma consisting of two or three consecutive 10- to 11-week Food and Wine certificate courses (Beginner's, Intermediate, Advanced), foundation and beginner's cookery certification, certificate course in wine. Nonprofessional courses. Established in

1975. Calendar: semester. Admission dates: October or January. Total enrollment: 96. 96 each admission period. 99% of applicants accepted. Student to teacher ratio 8:1. 90% of graduates obtain employment within six months. Facilities: Include 3 kitchens and a demonstration theatre.

Courses: Beginner's: basic cookery methods. Intermediate: butchery, exotic fish, commercial catering. Advanced: boned poultry, aspics, advanced patisserie, exotic canapes, costing, menu planning, market visits, business skills.

Faculty: 13 full-, 2 part-time. School founder and cookbook author Prue Leith is former Veuve Cliquot Business Woman of the Year. Principal is Caroline Waldegrave, vice-principal is A. Cavaliero.

Costs: Tuition ranges from £3,300-£3,600 per term; cost for all 3 is £9,650. Equipment fee is £280. All prices include VAT. The school assists in obtaining lodging, which ranges from £80-£120 per week.

Location: A refurbished Victorian building in Kensington, the center of London.

Contact: Judy Van DerSande, Registrar, Leith's School of Food and Wine, 21 St. Alban's Grove, London, W85 5BP England; (44) (0)171-229-0177, Fax (44) (0)171-937-5257, E-mail info@leiths.com, URL http://www.leiths.com.

THE MANOR SCHOOL OF FINE CUISINE

Widmerpool/Year-round *(See also page 150)*

This proprietary institution offers a 4-week Cordon Bleu certificate, 6-week Advanced Cordon Bleu certificate. Established in 1988. Curriculum: culinary. Admission dates: Rolling. Total enrollment: 12. 8 each admission period. 90% of applicants accepted. 5% receive financial aid. 60% enrolled part-time. Student to teacher ratio 8:1. 100% of graduates obtain employment within six months. Facilities: Include 7 kitchens, lecture room, large dining room, cooking library.

Courses: Include cuisine preparation, baking, basic nutrition, and menu planning.

Faculty: Principal Claire Tuttey, Cordon Bleu Diploma, head chef of noted restaurants, member of Cookery and Food Association, Craft Guild of Chefs, Chefs and Cooks Circle.

Costs: Resident (nonresident) tuition is £1,162.07 (£1,044.57). Residents are housed in The Manor (except weekends). Recommendations are provided for local lodging.

Location: The Manor, refurbished 17th century inn, is 9 miles from Nottingham City center and 80 minutes from London's Kings Cross train station.

Contact: The Manor School of Fine Cuisine, Old Melton Road, Widmerpool, Nottinghamshire, NG12 5QL England; (44) (0)1949-81371, Fax (44) (0)1949-81371.

THE MOSIMANN ACADEMY

London/Year-round *(See also page 288)*

This private school offers a 4-day certificate course (Anton Mosimann Food Experience) that covers menu planning, trends, financial management, kitchen concepts, planning and technique, wine tasting, leadership, cuisine naturelle, food presentation, winning strategies. Program started 1996. Curriculum: culinary. Student to teacher ratio 22-60:1. Facilities: New seminar and demonstration theater, library of Anton Mosimann's 6,000 cookery books.

Courses: Motivation and creativity, the Anton Mosimann philosophy (Cuisine Naturelle), which eschews the use of fats and alcohol. Tailor-made corporate courses.

Faculty: 3 full- and 3 part-time, including Anton Mosimann, author of 9 cookery books, Shaun Hill, John Burton-Race, Brian Turner, Jean-Christophe Novelli.

Costs: £995+VAT for 4-day course. Admission requirements: Background as professional chef is advisable.

Location: Centrally located in Battersea, London.

Contact: Emily Manson, Course Coordinator, The Mosimann Academy, 5 William Blake House, The Lanterns, Bridge Lane, London, SW11 3AD England; (44) (0)171-924-1111, Fax (44) (0)171-924-7187, E-mail academy@mosimann.com, URL http://www.mosimann.com.

PAUL HEATHCOTE'S SCHOOL OF EXCELLENCE

(See also page 288) **Manchester/Year-round**

This private school offers apprenticeships, supervisory and management training to NVQ level 4. Program started 1997. Curriculum: culinary. Facilities: Professionally-equipped kitchen with 6 individual work spaces, demonstration auditorium with projector and screen.

Faculty: Paul Heathcote, chef and owner of four restaurants, and his staff of instructors.

Location: Central Manchester, off Deansgate.

Contact: Administration Office, Paul Heathcote's School of Excellence, Jacksons Row, Deansgate, Manchester, M2 5WD England; (44) (0)161-839-5898, Fax (44) (0)161-839-5897, E-mail cookeryschool@heathcotes.co.uk, URL http://www.heathcotes.co.uk.

ROSIE DAVIES

Somerset/January-July, September-November

This culinary professional offers a 4-week Basics Plus course for professionals, A B Ski/A B Sea courses for running a chalet/chefs on yachts, shorter tailor-made courses, Basic Hygiene Certificate Course. Established in 1996. Curriculum: culinary. 4 each admission period. Student to teacher ratio 4:1. Facilities: Penny's Mill's farmhouse-style kitchen.

Courses: Basic to advanced techniques for preparing appetizers, soups, sauces, meat, fish, vegetables, vegetarian dishes, ethnic foods, desserts, breads.

Faculty: Rosie Davies trained at Oxford in Catering and Hotel Management, is a freelance cookery writer and editor, has 20+ years teaching experience at Redlynch Park College and Cookery at the Grange.

Costs: £1,690 per 4-week course includes meals and lodging at Penny's Mill, a 200-year-old converted water mill. Short courses are from £100 per day.

Location: The village of Nunney, in English West Country.

Contact: Rosie Davies, Rosie Davies, Penny's Mill, Nunney, Frome, Somerset, BA11 4NP England; (44) (0)1373-836210, Fax (44) (0)1373-836018, E-mail pennysmill@aol.com.

SQUIRES KITCHEN INTERNATIONAL SCHOOL

(See also page 289) **Farnham, Surrey/Year-round**

This private school offers a part-time 1-week certificate in sugarcraft and cake decorating plus diploma courses. Established in 1987. Calendar: trimester. Curriculum: core. Admission dates: Rolling. 100% of applicants accepted. 100% enrolled part-time. Student to teacher ratio 12:1/1:1. Facilities: A kitchen with specialized equipment and materials.

Courses: Include royal icing, sugarpaste, flowers, pastillage, chocolate.

Faculty: 17 full- and part-time. Members of the British Sugarcraft Guild. Guest tutors include Eddie Spence, Alan Dunn, Tombi Peck.

Costs: Range from £50 per day.

Location: A period building in a suburban area, a 2-minute walk to train station, 45 minutes from London.

Contact: Course Coordinator, Squires Kitchen Internat'l. School of Sugarcraft and Cake Decorating, 3 Waverley Lane, Farnham, Surrey, GU9 8BB England; (44) (0)1252-711749, Fax (44) (0)1252-714714, E-mail school@squires-group.co.uk, URL http://www.squires-group.co.uk.

TANTE MARIE SCHOOL OF COOKERY

(See also page 289) **Surrey/Year-round**

This private school offers a 36- or 24-week Intensive Tante Marie Cordon Bleu diploma. Established in 1954. Accredited by BACIFHE (British Accreditation Council). Calendar: trimester. Curriculum: culinary. Admission dates: January, April, September. Total enrollment: 84. 24-72 each admission period. 100% of applicants accepted. 20% receive financial aid. 0% enrolled part-time. Student to teacher ratio 12:1. 100% of graduates obtain employment within six months. Facilities: Include 5 modern teaching kitchens, a mirrored demonstration theatre, and a lecture room.

Courses: 36-week course for beginners: three 12-week terms of basic skills, labor-saving appliances, British cookery, French cuisine. Intensive 24-week course for experienced cooks: practical and theoretical elements. 4-day wine seminar prepares for Wine Certificate exam.

Faculty: 12 full- and part-time. Qualified to work in state schools and many have held catering positions. All undergo teacher training. Well-known TV cookery demonstrators, a noted wine expert, and local tradesmen also present.

Costs: £9,000 (£6,900) for 36-week (24-week) course, £3,300 for 12-week certificate course. Uniform and equipment £125. Overseas students qualify for 23% vocational training tax deduction from fees quoted. Admission requirements: English language fluency. Loans: yes.

Location: A turn-of-the-century country mansion near the center of Woking, a small country town approximately 25 minutes by train from London.

Contact: Margaret A. Stubbington, Registrar, Tante Marie School of Cookery, Woodham House, Carlton Rd., Woking, Surrey, GU21 4HF England; (44) (0)1483-726957, Fax (44) (0)1483-724173, E-mail info@tantemarie.co.uk, URL http://www.tantemarie.co.uk.

THAMES VALLEY UNIVERSITY
Berkshire/Year-round

This university offers a 3-year NVQ Level 2 international diploma in Culinary Arts. Program started 1992. Accredited by City and Guilds. Calendar: semester. Admission dates: September. Total enrollment: 40. 40 each admission period. 70% of applicants accepted. 10% receive financial aid. Student to teacher ratio 15:1. 100% of graduates obtain employment within six months. Facilities: Include 4 kitchens, demonstration kitchen, 2 science labs, 3 restaurants, computer lab.

Faculty: 60 full-time.

Costs: Annual tuition £4,846 plus fees and certificate costs.

Contact: David Foskett, Thames Valley University, Hospitality Studies, Wellington St., Berkshire, England; (44) (0)1753-697604, Fax (44) (0)7553-677682, URL http://www.tvu.ac.uk.

FRANCE

ÉCOLE GASTRONOMIQUE BELLOUET-CONSEIL
Paris/Year-round

This private school offers 1- to 5-day intensive seminars. Customized classes also available. Established in 1989. Calendar: semester. Student to teacher ratio 8-10 per class.

Courses: More than thirty individual courses including cakes, individual cakes, petits fours, chocolate, artistic sugar, catering, viennoiserie.

Faculty: G.J. Bellouet and J.M. Perruchon, both Meilleur Ouvrier de France.

Costs: Start at 1,400FF/day, includes lunch.

Contact: École Gastronomique Bellouet-Conseil, 304 & 306 LeCourbe, Paris, 75015 France; (33) (0)1-40-60-16-20, Fax (33) (0)1-40-60-16-21.

ÉCOLE LENÔTRE
Plaisir Cedex/Year-round

This French gastronomy school offers a 6-month intensive professional culinary training diploma course, more than thirty 1- to 4-day certificate courses. Established in 1970. Calendar: semester. Curriculum: culinary. Total enrollment: 3,000 per year. 60 per 6 classes each admission period. 90% of applicants accepted. Student to teacher ratio 12:1. Facilities: Part of LeNôtre, the school covers 198 acres with 6 specialized classrooms, a meeting room, and a boutique.

Courses: Classes are 90% participation. Several courses in each of the categories: cuisine, catering, buffet decoration, breads and pastry doughs, pastry plated desserts, ice cream and frozen desserts, chocolate and confectionery, work sugar, matching food and wine. Special a la carte courses for groups of 8 to 12 can also be arranged.

Faculty: 4 instructors are recipients of the Meilleur Ouvrier de France in pastry-confectionery, pork butchery, ice cream, and bakery-viennoiserie. Founded by Gaston Lenôtre and managed by Marcel Derrien, Meilleur Ouvrier de France in pastry-confectionery.

Costs: Tuition (non-French students), including breakfasts and lunches, ranges from 997FF (626FF) to 8,650FF (5,293FF). The school can reserve rooms in one of Plaisir's hotels, which range from 130FF-310FF. Scholarships: yes.

Location: About 20 miles from Paris, 8 miles from Versailles.

Contact: Marie-Anne Dufeu, Sales Manager, École Lenôtre, 40, rue Pierre Curie-BP 6, Plaisir Cedex, 787375 France; (33) (0)1-30-81-46-34, Fax (33) (0)1-30-54-73-70.

ÉCOLE SUPERIEURE DE CUISINE FRANCAISE GROUPE FERRANDI

Paris/September-June

This professional restaurant and culinary school offers a bilingual 9-month program, awarding a diploma issued by Paris' Chamber of Commerce and preparing qualified students for the C.A.P. certificate of the French Ministry of Education. Established in 1986. Accredited by French Ministry of Education, US Dept. of Education for Student Aid Programs. Calendar: trimester. Curriculum: culinary. Admission dates: February-May. Total enrollment: 200. 12-25 each admission period. 60-90% of applicants accepted. 50% receive financial aid. Student to teacher ratio 10-12:1. 100% of graduates obtain employment within six months. Facilities: Include more than 12 professional kitchens, tasting laboratory, auditorium, classrooms, 2 working restaurants.

Courses: General theoretical and practical courses, cooking, baking, pastry, butchery, delicatessen products, fish cookery. Other activities include a 3-day wine country excursion. The 9-month program includes visits to museums, markets, fine restaurant. 1- to 3-month restaurant apprenticeships in Paris and provinces arranged for top students.

Faculty: The curriculum is supervised by a Board of Advisors including well-known French chefs Joel Robuchon and Antoine Westermann.

Costs: The 9-month program costs 92,000FF. Off-campus lodging ranges from $500-$800 per month. Admission requirements: Proof of full medical and accident coverage, long-term student visa, certified birth certificate, and undergraduate transcript. Loans: yes.

Location: Rue Ferrandi in Paris's Latin Quarter, between St. Germain de Pres and Montparnasse. Convenient to 4 major metro stations.

Contact: Stephanie Curtis, Coordinator, ESCF Groupe Ferrandi, Bilingual Program, 10 rue de Richelieu, Paris, 75001 France; (33) (0)1-40-15-04-57, Fax (33) (0)1-40-15-04-58.

ÉCOLE DES ARTS CULINAIRES ET DE L'HÔTELLERIE, EACH

(See also page 298) **Lyon/Ecully/Year-round**

This private school offers programs in cooking and hotel and restaurant management, including the 16-week Cuisine and Culture program (taught in English), the 2-year Culinary Arts Management program (taught in French), short courses for professionals and cooking enthusiasts. Established in 1990. Calendar: semester. Admission dates: March for 16-week program, September and March for 2-year program. 100% of graduates obtain employment within six months. Facilities: 13 seminar rooms, 2 computer labs, 8 teaching kitchens, pastry/pantry facilities, video-equipped amphitheatre, sensory analysis lab, restaurants.

Courses: Include food preparation and processing, pantry, pastry and bakery workshop, basic cooking and catering, and restaurant cuisine. Seminars cover cheese, wine, French ingredients, and French culinary culture. Excursions and conversational French classes are provided.

Faculty: In addition to the Board of Trustees, headed by Paul Bocuse, the permanent teaching staff includes 2 Meilleurs Ouvriers de France, Pastry Chef Alain Berne, and Restaurant Chef Alain Le Cossec.

Costs: 16-week program is 56,000FF, which includes lunch, dinner, lodging. 2-year program is 59,000FF (47,000FF) the first (second) year, which includes lunch. Student residence has 114 rooms with private shower; lodging is 2,300FF for a single room. Admission requirements:

Students who complete the 16-week program and pass an exam may take the longer programs. Loans: yes.

Location: The restored 19th century Château du Vivier, in a 17-acre wooded park in the Lyon-Ecully University-Research Zone, 10 minutes from downtown Lyon.

Contact: Eleonore Vial, Dean, Angeline Phan, Enrollment Dept., École des Arts Culinaires et de l'Hôtellerie, EACH Lyon, Château du Vivier, B.P. 25, Ecully, Cedex, 69131 France; (33) (0)4-78-43-36-10, Fax (33) (0)4-78-43-33-51, E-mail information@each-lyon.com, URL http://www.each-lyon.com.

LA VARENNE

Burgundy/June-August *(See also pages 265 and 301) (See display ad page 301)*

This private school offers a 3- and 4-week residential program - French Cooking Today - (limit 15 students) that focuses on French cuisine. Established in 1975. Facilities: The school is in the 17th-century Château du Feÿ, a registered historic monument owned by founder Anne Willan.

Courses: Fundamentals, pastry, French regional cooking, contemporary cuisine, vineyard tours and fine dining, and such topics as classical cooking trends, pastry and chocolate, bistro and regional dishes. Students who pass a written and practical exam receive La Varenne's Diplôme d'Etudes Culinaire. Other activities include seminars and escorted excursions to northern Burgundy, Beaune, Paris.

Faculty: 2/3 of curriculum is taught by La Varenne's cuisine and pastry chefs, the rest by noted French restaurant chefs. Programs are directed by Anne Willan, author of the Look and Cook how-to series featured on PBS, Château Cuisine, and La Varenne Pratique.

Costs: Fee is $9,995 ($7,995) for the 4-week (3-week) program, which includes full board, shared twin lodging at the Château, planned excursions, transportation to/from Paris. Nonrefundable $1,000 deposit, balance 60 days prior to course.

Location: The Château du Feÿ is 90 minutes south of Paris. Amenities include a tennis court and outdoor swimming pool in season.

Contact: La Varenne, PO Box 25574, Washington, DC 20007 US; 800-537-6486/202-337-0073, Fax 703-823-5438, E-mail lavarenne@compuserve.com, URL http://www.lavarenne.com.

LE CORDON BLEU – PARIS

Paris/Year-round *(See also pages 139, 143, 148, 159, and 301) (See display ad page 149)*

This private school acquired by Andre J. Cointreau in 1984, re-opened in 1988, 5 locations worldwide, offers 9-months of basic, intermediate and superior level diploma programs in Pastry and Cuisine leading to Le Grand Diplome, 10-week International Culinary Summer Abroad Program, Buffet Techniques Program, 3- and 5-week programs. Established in 1895. Calendar: quinmester. Curriculum: culinary. Admission dates: Five times per year. Total enrollment: 150. 150 each admission period. 100% of applicants accepted. 10% receive financial aid. 20% enrolled part-time. Student to teacher ratio 10:1. 90% of graduates obtain employment within six months. Facilities: Professionally equipped kitchens; individual workspaces with refrigerated marble tops; demonstration rooms with video and overhead mirrors.

Courses: The Classic Cycle: 3 levels of Cuisine and 3 levels of Pastry taken consecutively or together. Courses cover basic to complex techniques; classic, regional, ethnic, contemporary cuisines; food and wine; catering; planning; presentation; decoration and execution. Students assist chef at least once and are assisted in finding a 3-month internship. Continuing education: Professional Chef and Buffet Techniques Catering Courses, 1-2 week intensives.

Faculty: 10 full-time Master Chefs, international staff. Director of Academics: Patricia Gastaud-Gallagher. Culinary staff: French Master Chefs from Michelin-starred restaurants and fine hotels, including two holding the title of Meilleur Ouvrier de France.

Costs: Tuition ranges from 220FF for a half-day demo, 750FF for a full-day hands-on workshop, to 4,590FF for a 4-day course. 10-week Certificate courses range from 22,950FF in pastry to 36,550FF in cuisine. 9-month Grand Diplome 172,750FF. Admission requirements: Personal statement, 2 letters of recommendation, resume. Scholarships: yes.

Location: Paris' 15th arrondissement, centrally located in the southwestern part of the city.

Contact: Sabine Bailly, Director of Admissions, Le Cordon Bleu, 8, rue Leon Delhomme, Paris, 75015 France; 800-457-CHEF, Fax (33) (0)1-48-56-03-96, E-mail info@cordonbleu.net, URL http://www.cordonbleu.net.

L'ÉCOLE DES CHEFS

(See also page 302) (See display ad page 303) **Paris, Lyon/Year-round**

Culinary professional Annie Jacquet-Bentley offers 5- to 6-day one-on-one internships with 2- and 3-star Michelin chefs. Certificate on completion. Pastry program at Ladurée with Pierre Herme, catering program with Potel & Chabot. Program started 1998. Curriculum: culinary. Total enrollment: 1 chef/restaurant. Student to teacher ratio 1:1. Facilities: On-site in French 2 and 3-star restaurant kitchens.

Courses: Immersion in a French kitchen with executive chef, garde-manger, saucier, saute cook, pastry chef. Visits to food market and kitchen equipment store. Programs specializing in pastry and catering available.

Faculty: 17 Michelin-starred chefs include Troisgros, Georges Blanc, Alain Passard, Pierre Herme of Ladurée, Jean-Pierre Biffi of Potel & Chabot.

Costs: Restaurant professionals: $3,900 ($4,900) for 2-star (3-star) restaurant; culinary school students: $1,950 ($2,450) for 2-star (3-star) restaurant. Admission requirements: Resume & references.

Location: Paris, Lyon, French Countryside (Provence, Brittany, Champagne, Alps, Burgundy).

Contact: Annie Jacquet-Bentley, President, L'École des Chefs, PO Box 183, Birchrunville, PA 19421 US; 610-469-2500, Fax 610-469-0272, E-mail info@leschefs.com, URL http://www.leschefs.com.

L'ÉCOLE DE PATISSERIE FRANCAISE

Uzes/Year-round

French pastry chef Didier Richeux offers 40 intensive 5-day private tuition master workshops per year. A school certificate is awarded on completion. Established in 1994. Total enrollment: 2 maximum. 2 each admission period. Student to teacher ratio 2:1. Facilities: Commercial patisserie kitchen with separate ice cream preparation kitchen.

Courses: Traditional and modern patisserie, including ice creams, sorbets, petits fours, chocolates, celebration cakes, plated desserts, croissants, brioches, croqu'embouches, individual pastries, catering. Individual needs are accommodated. Nonparticipant can visit olive oil mill, truffle farms, local markets, and feudal castles.

Faculty: Didier Richeux, co-founder, 20 years experience in pastry cooking, recipient of 18 int'l awards, including Gold Medal, Cordon Bleu de France; finalist Meilleur Ouvrier de France Glacier; Executive Pastry Chef at Le Cordon Bleu and The Savoy in London.

Costs: 5,500FF (Tuesday to Saturday), 9,900FF (2 weeks). Discount for 2 individuals booking together.

Location: 30 minutes north of Nimes Airport, 40 minutes southwest of Avignon, and an hour from Montpellier Airport.

Contact: Didier Richeux, L'École de Patisserie Francaise, 12 Rue de la Republique, Uzes, 30700 France; (33) (0)4-66-22-12-09, Fax (33) (0)4-66-22-26-36.

RITZ-ESCOFFIER ÉCOLE DE GASTRONOMIE FRANCAISE

(See also page 308) (See display ad page 157) **Paris/Year-round**

This culinary school in the Hotel Ritz, named for the hotel's first chef, Auguste Escoffier, offers a 30-week Grand Diploma course for future professionals. Segments: 1- to 6-week Cesar Ritz (beginner), 12-week Ritz-Escoffier (intermediate-advanced), 1- to 12-week Art of French Pastry (beginner-advanced). All segments may be taken separately. Established in 1988. Calendar: semester. Curriculum: core. Admission dates: Courses begin each week. Student to teacher ratio 8-10:1. Facilities: 2,000-square-foot custom-designed facility includes a main kitchen, pastry kitchen, conference room/library, changing rooms.

Courses: The Grand Diploma, César Ritz and Art of French Pastry courses consist of 25-28 hours per week of demonstrations, hands-on practice, and theory. Classes taught in French with simultaneous English translation. Students eat with chef after most practical classes. Diploma recipients may apply for internship in the Hotel Ritz's restaurant kitchens. One-week intensive classes on various topics including breadmaking, sauces, terrines.

Faculty: 7 full-time, 3 part-time. Instructors are Chefs de Cuisine Blaise Volckaert and Christophe Bellet and Chef Patissier Gilles Maisonneuve.

Costs: One-week classes start at 5,550FF per week, 6-week Cesar Ritz 34,000FF, 12-week Ritz-Escoffier 71,000FF, 12-week Art of French Pastry 63,000FF. Admission requirements: Ritz Escoffier course students must have Cesar Ritz diploma or equivalent experience; no experience required for Grand Diploma, César Ritz, Pastry.

Location: Central Paris, near 3 main subway entrances and most major department stores.

Contact: M. Jean-Philippe Zahm, Director, Ritz-Escoffier École de Gastronomie Francaise, 15, Place Vendome, Paris Cedex 01, 75041 France; (33) (0)1-43-16-30-50, Fax (33) (0)1-43-16-31-50, E-mail ecole@ritzparis.com. US toll-free phone: 800-966-5758.

ROBERT REYNOLDS, NORTHWEST FORUM (See page 104)

GREECE

ALPINE CENTER FOR HOTEL & TOURISM MANAGEMENT STUDIES

Athens/Year-round

The Swiss-managed Associate Institute of the International Hotel & Tourism Training Institutes offers a 2-year certificate in Culinary Arts for Chef Training. Accredited by Neuchatel Dept. of Public Economy, Switzerland. Curriculum: culinary. Admission dates: October.

Courses: Includes food and beverage, food production, kitchen organization, menu planning, nutrition, cooking methods, baking and pastry, buffet catering, international cuisine, professional development, language. Graduates can transfer to Johnson & Wales University, Providence, RI to complete BS degree in Culinary Arts.

Faculty: 6 instructors.

Costs: 9.980 (13.430, 16.930) Swiss Francs per year for commuter (shared room, single room) students. Admission requirements: High school diploma & placement test.

Contact: Alpine Center for Hotel & Tourism Management Studies, PO Box 17082, Athens, GR-10024 Greece; (30) 1-7251036/7, Fax (30) 1-8981189, E-mail admission@alpine.edu.gr, URL http://www.alpine.edu.gr.

IRELAND

BALLYMALOE COOKERY SCHOOL

Midleton/September-July *(See also page 311)*

This proprietary school offers a 12-week certificate course in Food and Wine, a variety of short courses. Established in 1983. Calendar: quarter. Curriculum: culinary. Admission dates: September, January. Total enrollment: 44. 44 each admission period. 90% of applicants accepted. Student to teacher ratio 6:1. 100% of graduates obtain employment within six months. Facilities: Include a specially-designed kitchen with gas and electric cookers, mirrored demonstration area, TV monitors, gardens that supply fresh produce.

Courses: Most courses are hands-on and emphasize traditional Irish, contemporary, and classic cookery, international cuisine, vegetarian, and seafood cooking.

Faculty: 4 full- and 4 part-time. Includes Principal Darina Allen, IACP-certified Teacher and Food Professional, her brother, Rory O'Connell, both trained in the Ballymaloe House restaurant kitchen, and her husband Tim Allen. Guest chefs are featured.

Costs: 12-week course is IR £3,575. Students live in cottages during the 3-month course and may

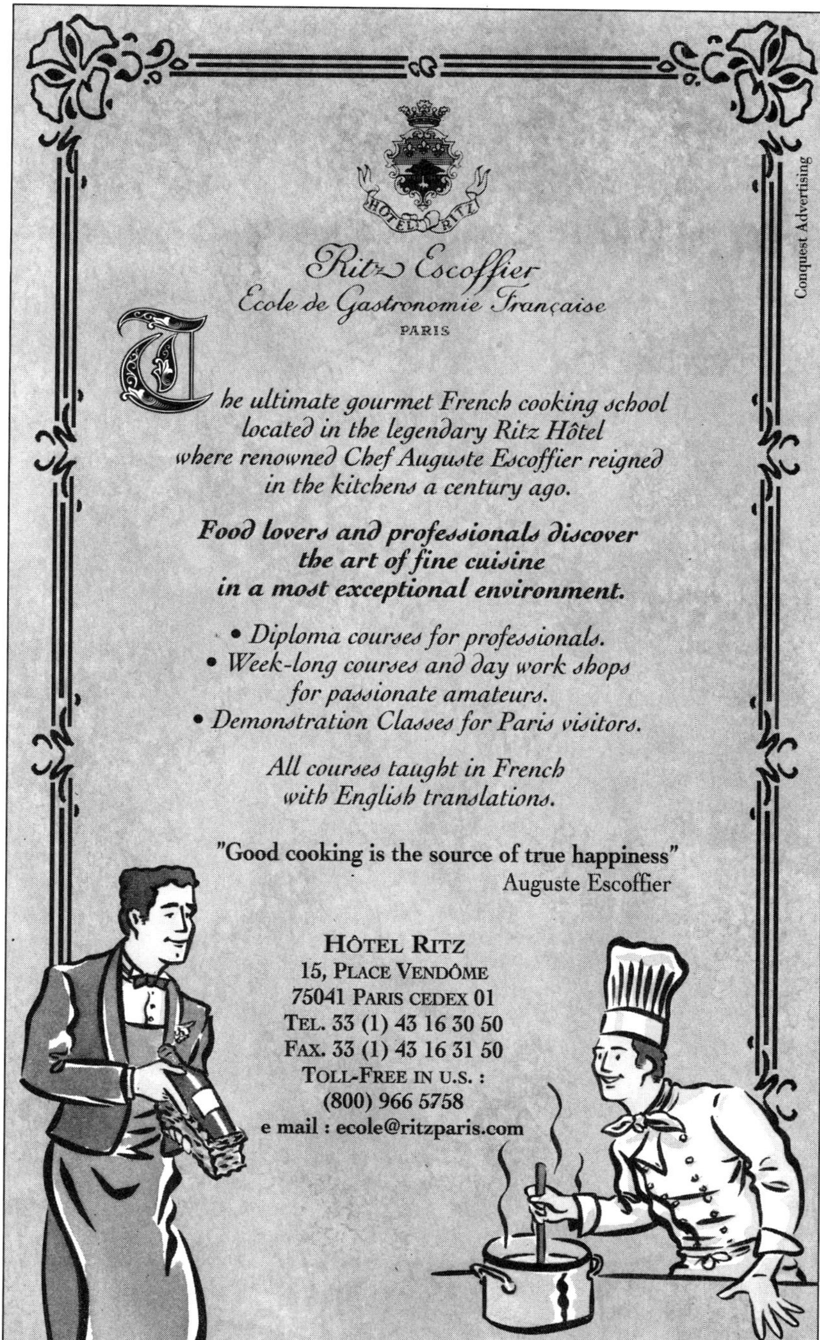
HÔTEL RITZ
Ritz Escoffier
Ecole de Gastronomie Française
PARIS
The ultimate gourmet French cooking school
located in the legendary Ritz Hôtel
where renowned Chef Auguste Escoffier reigned
in the kitchens a century ago.
Food lovers and professionals discover
the art of fine cuisine
in a most exceptional environment.
• Diploma courses for professionals.
• Week-long courses and day work shops
for passionate amateurs.
• Demonstration Classes for Paris visitors.
All courses taught in French
with English translations.
"Good cooking is the source of true happiness"
Auguste Escoffier
HÔTEL RITZ
15, PLACE VENDÔME
75041 PARIS CEDEX 01
TEL. 33 (1) 43 16 30 50
FAX. 33 (1) 43 16 31 50
TOLL-FREE IN U.S. :
(800) 966 5758
e mail : ecole@ritzparis.com
Conquest Advertising

assist in the restaurant. Self-catering cottage lodging is IR £35 per week double, £48 single.

Location: A mile from the sea, outside the village of Shanagarry in southern Ireland, 4 hours from Dublin. Airport and train station pick-up is available.

Contact: Tim Allen, Ballymaloe Cookery School, Kinoith, Shanagarry, County Cork, Midleton, Ireland; (353) 21-646785, Fax (353) 21-646909.

BALTIMORE INTERNATIONAL COLLEGE (See page 58)

ITALY

APICIUS – LORENZO DE'MEDICI INSTITUTE

Florence/Year-round *(See also page 313)*

This private school, a member of the Federation of European Schools and IACP, offers 1-year diploma and 2-semester programs that combine Italian language, art history, and culinary arts. Non-credit courses 1 week or more, private and group courses, programs for adults and children. Established in 1973. Authorized by the Italian government. Calendar: semester. Curriculum: core. Admission dates: January and August for semesters, any month for short courses. Total enrollment: 150. 20-25 each admission period. 80% of applicants accepted. 5% receive financial aid. 20% enrolled part-time. Student to teacher ratio 12:1. 30% of graduates obtain employment within six months. Facilities: 4-story building offering language, art, and cooking programs.

Courses: Renaissance culture through cooking, Italian food resources, Regional Italian food, restaurant management, food producers and sellers in Italy, food in art, entertaining in Europe, Italian vegetarian cooking, Italian-Jewish food tradition, wine appreciation. Internships as kitchen assistants in the school and local restaurants during diploma program. Special 10-day programs.

Faculty: Includes professional chefs Stefano Innocenti and Daniele Pescator, food historian Giuseppe Alessi, and Gabriella Ganugi, Director.

Costs: $2,900 tuition per semester, courses from $200. Admission requirements: Vary with the program. Scholarships: yes.

Location: In the San Lorenzo district, between the Duomo and the central train station.

Contact: Dr. Gabriella Ganugi, Director, Lorenzo de'Medici Institute, Via Faenza 43, Florence, 50123 Italy; (39) 055-287360, Fax (39) 055-2398920, E-mail LDM@dada.it, URL http://www.dada.it/ldm.

ETOILE INSTITUTE FOR ADVANCED CULINARY AND PASTRY ARTS

Venice/Year-round *(See also page 319)*

This private school offers weekly 3-day certificate programs in pastry, buffet, ice carving, chocolate sculpture. Program started 1985. Curriculum: culinary. 15 per class each admission period. Student to teacher ratio 15:1. Facilities: Modern, well-equipped labs, classrooms, library. Lodging at the adjacent Hotel Airone.

Courses: Advanced training based on regional Italian themes.

Faculty: Master Chef Rossano Boscolo, founder, author of 4 cookbooks for professionals, team leader of the Italian Culinary Olympic Team, partner of Boscolo Group, owner of hotels and restaurants. 40-chef faculty.

Costs: Average $1,000.

Location: 30 miles from Venice, near Chioggia, on the Venice lagoon across from the Adriatic beach.

Contact: Diana Place, U.S. Representative, Essence of Italy, PO Box 956, Boca Raton, FL 33429 US; 888-213-5678, 561-361-0301, Fax 561-361-0301, E-mail essenceofitaly@worldnet.att.net, URL http://www.etoile.org.

ITALIAN CULINARY INSTITUTE FOR FOREIGNERS

Piedmont region/Year-round

This professional culinary institute (ICIF) offers a 6-month certificate program consisting of 2 months classroom study and 4 months apprenticeship at an externship site in Italy. Program started 1991. Curriculum: culinary. Admission dates: January, June, September. 25 each admission period. 70% of applicants accepted. Student to teacher ratio 12:1. 100% of graduates obtain employment within six months. Facilities: High-tech equipment in an 18th-century castle.

Courses: Emphasis on product knowledge and developing an Italian palate. Externships at various locations in Italy.

Faculty: In-house instructors and visiting professional chefs.

Costs: $6,000, which includes room and board and NYC-Italy roundtrip airfare. Lodging in newly-constructed student housing facility. Admission requirements: Culinary school graduate or equivalent.

Location: Costigliole D'Asti.

Contact: Enrico Bazzoni, Director of Programs, ICIF c/o Enrico Bazzoni, 126 Second Place, Brooklyn, NY 11231 US; 718-875-0547, Fax 718-875-5856, E-mail Eabchef@aol.com, URL http://www.icif.com.

SCUOLA DI ARTE CULINARIA "CORDON BLEU"

(See also page 333) **Florence/Year-round**

This private school offers professional programs in spring and fall, nonvocational/vacation programs year-round. Professional curriculum consists of 11-21 one- to nine-session hands-on courses, a total of 150-230 hours of instruction. Established in 1985. Calendar: quarter. Admission dates: July, November for professional programs; year-round for other courses. Total enrollment: 150. 50-70 each admission period. 90% of applicants accepted. 90% enrolled part-time. Student to teacher ratio 6:1. 90% of graduates obtain employment within six months. Facilities: The school's 40-square-meter teaching kitchen.

Courses: Include basic to advanced cooking, pastry, bread, holiday menus, regional specialties, specific subjects, wine, history. More time devoted to Italian cuisine and traditions.

Faculty: 2 full-time instructors. Cristina Blasi and Gabriella Mari, 12 years experience, sommeliers and olive oil experts, authored a book on ancient Roman cooking, members Commanderie des Cordons Bleus de France, IACP, Italian Assn. of Cooking Teachers, AICI.

Costs: Tuition is 3,980,000 Lira (6,630,000 Lira) for the fall (spring) program, 7,910,000 Lira for both. 30% nonrefundable deposit 2 months prior, balance due at start of course. Off-campus housing available in apartments and hotels. Admission requirements: Anyone in culinary profession.

Location: Central Florence.

Contact: Gabriella Mari, Co-Director, Scuola di Arte Culinaria "Cordon Bleu", Via di Mezzo, 55/R, 50121 Firenze-Florence, Italy; (39) 055-2345468, Fax (39) 055-2345468, E-mail cordonbleu@aspide.it, URL http://www.cordonbleu-it.com.

JAPAN

LE CORDON BLEU – TOKYO

(See also page 154) (See display ad page 249) **Tokyo/Year-round**

This private school acquired by Le Cordon Bleu-Paris offers Cuisine and Pastry diplomas: 9-month programs of basic, intermediate, superior levels. Le Grand Diplome: 9-month program of cuisine and pastry. Introductory courses to cuisine, pastry, catering, bread baking. Daily demonstrations, evening sessions. Established in 1991. Accredited through international hospitality management schools. Calendar: quarter. Curriculum: culinary. Admission dates: Quarterly. Facilities: Professionally equipped kitchens; individual work spaces with refrigerated marble tables, convection ovens; specialty appliances.

Courses: Diploma and Certificate core curriculum consists of 3 levels of cuisine and pastry, taken consecutively or together. Covers basic to complex techniques, wine and food pairing, catering, presentation. Assistance in finding internships is provided. Offered at the Paris and London schools.

Faculty: 7 full-time French and English Master Chefs from Michelin-star restaurants and fine hotels.

Costs: Assistance is provided in finding lodging.

Location: Tokyo's Daikanyama district.

Contact: Yann Brochet, School Director, Le Cordon Bleu, Roob-1, 28-13 Sarugaku-cho, Daikanyama, Shibuya-ku, Tokyo, 150 Japan; (81) 3 5489 01 41, Fax (81) 3 5489 01 45, URL http://cordonbleu.net. US toll-free phone: 800-457-CHEF.

MEXICO

SEASONS OF MY HEART COOKING SCHOOL

Oaxaca/Year-round *(See also page 340)*

This private school offers a 1-week course twice a year, long weekend courses year-round for culinary professionals; 1-day to 1-week courses for nonprofessionals. Established in 1993. Curriculum: culinary. Total enrollment: 2-20. 100% of applicants accepted. 3% receive financial aid. 50% enrolled part-time. Student to teacher ratio 2-10:1. Facilities: Rancho Aurora, a working farm, has a handmade kitchen with 5 stations, outdoor kitchen with parilla, wood-fire and pre-hispanic cooking utensils.

Courses: Seven Days for Seven Regions, traditional foods, herbs, pre-hispanic foods, culinary and anthropological studies. Students come on a non-paid work study basis to be determined individually. Continuing education: archeological tours, herbal studies, farming, extended work-study.

Faculty: 1 full-time. Susana Trilling, teacher, writer, lecturer, chef, caterer, IACP member. Part-time teachers, cheese makers, bread bakers, chocolate makers, chefs, farmers, herbal healers.

Costs: $1,495 for 1-week course, which includes meals, lodging, planned activities. Lodging options are Oaxaca hotels or private casita at the ranch. Admission requirements: Student is working in or has studied culinary arts full-time. Scholarships: yes.

Location: Ranch surrounded by archeological sites in the mountains outside Oaxaca.

Contact: Susana Trilling, Director, Seasons of My Heart Cooking School, Rancho Aurora, PO Box AP 42, Admon 3, Oaxaca, Oaxaca, CP., 68101 Mexico; (52) 951-87726/954-83115, Fax (52) 951-87726/951-65280, E-mail seasons@antequera.com.

NEW ZEALAND

CENTRAL INSTITUTE OF TECHNOLOGY

Upper Hutt/Year-round

This trade school offers a 3-year diploma, 1-year certificate in Hospitality Operations-Management, 3-year bachelor's degree. Program started 1978. Accredited by State. Calendar: semester. Curriculum: core. Admission dates: August. Total enrollment: 300. 90 each admission period. 50% of applicants accepted. 40% receive financial aid. Student to teacher ratio 14:1. 50% of graduates obtain employment within six months. Facilities: 3 kitchens, 20 classrooms, 2 restaurants.

Faculty: 24 full-time.

Costs: Tuition in-country NZ$1,600, out-of-country NZ$10,500. On-campus housing cost: NZ$130 per week. Admission requirements: High school diploma or equivalent and admission test.

Contact: Mick Jays, Head of Department, Central Institute of Technology, Centre for Hospitality & Tourism Management, PO Box 40-740, Upper Hutt, New Zealand; (64) (0)4-527-6398 x6758, Fax (64) (0)4-527-6364, E-mail mick.jays@cit.ac.nz, URL http://www.cit.ac.nz.

NEW ZEALAND SCHOOL OF FOOD AND WINE

Christchurch/Year-round

This proprietary institution offers a 16-week full-time certificate in Foundation Cookery Skills, which includes New Zealand certificate in wine; 22-week full-time certificate in Restaurant-Cafe Management. Established in 1994. Accredited by New Zealand Qualifications Authority. Calendar: trimester. Curriculum: culinary. Admission dates: January, February, May, July, August. 12 each admission period. Student to teacher ratio 12:1. 90% of graduates obtain employment within six months. Facilities: Include 1 demonstration kitchen with overhead mirrors and 1 practical kitchen with commercial equipment, computer and seminar rooms.

Courses: Foundation skills based on Leith's Bible, concepts of French cuisine, food presentation, menu planning, costing, catering, wine. Restaurant and cafe management, computer skills to run a restaurant business.

Faculty: 10 full-time and part-time tutors.

Costs: Tuition is approximately NZ$4,650 (includes tax). Registered with the New Zealand Qualifications Authority (NZQA); students eligible for student visas. Loans: yes.

Location: Christchurch, a city of 300,000 in South Island.

Contact: Celia Hay, Director, The New Zealand School of Food and Wine, 63 Victoria St., Box 25217, Christchurch, So. Island, New Zealand; (61) (0)3-3797-501, Fax (61) (0)3-366-2302.

TAI POUTINI POLYTECHNIC

Greymouth/February-November

This trade-technical school offers NZQA 751 and 752 : One year full time. Established in 1988. Curriculum: culinary. Admission dates: November-February. Total enrollment: 24. 24 each admission period. 75% of applicants accepted. Student to teacher ratio 12:1. 95% obtain jobs.

Courses: Entry level (commis) into fine dining restaurants and hotels.

Faculty: 3 full-time catering and hospitalty.

Contact: Joseph Wellman, Chef Tutor, Tai Poutini Polytechnic, PO Box 469, Greymouth, SI, 3 New Zealand; (64) (0)3 7680411, E-mail jos@minidata.co.nz, URL http://www.geocities.com/NapaValley/6454/.

PHILIPPINES

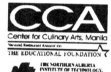

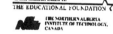

CENTER FOR CULINARY ARTS, MANILA

Quezon City/Year-round *(See display ad page 161)*

This private school offers a 2-year diploma program in Culinary Arts Technology Management, 1-year certificate program in Baking and Pastry Arts. 5 terms of 2 months per term. Program started 1995. Accredited by TESDA (Technical Education Skills Development Authority). Admission dates: June. Total enrollment: 50. 100 (50 first year, 50 second year) each admission period. 90% of applicants accepted. 0% receive financial aid. 0% enrolled part-time. Student to teacher ratio 6:1. Facilities: Modern lecture and lab facilities for institutional and small quantity cooking and baking; laboratory, computer, foodservice outlet.

Courses: Scientific background in principles, hands-on training in actual strategies/procedures with emphasis on positive work attitudes, values, and ethics. Practicum training in accredited institutions. Continuing education courses for chefs employed in the industry.

Faculty: 12 instructors.

Costs: 167,500 Philippine pesos (P39=US$1). Admission requirements: High school diploma, interview, written exam, medical exam. Scholarships: yes.

Location: Quezon City in the university belt. Part of Cravings Center, a full-service restaurant.

Contact: Corazon F. Gatchalian, Ph.D., Director, Center for Culinary Arts, Manila, 287 Katipunan Ave., Loyola Heights, Quezon City, Philippines; (632) 4264841, Fax (632) 4264840, E-mail culinary@mnl.sequel.net.

SCOTLAND

EDINBURGH COOKERY SCHOOL

Edinburgh/September-July

This private school offers full-time 10-week foundation, intermediate, advanced certificate courses; 22-week Intensive Certificate course, 1-year Cookery, French certificate course in conjunction with Basil Paterson College, 4-week Chalet Cook course, 4-week Host/Hostess course. Established in 1988. Admission dates: September, October, January, April, summer months. 24 each admission period. Student to teacher ratio 12 per class. 100% of graduates obtain employment within six months. Facilities: Include domestic and commercial gas cookers, electric ovens, and Aga cookers.

Faculty: Certificate/diploma.

Costs: Tuition, exclusive of VAT, is £3,975 for the Intensive Certificate course, £2,115 for the other certificate courses, £850 for 4-week course. Housing is £65 per week (space limited).

Contact: Ronnie Murray-Poore, Edinburgh Cookery School, The Coach House, Newliston, Kirkliston, Edinburgh, EH29 9EB Scotland; (44) (0)1764 655 440, Fax (44) (0) 764 655 440. Additional contact: (44) (0)131 662 9320.

TOP TIER SUGARCRAFT

Inverness/Year-round

This private school and mail order business offers 1-, 2-, and 3-week demonstration and participation courses for beginners and experts. Established in 1985. Student to teacher ratio 1-8:1. Facilities: 600-square-foot classroom with specialized equipment.

Courses: Sugarpaste, royal, floral, chocolate, marzipan, pastillage, personalized wedding cake instruction, cold porcelain work suitable for window display and centerpieces, individually tailored courses. Other activities include sightseeing in the Highlands.

Faculty: Principal teacher Diana Turner, British Sugarcraft Guild member and judge, has more than 10 years of experience in sugarcraft work.

Costs: 2-day (3-day, week-long) course £300 (£450, £750). Private tuition provided. Bed and breakfast lodging available on request.

Location: A mile from Culloden Battlefield, 5 miles from Inverness.

Contact: Diana Turner, Top Tier Sugarcraft, 10 Meadow Rd., Balloch, Inverness, IV1 2JR Scotland; (44) (0)1463-790456, Fax (44) (0)1463-790456.

SOUTH AFRICA

CHRISTINA MARTIN SCHOOL OF FOOD AND WINE

(See also page 266)
Durban/February-November

This private school offers a 1-year Intensive Diploma Chef's Course that covers food preparation, cooking and food and beverage service. Program started 1988. Calendar: semester. Curriculum: culinary, core. Admission dates: February. Total enrollment: 36. 36 each admission period. 80% of applicants accepted. 10% receive financial aid. 36% enrolled part-time. Student to teacher ratio 6:1. 100% of graduates obtain employment within six months. Facilities: 4 industrial kitchens with the latest equipment, auditorium, delicatessen, 60-seat restaurant, 80-seat conference venue, garden restaurant.

Courses: Development of sound basic culinary skills, honing skills and techniques, presenting food with flair. Cordon Bleu Part One and Part Two Courses, short courses for the public.

Faculty: Christina Martin is a Maitre Chef de Cuisine and Commandeur Associé de la Commanderie des Cordons Bleus de France. Instructors include vice-principal Michelle Barry, Chef de Cuisine Gerhard Van Rensburg, and Andrew White.

Costs: R47,880 ($7,366) for Intensive Diploma Course (payable in 3 installments), R900 for Cordon Bleu courses (6 lectures). Admission requirements: Senior School leavers certificate (matric). Scholarships: yes.

Location: 10 minutes from city center.

Contact: Christina Martin, Principal and Owner, Christina Martin School of Food and Wine, PO Box 4601, Durban, 4000 South Africa; (27) 31-3032111, Fax (27) 31-233342, E-mail chrismar@iafrica.com.

SILWOOD KITCHEN CORDONS BLEUS COOKERY SCHOOL

(See also page 267)
Rondebosch Cape/Year-round

This private school offers three 1-year culinary career courses that begin each January: the certificate course, the diploma course, the Grande Diploma. Established in 1964. Accredited by HITB-Hospitality Industries Training Board. Calendar: quarter. Curriculum: core. Admission dates: Last Thursday in January. Total enrollment: 44. 44, divided into groups of 11 each admission period. 60% of applicants accepted. Student to teacher ratio 11:1. 100% of graduates obtain employment within six months. Facilities: A 200-year-old coach-house converted into a demonstration and experimental kitchen, 3 additional kitchens, demonstration hall, and a library.

Courses: Cooking and baking, icing, wine, table art, mise en place, floral art.

Faculty: 11-member faculty. Includes school principal Alicia Wilkinson, nutrition instructors Jeanette Rietmann, menu reading instructor Alisa Smith, and practical supervisors Louise Faull, Peggy Loebenberg, Carianne Wilkinson, Pauline Copus and Shirley Henderson.

Costs: Tuition is R20,680 for the Certificate course, R8,470 for the Diploma course, R825 for the Grande Diploma. Housing is R500 to R1,000 per month.

Location: Cape Town.

Contact: Mrs. Alicia Wilkinson, Silwood Kitchen, Silwood Rd., Rondebosch Cape, South Africa; (27) 21-686-4894, Fax (27) 21-685-4378.

SPAIN

EL TXOKO DEL GOURMET

(See also page 345)
San Sebastian/Year-round

This specialty food shop offers a 12-month vocational course in Basic Cuisine consisting of 160 hours per month. Instruction in Spanish. Curriculum: culinary. Facilities: Basement workroom.

Courses: Internships at leading local restaurants on weekends and during summer months.

Faculty: Pepa Armendáriz.

Costs: $340.

Location: Opposite San Sebastian's covered market.

Contact: Pepa Armendáriz, El Txoko del Gourmet, Aldamar, 4 Bajo, San Sebastian, 20003 Spain; (34) 943-422-218, Fax (34) 943-427-641.

ESCOLA DE RESTAURACIO I HOSTALAGE A BARCELONA

Barcelona/Year-round

This private school in collaboration with the local government, La Generalitat de Catalunya offers a 2-year program to become a waiter, bartender, or cook, 3rd year part- or full-time training to become a chef or restaurant manager. Curriculum: culinary.

Costs: $4,000 per year.

Contact: Escola de Restauracio i Hostalage a Barcelona, C/Muntaner 70-72, Barcelona, 08011 Spain; (34) 934-532-904, Fax (93) 323-7423.

ESCUELA DE COCINA LUIS IRÍZAR

San Sebastian/September-July *(See also page 345)*

This private school offers a 2-year diploma program plus 15-day summer courses for professionals and nonprofessionals. Instruction in Spanish. Program started 1992. Curriculum: culinary. Admission dates: Year-round, selection in May. Total enrollment: 28 per year. 28 each admission period. 80% of applicants accepted. Student to teacher ratio 14:1. 100% of graduates obtain employment within six months. Facilities: Fully-equipped kitchen, separate classroom, TV and video.

Courses: Basque cuisine. Apprenticeships at leading restaurants. Wine tasting, nutrition, cocktails, visits to food and wine producers.

Faculty: Founder Luis Irízar has served as chef in Spain's leading restaurants. His staff includes 3 full-time instructors plus part-time teachers for continuing ed.

Costs: $450 per month. Admission requirements: Entry exam and interview.

Location: Old part of San Sebastian, facing the port and beach.

Contact: Virginia Irízar, Escuela de Cocina Luis Irízar, c/Mari, #5, Bajo, San Sebastian, 20003 Spain; (34) 43-431540, Fax (34) 43-423553.

ESCUELA DE HOSTELERIA ARNADI

Barcelona/Year-round *(See also page 345)*

This private school offers a 28-month Cooking and Pastry Chef course, 6-month Intensive Cooking course, 1-year Complete Cooking course, 1-year Basic Pastry course, 6-month Advanced Pastry course. Instruction in Spanish. Established in 1982. Curriculum: culinary.

Courses: Cooking basics, nutrition, baking and pastry.

Contact: Mey Hoffman, Escuela de Hosteleria Arnadi, C/Argenteria 74-78, Barcelona, 08003 Spain; (34) 933-195-889/882, Fax (34) 933-195-859.

ESCUELA DE HOSTELERIA Y COCINA DE VITORIA-GASTEIZ

Alava/Year-round

This private school with government support offers a 2-year program that prepares chefs to work in the hostelry field. Established in 1981. Curriculum: culinary. 25 each admission period.

Faculty: 10 instructors.

Costs: $200 per year. Admission requirements: High school diploma or entry exam, 18 years of age, one year professional experience.

Contact: Rodolfo Villate Ruiz de Gordexola, Director of School of Hostelry, Escuela de Hosteleria y Cocina de Vitoria-Gasteiz, Frontones de Mendizorrotza, Pza. Amadeo Salazar, Vitoria-Gasteiz (Alava), 01007 Spain; (34) 945-132-577, Fax (34) 945-132-083.

FUNDACIÓ ESCOLA DE RESTAURACIÓ I HOSTALARIA DE BARCELONA

Barcelona/Year-round

This private school offers a 3-year chef training program. Instruction in Spanish. Established 1985.

Costs: $4,000 per year. Admission requirements: High school diploma.

Contact: Gloria López de Padilla, Fundació Escola de Restauració i Hostalaria de Barcelona, C/Muntaner, 70-72, Barcelona, 08011 Spain; (34) 934-451-6982.

LA TABERNA DEL ALABARDERO

(See also page 346) **Seville/Year-round**

This private school offers a 3-year Professional Chef diploma program, 2-year Hotel Management diploma program. Instruction in Spanish. Program started 1993. Curriculum: culinary. Total enrollment: 60.

Costs: $5,000 for the first year, 5% more for each additional year.

Contact: Pedro Olives, La Taberna del Alabardero, Pza. de Molviedro, 4, Seville, 41001 Spain; (34) 954-560-637, Fax (34) 954-215-555.

LA TAHONA DEL MAR

Barcelona/January-May, September-November

This bakery offers 1-week (40-hour) courses in artisanal European breadmaking and Spanish and Catalan breads and pastries. Started 1998. Curriculum: culinary. Admission dates: spring and fall. Admission closes 2 weeks before class date. Total enrollment: 6. 6 enrollees each admission period. Facilities: A modern culinary classroom attached to a working bakery.

Courses: Traditional baking of Spain, in particular Catalonia, using natural starters and traditional formulas modernized for commercial production.

Faculty: Joaquín Llarás, professional baker & owner of La Tahona del Mar, studied at LeNôtre in France & with traditional bakers in Spain & Italy. His assistants are graduates of Spanish professional culinary schools and have industry experience.

Costs: 80,000 Spanish pesetas (about $650). Instruction is in Spanish. English interpretation is $200 additional. Admission requirements: Professional experience or training in bread or pastry making.

Contact: Elizabeth Duran, English Language Representative, La Tahona del Mar, 605 W. 111th St., #63, New York, NY 10025 US; 212-222-9062, Fax 212-222-6613, E-mail eduran@pipeline.com, URL http://www.sinix.net/paginas/tahona/ingles.htm. In Barcelona: Joaquín Llarás, (34) 93-307-1566.

THAILAND

ROYAL THAI SCHOOL OF CULINARY ARTS

(See also page 347) **Bangkok/Year-round**

This private school offers six 3- to 6-week Royal Thai cooking certificate courses leading to a diploma in Royal Thai Cooking. Nine Thai cooking courses (5-week intensive) leading to a Grand Diploma in Thai Cooking. Classes in Fusion and Burmese cooking. Program started 1997. Curriculum: culinary. Total enrollment: 32. 8 per class each admission period. 100% of applicants accepted. Student to teacher ratio 4:1. Facilities: Demonstration and 2 practical kitchens, each for 8 students.

Courses: Introductory to advanced Royal Thai (6 courses), regional Thai (3 courses), Fusion (3 courses), and Burmese (3 courses) cooking. Emphasis on feel, taste, and presentation of each cuisine.

Faculty: Le Cordon Bleu Grand Diplome graduates.

Costs: Ranges from $1,700-$12,000, includes single lodging in ocean-view room, two meals daily. Admission requirements: Culinary professional or serious home chef. Scholarships: yes.

Location: On the beach in Bang Saen about 88 km from the center of Bangkok.

Contact: Chris Kridakorn-Odbratt, Head Chef, Royal Thai School of Culinary Arts, 53 Sukhumvit 33 Rd., Bangkok, 10110 Thailand; 66 - 38 - 748 404, Fax 66 - 38 - 748 405, E-mail lechef@chef.net, URL http://www.rtsca.com.

National Apprenticeship Training Program for Cooks

American Culinary Federation Educational Institute (ACFEI)

Culinary Apprenticeship is a three-year on-the-job paid training program complemented by related instruction from an educational institution. The program began under the Carter Administration in 1976 with a grant from the U.S. Government and is now the 7th largest apprenticeship program in the U.S., with over 18,500 cooks being trained since its inception.

The apprenticeship program offers career-oriented cooks an alternative to private culinary institutions, community colleges, and vocational-technical schools. Apprentices, who generally range in age from 18 to 40 (average age 24), receive three years (6,000 hours total) of on-the-job training while earning an income. The first 500 hours are a probationary period, after which the apprentice is required to join the ACF and become registered with the Department of Labor. In addition to a 40-hour work week, the apprentice attends school part-time (a minimum of 192 hours per year) and may also have the opportunity to earn an associate degree. The average cost for school is between $500 and $3,000 per year.

To qualify for the program, applicant must be at least 17 years of age, have a high school diploma or equivalent, and have passed all entry-level academic and aptitude examinations as prescribed by the Apprenticeship Committee of the ACF. Consideration is given to those who have had high school foodservice training or on-the-job experience. A multiple-step screening process includes an orientation seminar, documentation of prior experience, and personal interviews.

The program is planned in six semi-annual stages, which can be shortened or lengthened according to the individual's ability. The apprentice keeps a weekly Log Book in which recipes and food preparation techniques are recorded. Those who complete the apprenticeship can: prepare, season, and cook soups, sauces, salads, meats, fish, poultry, game, vegetables, and desserts; produce baked goods and pastries; fabricate meat portions from primal cuts; prepare a buffet dinner; select and develop recipes; plan, write, and design complete menus; plan food consumption, purchasing, and requisitioning; operate a working budget in food and labor costing; recognize quality standards in fresh vegetables, meats, fish, and poultry; demonstrate basic artistic culinary skills, including ice carving, tallow sculpturing, cake decorating, and garniture.

In addition to work skills, the apprentice completes 30 hours minimum class time at an accredited postsecondary institution in each of 12 areas of related instruction; 1) Introduction to Food Service (Industry Survey); 2) Sanitation and Safety; 3) Basic Food Preparation (Introduction to Cooking); 4) Business Math (Food Cost Accounting); 5) Food and Beverage Service; 6) Nutrition; 7) Garde Manger; 8) Menu Planning and Design; 9) Baking; 10) Purchasing; 11) Supervisory Management; 12) Advanced Food Preparation.

On completion of the program and successful completion of a written exam, the apprentice is awarded the status of Certified Cook and may be offered employment at the training establishment or recommended for job placement.

Contact: ACF Apprenticeship Coordinator, American Culinary Federation, P.O. Box 3466, 10 San Bartola Rd., St. Augustine, FL 32085; (800) 624-9458, (904) 824-4468, Fax (904) 825-4758.

ALABAMA

ACF BIRMINGHAM CHAPTER

This chapter has 25 apprentices, 15 under age 25, 10 age 25 or over. Costs are $3,500 in-state, $5,000 out-of-state. Beginning salary is $5 per hour with 40 cent increases every 6 months. The 15 locations are country club, restaurant, hotel, corporate dining room, private club. Most desirable settings are local fine-dining restaurants and country clubs. Housing cost is $350-$500 per month. Degree program through Jefferson State Community College.

Contact: George White, Jefferson State Community College, 2601 Carson Rd., Birmingham, AL 35215; 205-856-7898.

ACF GREATER MONTGOMERY CHAPTER

This chapter has 60+ apprentices, 30 under age 25, 30 age 25 or over. Costs are $42 per semester hour. Beginning salary is $20,000+. The 25 locations are unlimited from healthcare to fine dining. Housing cost is $200 per month and up. Degree program through Trenholm State Technical College. Baking apprenticeship available.

Contact: Mary Ann Ward Campbell, CEC, CCE, 1225 Air Base Blvd., Montgomery, AL 36108; 205-262-4728, Fax 334-832-CHEF(2433).

ARIZONA

CHEFS ASSN. OF GREATER PHOENIX

This chapter has 8 apprentices, 4 under age 25, 4 age 25 or over. The 9 locations are resorts. Degree program through Scottsdale and Phoenix Community Colleges. Baking apprenticeship available.

Contact: Camron Clarkson, CWC, 2210 E. Sunnyside Dr., Phoenix, AZ 85028.

CHEFS ASSN. OF SOUTHERN ARIZONA, TUCSON

Contact: Ed Doran, 4123 E. Glenn Street, Tucson, AZ 85712; 520-881-3408.

RESORT & COUNTRY CLUB CHEFS OF THE SW ACF

Contact: Robert J. Chantos, CEC, AAC, Resort & Country Club Chefs Association, PO Box 12784, Scottsdale, AZ 85267-2784; 602-947-9295.

ARKANSAS

ACF CENTRAL AR CULINARY SCHOOL OF APPRENTICESHIP

This chapter has 40 apprentices. Classes begin August, January. Cost is $2,000 per year (payable by semester or financed monthly) plus books, uniform, knife kit. Beginning salary is minimum wage or determined by employer. The 12 locations include hotels, hospitals, owner-operated small restaurants, foodservice management companies.

Contact: Sharon B. McCone, RD LD, Executive Director, ACF Central Arkansas Culinary School of Apprenticeship, 24 Fairway Woods Circle, Maumelle, AR 72113; 501-831-CHEF, Fax 501-851-6684, E-mail acfark@juno.com.

ACF GREATER MONTGOMERY CHAPTER

This chapter has 11 apprentices, 4 under age 25, 7 age 25 or over. Costs are $650 per year. Beginning salary is $4.75 per hour with 5% increases every 6 months. The 8 locations include hotel, restaurant, country club.

Contact: Edward Hornyak, CEC, 6 Silver Birch Court, Little Rock, AR 72212; 501-225-5622.

CALIFORNIA

ACF CHEFS ASSN. SAN JOAQUIN

Contact: Bill McComas, 1175 E. Alluvial, Fresno, CA 93720.

CALIFORNIA CAPITOL CHEFS ASSN.

This chapter has 43 apprentices, 33 under age 25, 10 age 25 or over. Wait list is 1 year. Costs are $15 for books and materials. Beginning salary is $5.40 per hour with increases every 6 months. The 26 locations are restaurants and convalescent homes. Most desirable settings are Red Lion Hotel and El Paso Country Club. Housing cost is $425 per month. Degree program through Sierra College. Baking apprenticeship available.

Contact: Jon Greenwalt, CEC, AAC, 5475 Asby Lane, Granite Bay, CA 95746; 916-791-2554.

CHEFS DE CUISINE ASSN. OF CALIFORNIA

This chapter has 15 apprentices, 14 under age 25, 1 age 25 or over. Costs are $150. Beginning salary is variable with increases every 6 months. The locations are hotel, private club, restaurant. Most desirable settings are hotels. Housing cost is variable. Degree program through California Polytechnic. Baking apprenticeship available.

Contact: LeRoy Blanchard, CEC, Culinary Arts Dept., 400 W. Washington Blvd., Los Angeles, CA 90015; 213-744-9480, Fax 213-748-7334, E-mail LeRoy_S._Blanchard@laccd.cc.ca.us, LeRoy_S._Blanchard@laccd.cc.ca.us.

NORTHERN CALIFORNIA CHEFS ASSN.

Contact: Michael Piccinino, CEC, CCE, AAC, 6945 Pine Dr., Anderson, CA 96007; 530-225-4829.

ORANGE EMPIRE CHEFS ASSN.

This chapter has 25 apprentices. Costs are $100 one time ACF fee, $55 annually and $13 per unit. Beginning salary is negotiable with negotiable increases. The 30 locations are hotel, country club and restaurant. Most desirable settings are Ritz Carlton, Laguna Niguel, Sutton Place Hotel, and Westin. Housing cost is $400-$600 per month. Degree program through Orange Coast College. Baking apprenticeship available.

Contact: Bill Barber, CWC, Orange Coast Community College, 2701 Fairview Rd., P.O. Box 5005, Costa Mesa, CA 92628-5005; 714-432-5835.

SAN FRANCISCO CULINARY/PASTRY PROGRAM

This chapter has 15 apprentices, all age 25 or over. Wait list is 12-18 months. Beginning salary is 55% of journeyman wage with 5% increases every 6 months. The 5-7 locations are full-service hotels and restaurants. Degree program through City College. Baking apprenticeship available.

Contact: Joan Ortegea, 760 Market St., Suite 1066, San Francisco, CA 94102; 415-989-8726, Fax 415-989-2920.

SANTA CLARA COUNTY CHEFS ASSN.

Contact: Eric Carter, 1572 Monteval Lane, San Jose, CA 95120; 408-479-5012.

COLORADO

ACF COLORADO CHEFS DE CUISINE ASSOCIATION DENVER

Contact: Michelle Wise, 820 16th St., Ste. 421, Denver, CO 80202; 303-575-4808, Fax 303-575-4840.

ACF CULINARIANS OF COLORADO

This chapter has 35 apprentices, 17 under age 25, 17 age 25 or over. Costs are $2,000 approximately. Beginning salary is 70% of journeyman wage with 5% increases every 6 months. The 30 locations are high-end hotels, country clubs, and restaurants. Housing cost is $300+ per month. Degree program through Community College of Denver.

Contact: Michelle Wise, 820 - 16th St., Ste. 421, Denver, CO 80202; 303-571-5653, Fax 303-571-5653.

ACF PIKES PEAK CHAPTER, INC.

Contact: Siegfried Eisenburger, CEC, AAC, Broadmoor Hotel, 1 Lake Ave., Colorado Springs, CO 80909; 719-634-7711 #5336.

FLORIDA

ACF CENTRAL FLORIDA CHAPTER

This chapter has 300 apprentices, 80 under age 25, 20 age 25 or over. Costs are $615. Beginning salary is $6.50 per hour with 25 cent increases annually. The 50 locations are restaurant, corporate and resorts. Most desirable settings are Swan(Westin) Hotel, Peabody Hotel, Sheraton Plaza. Baking apprenticeship available.

Contact: Dale Pennington, Mid Florida Tech, Culinary Arts O/TEC, 2900 W. Oak Ridge Road, Orlando, FL 32809; 407-855-5880 #286.

ACF FIRST COAST CHAPTER

This chapter has 25 apprentices, 15 under age 25, 10 age 25 or over. Wait list is 3 months. Costs are $300 per year. Beginning salary is $5.50 per hour with 25 cent increases every 6 months. The 20 locations are restaurant, hotel, resort, private club, country club. Most desirable settings are River Club, Ritz Carlton Amelia Island, Omni Hotel. Housing cost is $400 per month. Degree program through Florida Community College at Jacksonville. Baking apprenticeship available.

Contact: John Wright, CEPC, CEC, CCE, AAC, 5437 Calloway Ct., Jacksonville, FL 32209; 904-765-2140.

ACF GREATER FT. LAUDERDALE CHAPTER

This chapter has 22 apprentices, 6 under age 25, 16 age 25 or over. Costs are $200. Beginning salary is $6-$8 with 25-50 cent increases every 3-6 months. The 12 locations are restaurant, resort, hospital. Most desirable settings are Marriott Harbor Beach Resort, Bonaventure Hotel Resort, Lauderdale Yacht Club. Degree program through Atlantic Vocational Technical Center. Baking apprenticeship available.

Contact: Thomas Ferrel, 5600 NW 59th St., #1, Tamarac, FL 33319; 954-848-7133 (bpr).

ACF GULF TO LAKES CHEFS AND COOKS CHAPTER

This chapter has 45 apprentices, 18 under age 25, 27 age 25 or over. Costs are paid by school. Beginning salary is $6-$8, subject to wage scales with 25 cent increases every 6 months. The 18 locations are restaurant, resort, hotel, inn, nursing care home. Most desirable settings are full service resorts, hotels and restaurants. Housing cost is $350-$550 per month. Degree program through Lake County Vocational Tech Center.

Contact: Steve Neverman, c/o Gulf to Lakes Chapter, P.O. Box 1179, Eustis, FL 32727; 352-742-6486, E-mail gulfchefs@aol.com, gulfchefs@aol.com.

ACF PALM BEACH COUNTY CHEFS

This chapter has 5 apprentices, all under age 25. Costs are $300 per year. Beginning salary is $8 per hour with 50 cent increases every 6 months. The 10 locations are restaurant, hotel, resort, private club, country club, corporate dining room. Most desirable settings are hotel, private country club. Housing cost is $400 per month. Degree program through Palm Beach Community College.

Contact: Roderick G. Smith, CEC, AAC, 6615 Lawrence Woods Ct., Lantana, FL 33462; 561-732-2520, Fax 561-732-7400.

ACF SOUTHWEST FLORIDA CHEFS ASSN.

Contact: , 1625 NE 6th St., Cape Coral, FL 33909; 941-574-7393.

ACF TAMPA BAY CHEFS AND COOKS ASSN.

This chapter has a variable number of apprentices. Costs are $1,200 approx., $34 per credit. Beginning salary is open. Degree program through Hillsborough Community College. Baking apprenticeship available.

Contact: George Pastor, CEC, CCE, AAC, 11722 Spanish Lake Dr., Tampa, FL 33635-6307; 813-932-8612.

ACF TREASURE COAST CHAPTER

This chapter has 30 apprentices, 10 under age 25, 20 age 25 or over. Costs are $160 plus texts and materials. Beginning salary is $5-$6 per hour with increases every 3 months. The 15 locations are country club, resort, corporate, institutional and restaurant. Most desirable settings are Indian

River Plantation, Harbor Ridge CC, Monarch CC. Baking apprenticeship available.

Contact: Thomas Boehm, CWC, 2801 S. Kanner Hwy., Stuart, FL 34994; 561-287-0710 x315.

DISNEY'S CULINARY ACADEMY

This chapter has 55 apprentices. Costs are $2,500 per year, payroll deduction 2nd and 3rd years. Beginning salary is $6.20 per hour with full benefits. The 20+ locations are Walt Disney World's resort hotels, theme parks, Downtown Disney West Side & Marketplace.

Contact: Carolyn Tremblay, Walt Disney World Co., Culinary Development, Box 10,000, Lake Buena Vista, FL 32830-1000; 407-827-4706, Fax 407-827-4709, E-mail Carolyn_Tremblay@wda.disney.com, Carolyn_Tremblay@wda.disney.com.

GULF COAST CULINARY ASSOCIATION

Contact: Gus Silivos, CEC, 670 Scenic Hwy., Pensacola, FL 32503; 850-432-6565.

TC CENTRAL FLORIDA-TECOM

Contact: Rick Petrello, CEC, CCE, AAC, 2900 West Oakridge Rd., Orlando, FL 32809; 407-344-5080.

VOLUSIA COUNTY CHEFS & COOKS

This chapter has 28 apprentices, 12 under age 25, 16 age 25 or over. Costs are $700 in-state, $900 out-of-state. Beginning salary is $5-$6 per hour with increases annually. The 30 locations are restaurant, hotel, country clubs. Most desirable settings are country clubs, hotels, and gourmet restaurants.

Contact: Jeff Conklin, Daytona Beach Community College, 1200 West International Blvd., Daytona Beach, FL 32120; 904-255-8131 #3735.

GEORGIA

ACF AUGUSTA CHAPTER

Contact: Harry Sayles, CEC, 2010 Bald Eagle Dr., Hephzibah, GA 30815; 706-722-3008.

ACF GOLDEN ISLES OF GEORGIA

Contact: Charles Miller, 504 E. Island Square, St. Simons Island, GA 31522; 912-638-3611.

ACF INC. CHEFS ASSN. OF GREATER ATLANTA

Contact: John Brantley, CEC, 3571 Forrest Glen Trail, Lawrenceville, GA 30244; 770-495-0997.

HAWAII

CHEFS DE CUISINE ASSN. OF HAWAII

Contact: Randy Francisco, Kapiolani Community College OCS, 1424 Diamond Rd., Honolulu, HI 96816; 808-734-9457.

MAUI CHEFS ASSN.

This chapter has 6 apprentices, 4 under age 25, 2 age 25 or over. Wait list is 1 year. Costs are $468 resident, $2,856 nonresident per semester. Beginning salary is $10.50-$11 per hour with variable increases every 6 months. The 3 locations are resort and restaurant. Most desirable settings are Westin Maui and Kea Lani Resort. Housing cost is $600-$800 per month. Degree program through Maui Community College.

Contact: Christopher Speere, Maui Community College, 310 Kaahumanu Ave., PO Box 1284, Kahului, HI 96732; 808-242-1210, Fax 808-242-1210, E-mail speere@mccada.mauicc.hawaii.edu.

ILLINOIS

ACF CHICAGO CHEFS OF CUISINE, INC.

Contact: Jeff M. Lemke, CWC, 23808 Sunset Drive, Lake Zurich, IL 60047; 847-854-3010.

INDIANA

ACF GREATER INDIANAPOLIS CHAPTER

Contact: Allen Elsesy, CEC, PO Box 355, St. Paul, IN 47242; 317-615-1510 x220.

ACF SOUTH BEND CHAPTER

This chapter has 20 apprentices, 7 under age 25, 13 age 25 or over. Costs are $3,825 for 3 years(6,000 hours). Beginning salary is $7.50 per hour with 8% increases every 1,000 hours. The 14 locations are restaurant, hotel, private club, country club, college, hospital. Most desirable settings are restaurant, hotel, country club. Degree program through Lake Michigan College.

Contact: Denis F. Ellis, CEC, AAC, University of Notre Dame, South Dining Hall, Notre Dame, IN 46556; 219-631-5416.

IOWA

ACF CHEF DE CUISINE/QUAD CITIES (IOWA)

This chapter has 25 apprentices, 13 under age 25, 12 age 25 or over. Wait list is 1 year. Costs are $5,100 for 3 years. Beginning salary is $5 per hour with 25 cent increases every 6 months. The 22 locations are country club, hotel and restaurant. Housing cost is $300-$350 per month. Degree program through Scott Community College.

Contact: Brad Scott, Scott Community College, 500 Belmont Rd., Bettendorf, IA 52722; 319-359-7531 x278.

ACF GREATER DES MOINES CULINARY ASSN.

Contact: Robert Anderson, Des Moines Area Community College, 2006 S. Ankeny Blvd., Ankeny, IA 50021; 515-964-6200 x6566.

KANSAS

ACF GREATER KANSAS CITY CHEF'S ASSOCIATION

Degree program through Johnson County Community College.

Contact: Patrick Sweeney, CEC, AAC, Johnson County Community College, 12345 College at Quivira, Overland Park, KS 66210; 913-469-8500 x3611.

LOUISIANA

ACF NEW ORLEANS CHAPTER

This chapter has 190 apprentices, 123 under age 25, 67 age 25 or over. Costs are $3,600. Beginning salary is with 25 cent increases every 6 months. The 75 locations are restaurant, hotel, country club, private club. Housing cost is $350-$550. Degree program through Delgado Community College.

Contact: Iva Bergeron, CCE, Delgado Community College, 615 City Park Ave., Bldg. 11, New Orleans, LA 70119-4399; 504-483-4208.

MARYLAND

CENTRAL MARYLAND CHEFS ASSN.

This chapter has 26 apprentices, 24 under age 25, 1 age 25 or over. Beginning salary is $5 per hour with increases every 6 months. Most desirable settings are hotel, country club, restaurant. Degree program through Anne Arundel Community College. Baking apprenticeship available.

Contact: Terry Green, CCE, Western School, 100 Kenwood, Baltimore, MD 21228; 410-887-0852.

MASSACHUSETTS

EPICUREAN CLUB OF BOSTON

This chapter has 5 apprentices, 1 under age 25, 5 age 25 or over. Beginning salary is variable with variable increases annually. Housing cost is $700 and up. Degree program through Bunker Hill Community College.

Contact: Christoph Leu, The Westin Hotel, 10 Huntington Ave., Boston, MA 02116-5798; 617-424-7524.

MASSACHUSETTS CULINARY ASSOCIATION

Contact: Stanley Nicas, CEC, AAC, 1230 Main St., Leicester, MA 01524; 508-892-9090.

MICHIGAN

ACF BLUE WATER CHEFS ASSN.

Contact: David F. Schneider, CEC, CCE, Macomb Community College, 44575 Garfield Rd., Clinton Township, MI 48038-1139; 810-286-2088, Fax 810-286-2038.

ACF MICHIGAN CHEFS DE CUISINE ASSN.

This chapter has 65 apprentices, 30 under age 25, 30 age 25 or over. Costs are $1,500. Beginning salary is $6 per hour with increases every 6 months. The 60 locations are hotel, restaurant, country and city club, hospital. Most desirable settings are In Oakland County. Degree program through Oakland Community College.

Contact: Kevin Enright, CEC, CCE, Oakland Community College, 27055 Orchard Lake Rd., Farmington Hills, MI 48334-4579; 248-471-7785, Fax 248-471-7553, E-mail kmenrigh@occ.cc.mi.us.

ACF OF NORTHWESTERN MICHIGAN

This chapter has 36 apprentices. Costs are $2,800 for 3 years. Beginning salary is $6 per hour with increases every 6 months. The 22-25 locations are institutional, hotel, country club, restaurant. Housing cost is $280 per month. Degree program through Lake Michigan College.

Contact: Andrew Colvin, CC, Shanty Creek-Schuss Mountain, Bellair, MI 49615; 616-533-8621.

MISSOURI

ACF CHEFS & COOKS OF SPRINGFIELD/OZARK

This chapter has 19 apprentices, 19 under age 25, age 25 or over. Beginning salary is minimum wage with increases every 6 months. The 6-8 locations are hotel, country club, restaurant.

Contact: James Lekander, CEC, 707 McLean Ct., Nixa, MO 65714; 417-724-0968.

CHEFS DE CUISINE OF ST. LOUIS

This chapter has 14 apprentices, 13 under age 25, 1 age 25 or over. Wait list is 6 months. Degree program through St. Louis Community College. Baking apprenticeship available.

Contact: Michael Downey, St. Louis Community College, 5600 Oakland Ave., HRM Dept., St. Louis, MO 63110; 314-664-9100.

NEBRASKA

ACF PROFESSIONAL CHEFS OF OMAHA

This chapter has 20 apprentices, 10 under age 25, 10 age 25 or over. Costs are $4,000. Beginning salary is $7 per hour with variable increases annually. The 15 locations are hotels, casinos. country clubs, upscale restaurants. Degree program through Metro Community College.

Contact: Jim Trebbien, CCE, Metropolitan Community College, Bldg. 10, P.O. Box 3777, 30th & Fort Sts, Omaha, NE 68103-0777; 402-457-2510, Fax 402-457-2515, E-mail jtrebbien@metropo.mccneb.edu.

NEVADA

HIGH SIERRA CHEFS ASSN.

This chapter has 8 apprentices, 6 under age 25, varies age 25 or over. Wait list is 3 months. Costs are $1,200-$1,500 with partial reimbursement. Beginning salary is $6.87-$7.37 per hour (varies with hotel) with 3-5% increases annually. The 2+ locations are hotel and casino. Most desirable settings are Harrah's Tahoe, Harrah's Reno. Housing cost is $300-$500 per month. Degree program through Lake Tahoe and Truckee Meadows Community Colleges. Baking apprenticeship available.

Contact: Paul J. Lee, CEC, Harrah's Lake Tahoe, PO Box 8, Lake Tahoe, NV 89449; 702-588-6611 x2205, Fax 702-586-6643, E-mail plee@laketahoe.harrahs.com, plee@laketahoe.harrahs.com.

THE FRATERNITY OF EXECUTIVE CHEFS OF LAS VEGAS ACF

This chapter has 11 apprentices, 11 under age 25, age 25 or over. Wait list is variable. Costs are $165. Beginning salary is 80% of cook's helper's wages with increases annually. The 4 locations are resort. Most desirable settings are Mirage, Caesar's, Hilton. Housing cost is $350-$500. Degree program through University of Las Vegas. Baking apprenticeship available.

Contact: Joseph Mulligan, CEC, 132 Villaggio, Henderson, NV 89014; 702-435-3206.

NEW HAMPSHIRE

GREATER NORTHERN NEW HAMPSHIRE

This chapter has 13 apprentices, 12 under age 25, 1 age 25 or over. Beginning salary is minimum wage plus room and board with 85 cent increases every 2,000 hours. The 7 locations are Balsams Grand Resort Hotel, Gasparilla Inn, The Cloister, American Club, Wigwam, Broadmoor, Hyatt Regency, Grand Cypress. Housing cost is free. Degree program through New Hampshire Technical College.

Contact: Phil Learned, CEC, AAC, The Balsams Resort Hotel, Box 112, Dixville Notch, NH 03576; 603-255-3861, Fax 603-255-4670.

NEW JERSEY

ACF NORTHERN NEW JERSEY CHAPTER

Contact: Joe Amabile, CEC, Bergan County Tech High School, 200 Hackensack Avenue, Hackensack, NJ 07601; 201-343-6000 x2255.

PROFESSIONAL CHEFS GUILD OF CENTRAL NEW JERSEY

Degree program through Mercer County Community College.

Contact: Doug Fee, Mercer County Community College, 1200 Old Trenton Rd., Trenton, NJ 08690; 609-586-4800 x3476.

PROFESSIONAL CHEFS OF SOUTH JERSEY

Contact: John Carbone, CCE, CEC, AAC, P.O. Box 157, Port Republic, NJ 08241; 609-646-4950.

NEW MEXICO

ACF CHEFS OF SANTA FE

Contact: Maurice Zeck, CEC, AAC, 100 E. San Francisco St., Santa Fe, NM 87501; 505-982-5511.

ACF RIO GRANDE VALLEY CHAPTER

Contact: Diane Ciampolillo, CEC, Albuquerque Tech-Vocational Inst., 920 River View Dr. SE, Rio Rancho, NM 87124; 505-896-3000, Fax 505-896-4205.

ACF S.W. NEW MEXICO & TEXAS

Contact: Tatsuya Miyazaki, 930 El Paseo Road, Las Cruces, NM 88001; 505-526-7144.

NEW YORK

ACF CAPITAL DISTRICT OF CENTRAL NEW YORK

Costs are $3,045 for 3 years. Beginning salary is $6 with sliding scale increases. The 5 locations are full-service restaurants, country clubs. Housing cost is $250 per month. Degree program through Schenectady County Community College.

Contact: Scott A. Vadney, 43 Berwyn Street, Schenectady, NY 12304; 518-388-6345.

ACF OF GREATER BUFFALO NEW YORK

This chapter has 3 apprentices, 3 under age 25, age 25 or over. Beginning salary is negotiable. The locations are private clubs. Most desirable settings are full service with extensive catering. Degree program through Niagara County Community College or Erie Community College. Baking apprenticeship available.

Contact: Samuel J. Sheusi, CEC, CCE, 5084 Dana Dr., Lewiston, NY 14092; 716-731-4101.

MID-HUDSON CULINARY ASSN.

Contact: Brent Wertz, CWC, Mohonk Mountain House, New Paltz, NY 12561; 914-256-2070, Fax 914-256-2161.

NORTH CAROLINA

ACF INC. CHARLOTTE CHAPTER

Contact: Chris Jones, PO Box 471053, Charlotte, NC 28248; 704-597-0414.

ACF SANDHILLS/CROSS CREEK CHEFS ASSN.

Contact: Gary Kowal, CEC, Pinehurst Resort, PO Box 4000, Carolina Vista, Village of Pinehurst, NC 28374; 919-295-6565.

TRIAD PROFESSIONAL CHEFS ASSN.

Contact: S. Mitchell Mack, c/o HIFS, 3121 High Point Rd., Greensboro, NC 27407; 919-242-9161 x4112.

WESTERN NORTH CAROLINA CULINARY ASSN.

Contact: Dennis R. Trantham, CC, Route 4, Box 256A, Canton, NC 28716; 704-648-7195.

OHIO

ACF CLEVELAND CHAPTER

This chapter has 10-20 apprentices, 50% under age 25, 50% age 25 or over. Costs are $140 start up for membership in ACF and books. Beginning salary depends on skill level. The numerous locations are clubs, country clubs, restaurants. Degree program through Cuyahoga Community College. Baking apprenticeship available.

Contact: Richard Fulchiron, CEC, CCE, Cuyahoga Community College, Hospitality Mgmt. Dept., 2900 Community College Ave., Cleveland, OH 44115; 216-987-4087, Fax 216-987-4096, E-mail Richard.Fulchiron@TRI-C.CC.OH.US, Richard.Fulchiron@TRI-C.CC.OH.US.

ACF COLUMBUS CHEFS CHAPTER

This chapter has 75 apprentices, 50 under age 25, 25 age 25 or over. Costs are $7,000 for 3 years. Beginning salary is $6 per hour and up with variable increases. The 45 locations are hotel, country club, private club, restaurant. Most desirable settings are restaurant, hotel, club. Degree program through Columbus State Community College.

Contact: Carol Kizer, CCE, Columbus State Community College, 550 E. Spring St., Columbus, OH 43215; 614-227-2579, Fax 614-227-5973, E-mail ckizer@cscc.edu, ckizer@cscc.edu.

OKLAHOMA

ACF CULINARY ARTS OF OKLAHOMA

This chapter has 12 apprentices. Beginning salary is minimum wage. The 6 locations are restaurant, hotel, resort, country club. Most desirable settings are country clubs. Degree program through OSU-Oklahoma City. Baking apprenticeship available.

Contact: Genni Thomas, CEPC, CEC, CCE, AAC, 4337 Dahoon Dr., Oklahoma City, OK 73120; 405-755-0550 x155, Fax 405-751-8971.

ACF TULSA CHAPTER

Contact: Robert M. Boyce, CWC, 5531 S. Toledo Pl., Tulsa, OK 74135-4325; 918-486-6575, Fax 918-486-6576.

PENNSYLVANIA

ACF LAUREL HIGHLANDS CHAPTER

This chapter has 74 apprentices, 56 under age 25, 18 age 25 or over. Costs are $3,216 for 3 years. Beginning salary is minimum wage with increases every 1,000 hours. The 35 locations are restaurant, hotel, club, resort, institution. Most desirable settings are club, hotel, restaurant. Degree program through Westmoreland County Community College.

Contact: Mary Zappone, CCE, Westmoreland Community College-Culinary Arts, Armbrust Rd., Youngwood, PA 15697-1895; 724-925-4016, Fax 724-925-4293, E-mail zappinm@astro.westmoreland.cc.pa.us, zappinm@astro.westmoreland.cc.pa.us.

ACF PITTSBURGH CHAPTER

This chapter has 60 apprentices, 40 under age 25, 20 age 25 or over. Costs are $4,000. Beginning salary is negotiable with 25 cent increases every 6 months. Most desirable settings are hotels, clubs, restaurants. Degree program through Community College of Allegheny Cty. Baking apprenticeship available.

Contact: Paul Passafume, 274 Washington Street, Whitacre, PA 15120; 412-578-5513.

DELAWARE VALLEY CHEFS ASSN.

Contact: William Tillinghast, 3191 Janney Street, Philadelphia, PA 19134; 215-895-1143.

SOUTHERN ALLEGHENY CHEFS ASSN.

This chapter has 12 apprentices, 9 under age 25, 3 age 25 or over. Costs are $2,500 per year. Beginning salary is $6 with 50 cent increases every 6 months. The 8 locations are catering- white tablecloth.

Contact: Andrew G. Iannacchione, CEC, AAC, Hospitality Referrals/Consultants, RD 1 Box 264, Bedford, PA 15522; 814-623-9029, Fax 814-623-9029, E-mail agicec@aol.com, agicec@aol.com.

PUERTO RICO

CARIBBEAN CULINARY FEDERATION

Contact: William Moore, CEC, AAC, Caribbean Culinary Association, 18 Marseilles Street, Suite 2B, San Juan, PR 00907; 787-725-9139.

SOUTH CAROLINA

ACF MIDLAND CHAPTER

Contact: Bryant Withers, CEC, 300 Spring Valley Rd., Columbia, SC 29223; 803-788-3080.

TENNESSEE

ACF MIDDLE TENNESSEE CHAPTER, OPRYLAND HOTEL

This chapter has 50 apprentices, 25 under age 25, 25 age 25 or over. Wait list is 1 year. Costs are $500. Beginning salary is $5.25 per hour with 25 cent increases every 6 months. The locations are restaurant, hotel, club kitchens, in-house butcher shop, bakery, pastry shop of the Opryland complex. Degree program through Volunteer State Community College.

Contact: Dina D. Starks, RD, 2800 Opryland Dr., Nashville, TN 37214; 615-871-7765, Fax 615-871-6942, E-mail dstarks@oprylandusa.com, dstarks@oprylandusa.com.

TEXAS

ACF CAPITOL OF TEXAS CHEFS

This chapter has 6 apprentices, 4 under age 25, 2 age 25 or over. Costs are $500 per year. Beginning salary is $6-$7 per hour with 25-40 cent increases every 6 months. The 12 locations are hotel, resort, country club, conference centers. Most desirable settings are Hill country, downtown, university area. Degree program through Le Chef College Hospitality of Careers.

Contact: Matt Collins, CWC, 6020 Dilliard Circle, Austin, TX 78752; 512-323-2511.

ACF PROF. CHEFS ASSN. OF HOUSTON

Contact: Dan Capello, The Houston Country Club, 1 Potomac Dr., Houston, TX 77057; 713-465-8381 x246, Fax 713-465-7455.

ACF TEXAS CHEFS ASSN.

This chapter has 80 apprentices. Costs are less than $1,000 per year. The locations are hotels, restaurants, and country clubs of Dallas and Houston areas. Degree program through San Jacinto Community in Houston and El Centro Community in Dallas. Baking apprenticeship available.

Contact: Apprenticeship Coordinator, 2161 N.W. Military Hwy., Ste. 305, San Antonio, TX 78213; 210-377-1092, Fax 210-377-1093, E-mail texchef@swbell.net, texchef@swbell.net.

TCA-BRAZOS VALLEY CHAPTER

This chapter has 13 apprentices. Wait list is 30 applicants long. Costs are none. Beginning salary is $6 per hour with state mandated increases annually. The 5 locations are educational and institutional. Most desirable settings are Texas A&M University. Housing cost is $450 per month. Baking apprenticeship available.

Contact: Vicki Beck, Texas A&M University, FoodService Dept., College Station, TX 77843; 409-845-9312.

TCA-DALLAS

Contact: James Goering, CCE, CEC, El Centro College, Main at Lamar, Dallas, TX 75202-9604; 214-746-2217.

TCA-HOUSTON

Contact: George Messinger, CEC, CCE, 19826 Atascocita Pines Dr., Humble, TX 74346; 713-459-7150.

VERMONT

NORTH VERMONT CHEFS & COOKS ASSN.

Costs are $835 per year. 15 locations. Most desirable settings are Marriott, Perry Restaurant Group, restaurants in Stowe locations. Baking apprenticeship available.

Contact: Patrick R. Miller, CEC, 2575 Weeks Hill Rd., Stowe, VT 05672; 802-253-4236, Fax 802-253-4236.

VIRGINIA

ACF NATIONS CAPITOL CHEFS

This chapter has 28 apprentices, 18 under age 25, 10 age 25 or over. Costs are $4,500 for 3 years. Beginning salary is $6 per hour with increases every 6 months. The 28 locations are restaurant, hotel, country club. Most desirable settings are Radisson Mark Plaza, Congressional Country Club, Vista International Hotel. Degree program through North Virginia Community College.

Contact: Forest Bell, 6289 Dunaway Ct., McLean, VA 22101; 301-469-2018, Fax 301-469-2035.

BLUE RIDGE CHEFS ASSN.

Contact: Bill King, CEC, 3425 Pippin Ln., Charlottesville, VA 22903; 703-894-5436, Fax 703-894-0534.

VIRGINIA CHEFS ASSN.

This chapter has 20 apprentices, 19 under age 25, 1 age 25 or over. Costs are $3,000 in-state plus $1,250 for books and materials. Beginning salary is $6.50 per hour with 25 cent increases quarterly. The 15 locations are restaurant, country club, hotel, grocery store, central commissary and hospital. Most desirable settings are Tobacco Company Restaurant, country clubs, several independent restaurants. Housing cost is $300-$500 per month. Degree program through J. Sargeant Reynolds Community College.

Contact: Bruce Clarke, 3204 Old Gun Rd. East, Midlothian, VA 23113; 804-272-0761, Fax 804-786-0655, E-mail stnfdhill@aol.com, stnfdhill@aol.com.

WASHINGTON

WASHINGTON STATE CHEFS ASSN.

Contact: Jamie Callison, CC, 8720 - 231st St. SW, Edmonds, WA 98026; 206-363-2169.

WEST VIRGINIA

ACF WEST VIRGINIA CHAPTER

Contact: Dan Ferguson, CWC, 216 Rockledge Drive, Nitro, WV 25143.

WISCONSIN

ACF CHEFS OF MILWAUKEE, INC.

This chapter has 30 apprentices, 10 under age 25, 20 age 25 or over. Costs are $2,500. Beginning salary is $6 per hour with 5%-10% increases every 6 months. The 48 locations are restaurant, hotel, private club, country club, catering. Most desirable settings are restaurant, hotel, country club. Degree program through Milwaukee Area Technical College.

Contact: John Reiss, CCC, CCE, Milwaukee Area Technical College, 700 W. State St., Milwaukee, WI 53233; 414-297-6861, Fax 414-297-7990, E-mail reissj@milwaukee.tec.wi.us, reissj@milwaukee.tec.wi.us.

ACF FOX VALLEY CHAPTER

This chapter has 27 apprentices. Costs are $1,400 for 3 years. Beginning salary is $5 per hour with increases each semester. The 19 locations are restaurant, hotel, resort, convention center, country club. Most desirable settings are American Club Resort, Paper Valley Hotel, Oneida Golf and Riding Club. Housing cost is $300 per month. Degree program through Fox Valley Technical College.

Contact: Albert Exenberger, CEC, CCE, AAC, Fox Valley Technical Institute, 1825 N. Bluemound Dr., PO Box 2277, Appleton, WI 54913-2277; 414-735-5600 x735.

ACF MIDDLE WISCONSIN CHEFS

Contact: Gregory Krzyminski, Mid-State Technical College, 500 32nd Street N., Wisconsin Rapids, WI 54494; 715-423-5650.

BAHAMAS

BAHAMAS CULINARY ASSN.

This chapter has 53 apprentices, 90% under age 25, 10% age 25 or over. Wait list is 1 year. Costs are $150. Beginning salary is $120 per week with increases annually. The 208 locations are resort, hotel, restaurants. Most desirable settings are Princess Towers Hotel, Sun International, and Carnival's Crystal Palace Hotel. Degree program through Bahamas Hotel Training College. Baking apprenticeship available.

Contact: Linda Mortimer, Bahamas Hotel Training College, PO Box N 4896, Nassau, Bahamas; 242-326-5860, Fax 242-325-2459.

BAHAMAS HOTEL TRAINING COLLEGE

This chapter has 25 apprentices, 90% under age 25, 10% age 25 or over. Wait list is 1 year. Costs are $1000 per year. Beginning salary is $120 per week with increases annually. The 8 locations are resort, hotel, restaurant. Most desirable settings are Bahamas Princess Country Club, Lucaya Beach Hotel, Princess Casino. Degree program through Bahamas Hotel Training College. Baking apprenticeship available.

Contact: Paul Pelfrense, PO Box F-1679, Freeport, Bahamas; 242-352-2896, Fax 242-352-9002.

2

Non-Vocational and Vacation Programs

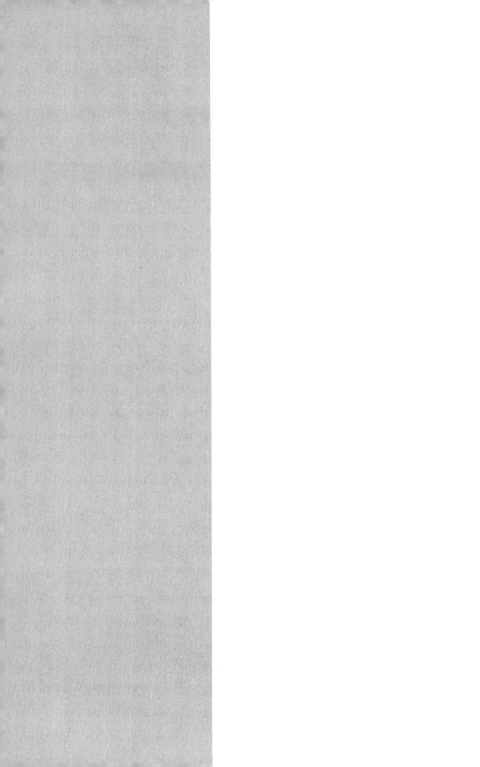

ALABAMA

COOK'S GARDEN

Mobile/January-May, August-November

This restaurant offers series of hands-on and demonstration classes that cover a variety of topics. Established in 1970. Maximum class/group size: 25 demo, 14 hands-on. 4 programs per year. Facilities: Restaurant kitchen.

Emphasis: Teaching the fundamental techniques and then the frills.

Faculty: Local and guest chefs.

Costs: $35-$85/class.

Location: Alabama's Gulf Coast, 2 hours from New Orleans.

Contact: Priscilla Gold-Darby, CCP, Owner/Director, Cook's Garden, 61 Westwood St., Mobile, AL 36606 US; 334-476-6184, Fax 334-450-2245.

GREGERSON'S SCHOOL OF COOKING

Rainbow City/September-May

Gregerson's Fine Foods of Rainbow City offers demonstration classes on a variety of topics biweekly, hands-on junior chef cooking classes once monthly. Established in 1996. Maximum class/group size: 16 adult, 10 children. 30+ programs per year. Facilities: Fully-equipped cooking school located within a new upscale grocery store. Also featured: Classes for private groups.

Emphasis: A variety of topics, including the healthy gourmet.

Faculty: Director Cynthia Sirna, Registered Dietitian, teaches classes on the healthy gourmet. A variety of chefs and cookbook authors from nearby.

Costs: $12 Junior Chef, $25 adult.

Location: Rainbow City, approximately 5 miles from downtown Gadsden.

Contact: Cynthia Sirna, RD, Director, Gregerson's School of Cooking, 115 West Grand Ave., Rainbow City, AL 35906 US; 256-442-1940, Fax 256-442-1980.

SOUTHERN LIVING COOKING SCHOOL

Southern cities/March-May, September-November

Southern Living magazine sponsors cooking shows, co-sponsored by nationally-known food brands and local newspapers, that feature the demonstration of more than a dozen recipes containing sponsors' products. Established in 1975. Maximum class/group size: 1500 demo.

Emphasis: Southern cuisine.

Faculty: Professional foods and entertaining specialists from Southern Living.

Costs: $1-$10/session.

Location: Auditoriums and community centers in southern cities.

Contact: Kathryn Dunlap, Internal Coordinator, Southern Living Cooking School, P.O. Box 2581, Birmingham, AL 35202 US; 205-877-6000, Fax 205-877-6200, E-mail Kathy_Dunlap@spc.com.

ARIZONA

CULINARY CONCEPTS

Tucson/Year-round

This retail kitchenware store/school offers mostly participation classes that include a certificate course, series for youngsters, dinner workshops, weekends in the Southwest, guest chefs. Established in 1994. Maximum class/group size: 32. 400+ programs per year. Facilities: 900-square-foot teaching kitchen with 8 workspaces. Also featured: Wine appreciation, private classes, bridal showers.

Emphasis: A variety of topics, guest chef specialties.

Faculty: Proprietor Judith Berger, CCP.

Costs: $40-$80/session.

Location: Shopping plaza in north-central Tucson.

Contact: Judith B. Berger, CCP, Culinary Concepts, 2930 N. Swan, #126, Tucson, AZ 85712 US; 520-321-0968, Fax 520-321-0375.

FOOD FESTS (See page 224)

THE HOUSE OF RICE STORE

Scottsdale/Year-round

This retail store/school offers single-session participation classes. Established in 1977. Maximum class/group size: 13. 130 programs per year. Facilities: Kitchen with large U-shaped counter. Also featured: Private group classes.

Emphasis: Chinese, Japanese, Vietnamese, and Thai.

Faculty: Owner Kiyoko Goldhardt, Chau Liaw, Lan Nguyen Altman, and Mark Gerding.

Costs: $17-$30/class.

Location: A mile east of downtown Scottsdale.

Contact: Kiyoko Goldhardt, Owner, The House of Rice Store, 3221 N. Hayden Rd., Scottsdale, AZ 85251 US; 602-949-9681/602-947-6698, Fax 602-947-0889.

KITCHEN CLASSICS

Phoenix/Year-round

This kitchen store offers demonstration and hands-on classes four times a week. Established in 1987. Maximum class/group size: 12-32. 200+ programs per year. Facilities: Retail store with full kitchen and dining areas. Also featured: Classes for youngsters.

Faculty: Local restaurant chefs and in-store chefs.

Costs: $25/class.

Contact: Shauna Pyland, Manager, Kitchen Classics, 4041 E. Thomas Rd., Phoenix, AZ 85018 US; 602-954-8141, Fax 602-954-6828, URL http://www.kitchen-classics.com.

LES GOURMETTES COOKING SCHOOL

Phoenix/September-May

This private school offers demonstration classes and series. Established in 1982. Maximum class/group size: 15. 40-50 programs per year. Facilities: Private home in central Phoenix. Also featured: Summer classes for children, culinary travel.

Emphasis: French, Southwest, and other cuisines.

Faculty: School proprietor Barbara Fenzl, CCP, studied at Le Cordon Bleu and École LeNôtre. Guest instructors have included Giuliano Bugialli, Hugh Carpenter, Lydie Marshall, Jacques Pépin, Anne Willan.

Costs: $45-$85/class.

Location: Central Phoenix.

Contact: Barbara Fenzl, Owner, Les Gourmettes Cooking School, 6610 N. Central Ave., Phoenix, AZ 85012 US; 602-240-6767, Fax 602-266-2706, E-mail tfenzl@primenet.com.

LES PETITES GOURMETTES CHILDREN'S COOKING SCHOOL

Scottsdale/December-January, March, June-August

Culinary professional Linda Hopkins offers 1- and 4-day hands-on courses for youngsters ages 6-15. Established in 1994. Maximum class/group size: 10-12. 14-20 programs per year. Facilities: Home kitchen.

Emphasis: Beginning and advanced techniques, international holiday cuisines and specialties.

Faculty: Linda Hopkins has assisted for ten years at Les Gourmettes in Phoenix and worked with noted professionals, including Jacques Pépin, Ken Hom, Anne Willan.

Costs: $30/session, $125 for a 4-session course.

Contact: Linda Hopkins, Owner/Teacher, Les Petites Gourmettes Children's Cooking School, 14402 N. 60th Pl., Scottsdale, AZ 85254 US; 602-991-7648, Fax 602-991-4516, E-mail LPGourmett@aol.com.

SCHOOL OF NATURAL COOKERY (See page 199)

SWEET BASIL GOURMETWARE & COOKING SCHOOL

Scottsdale/Year-round

This cookware store offers demonstration classes and 1- to 3-session participation courses. Established in 1993. Maximum class/group size: 10 hands-on/25 demo. Facilities: 400-square-foot kitchen with 6 workspaces, gas and electric appliances. Also featured: Field trips to herbfarms and other food related sites.

Emphasis: Low fat, ethnic, regional cuisines, specific subjects.

Faculty: 8 instructors include school director Mary Schlueter, IACP award winner Barbara Colleary, nutritionist TJ Majeras.

Costs: $30-$85/course.

Location: 2 miles from Phoenix.

Contact: Martha Sullivan, Owner, Sweet Basil Gourmetware & Cooking School, 10701 N. Scottsdale Rd., #101, Scottsdale, AZ 85260 US; 602-596-5628, Fax 602-596-5629.

THE TASTING SPOON

Tucson/September-May

This private school offers classes, Lunch and Learn sessions, diploma courses in spring and fall, 10-session international series, regional series, basics and advanced series. Established in 1978. Maximum class/group size: 10-12. Also featured: Wine tastings.

Emphasis: Techniques, beginning to advanced recipes, international cuisines.

Faculty: Chefs Jeff Azersky, Doug Levy, James Murphy, Donna Nordin.

Costs: $35-$40/class, $15 for Lunch and Learn, $375 for series.

Location: Northwest Tucson.

Contact: Virginia Selby, Director, The Tasting Spoon, P.O. Box 44013, Tucson, AZ 85733-4013 US; 520-327-8174.

ARKANSAS

HARRIET NEIMAN

Fayetteville/Year-round

Caterer Harriet Neiman conducts demonstration and participation classes. Established in 1992. Maximum class/group size: 8-13. 50-60 programs per year. Facilities: 36-square-foot demonstration station. Also featured: Children's classes, private instruction.

Emphasis: Ethnic, especially Mediterranean, easy techniques.

Faculty: Harriet Neiman studied at La Varenne, Linda Gaddy studied at Le Cordon Bleu.

Location: 2 hours from Tulsa, 4 hours from Little Rock.

Contact: Harriet Neiman, 40 N. Crossover Rd., Fayetteville, AR 72701 US; 501-521-3739, Fax 501-443-9504.

SCHOOL OF SOUTHERN BBQ

De Witt/February-September

Culinary professional Jerry Roach offers 1-day courses in various cities. Instruction ranges from basic barbeque principles to advanced techniques, includes meat preparation, sauces, rubs, temperature control. Established in 1992. Maximum class/group size: 50. 11 programs per year. Facilities: Hotel kitchens.

Emphasis: Barbeque and grilling techniques for the novice or experienced chef.

Faculty: Instructors have won over 300 awards at barbeque competitions.

Costs: $125/class.

Location: Cities where major airport service is available or close to metropolitan areas.

Contact: Jerry Roach, J-R Enterprises-School of Southern BBQ, 2055 Hwy 165 So., De Witt, AR 72042 US; 800-432-8187, Fax 870-946-3682, E-mail jr@jrenterprises.com, URL http://www.jrenterprises.com.

CALIFORNIA

ACADEMY OF COOKING – BEVERLY HILLS

Beverly Hills/February-July, September-December

Meredith's Marvelous Morsels catering firm offers participation classes. Established in 1990. Maximum class/group size: 10. 12 programs per year. Facilities: Restaurant kitchen with 10 work stations. Also featured: Children's, private, corporate and group classes, culinary tours in southern California.

Emphasis: Afternoon tea, buffet, brunch menus; California, international, vegetarian cuisines.

Faculty: Meredith Jo Mischen studied with chefs at New York's Plaza and Waldorf-Astoria Hotels.

Costs: $50/class.

Contact: Meredith Jo Mischen, Director, Academy of Cooking - Beverly Hills, 400 S. Beverly Dr., #214, Beverly Hills, CA 90212 US; 310-284-4940.

ALL SEASONS COOKING SCHOOL

Beverly Hills/Year-round

Caterer and party consultant Dahlia Haas conducts sessions that utilize seasonal produce and include holiday menus. Established in 1994. Maximum class/group size: 15+. 12+ programs per year. Facilities: Newly equipped kitchen in a country setting. Also featured: Classes for youngsters, private instruction.

Emphasis: Cooking for family and everyday, healthy menus, party menus.

Faculty: Dahlia Haas, a caterer and party consultant in Los Angeles for 12+ years.

Costs: $50-$75, depending upon the menu; holiday meals are $75.

Location: Central Beverly Hills.

Contact: Dahlia Haas, Owner, All Seasons Cooking School, 1109 Tower Rd., Beverly Hills, CA 90210 US; 310-276-3110, Fax 310-201-7701.

AMY MALONE SCHOOL OF CAKE DECORATING

La Mesa/Year-round

Cake decorating professional Amy Malone conducts morning and evening participation and demonstration classes. Established in 1977. Maximum class/group size: 14 hands-on/30 demo. 150+ programs per year.

Emphasis: Cake decorating, candy-making, creative garnishes, desserts, and food presentation.

Faculty: Amy Malone is a graduate of the Wilton, Betty Newman May, John McNamara, and Frances Kuyper schools of cake decorating and was guest instructor at L'Academie de Cuisine.

Costs: $15-$75/class.

Location: 10 miles east of San Diego.

Contact: Amy Malone, Amy Malone School of Cake Decorating, 4212 Camino Alegre, La Mesa, CA 91941 US; 619-660-1900, E-mail amymalone@aol.com.

THE APPLE FARM

Philo/Year-round

This 30-acre apple farm offers twice monthly hands-on farm weekend cooking courses and single session demonstration classes. Established in 1995. Maximum class/group size: 8 hands-on/12 demo. Facilities: New kitchen.

Emphasis: Meals prepared using the farm's produce and other seasonal foods.

Faculty: Sally Schmitt, former chef-owner of the French Laundry in Yountville.

Costs: Farm weekends are $200, which includes most meals. Lodging recommendations are provided.

Location: Anderson Valley in Mendocino County, on Hwy. 128.

Contact: Sally Schmitt, Owner, The Apple Farm, 18501 Greenwood Rd., Philo, CA 95466 US; 707-895-2461, Fax 707-895-2461. Also: Karen Bates 707-895-2333.

THE ART OF THAI COOKING

Oakland/February-October

Cookbook author Kasma Loha-unchit offers hands-on programs that include 4-week beginner and intermediate courses and on-going advanced classes. Week-long summer intensive for out-of-towners. Established in 1985. Maximum class/group size: 12. Facilities: Fully-equipped private kitchen. Also featured: Private instruction, market visits, classes in private homes, food and cultural tours to Thailand.

Emphasis: Thai, Chinese, Southeast Asian.

Faculty: Kasma Loha-unchit is a Thai chef and author of It Rains Fishes: Legends, Traditions and the Joys of Thai Cooking, Julia Child Award winner.

Costs: $30-$35/session for classes, $400 for week-long summer intensive. 18- to 27-day tours of Thailand are $2,850-$3,450, including airfare, meals, and lodging.

Location: 20 minutes from San Francisco.

Contact: Kasma Loha-unchit, The Art of Thai Cooking, P.O. Box 21165, Oakland, CA 94620 US; 510-655-8900, Fax 510-655-8900, E-mail kasma@lanminds.com, URL http://users.lanminds.com/~kasma.

BORDER GRILL

Los Angeles/Year-round

This restaurant offers quarterly demonstration classes. Established in 1992. Maximum class/group size: 65.

Emphasis: Latin cuisines.

Faculty: Susan Feniger and Mary Sue Milliken, chef-owners of the Border Grill and authors of City Cuisine, Mesa Mexicana, Cantina, and Cooking With Too Hot Tamales.

Costs: $75/session includes full lunch and bar.

Contact: Andrea Uyeda, Catering Manager, Border Grill, 1445 Fourth St., Santa Monica, CA 90401 US; 310-451-1655, Fax 310-394-2049, E-mail mail@bordergrill.com, URL http://www.bordergrill.com.

BRISTOL FARMS COOKING SCHOOL

Manhattan Beach/Year-round

This gourmet specialty foods and cookware store offers 1- to 6-session demonstration and participation courses. Established in 1985. Maximum class/group size: 20 hands-on/40 demo. 200+ programs per year. Facilities: Kitchen with 6-burner stove, grill, oven. Also featured: Children's and private classes, field trips, tours.

Emphasis: International and regional cuisine, baking, low fat cooking, basic cooking techniques.

Faculty: Director Grace-Marie Johnston. Guest instructors have included Graham Kerr, Paul Prudhomme, Stephen Pyles, Patricia Wells, Jacques Pépin, Tommy Tang.

Costs: $40-$55/session.

Location: Upstairs classroom in Manhattan Beach Bristol Farms Market, 30 minutes from Los Angeles, 5 minutes from the beach.

Contact: Grace-Marie Johnston, Cooking School Director, Bristol Farms Cooking School, 1570 Rosecrans Ave., Manhattan Beach, CA 90266 US; 310-726-1350, Fax 310-726-1341.

CAKEBREAD CELLARS

Napa Valley/January, July, November

This winery offers demonstration and participation classes. Maximum class/group size: Demonstration 25, hands-on 16. 4 programs per year. Facilities: Winery with kitchen.

Emphasis: Seasonal specialties.

Faculty: Resident chef Brian Streeter, a New England Culinary Institute graduate, and guest chefs.

Costs: $125/class.

Location: 1 hour from San Francisco.

Contact: Pat Kincaid, Cakebread Cellars, 8300 St. Helena Hwy., Box 216, Rutherford, CA 94573-0216 US; 707-963-5221, Fax 707-963-1034, E-mail cellars@cakebread.com, URL http://www.cakebread.com. Additional contact: Karen Cakebread.

CALIFORNIA CULINARY ACADEMY

San Francisco/Year-round *(See also page 9) (See display ad page 10)*

This career culinary school offers 4-session courses and 6-hour single topic classes. Established in 1977. Maximum class/group size: 20. 40-50 programs per year. Facilities: The Academy kitchens.

Emphasis: Culinary fundamentals, specialties.

Faculty: San Francisco Bay area chefs.

Costs: $500 for 4 sessions (includes books, tools, uniforms), $130/6-hour class.

Location: Civic Center area.

Contact: Weekend Program Manager, California Culinary Academy, 625 Polk St., San Francisco, CA 94102 US; 800-229-2433, URL http://www.baychef.com.

CHEZ LINDA COOKING

Los Gatos/Year-round

Culinary professional Linda Vandemarliere offers 4-session demonstration and participation courses, 2 one-week vacations to France/year that feature demonstrations and visits to wineries, food producers, and markets. Established in 1995. Maximum class/group size: 10 vacation/12-15 demo and hands-on. 100+ programs per year. Facilities: 500-square-foot home kitchen. Also featured: Food and wine pairing classes.

Faculty: Linda Vandermarliere, who graduated from La Varenne and studied with Madeleine Kamman.

Costs: $39/class. Trip to France is $3,000, which includes meals, lodging, planned activities.

Location: 5 miles from the San Jose Airport.

Contact: Linda Vandermarliere, Chez Linda Cooking, 167 Teresita Way, Los Gatos, CA 95032 US; 408-358-3169, Fax 408-358-2432, E-mail chezlinda@aol.com.

COOKING AT TOBY'S

Ventura/Year-round

Toby's Kitchen Store offers classes that feature local chefs. Established in 1996. Maximum class/group size: 18 demo, 10 hands-on. 120 programs per year. Facilities: Classroom in a cooking store. Also featured: Private instruction, classes for youngsters.

Faculty: Peggy Carr, caterer for 20 yrs, owned a specialty market, had food column in paper; Linda Hale, studied in Italy; Tina Reynolds, kitchen designer and has a service specializing in high teas.

Costs: $25-$40/class.

Location: 25 miles south of Santa Barbara, 60 miles north of Los Angeles.

Contact: Peggy Carr, Owner, Toby's Kitchen Store, 2721 E. Main St., Ventura, CA 93003 US; 805-644-5608 (Peggy), E-mail peggycarr@msn.com. Store phone: 805-643-4577.

COOKS AND BOOKS COOKING SCHOOL

Danville/Year-round

This cookbook and cookware store, wine shop, and school offers demonstration and participation courses, 4- and 5-part hands-on series, single topic classes covering a variety of cuisines. Established in 1991. Maximum class/group size: 10-30 demo and hands-on. 100+ programs per year. Facilities: 1,600-square-foot teaching area and 600-square-foot commercial kitchen. Also featured: Local shopping excursions, culinary tours.

Emphasis: International cuisines, seasonal and holiday menus, nutritious foods, wine and food pairing, guest chef specialties, and basic series.

Faculty: In-house instructors D.J. Rae, CCA graduate and Kent Nielsen. Other instructors are guest chef/instructors, culinary teachers, and cookbook authors.

Costs: $40-$50/class.

Location: 30 miles east of San Francisco.

Contact: D.J. Rae, Chef/Owner, Cooks and Books, 148 E. Prospect Ave., Danville, CA 94526 US; 925-831-0708, Fax 925-831-0741, E-mail ckbkcrk@silcon.com.

CUCINA CASALINGA (See page 200)

CULINARY ARTS, INTERNATIONAL (See page 318)

DEPOT

Torrance/Year-round

This restaurant offers 1-2 Saturday afternoon demonstrations/month. Established in 1992. Maximum class/group size: 60. 15-20 programs per year. Facilities: Private dining room of the restaurant. Also featured: Wine tastings.

Emphasis: Italian, grilling, holiday meals, soups, chef specialties, wine pairing.

Faculty: Michael S. Shafer, CEC, chef and general manager of Depot, an Urban Grill Room and Bar, also oversees operations in Fino, Misto, and Chez Melange. He received the Gold Medal in the 1988 Culinary Olympics.

Costs: $35/session.

Location: Los Angeles suburb.

Contact: Michael S. Shafer, Owner/Chef, Depot, 1250 Cabrillo, Torrance, CA 90501 US; 310-787-7501, Fax 310-787-9647.

THE DEPOT HOTEL COOKING SCHOOL

Sonoma/January-July, October

The Depot Hotel Restaurant offers monthly demonstration and hands-on classes in Mediterranean, Italian, French cuisine, classic sauce work. Established in 1987. Maximum class/group size: 23. 12 programs per year. Facilities: Depot Hotel restaurant kitchen. Also featured: Private group classes.

Emphasis: Mediterranean and Northern Italian cuisine, cucina rustica style of cooking.

Faculty: Michael Ghilarducci, chef-owner of the Depot Hotel Restaurant, has 30+ yrs experience in French and Italian cuisine. Pastry Chef Gia Ghilarducci also teaches.

Costs: $50-$75/class.

Location: A block north of the historic plaza in Sonoma, an hour north of San Francisco.

Contact: Gia Ghilarducci, Pastry chef-owner, The Depot Hotel Cooking School, 241 First St. West, Sonoma, CA 95476 US; 707-938-2980, Fax 707-938-5103, E-mail depotel@interx.net, URL http://www.depotel.com.

DRAEGER'S CULINARY CENTERS

Menlo Park, San Mateo/Year-round

Draeger's Market Place offers demonstration and hands-on classes on a variety of topics. Established in 1991. Maximum class/group size: 35. 525 programs/location per year. Facilities: Menlo Park: 38-seat classroom with kitchen and overhead mirror; San Mateo: 38-seat classroom with closed circuit video monitors. Also featured: Wine classes and dinners, market tours, private classes.

Emphasis: Ethnic and regional cuisines, fundamentals, baking, vegetarian and healthful foods, entertaining menus, food history.

Faculty: Guest instructors include well-known chefs, cookbook authors, and culinary professionals.

Costs: Range from $40-$75/session.

Location: Menlo Park and San Mateo, ~35 miles and~15 miles south of San Francisco respectively.

Contact: Pamela Keith, Culinary Director, Draeger's Culinary Centers, 222 E. Fourth Ave., San Mateo, CA 94401 US; 650-685-3795, Fax 650-685-3728, E-mail cookschool@aol.com.

ELDERBERRY HOUSE COOKING SCHOOL

Oakhurst/March, November

The Château du Sureau (Estate by the Elderberries) offers 3-day participation programs. 8 hours of daily cooking instruction are devoted to preparing a 6-course menu. Wine pairing instruction. Established in 1985. Maximum class/group size: 12. 6 programs per year. Facilities: Erna's Elderberry Restaurant's full commercial kitchen, herb garden, local organic vegetable farm. Also featured: Bass fishing, golf, hiking, tennis, and visits to Yosemite National Park.

Emphasis: Sauces, soups, seafood and meat cookery, desserts.
Faculty: Chef-Proprietor Erna Kubin-Clanin has 30 years of culinary and restaurant experience together with Executive Chef James Overbaugh.
Costs: $700 ($250/day), which includes some meals. Nine 2-person guest rooms at Château du Sureau range from $310-$410 including breakfast. 10% student discount.
Location: The château, a member of Relais & Châteaux, is in a mountain village near Yosemite, 45 minutes from Fresno, 4 hours north of Los Angeles.
Contact: Erna Kubin-Clanin, Proprietor, Elderberry House Cooking School, 48688 Victoria Ln., Box 2413, Oakhurst, CA 93644 US; 209-683-6800, Fax 209-683-0800, E-mail chateau@sierranet.net.

EPICUREAN SCHOOL OF CULINARY ARTS

Los Angeles/Year-round *(See also page 14)*

This private school offers participation classes in fish, chicken, and other specialties. Established in 1985. Maximum class/group size: 15. Facilities: Teaching kitchen with 5 work stations.
Emphasis: Beginners' classes, specific cuisines.
Faculty: CIA and CCA graduates.
Costs: $65/class.
Location: West Hollywood.
Contact: Epicurean School of Culinary Arts, 8759 Melrose Ave., Los Angeles, CA 90069 US; 310-659-5990, Fax 310-659-0302, E-mail epicureans@aol.com.

THE GREAT CHEFS AT ROBERT MONDAVI WINERY

Oakville/April, November

Robert Mondavi Winery offers 2- to 3-day weekend and 1-day Monday programs that feature cooking demonstrations by noted chefs, wine seminars, private winery tours, and theme lunches and dinners. Established in 1976. Maximum class/group size: 28 for weekend programs, 35 for 1-day programs.
Emphasis: International cuisines, table setting, flower arranging, food and wine pairing.
Faculty: Julia Child, Jacques Pépin, Pierre Gagnaire, and Marc Meneau.
Costs: One-day sessions range from $150-$180; two-day program is $750; three-day program is $1,650, including transportation and lodging.
Location: Napa Valley; 90 minutes from San Francisco.
Contact: Valerie Varachi, Administrative Secretary, The Great Chefs at Robert Mondavi Winery, P.O. Box 106, Oakville, CA 94562 US; 707-968-2100, Fax 707-968-2174, E-mail valerie.varachi@robert-mondavi.com, URL http://www.robertmondavi.com.

GREAT NEWS!

San Diego/Year-round

This cookware store offers hands-on and demonstration classes in basic techniques, ethnic cooking, individual subjects. Established in 1996. Maximum class/group size: 25 hands-on, 50 demo. 240+ programs per year. Facilities: Teaching kitchen with 5 big-screen TV monitors. Also featured: Market visits.
Emphasis: Classes for the home chef.
Faculty: Cookbook authors and local restaurant chefs.
Costs: $25-$50/class.
Location: Pacific beach area, near the ocean.
Contact: Catherine Kerulis, Manager, Great News!, 1788 Garnet, San Diego, CA 92109 US; 619-270-1582, Fax 619-270-6815, E-mail greatnews@great-news.com, URL http://www.great-news.com. Additional contact: Megan Barnet, Assistant Manager.

HOMECHEF COOKING SCHOOL

Eight locations in California/Year-round

This kitchen store and cooking school offers an 8- and 12-week Essential Cooking Series adapted from professional curriculum plus demonstration and participation courses. Established in 1976.

Maximum class/group size: 20 hands-on/45 demo. Facilities: Approximately 1,000-square-foot classrooms with kitchens. Also featured: Private events, free sampler classes.
Emphasis: Fine cuisine in home kitchens by home chefs.
Faculty: Founder Judith Ets-Hokin, CCP, author of the The Dinner Party Cookbook, The Homechef, Fine Cooking Made Simple, Great Cooking in Minutes, holds certificates from cooking schools in England, France, and Italy.
Costs: The Essential Cooking Series is $270 for 8 weeks, $360 for 12 weeks; single classes are $45 for demonstrations, $69 for workshops.
Location: Eight locations in California: San Francisco, Corte Madera (headquarters), Palo Alto, San Jose, Sacramento, Newport Beach, Walnut Creek, Pasadena.
Contact: Homechef Cooking School, 5725 Paradise Drive, Ste. 360, Corte Madera, CA 94925 US; 415-927-3290, Fax 415-927-4164, E-mail info@homechef.com, URL http://www.homechef.com.

HUGH CARPENTER'S CAMP NAPA CULINARY
Oakville, Napa Valley/May-October

Chef and cookbook author Hugh Carpenter offers 6-day food and wine tours that feature participation classes. Established in 1992. Maximum class/group size: 18. Facilities: Cakebread Cellars Winery kitchen. Also featured: Dining in fine restaurants, private winery tours, seminars on food and wine pairing, a croquet tournament. Hot-air ballooning, Calistoga spa, golf, and tennis also available.
Emphasis: California-Asian and cross-cultural cuisine; winery chef specialties.
Faculty: Hugh Carpenter, founding chef of 6 Chopstix restaurants in Los Angeles, is author of the IACP-award-winning Pacific Flavors, Chopstix, Hot Wok, Hot Chicken, Hot Pasta, Hot Barbecue, Hot Vegetables, and Quick Cooking with Pacific Flavors.
Costs: Cost is $1,440, which includes meals and planned itinerary. A list of recommended lodging is available.
Location: Napa Valley, 50 miles northeast of San Francisco.
Contact: Hugh Carpenter, Hugh Carpenter's Camp Napa Culinary, P.O. Box 114, Oakville, CA 94562 US; 707-944-9112/888-999-4844, Fax 707-944-2221.

THE JEAN BRADY COOKING SCHOOL
Santa Monica/September-June

Culinary professional Jean Brady offers a 7-session demonstrations. Established in 1973. Maximum class/group size: 6-8 hands-on/15 demo. Facilities: Commercially-equipped home kitchen featured in Bon Appètit, guest classes with restaurant chefs in their kitchens. Also featured: Children's classes, market visits, 1-week seminars for private groups, culinary tours to Europe.
Emphasis: A variety of topics; low-fat savories; menus for easy entertaining; guest chef specialties.
Faculty: Proprietor Jean Brady studied with Lydie Marshall, Jacques Pépin, and Paula Wolfert and attended the Cordon Bleu and La Varenne. Guest chefs include Lydie Marshall, Jacques Pépin, Paula Wolfert, and top local chefs in their restaurant kitchens.
Costs: Guest chef classes range from $60-$90; 7-session classes are $300.
Location: 20 minutes from Beverly Hills, 30 minutes from downtown LA.
Contact: Jean Brady, The Jean Brady Cooking School, 680 Brooktree Rd., Santa Monica, CA 90402 US; 310-454-4220, Fax 310-454-4220.

KITCHEN WITCH GOURMET SHOP
Encinitas/Year-round

This gourmet shop and school offers demonstration sessions. Established in 1981. Maximum class/group size: 14. 400+ programs per year. Also featured: After-school classes for children, private group lessons.
Emphasis: Ethnic and regional cuisines, nutrition, vegetarian, macrobiotic, breads, holiday menus, pastries, chocolate, microwave and food processor techniques.
Faculty: Includes Carole Bloom, Phillis Carey, Suzy Eisenman, Kay Pastorius, Dee Biller, Nadia Frigeri, Nancy Brown, Stella Fong, George Geary, Janet Chatfield, Nancie McDermott, Liz Strahle,

Evelyn Sudora, and Heath Fox.

Costs: Range from $16-$27.

Location: North of San Diego, on the Pacific Coast.

Contact: Marie Santucci, Owner, Kitchen Witch Gourmet Shop, 127 N. El Camino Real, Ste. D, Encinitas, CA 92024 US; 760-942-3228.

LET'S GET COOKIN'

Westlake Village/Year-round *(See also page 22, Westlake Culinary Institute)*

This private school offers 1- to 6-session demonstration and participation classes. Maximum class/group size: 10-30. 150+ programs per year. Facilities: 1,000-square-foot combination demonstration/participation facilities, cookware store. Also featured: Classes for young people, day trips, travel abroad.

Emphasis: Basic and advanced techniques for the home cook.

Faculty: Includes cookbook authors and guest chefs.

Costs: $45-$75/session ($25 for children's classes).

Location: North of Malibu, 30 minutes from Los Angeles.

Contact: Phyllis Vaccarelli, Owner/Director, Let's Get Cookin', 4643 Lakeview Canyon Rd., Westlake Village, CA 91361 US; 818-991-3940, Fax 805-495-2554.

LOCANDA VENETA COOKING CLASSES

Los Angeles/Year-round

The Locanda Veneta restaurant offers demonstration classes in Italian cuisine the last Saturday of each month, followed by lunch and discussion. Established in 1998. Maximum class/group size: 15. 12 programs per year. Facilities: Small restaurant with open kitchen. Also featured: Private instruction available.

Emphasis: Tuscan cuisine, including soups, pastas, and main courses.

Faculty: Executive Chef Massimo Ormani.

Costs: $90/class.

Location: West Los Angeles.

Contact: Massimo Ormani, Executive Chef, Locanda Veneta Cooking Classes, 8638 West Third St., Los Angeles, CA 90048 US; 310-246-8892, Fax 310-286-1043, E-mail Psycook@aol.com.

LUCY'S KITCHEN

Albany/Year-round

Culinary professional Lucy Seligman offers monthly (and/or weekly) cooking classes that are either demonstration or hands-on in English or Japanese. Established in 1994. Maximum class/group size: 6-12. Facilities: Lucy's kitchen or students' home kitchens. Also featured: Children's classes, event planning, kitchen organization, private meal preparation, small dinner party catering, menu planning and consulting.

Emphasis: A variety of topics, including Japanese, Italian, Russian, Turkish, and American regional cuisines.

Faculty: Lucy Seligman graduated from Boston University's Culinary Arts program, owned a cooking school in Japan and Ann Arbor, and studied cooking in Paris, Bangkok, Tokyo, Florence, and Los Angeles. She published a Japanese cuisine newsletter for over 7 years.

Costs: Vary.

Location: Near Berkeley.

Contact: Lucy Seligman, Lucy's Kitchen, 753 Taft St., Albany, CA 94706 US; 510-524-9504, E-mail ls@impactfund.org.

MANDOLINE COOKING SCHOOL

Sunnyvale/Year-round

Culinary professional Paula Barbarito-Levitt offers afternoon and evening participation classes. Established in 1992. Maximum class/group size: 6-7. 50 programs per year. Facilities: Large kitchen with 5 work stations. Also featured: Private classes for groups/organizations.

Emphasis: Regional Italian, Mediterranean, regional American, vegetarian, and French bistro cuisine, pastry and desserts, bread making, wine pairing, techniques, seasonal ingredients, equipment selection.
Faculty: Paula Barbarito-Levitt, member of the IACP and WCR, studied at Le Cordon Bleu, the New York Restaurant School, California Culinary Academy, with Giuliano Bugialli and Lydie Marshall.
Costs: Range from $35-$150.
Location: A 50-minute drive south of San Francisco and 10 minutes north of San Jose.
Contact: Paula Barbarito-Levitt, Owner/Instructor, Mandoline Cooking School, 1083 Robbia Dr., Sunnyvale, CA 94087 US; 408-733-4224, Fax 408-773-1863, E-mail mandoline@aol.com.

MON CHERI COOKING SCHOOL/UCSC EXTENSION

Santa Cruz/Year-round

This university extension-private school cooperative program offers 1- to 4-session participation workshops and courses. Established in 1983. Maximum class/group size: 18-20. 20+ programs per year. Facilities: Historic house with modern commercial kitchen.
Emphasis: Stress relief cooking and a variety of other topics.
Faculty: Director Sharon Shipley, an IACP member who received certificates from La Varenne and Le Cordon Bleu; noted guest chefs.
Costs: Range from $85 for a single session to $165 for four.
Location: Silicon Valley, 40 miles south of San Francisco.
Contact: Mon Cheri Cooking School/UCSC Extension, 740 Front St., #155, Santa Cruz, CA 95060 US; 408-427-6695, Fax 408-427-6608/736-0932, E-mail sship25521@aol.com, URL http://www.monchericaterers.com.

NAPA VALLEY COOKING SCHOOL

(See also page 16) (See display ad page 17) **St. Helena/Year-round**

This college offers 1- to 6-session demonstration and participation courses. Established in 1990. Maximum class/group size: 12-28. Facilities: New kitchen with 18 burners, 4 ovens, demonstration counter, outdoor dining area. Also featured: Wine and food classes, farmer's market visits, catering seminars.
Emphasis: Various topics, including cooking basics, bread baking, Indian, Asian, Italian, and Mediterranean cuisines, wine appreciation.
Faculty: Guest chefs have included: Bruce Aidells, Steven Levine, Michael Chiarello, Gary Danko, Carlo Middione, John Ash, and Jeremiah Tower.
Costs: $50-$75/session.
Location: Napa Valley, 75 minutes from San Francisco.
Contact: Eric Lee, Program Coordinator, Napa Valley College, 1088 College Ave., St. Helena, CA 94574 US; 707-967-2900 x2930, Fax 707-967-2909.

NATURAL FOODS COOKING SCHOOL

Woodland Hills/Year-round

This private school offers a 12-month basic natural foods curriculum that includes two group classes/month and weekend retreats year-round. Established in 1985. Maximum class/group size: 6-15.
Emphasis: Grains, pasta, vegetables, breads, fermented foods, catering, food and healing.
Faculty: Donna Wilson, who also owns the Ginkgo Leaf Bookstore, has operated natural foods stores and restaurants in southern California since 1978.
Costs: Individual evaluation is $25, nonrefundable; group classes are $25 each, $75 for a series of 5; private classes are $50/hour.
Location: Woodland Hills.
Contact: Donna Wilson, , 21109 Costanso St., Woodland Hills, CA 91364 US; 818-716-6332, Fax 818-716-6332, E-mail ginkgo@earthlink.net.

PALATE & SPIRIT

Los Angeles/Year-round

This private school (formerly Montana Mercantile) offers private and small group instruction, demonstration and participation classes, including a series for beginning cooks. Established in 1976. Maximum class/group size: Varies. Facilities: Professional home kitchens. Also featured: Classes for household staff cooks (also taught in Spanish), California wine seminars, home kitchen design and organization.

Emphasis: International cuisines with emphasis on health concerns and easy preparation.

Faculty: Rachel Dourec and associates.

Contact: Rachel Dourec, Director, Palate & Spirit, P.O. Box 17178, Beverly Hills, CA 90209 US; 310-472-3220, Fax 310-472-8846.

PATINA AND PINOT RESTAURANTS

Los Angeles/Year-round

These restaurants offer Saturday morning demonstration and participation classes and a full day in the kitchen. Established in 1989. Maximum class/group size: 12. Facilities: Patina Restaurant, Pinot Bistro, Cafe Pinot, Pinot Hollywood, Pinot at the Chronicle. Also featured: Full-day class includes working in the pastry kitchen in the morning and in the afternoon preparing a 5-course dinner for the student's seven dinner guests.

Emphasis: Pastries, bistro cooking, spa, vegetarian, potatoes.

Faculty: Joachim Splichal, chef/owner of Patina and Pinot Restaurants; Octavio Becerra, executive chef of Pinot Bistro; Jon Ferow, executive chef of Pinot Hollywood; Bernhard Renk, executive chef of Cafe Pinot.

Costs: Saturday classes are $55-$100. Day in the Kitchen is $950-$1,250.

Location: Downtown Los Angeles, Valley, Hollywood, and Pasadena locations.

Contact: Susan Goodwin, Special Events Coordinator, Patina and Pinot restaurants, 5735 Melrose Ave., Los Angeles, CA 90038 US; 213-960-1762, Fax 213-467-1924, URL http://www.patina-pinot.com.

PEGGY RAHN COOKS

Pasadena/Year-round

This private school in a 1918-vintage home offers 1- and 2-session demonstration and participation workshops. Established in 1974. Maximum class/group size: 10 hands-on/20 demo. 50 programs per year. Facilities: Well-equipped, home kitchen with overhead mirror. Also featured: Private classes, small group excursions to markets, party classes, and culinary trips.

Emphasis: Ethnic cuisines, technique classes, healthful eating.

Faculty: Peggy Rahn, CCP, food and travel columnist, cookbook/restaurant reviewer, co-host of CBS's Meet the Cook, teacher at UCLA, studied at La Varenne, Le Cordon Bleu, and the Ritz Escoffier. Guest faculty has included Giuliano Bugialli and Madeleine Kamman.

Costs: Range from $50-$75/course.

Location: Ten minutes from downtown Los Angeles, 20 minutes from Burbank.

Contact: Peggy Rahn, Owner, Peggy Rahn Cooks, 484 Bellefontaine St., Pasadena, CA 91105 US; 818-441-2075, Fax 818-441-5286, E-mail prahn@earthlink.net.

RAMEKINS SONOMA VALLEY CULINARY SCHOOL

Sonoma/Year-round

This private school, cookware store, and restaurant offers one- and two-day classes that include basic and general cooking instruction, ethnic and skill-specific sessions, and seasonal menus. Established in 1998. Maximum class/group size: 36 demo, 18 hands-on. Facilities: Two high-tech teaching kitchens: one primarily demonstration with TVs, mirrors and residential equipment, the other a full-service commercial kitchen. Also featured: Private instruction, classes for youngsters, visits to markets, food producers, and wineries.

Emphasis: Ethnic cuisines and wine-country menus.

Faculty: 40+ instructors include noted chefs, cookbook authors, and other culinary professionals.

Costs: Demos $35-45, hands-on classes $50-75.
Location: 45 minutes from San Francisco, 20 minutes from Napa Valley.
Contact: Bob Nemerovski, Culinary Director, Ramekins Sonoma Valley Culinary School, 450 West Spain St., Sonoma, CA 95476 US; 707-933-0450, Fax 707-933-0451, E-mail info@ramekins.com, URL http://www.ramekins.com.

THE SEASONAL TABLE COOKING SCHOOL

Los Angeles/October-June

Culinary professionals Karen Berk and Jean Brady offers classes. Established in 1994. Maximum class/group size: Up to 20. 50 programs per year. Facilities: Local restaurant kitchens and private homes. Also featured: Market visits, culinary tours and wine instruction, private classes for individuals and groups, and special events.
Emphasis: A variety of topics, including seasonal and entertaining menus, techniques in healthful and vegetarian cooking, baking, wine, restaurant specialties, ethnic cuisines.
Faculty: Co-owners Karen Berk, founder of Incredible Edibles Cooking School and co-editor of Southern California Zagat Restaurant Survey, and Jean Brady of The Jean Brady Cooking School; noted restaurant chefs, cookbook authors, and other culinary professionals.
Costs: $55-$75/class.
Location: Westside of Los Angeles.
Contact: Karen Berk, Co-owner, The Seasonal Table Cooking School, 12618 Homewood Way, Los Angeles, CA 90049 US; 310-472-4475, Fax 310-471-3904, E-mail kjberk@aol.com. Other contact: Jean Brady, 310-454-4220.

SIGNATURE FOOD DESIGN-INNSPIRED COOKS

Northern California/Year-round

This private school offers single sessions, weekend retreats, and 3- and 5-day courses that cover a variety of topics, including essential skills, ethnic and seasonal cuisines, desserts, food and wine pairing. Established in 1996. Maximum class/group size: Demo 20, hands-on 8. 40-50 programs per year. Facilities: Country inns and bed & breakfasts. Also featured: Visits to specialty and farmers' markets, food producers, winemaker.
Emphasis: Northern California B&B and country inn specialties.
Faculty: Signature owner Lana Richardson, an honor graduate of the California Culinary Academy with catering experience; local chefs and authors.
Costs: Ranges from $50 for a class to $800 for a 5-day course, which includes 20-36 hours of instruction and meals.
Location: Calistoga, in the Napa Valley wine region north of San Francisco.
Contact: Lana Richardson, Owner, Signature Food Design, 3669 Chucker Ct., Walnut Creek, CA 94598 US; 510-256-0415, Fax 510-932-4870, E-mail Lana@value.net, URL http://pwp.value.net/signaturefood.

SOUTHERN CALIFORNIA SCHOOL OF CULINARY ARTS

(See also page 20) (See display ad page 19) **South Pasadena/Year-round**

This private school offers 12-week series and individual classes, celebrity chef demos. Established in 1994. Maximum class/group size: 12-24. Facilities: 8-station modern kitchen with modular design.
Emphasis: A variety of topics, including international and vegetarian cuisines, entertaining menus, low-fat cooking, wine and food pairing.
Faculty: 5 staff instructors, guest chefs.
Costs: $45-$60/class.
Location: The 12,000-square-foot school facility is in an historic suburb, 12 minutes from downtown Los Angeles.
Contact: Southern California School of Culinary Arts, 1420 El Centro St., S. Pasadena, CA 91030 US; 888-900-CHEF, Fax 626-403-8494, E-mail scsca@earthlink.net, URL http://www.scsca.com.

A STORE FOR COOKS

Laguna Niguel/Year-round

This cookware store and school offers morning and evening demonstration classes and Lunch & Learn classes. Established in 1981. Maximum class/group size: 25 demo. 100+ programs per year. Also featured: Classes for private groups.

Emphasis: Ethnic and regional cuisines, holiday and seasonal foods, guest chef specialties.

Faculty: Proprietor and cookbook author Susan Vollmer, Hugh Carpenter, Phillis Carey, Tarla Fallgatter, cookbook authors, and local chefs.

Costs: Lunch and Learn classes are $15; demonstrations range from $40-$75.

Location: On the Pacific coast, 55 miles south of Los Angeles.

Contact: Susan Vollmer, Owner, A Store for Cooks, 30100 Town Center Dr., Ste. R, Laguna Niguel, CA 92677 US; 714-495-0445, Fax 714-495-2139, E-mail 73571.3511@compuserve.com.

SUGAR 'N SPICE CAKE DECORATING SCHOOL

Daly City/Year-round

This gourmet bakeware, cake decorating and candy making supply store and school offers 1- to 5-session participation courses. Established in 1973. Maximum class/group size: 10-15. 50 programs per year. Facilities: Classroom and retail store.

Emphasis: Cake decorating.

Faculty: Jeanne Lutz is a graduate of Edith Gate's Cake Decorating School and has studied with many other professionals. Guest instructors from around the world are featured.

Costs: Range from $30-$250/course. Supplies vary with each course.

Location: The school is in Daly City, 20 minutes from downtown San Francisco.

Contact: Jeanne Lutz, Owner, Sugar 'n Spice Cake Decorating School, 2965 Junipero Serra Blvd., Daly City, CA 94014-2549 US; 650-994-4911, Fax 650-994-4912, E-mail SugNSpiz@aol.com.

SUR LA TABLE

Four locations/Year-round *(See also page 264)*

This cookware store offers demonstrations and hands-on classes covering a variety of cuisines, the basics, special subjects, chef specialties. Established in 1996. Maximum class/group size: 16 hands-on, 36-40 demo. 100-150 programs/location per year. Facilities: Full demonstration kitchens with TV monitors and overhead mirrors over teaching islands, hands-on tables. Also featured: Professional classes, corporate team-building classes, culinary walking tours, market visits, programs for youngsters.

Emphasis: Basic cooking and baking, topics of current interest featuring popular restaurant chefs and authors.

Faculty: Local and nationally known chefs, restaurateurs, and cookbook authors.

Costs: $35-$100/class, includes a 10% merchandise discount coupon.

Location: Four locations in California (San Francisco, Berkeley, Santa Monica, Newport Beach); Kirkland, WA, 10 miles northeast of Seattle on Lake Washington.

Contact: Doralece Dullaghan, Culinary School Manager, Sur La Table, 77 Maiden Lane, San Francisco, CA 94108 US; 415-732-7900, Fax 415-732-7797. Berkeley, CA: 1806-4th St., 510-849-2252.

TANTE MARIE'S COOKING SCHOOL

San Francisco/Year-round *(See also page 21) (See display ad page 21)*

This small private school offers 1-week, weekend, 6-session evening, and single-session morning participation courses, afternoon and weekend demonstrations, and party classes. Established in 1979. Maximum class/group size: 16-38. Facilities: Store front. Also featured: 1-week courses that include shopping at the Farmer's Market, visits to bread bakeries and cheese makers, winery tours, and dining in fine restaurants.

Emphasis: General and specific topics, including pastries and regional cuisines.

Faculty: Founder Mary Risley studied at Le Cordon Bleu and La Varenne; Catherine Pantsios, former chef/owner of Zola's; Cathy Burgett, former pastry chef of Campton Place; guest instructors.

Costs: 1-week course $550, weekend courses $40-$125, 6-session courses $450, morning classes $75, demonstrations $40 (5 for $150). Hotel lodging available.
Location: On San Francisco's Telegraph Hill, near Fisherman's Wharf and public transportation.
Contact: Peggy Lynch, Administrator, Tante Marie's Cooking School, 271 Francisco St., San Francisco, CA 94133 US; 415-788-6699, Fax 415-788-8924, URL http://www.tantemarie.com.

UCLA EXTENSION, HOSPITALITY/FOODSERVICE MANAGEMENT

(See also page 22) **Los Angeles/Year-round**

This continuing education provider offers a variety of demonstration and participation courses and 1- to 5-session seminars. Maximum class/group size: 15+ demo, 15 hands-on. Facilities: Commercial kitchens.
Emphasis: General interest and professional development.
Faculty: CIA- and CCA-trained chefs, restaurant owners.
Costs: $95-$325.
Location: Westside Los Angeles.
Contact: UCLA Extension, 10995 Le Conte Ave., Room 515, Los Angeles, CA 90024-0901 US; 310-206-8120, Fax 310-206-7249.

WEIR COOKING

San Francisco/Year-round

Culinary professional Joanne Weir offers weekend and 5-day participation courses, hands-on and demo. 1 week programs abroad. Established in 1989. Maximum class/group size: 8-9. Facilities: Newly-designed professional commercial kitchen with wood-fired Tuscan oven and 4 work stations. Also featured: Courses include Napa and Sonoma Valley tours and dining at fine restaurants. Private classes and visits to restaurants, wineries, and markets.

Emphasis: French, Italian, Mediterranean, and American regional cuisines.
Faculty: Joanne Weir cooked at Berkeley's Chez Panisse, studied with Madeleine Kamman, received the Julia Child/IACP Cooking Teacher Award of Excellence, and is author of From Tapas to Meze, the Williams Sonoma Seasonal Celebrations series, and You Say Tomato.
Costs: \$80-\$100/class in US, \$2,500-\$2,950 in Italy and France.
Location: Pacific Heights, San Francisco.
Contact: Joanne Weir, Weir Cooking, 2107 Pine Street, San Francisco, CA 94115 US; 415-776-4200, Fax 415-776-0318, E-mail weircook@aol.com.

WOK WIZ WALKING TOURS & COOKING CENTER
San Francisco/Year-round

This private school offers daily walking tours of Chinatown, weekend Walk 'n Wok Workshop, which includes shopping + cooking class. Custom tours and classes for groups. Established in 1986. Maximum class/group size: 10-15. 300+ programs per year. Facilities: 1,100-square-foot 2-story building with demonstration kitchen built in 1996. Also featured: 10-day epicurean tour to Hong Kong and Thailand.
Emphasis: Chinese cuisine.
Faculty: Shirley Fong-Torres is author of the Wok Wiz Chinatown Cookbook, In the Chinese Kitchen, and San Francisco Chinatown, A Walking Tour.
Costs: Walking tours \$37-\$65 including lunch, shopping and cooking workshop \$75.
Location: A half-block from Chinatown.
Contact: Shirley Fong-Torres, Owner, Wok Wiz Walking Tours & Cooking Center, 654 Commercial St., San Francisco, CA 94111-2504 US; 800-281-9255/415-981-8989, Fax 415-981-2928, E-mail wokwiz@aol.com, URL http://www.mim.com/wokwiz.

YANKEE HILL WINERY – WHAT'S COOKING AT THE WINERY
Columbia/Year-round

This winery offers weekly demonstration and participation cooking and wine classes. Established in 1970. Maximum class/group size: 12 hands-on/30 demo. Facilities: 2,500-square-foot area with 12 workspaces, Swiss baking ovens, pizza oven, candy stove, sausage and salami-making equipment, smoker, wine-making equip. Also featured: Classes for youngsters, market visits, private classes, facility rental.
Emphasis: Baking, international cuisines taught by native instructors, instruction in a relaxed environment.
Faculty: The 10 instructors include Yankee Hill Winery owner Ron Erickson, who teaches at the community college and has owned several Bay area restaurants; Jerry Phillips, Gretchen Erickson, and Nestor Ramirez; guest chefs.
Costs: \$30-\$100/class.
Location: Two hours south of Sacramento, in the grape-growing regions of Tuolumne and Calaveras Counties. Overlooking the Sierra Nevada foothills.
Contact: Ron Erickson, Owner, Yankee Hill Winery, P.O. Box 330, Columbia, CA 95310 US; 209-533-2417/800-497-WINE, Fax 209-533-2417, E-mail Columbia@mlode.com.

YOSEMITE CHEFS' & VINTNERS' HOLIDAYS
Yosemite National Park/November-February

Yosemite Concession Services Corporation offers a series of seven 2-day/3-night Chefs' and Vintners' Holidays that feature 3 cooking demonstrations or 4 wine seminars and a concluding banquet. Established in 1982. Maximum class/group size: 180. Facilities: Great Lounge of The Ahwahnee Hotel.
Emphasis: Cuisines of Western chefs, California wines.
Faculty: Each program features three noted cooking instructors or four wineries. Executive Chef Robert Anderson and his staff prepare the vintner's banquet, visiting chefs prepare the chefs' banquet.
Costs: Lodging at the Ahwahnee Hotel is \$215/night. Chefs'/vintners' banquets are \$75/\$80.

Location: Yosemite National Park, 90 miles from Fresno and 175 miles from San Francisco.
Contact: Yosemite Reservations, 5410 East Home Ave., Fresno, CA 93727 US; 209-252-4848, Fax 209-372-1362, URL http://www.yosemitepark.com/events/chef.html.

ZOV'S BISTRO
Tustin/Year-round

This bakery/cafe, bistro, catering service, cooking school offers demonstration classes on a variety of topics. Established in 1987. Maximum class/group size: 25-40. 19 programs per year. Facilities: Front kitchen of Zov's Bistro.
Emphasis: Chef specialties.
Faculty: Noted chefs, TV personalities, cookbook authors, including John Ash, Hugh Carpenter, George Geary, Joyce Goldstein, David Rosengarten, Julie Sahni, Martin Yan, and Zov Karamardian, caterer and teacher for over 20 years.
Costs: $50-$100/class.
Location: Orange County, near Anaheim and Newport Beach.
Contact: Zov Karamardian, Chef/Owner, Zov's Bistro, 17440 E. 17th St., Tustin, CA 92780 US; 714-838-8855 x5, Fax 714-838-9926, E-mail zov@zovs.com, URL http://www.zovs.com.

COLORADO

ASIAN COOKERY
Colorado Springs/Year-round

Culinary professional Peng Jones offers classes. Established in 1989. Maximum class/group size: 8-12. 18 programs per year. Facilities: Specially-designed teaching kitchen. Also featured: Private classes, specialty classes, dinner parties.
Emphasis: Chinese, Malaysian, Thai, Vietnamese, Indian, low-fat, and vegetarian cuisines.
Faculty: Peng Jones, CCP, studied at the International School of Home Cookery in Malaysia and trained in Oriental food and vegetable carving.
Costs: $25-$35/session.
Location: A 20-minute drive from the Colorado Springs airport; ~ 60 miles south of Denver.
Contact: Peng Jones, Owner, Asian Cookery, P.O. Box 62674, Colorado Springs, CO 80962 US; 719-590-7768, E-mail TedJones@aol.com.

COOKING SCHOOL OF ASPEN
Aspen/Year-round

This private school and cookware shop offers demonstration classes that include American regional and international cuisines, seasonal recipes, and guest chef specialties. Established in 1998. Maximum class/group size: 20 max. 200+ programs per year. Facilities: In-store demonstration kitchen. Also featured: Classes for youngsters, private instruction, classes in French technique.
Emphasis: A variety of topics and guest chef specialties.
Faculty: More than 20 instructors, includes chefs, caterers, and cookbook authors.
Costs: $95-$135/class.
Location: Downtown Aspen in Hyman Ave. Mall.
Contact: Rob Seideman, Cooking School of Aspen, 414 Hyman Ave. Mall, PO Box 11424, Aspen, CO 81612 US; 800-603-6004/970-920-1879, Fax 970-920-2188, E-mail cookaspn@sopris.net, URL http://www.aspen.com/aspenonline/dir/service/sponsors/cooking.

COOKING SCHOOL OF THE ROCKIES
(See also page 24) (See display ad page 24)
Boulder/Year-round

This private school offers individual classes, short courses, and 5-day basic techniques intensives that emphasize creativity, organization, and presentation. Established in 1991. Maximum class/group size: 32 demos, 12-16 hands-on. Facilities: Modern, fully-equipped kitchen with overhead mirror. Also featured: Corporate training, private parties, bridal showers, retail cookware store, diploma/professional program.

Emphasis: Basic French techniques, Italian cuisine, pastry, baking, ethnic cuisines, wine appreciation.

Faculty: Revolving visiting instructors program featuring local and national restaurant/bakery chefs, guest chefs, cookbook authors.

Costs: Classes range from $35-$75 each; intensives are $425. A list of bed & breakfasts and lodgings is available.

Location: 25 miles northwest of Denver, at the base of the Rocky Mountain Foothills.

Contact: Joan Brett, Director, Cooking School of the Rockies, 637 S. Broadway, Ste. H, Boulder, CO 80303 US; 303-494-7988, Fax 303-494-7999, E-mail csrockies@aol.com, URL http://www.cookingschoolrockies.com.

FLAVORS OF MEXICO – CULINARY ADVENTURES (See page 339)

FOOD & WINE MAGAZINE CLASSIC AT ASPEN

Aspen/June

Food & Wine Magazine sponsors an annual 3-day weekend festival featuring a variety of events for food and wine enthusiasts and professionals. The 20-hour program offers over 80 lectures, demonstrations, panels, and tastings; a benefit auction; and fine dining. Established in 1983. Maximum class/group size: 70-800. 1 programs per year. Facilities: Hotel and tented park area. Also featured: Winemaker dinners.

Emphasis: Trade seminars for chefs and restaurateurs cover employee relations, marketing, direct mail, customer relations, insurance; consumer events include chef demonstrations and tastings from over 250 vintners.

Faculty: Has included Julia Child, Marcella Hazan, Jacques Pépin, Emeril Lagasse, Robert M. Parker, Jr., Frank Prial.

Costs: Three-day tickets are ~$495. Reserve tastings are $75 to $200 extra. Deluxe hotel and condominiums available.

Location: 3 hours from Denver by car, 45 minutes by plane.

Contact: Laura Powers, Associate Event Marketing Manager, Food & Wine Magazine Classic at Aspen, 425 Rio Grande Plaza, Aspen, CO 81611-9938 US; 888-7WI-NE97, Fax 970-925-9008, E-mail arone@rof.net. Also: Food & Wine Magazine, 1120 Sixth Ave., New York, NY 10036; 212-382-5627.

KEYSTONE COOKING SCHOOL

Keystone/April-May, September-November

Keystone Resort offers 3-day hands-on programs that feature fine dining and classes by the Resort's chefs. Established in 1996. 4-5 programs per year. Facilities: Conference facilities. Also featured: Recreation options include skiing, snowboarding, snowshoeing, cross-country skiing, hiking, mountain biking, golf.

Emphasis: Hot and chilled appetizer preparation, entrees, desserts.

Faculty: Chefs of Keystone Resort's restaurants.

Costs: $775/person, $1,150/couple includes lodging at Ski Tip Lodge and meals. Price without lodging is $575/person.

Location: In the mountains 90 miles west of Denver.

Contact: Jenny McCabe, Culinary Public Relations Manager, Keystone Cooking School, 730 Burbank St., Broomfield, CO 80020 US; 303-404-2780, Fax 303-404-2768, E-mail jcomm@jcomm.com.

LEGENDS PARK COOKING SCHOOL

La Veta/Year-round

This private school offers classes that include basic techniques and specialized topics. Established in 1996. Maximum class/group size: 20 demo, 12 hands-on. 60-75 programs per year. Facilities: Restaurant kitchen. Also featured: Outings to gather edible plants, market visits, sightseeing.

Emphasis: Low fat gourmet.

Costs: $20-$50/class.

Location: 3 miles south of Denver, 2 hours south of Colorado Springs.

Contact: Karen Briggs and Sally Greer, Owners, Legends Park Cooking School, P.O. Box 788, 902 S. Oak St., La Veta, CO 81055 US; 719-742-3147, Fax 719-742-3762, E-mail Legends@designelk.com, URL http://www.designelk.com/legends.html.

MIJBANI INDIAN RESTAURANT

Boulder/Year-round

Jessica Shah offers 1- to 3-session demonstration and participation courses. Established in 1993. Maximum class/group size: 6-16 hands-on/20 demo. 10-15 programs per year. Also featured: A newsletter, private classes.

Emphasis: Homestyle cuisine, vegetarian, quick and easy to prepare, techniques, spices, flat breads, chutneys, appetizers, fast foods.

Faculty: Jessica Shah, a native of Bombay, has 14 years cooking experience, owns MijBani Restaurant, and writes for local and national publications.

Costs: $35-$100.

Location: Boulder, 45 minutes from Denver.

Contact: Jessica Shah, MijBani Indian Restaurant, 2005 18th St., Boulder, CO 80302 US; 303-442-7000, Fax 303-442-5380.

SCHOOL OF NATURAL COOKERY – THE MAIN COURSE

(See also page 208, The Natural Cook) **Boulder, Denver/Year-round**

This vegetarian cooking school offers intensives and weekly courses, no recipes, hands-on, the language of chefs. Established in 1985. Maximum class/group size: 4-14. Facilities: Vary. Also featured: In Boulder, nearby activities include mountain biking, hiking, skiing, music, dance, theater, festivals.

Emphasis: Theory and techniques for preparing whole grains, beans, vegetables.

Faculty: Mary Bowman, Michael Thibodeaux, Vivian Gold, Elizabeth Archerd, Jane Angulo, and Vicki Johnston are certified Main Course instructors.

Costs: Tuition is $250 and materials $55 for Parts I and II, $125 for Part III for weekly classes. Intensives are $800 for all three parts, $400/weekend.

Location: North shore of the Big Island of Hawaii, Boulder, CO (35 minutes from Denver), central locations in Denver, Minneapolis, Atlanta, and Seattle.

Contact: Joanne Saltzman, Director, School of Natural Cookery, P.O. Box 19466, Boulder, CO 80308 US; 303-444-8068, E-mail snc@sprynet.com.

THE SEASONED CHEF

Denver/Year-round

This private school offers demonstration and participation classes. Established in 1993. Maximum class/group size: 12-15 hands-on/35 demo. 100+ programs per year. Facilities: Well-equipped home kitchen. Also featured: Classes for youngsters, wine appreciation, market visits, private instruction.

Emphasis: Includes basic techniques, healthful cooking, ethnic cuisines, menu planning.

Faculty: Area cooking school instructors and restaurant chefs, guest chefs and cookbook authors.

Costs: $35-$45/class.

Contact: Susan Stevens, Director, The Seasoned Chef, 999 Jasmine St., #100, Denver, CO 80220 US; 303-377-3222, E-mail swstevens@mindspring.com.

THE TELLURIDE WINE FESTIVAL

Telluride/June

The City of Telluride sponsors luncheon programs with guest chefs, seminars, tastings of over 200 wines, and a cooking class. Established in 1981.

Emphasis: Wine and food pairing.

Contact: Keith Hampton, Program Director, Telluride Wine Festival, 747 W. Pacific, #324, Box 1677, Telluride, CO 81435 US; 970-728-3178, Fax 970-728-4865.

CONNECTICUT

BELLA CUCINA

New Canaan/Year-round *(See also page 315)*

Culinary professional Carol Borelli offers 1- to 6-session courses that cover basic techniques, seasonal and regional menus, fresh pasta, soups, risotto, dinner party menus, breads. Established in 1996. Maximum class/group size: 8-20. 15-20 programs per year. Facilities: Fully equipped home kitchen. Also featured: Excursions to markets and restaurants, travel programs to Italy.

Emphasis: Fine cooking in the seasonal traditions of Italy's regions.

Faculty: Carol Borelli, artist and educator, studied at Le Cordon Bleu and with chefs worldwide. Guest teachers from the US and abroad.

Costs: Classes and courses are $45-$270.

Location: 50 miles northeast of New York City.

Contact: Carol Borelli, Owner, Bella Cucina, P.O. Box 421, New Canaan, CT 06840 US; 203-966-4477, Fax 203-966-8781, E-mail cborelli@earthlink.net.

THE COMPLETE KITCHEN COOKING SCHOOL

Darien/March-May, September-November

This school in a kitchenware store offers morning and evening demonstrations. Established in 1980. Maximum class/group size: 20. 50 programs per year.

Faculty: School Director Sigrid Laughlin and guest instructors, including Julia della Croce, Nicole Routhier, Stephen Schmidt, Patricia Wells.

Costs: $45-$75/session.

Location: 40 miles from New York City and New Haven.

Contact: Sigrid Laughlin, Director, The Complete Kitchen Cooking School, 863 Post Rd., Darien, CT 06820 US; 203-655-4055, Fax 203-655-0121.

CUCINA CASALINGA

Wilton/March-May, October-November

This private cooking school offers demonstration and hands-on classes in Connecticut for adults and children, two 10-day culinary and cultural tours to Italy each year, and a food and wine tour of California's wine country. Established in 1981. Maximum class/group size: 12 hands-on/20 demo/16 travel programs. Facilities: Open-plan home kitchen in Connecticut. Also featured: Private classes, children's summer camp, wine tasting, tours of Arthur Ave. in the Bronx include dining at fine restaurants and local trattorias.

Emphasis: Italian regional cuisine, California cuisine and wines.

Faculty: Owner/instructor Sally Maraventano graduated Georgetown University, studied at the University of Florence, and learned to cook from her mother and Sicilian grandfather, who owned an Italian bakery. Guest instructors include European chefs and American culinarians.

Costs: Adult (children's) classes $70 ($45)/session, $200 ($125) for a series of 3. Cost of Italy trip is $5,500, including shared deluxe lodging, most meals, planned activities, airfare and ground transport.

Location: CT: Lower Fairfield County, 1 hour north of Manhattan. Italy: Tuscany and Piedmont regions, Lake Country, Venice. California: Napa and Sonoma Valleys.

Contact: Sally Maraventano, Owner, Cucina Casalinga, 171 Drum Hill Rd., Wilton, CT 06897 US; 203-762-0768, Fax 203-762-0768, E-mail cucinacasa@aol.com, URL http://members.aol.com/cucinacasa/index.htm. Additional contact: Miriam Luck 203-226-1033.

FOODSEARCH PLUS, INC.

Ridgefield/Year-round *(See also page 309, The Vacationing Table)*

This private school offers 4- and 8-session hands-on courses that focus on culinary techniques. Established in 1990. Maximum class/group size: 10. 10 programs per year. Facilities: Commercially equipped kitchen in converted barn on pond. Also featured: Market visits, private instruction.

Emphasis: Cooking fundamentals; planning healthy, seasonal menus for family and entertaining.
Faculty: IACP member Karen Hanson, culinary educator, author, food consultant.
Costs: $300/4 classes.
Location: 75 minutes from New York City, 30 minutes from Greenwich, CT.
Contact: Karen Hanson, Foodsearch Plus, Inc., 258 Florida Rd., Ridgefield, CT 06877 US; 203-438-0422.

HAY DAY COOKING SCHOOL

Four locations/Year-round

This private school offers 12-18 demonstration classes/session, 3 sessions/year, in each of its locations. Established in 1982. Maximum class/group size: 45. Facilities: Professional demonstration kitchen-classroom with seating for 45 students, overhead mirror, stovetop, convection oven, P.A. system.
Emphasis: Regional and ethnic cuisines, classic techniques, methods and food history.
Faculty: Guest chefs include Bobby Flay, Bradley Ogden, Marcella Hazan, Jacques Pépin, Michael Romano, Steven Raichler, Chris Schlesinger, Gordon Hamersley.
Costs: $65-$75/class.
Location: Greenwich, CT, 45 minutes from NYC; Ridgefield, CT, 70 minutes from NYC; Westport, CT, 1 hour from NYC; Scarsdale, NY, 30 minutes from NYC.
Contact: Nicole J. Courtemanche, Director, Sutton Hay Day, Inc., 1385 Post Rd. East, Westport, CT 06880 US; 203-319-2777/454-6649, Fax 203-319-2772.

MYSTIC COOKING SCHOOL

Mystic, Stonington/Year-round

This private school offers classes on a variety of topics, including specific techniques and regional and ethnic cuisines. Established in 1994. Maximum class/group size: 16 hands-on, 20 demo, 25 travel. 50+ programs per year. Facilities: Stonington Vineyard. Also featured: Winery tours and trips to New Orleans and Cajun Country are also scheduled.
Faculty: Restaurant chefs and cookbook authors, including Charles van Over, Martha Murphy, Patrick Boisjot.
Costs: $40-$50/class.
Location: 50 miles from Hartford, 120 miles from New York City, 100 miles from Boston.
Contact: Annice Estes, Owner, Mystic Cooking School, P.O. Box 611, Mystic, CT 06355 US; 860-535-1151, Fax 860-535-1151.

PRUDENCE SLOANE'S COOKING SCHOOL

Hampton/Year-round

Culinary professional Prudence Sloane offers participation workshops, demonstrations, and dinner demonstrations. Established in 1993. Maximum class/group size: 8 hands-on/14 demo/14 dinner demo. 25-30 programs per year. Facilities: Well-equipped teaching kitchen. Also featured: Private party classes, food styling, kitchen design, knife skills workshops.
Emphasis: Ethnic and regional cuisines, techniques, theory, food history and flavoring principles, seasonal and holiday menus, wine selection.
Faculty: Prudence Sloane, an IACP member, was awarded the Blue Ribbon Professional diploma from Peter Kump's New York Cooking School, hosts a food radio show, TV cooking demos, food styling and kitchen design.
Costs: $30-$65/session, $210-$300 for intensive techniques series.
Location: Northeastern Connecticut, 90 minutes from Boston and 3 hours from New York City.
Contact: Prudence Sloane, Owner, Prudence Sloane's Cooking School, 245 Main St., Hampton, CT 06247 US; 203-455-0596, E-mail prudence.sloane@snet.net.

RONNIE FEIN SCHOOL OF CREATIVE COOKING

Stamford/Year-round

Culinary professional Ronnie Fein offers demonstration and participation classes that emphasize ingredients, techniques, and menus. Established in 1971. Maximum class/group size: 8 hands-on/16 demo. Facilities: Fully-equipped home teaching kitchen. Also featured: Children's classes, private instruction, year-round.

Emphasis: Regional cuisines, seasonal and holiday menus, food gifts, low-fat cuisine, ethnic foods, use of fresh herbs, and menu structure.

Faculty: Ronnie Fein writes for food publications (newspapers and magazines) and attended the China Institute and Four Seasons Cooking School. She is author of The Complete Idiot's Guide to Cooking Basics.

Costs: $45-$55.

Location: North Stamford, 45 minutes from New York City.

Contact: Ronnie Fein, Owner, Ronnie Fein School of Creative Cooking, 438 Hunting Ridge Rd., Stamford, CT 06903 US; 203-322-7114, Fax 203-329-3366, E-mail ronskie@aol.com.

SANDY'S BRAZILIAN & CONTINENTAL CUISINE

Oxford/October-June

This private school offers monthly hands-on classes in international cuisine and culinary tours to Brazil. Established in 1987. Maximum class/group size: 8 hands-on, 25 travel. 12-15 programs per year. Facilities: Modified home kitchen. Also featured: Private instruction, market visits, sightseeing, visits to food producers, coffee farm, brewery.

Emphasis: International menus, including European, Middle Eastern, Far Eastern, Brazilian, Oceania.

Faculty: Brazilian-born Sandra N. Allen, CCP, is a member of the IACP with 20+ yrs teaching experience in the US and Brazil, including Peter Kump's NY Cooking School, the New School for Social Research, and on TV cooking shows.

Costs: $60/class includes dinner. 9-day Brazil tour $2,500 includes 4-star hotel lodging, 5 cooking workshops, breakfasts and some lunches, 3 dinners in noted restaurants, planned activities.

Location: Country location between Southbury and Oxford,CT about 15 miles east of Danbury and 15 miles west of Waterbury off I-84.

Contact: Sandra Allen, Director, Sandy's Brazilian & Continental Cuisine, 222 Maple Tree Hill Rd., Oxford, CT 06478-1545 US; 203-264-0374, Fax 203-264-0374, E-mail Sandyna@juno.com, URL http://www.sbcc-cooking.com.

SILO COOKING SCHOOL

New Milford/March-December

This gourmet foods store and art gallery offers demonstration and participation courses. Established in 1972. Maximum class/group size: 14 hands-on/30-35 demo. 70 programs per year. Facilities: Well-equipped teaching kitchen. Also featured: Custom group and children's classes.

Emphasis: Ethnic and regional cuisines, holiday menus, baking, guest chef specialties, wine selection.

Faculty: Has included Giuliano Bugialli, Michael Romano, Daniel Leader, Jacques Pépin, Madeleine Kamman. School is owned by New York Pops founder Skitch Henderson and wife Ruth.

Costs: About $75-$85 for master chef classes, $65-$100 for others. Discount for 2 people registering for same class.

Location: About 80 miles from New York City on the Henderson's Hunt Hill Farm in the Litchfield Hills.

Contact: Sandra Daniels, Director, Silo Cooking School, Upland Rd., New Milford, CT 06776 US; 860-355-0300, Fax 860-350-5495, E-mail sales@thesilo.com, URL http://www.thesilo.com.

DELAWARE

WHAT'S COOKING AT THE KITCHEN SINK

Hockessin/September-June

This school in a kitchenware store offers demonstration classes. Established in 1991. Maximum class/group size: 16. 120+ programs per year. Facilities: 300-sq-ft, 16-seat teaching area with overhead mirror. Also featured: Children's workshops, private and party classes.

Emphasis: Special occasion menus, guest chef specialties, specific subjects.
Faculty: Director Lee Wooding, an IACP member, and traveling guest chefs/authors.
Costs: $26-$45/class. For lodging, the school recommends The Inn at Montchanin Village.
Location: Hockessin, a Wilmington suburb, is 40 miles from Philadelphia and 75 miles from Baltimore.
Contact: Wendy Ketcham, Director, What's Cooking at the Kitchen Sink, 425 Hockessin Corner, Hockessin, DE 19707 US; 302-239-7066, Fax 302-239-7665, E-mail cook@thekitchensink.com, URL http://www.thekitchensink.com.

DISTRICT OF COLUMBIA

WHAT'S COOKING!
Washington/September-May

Culinary professional Phyllis Frucht offers limited participation and/or hands-on classes on a variety of topics. Series include The International Gourmet, Asian Cooking, Techniques, Vegetarian, and Contemporary Cooking. Established in 1976. Maximum class/group size: 16. Facilities: Newly renovated townhouse kitchen.
Faculty: Phyllis Frucht has taught cooking 30+ yrs at home, in adult ed, and at the former What's Cooking! cookware store/cooking school in Rockville, MD, where she was chef/owner.
Costs: Classes are $35 each or 5/$150.
Location: The Dupont Circle area of Washington, D.C., two blocks from a metro station.
Contact: Phyllis Frucht, Teacher, What's Cooking!, 1917 S Street NW, Washington, DC 20009 US; 202-483-7282, Fax 202-483-7284, E-mail whatsckng@aol.com.

FLORIDA

ARIANA'S COOKING SCHOOL
Miami/Year-round

This cookware store and school offers demonstration and participation classes. Established in 1976. Maximum class/group size: 28. 150+ programs per year. Facilities: 400-square-foot kitchen with overhead mirror. Also featured: Private bridal showers, birthday parties for children and adults.
Faculty: The 15+-member faculty includes Wendy Kallergis, Paul Galadga, Ariana Kumpis, Allen Susser, Mark Militello, Sarah Benson.
Costs: $25-$100/session.
Contact: Ariana M. Kumpis, Director, Ariana's Cooking School, 7251 S.W. 57th Ct., Miami, FL 33143 US; 305-667-5957, Fax 305-665-7763.

CHEF ALLEN'S
North Miami Beach/Year-round

Chef Allen Susser offers demonstration and participation classes the second Wednesday of every month; one-on-one sessions in which the student works along with the restaurant staff. Established in 1986. Maximum class/group size: 1-25. 12 programs per year. Facilities: Chef Allen's restaurant in North Miami Beach, Allen's 2 Go gourmet market.
Emphasis: New World cuisine, local fish, tropical fruits, Latin root vegetables.
Faculty: Chef Susser, graduate and on faculty of Florida International University School of Hospitality & Restaurant Management. Author of Allen Susser's New World Cuisine and Cookery and The Great Citrus Book, he studied at Le Cordon Bleu and was chef at Paris' Bristol Hotel and Le Cirque in New York City.
Costs: Group classes range from $35-$50. Individual session is $195.
Contact: Chef Allen Susser, Chef/Owner, Chef Allen's, 19088 N.E. 29th Ave., N. Miami Beach, FL 33180 US; 305-935-2900, Fax 305-935-9062, E-mail ChefAllen@aol.com, URL http://www.chefallen.com.

CREATIVE CUISINE COOKING SCHOOL (See page 265)

DAMIANO'S AT THE TARRIMORE HOUSE
Delray Beach/October-August

This restaurant offers Wednesday day and evening theme classes. Established in 1992. Maximum class/group size: 25. Facilities: Demonstration kitchen in the restaurant dining room.
Emphasis: Low-fat, or fat-free cooking; Italian, Southwest, and Asian cuisines.
Faculty: Chef Anthony Basil Damiano.
Costs: $100 ($65) for 3 evening (day) classes. Lodging is available at the Seagate Beach Club.
Contact: Lisa Damiano, Damiano's at the Tarrimore House, 52 N. Swinton Ave., Delray Beach, FL 33444 US; 561-272-4706, Fax 561-272-4796.

DISNEY INSTITUTE CULINARY ARTS PROGRAMS
Lake Buena Vista/Year-round

Walt Disney World Resort offers culinary programs that range from the basics to entertaining menus. Established in 1996. Maximum class/group size: Approximately 15. Facilities: 3 participatory kitchens, each with 14 newly-equipped work stations. Also featured: Programs in Disney animation, photography, sports and fitness, gardening, golf and tennis, spa, special programs for youngsters, resort amenities.
Emphasis: Includes celebrations, techniques, healthy cooking, baking, international cuisines, wine.
Faculty: Programs managed by Chef Frank Brough, former executive chef at Ariel's restaurant at Disney's Yacht Club Resort and chef de cuisine at Victoria & Albert's at Disney's Grand Floridian Resort & Spa.
Costs: Vacation packages begin at $539 for 3-night/4-day vacations that include shared bungalow lodging at the Villas of the Disney Institute, 4 hands-on cooking classes, a personal session with a chef, and a bonus, e.g. dinner, golf lesson, theme park ticket.
Location: Walt Disney World Resort.
Contact: Frank Brough, Disney Institute, P.O. Box 10,000, Lake Buena Vista, FL 32830-1000 US; 407-827-1100, Fax 407-827-4586, URL http://www.disneyinstitute.com.

GOING SOLO IN THE KITCHEN
Dog Island/Year-round

Food and travel writer Jan Doerfer offers bimonthly 5-day participation courses that are geared to the needs of the solo cook. Instruction is scheduled mornings and evenings, afternoons are free. Maximum class/group size: 12. 6 programs per year. Facilities: The Pelican Inn resort on the beach.
Faculty: Going Solo newsletter publisher Jane Doerfer's cookbook credits include The Victory Garden Cookbook (collaborator), The Legal Sea Foods Cookbook, and Going Solo in the Kitchen.
Costs: $975 includes lodging and meals.
Location: Dog Island, a barrier island accessible by ferry 4 miles from the mainland, near Carrabel in Florida's Panhandle.
Contact: Jane Doerfer, Going Solo in the Kitchen, P.O. Box 123, Apalachicola, FL 32329 US; 850-653-8848, E-mail Jdoerfer@digitalexp.com.

HARRIET'S KITCHEN WHOLE FOODS COOKING SCHOOL
Winter Park/September-June

This private school offers demonstration and participation classes. Established in 1987. Maximum class/group size: 16 hands-on/35 demo. 100+ programs per year. Facilities: 500-square-foot teaching kitchen. Also featured: Classes for youngsters, a 9-session Healing Macrobiotic series, sourdough whole grain bread classes, and weekend and week-long cooking intensives.
Emphasis: Macrobiotic and gourmet vegetarian cuisines.
Faculty: Director Harriet McNear, a Kushi certified teacher and licensed nutrition counselor, studied at the Kushi Institute and the Natural Gourmet Cookery School. Local chefs include Mario Martinez, Clair Epting, Bruno Ponsot, Marc van Couwenberghe, M.D.
Costs: Classes range from $20-$40, 5-day retreat $750-$900, the 9-session course $225. Spouses

receive a 30% discount. Work-study and assistantship positions are available.
Location: Near Walt Disney World, 15 miles from Orlando International Airport.
Contact: Harriet McNear, Director, Harriet's Kitchen, 1136 Oaks Blvd., Winter Park, FL 32789 US; 407-644-2167, Fax 407-644-2187, E-mail harkit@mindspring.com.

THE KITCHEN HEARTH

Miami Beach/Year-round

This cookware store offers demonstration and participation classes. Established in 1994. Maximum class/group size: 15 hands-on/20 demo. 60+ programs per year. Facilities: Kitchen with 15 workspaces. Also featured: Classes for youngsters, market visits.
Emphasis: A variety of topics, including Italian, Thai, Chinese, and Indian cuisines; baking and cake decorating; low-fat cooking.
Faculty: Guest chefs from noted local restaurants.
Costs: $30 for full-length classes, $18 for mini-classes, $18 for children's classes.
Location: Two miles from the historic South Beach district and 15 minutes from Miami International Airport.
Contact: Gail Fix, The Kitchen Hearth, 456 Arthur Godfrey Rd., Miami Beach, FL 33140 US; 305-538-3358, Fax 305-538-3431.

THE PALM BEACH SCHOOL OF COOKING, INC.

Delray Beach/Year-round

This cookware store, gourmet take-out food retailer, and private school offers half-day workshops and multi-session courses that cover a variety of topics including Caribbean, Pacific Rim, sushi making, and vegetarian dishes. Established in 1998. Maximum class/group size: 10 hands-on, 20 demo. 30 programs per year. Facilities: Professional cooking equipment, individual work stations.
Emphasis: Asian gourmet low-fat and other ethnic cuisines, professional techniques for nonprofessionals.
Faculty: Professionally trained experienced teachers.
Costs: $45/class.
Location: Downtown Delray Beach near Restaurant Row in a historic renovated shopping village. 17 miles south of Palm Beach International Airport.
Contact: Doreen N. Moore, Director/Executive Chef, The Palm Beach School of Cooking, Inc., 25 N.E. 2nd Ave., #112, Delray Beach, FL 33444 US; 561-279-4707, Fax 561-279-8679, E-mail cybrcook@pb.seflin.org.

THE RITZ-CARLTON, AMELIA ISLAND COOKING SCHOOL

Amelia Island/Year-round

The Ritz-Carlton resort hotel offers monthly 2-day participation courses that focus on a theme. Established in 1994. Maximum class/group size: 15. 12 programs per year. Facilities: The Grill kitchen. Also featured: A tour of the food preparation facilities, notebook with recipes, champagne graduation ceremony.
Emphasis: Seasonal and entertaining menus, regional and ethnic cuisines, macrobiotic recipes.
Faculty: Kenneth Gilbert, AAA 5-Diamond chef of The Grill; the hotel's food and beverage staff.
Costs: $650/person, $965/couple, which includes lodging, 2 meals daily, and Mobil 4-star oceanfront resort amenities, which include 18-hole golf course, pools, tennis.
Location: On the Atlantic Ocean, 25 minutes north of Jacksonville.
Contact: Frank Cavella, Director of Sales & Marketing, The Ritz-Carlton, Amelia Island, 4750 Amelia Island Pkwy, Amelia Island, FL 32034 US; 800-241-3333/904-277-1100, Fax 904-277-1145, URL http://www.ritzcarlton.com.

SARASOTA FOOD AND WINE ACADEMY

Sarasota/Year-round

This wine and gourmet food store offers demonstration and participation courses. Established in 1995. Maximum class/group size: 15 hands-on/40 demo. 20 programs per year. Facilities: 350-

square-foot central teaching station and kitchen. Also featured: Wine courses.
Faculty: Special events coordinator Anthony Blue; guest chefs.
Costs: $25-$100.
Location: Adjacent to Michael's On East restaurant, 60 miles south of Tampa.
Contact: Michael Klauber, Director, Sarasota Food and Wine Academy, 1212 East Ave. S., Sarasota, FL 34239 US; 941-366-0007 #236, Fax 941-955-1945, E-mail michael@bestfood.com, URL http://www.bestfood.com. Other contact: Gayle Guynup, Coordinator.

GEORGIA

ART INSTITUTE OF ATLANTA – SCHOOL OF CULINARY ARTS

Atlanta/Year-round *(See also page 36) (See display ad page 8)*

This private career institution offers Saturday AM hands-on and demonstration classes. Established in 1991. Maximum class/group size: 16. Facilities: Include 4 kitchens and 3 classrooms.
Emphasis: Cooking basics.
Faculty: 17 full- and part-time.
Costs: $85/class.
Location: Atlanta's Buckhead section.
Contact: June Fischer, Director, Smart Fun Workshops, Art Institute of Atlanta-School of Culinary Arts, 3376 Peachtree Rd., Atlanta, GA 30326 US; 800-275-4242 x420, Fax 404-266-1383, URL http://www.aii.edu.

THE COOK'S WAREHOUSE, INC.

Atlanta/Year-round

This cookware store offers 2-hour demonstration and hands-on classes on a variety of topics, including basics, seasonal and holiday recipes, ethnic cuisines, chef specialties. Established in 1995. Maximum class/group size: 25 demo, 14 hands-on, 16 travel. Facilities: 5-Star gas range and stove, Sub-Zero refrigerator, Bosch dishwasher, 12-square-foot overhead mirror, Brazilian granite countertops. Also featured: Classes for youngsters, wine classes, private instruction, dining in private homes, quarterly volunteer day at the Atlanta Food Bank.
Emphasis: Informal non-intimidating learning atmosphere.
Faculty: ~30 restaurant chefs and other culinary professionals.
Costs: $35/class.
Location: Midtown Atlanta, 10 minutes from downtown, 20 minutes from Hartsfield International Airport.
Contact: Mary Moore, Owner/administrator, The Cook's Warehouse, Inc., 549-1 Amsterdam Ave. NE, Atlanta, GA 30306 US; 404-815-4993, Fax 404-815-0543, E-mail cookware@mindspring.com, URL http://www.cookswarehouse.com.

DIANE WILKINSON'S COOKING SCHOOL

Atlanta/Year-round

Culinary professional Diane Wilkinson offers 5-day intensive techniques courses and short courses. Established in 1974. Maximum class/group size: 8 for hands-on classes. Facilities: Remodeled Mediterranean-style kitchen with 2 fireplaces, one built for open-hearth cooking. Also featured: Private classes for individuals and groups.
Emphasis: French and Italian techniques, seasonal foods, reduced fat recipes.
Faculty: Diane Wilkinson, CCP, studied at Le Cordon Bleu-Paris, La Varenne, with Marcella Hazan, and has worked in kitchens in France and Italy, including those of Michael Guerard, Claude Deligne, Guenther Seeger, and L'Oustau de Baumaniere.
Costs: Classes are $50, five sessions are $225, the intensive is $650.
Location: Northwest Atlanta, between Buckhead and Cumberland Mall.
Contact: Diane Wilkinson, Diane Wilkinson's Cooking School, 4365 Harris Trail, Atlanta, GA 30327 US; 404-233-0366, Fax 404-233-0051.

RAY OVERTON'S LE CREUSET COOKING SCHOOL

Atlanta/Year-round

Culinary professional Ray Overton (formerly of The Cooking Scene) offers 1- to 3-session demonstration and participation classes. Established in 1993. Maximum class/group size: 8 hands-on/30 demo-lecture. 200+ programs per year. Facilities: 30-seat demonstration kitchen with overhead mirror and audio/video system, 8 separate work stations. Also featured: Cooking camp for youngsters, private classes, lectures, book signings, and culinary trips.

Emphasis: Ethnic and regional cuisines, special occasion dishes, baking, pastry, low fat, healthy and vegetarian, wine tasting, specific subjects.

Faculty: Cookbook author, IACP and AIWF member and cooking show host Ray Overton trained with Nathalie Dupree. Guest instructors may include Lydie Marshall, Shirley Fong-Torres, Fabrizio Bottero, Virginia Willis, Hugh Carpenter, Mara R. Rogers, Martin Yan, Pat Wells.

Costs: Demonstrations are $25/class; participation classes begin at $50. Lodging is available at hotels located across Perimeter Mall.

Location: Park Place Shops across from Perimeter Mall.

Contact: Ray L. Overton, III, Culinary Director/Owner, Culinary Concepts, 4505 Ashford Dunwoody Rd., Atlanta, GA 30346 US; 770-396-5925, Fax 404-733-6002, E-mail roverusa@aol.com. Also: Culinary Concepts, 89 26th St. NW, #2, Atlanta, GA 30309-2004; 404-875-7532.

SCHOOL OF NATURAL COOKERY (See page 199)

URSULA'S COOKING SCHOOL, INC.

Atlanta/September-May

Culinary professional Ursula Knaeusel offers 4-session demonstration courses. Established in 1966. Maximum class/group size: 40. 3 programs per year. Facilities: 3-level classroom with 18-foot mirror over a 22-foot granite counter. Also featured: Gingerbread house, cutting and decorating classes, bridal shower classes, couples classes.

Emphasis: Nouvelle cuisine, time-saving methods and advance preparation.

Faculty: Ursula Knaeusel's 40+ years of experience include supervising kitchens and operating restaurants in Europe and the U.S., teaching in Central America, the Caribbean, and the U.S. She hosts PBS' Cooking With Ursula and is author of the same.

Costs: $90 for the 4-session course.

Location: One mile from Interstate 75 and 85 and 4 miles from downtown Atlanta.

Contact: Ursula Knaeusel, President, Ursula's Cooking School, Inc., 1764 Cheshire Bridge Rd., N.E., Atlanta, GA 30324 US; 404-876-7463, Fax 404-876-7467, E-mail UrsulaKAtl@aol.com, URL http://www.angelfire.com/ga/ursulascookingschool/.

HAWAII

CELEBRATIONS OF HAWAII REGIONAL CUISINE

All Islands/Year-round

This full-service travel company offers culinary vacation programs featuring signature chefs, wine makers, and luxury resorts. Hands-on classes, demonstrations, evening dining events, wine tastings, seminars, off-site tours are all included. Established in 1994. Maximum class/group size: 15-30. Also featured: Spa treatments, golf, tennis, cultural and art tours, and enrichment seminars that include floral design, herb gardening, decorative displays, cooking with condiments, and star gazing.

Emphasis: Hawaiian and American regional cuisine, Pacific Rim, other ethnic cuisines, wine appreciation, vegetarian. Guest chefs.

Faculty: The founding chefs of Hawaii Regional Cuisine: Peter Merriman (Hula Grill), Mark Ellman (Avalon), Jean-Marie Josselin (A Pacific Cafe), Roy Yamaguchi (Roy's Kahana Bar & Grill), Beverly Gannon (Haliimaile General Store). Guest chefs.

Costs: Land packages begin at $1,000, which includes most meals, ground transport, and first class lodging.

Location: Throughout the Hawaiian Islands.
Contact: Noelle Rutter, Interactive Events, 155 Wailea #14, Wailea, HI 96753 US; 800-961-9196, Fax 808-875-1565, E-mail events@maui.net, URL http://www.maui.net/~events. Other contact: General information, 808-875-8808.

COOKING SCHOOL AT AKAKA FALLS INN

Honomu/June

This bed & breakfast establishment offers 5-day hands-on classes in regional cuisine that include visits to markets and food producers, winery tours, dining in a local restaurant, and sightseeing. Established in 1998. Maximum class/group size: 4. 20-26 programs per year. Facilities: Large, fully-equipped home-style kitchen. Also featured: Private instruction, classes for youngsters.
Emphasis: Local food products, Pacific cuisine.
Faculty: IACP member Sonia R. Martinez and Vicki Soule.
Costs: $950 includes meals, lodging at the Inn, all planned activities.
Location: Old sugar plantation village 13 miles north of Hilo on the Big Island of Hawaii.
Contact: Sonia R. Martinez, Co-owner, Cooking School at Akaka Falls Inn, PO Box 190, Honomu, HI 96728 US; 888-757-0924/808-963-5468, Fax 808-963-6353, E-mail akakainn@gte.net.

CUISINES OF THE SUN

Kohala Coast/August

The Mauna Lani Bay Hotel sponsors an annual 4-day culinary vacation that features daily demonstrations of tropical recipes. Established in 1990. Maximum class/group size: 200. 1 programs per year. Facilities: On-stage demonstration kitchen. Also featured: Resort amenities, including golf, tennis, spa.
Emphasis: Tropical cuisines. Theme and region changes each year.
Faculty: Noted chefs and beverage makers from the regions featured.
Costs: Approximately $1,795 single, $2,590 double occupancy, which includes some meals, planned activities, and lodging. Daily and individual event options available.
Location: The AAA 5-Diamond Mauna Lani Bay Hotel and Bungalows is on the Kohala Coast of Hawaii's Big Island, 20 miles from the airport.
Contact: Sharon Bianco, Director of Catering, The Mauna Lani Bay Hotel and Bungalows, One Mauna Lani Dr., Kohala Coast, HI 96743 US; 808-885-6622, Fax 808-885-4556, E-mail maunalani@maunalani.com, URL http://www.maunalani.com.

GRAND CHEFS ON TOUR

Wailea, Maui

The Kea Lani Hotel offers a 3-day culinary program that features lunch demonstrations and dinners with wine tastings. Established in 1996. Maximum class/group size: 50. Facilities: Kea Lani Hotel open-air classroom and kitchen. Also featured: Live fusion jazz music, island style.
Emphasis: Cuisines of the Pacific Rim and noted guest chefs; wine and food pairing.
Faculty: Each program features a Pacific Rim chef, a noted mainland U.S. chef, and experts from California wineries. Has included Emeril Lagasse, Dean Fearing, Bradley Ogden, Martin Yan.
Costs: Demonstration and lunch $35, dinner $85.
Location: Wailea Resort, South Maui, 30 minutes from Kahului Airport
Contact: Tom Risko, Public Relations Director, Classic Custom Vacations and Avis Rent-A-Car, 4100 Wailea Alanui Dr., Maui, HI 96753 US; 800-659-4100, Fax 808-875-1200, E-mail sales@kealani.com, URL http://www.kealani.com/.

THE NATURAL COOK

Pahoa/May-November *(See also page 199, School of Natural Cookery)*

The Kalani Oceanside Retreat resort offers 2-week Main Course and 1-week Natural Sweets baking intensives that feature daily demos &/or hands-on classes that progress from techniques to whole meals, no animal products. Established in 1996. Maximum class/group size: 8. 6-9 programs per year. Facilities: Open air teaching kitchen at Kalani Oceanside Retreat and other locations.

Emphasis: Non-recipe based method that combines creativity, cooking theory and technique, flavor and visual rhythm.
Faculty: Michael Thibodeaux, certified by the School of Natural Cookery in Boulder, CO.
Costs: Main Course (2 weeks) is $1,960, Natural Sweets (1 week) is $890, including lodging and meals at Kalani Oceanside Retreat.
Location: Island of Hawaii.
Contact: Michael Thibodeaux, Culinary Instructor, Kalani Oceanside Retreat, RR2 Box 4500, Pahoa, HI 96778 US; 800-800-6886, Fax 888-337-2177, E-mail mthib@usa.net, URL http://www.naturalcook.com.

ILLINOIS

A TASTE OF ELEGANCE FROM THE HEART OF MEXICO

Chicago/Year-round

Chef Geno Bahena offers hands-on classes in Mexican cuisine. Travel programs include market visits, winery tours, sightseeing, visits to food producers, and dining in private homes. Established in 1996. Maximum class/group size: 16 max. 8 programs per year. Facilities: Home, restaurant, cottage, hotels.
Emphasis: Gourmet Mexican cuisine, including Guerrero, a province on Mexico's Pacific Coast.
Faculty: Chef Geno Bahena, managing chef of Frontera Grill and Topolobampo restaurants.
Costs: $35/class.
Contact: Geno Bahena, Chef-Instructor, A Taste of Elegance from the Heart of Mexico, 300 N. State St., Chicago, IL 60610 US; 312-464-0399, Fax 312-464-0436, E-mail tasteoeleg@aol.com.

BAKING ANGELS COOKING SCHOOL

Carlinville/Year-round

This private school offers 2-hour, full-day and week-long classes in traditional midwest farm cooking including German Heritage, cookies and pies, breads, cake decorating, candy. Established in 1997. Maximum class/group size: 36 demo, 11 hands-on. Facilities: Wooded kitchen close to orchard. 1,600-square-foot kitchen with 640-square-foot deck and cooking fireplace, individual work stations, computer/business facilities. Also featured: Children's classes, private instruction, visits to markets and food producers.
Emphasis: Traditional midwest farm meals, homemade bread from fresh grains, cookies, and pies from farm-grown products, lean meat cookery with pork and beef products.
Faculty: Sharon Behme, home economist; Luke Behme, state meat and dairy evaluations team; Rachel Hodges, pastry chef; Gil Hodges, chef.
Costs: Adult specialty classes $50-$150, Lunch & learn demos and taste testing $30-$60, youth participation classes $15-$60.
Location: 38 miles south of Springfield, IL; 60 miles northeast of St. Louis, MO on US Rt 55; outside Carlinville city limits a half mile east of the Orchards.
Contact: Sharon Behme, Baking Angels Cooking School, RR 4 Box 242, Carlinville, IL 62626 US; 217-854-2334, Fax 217-854-4002, E-mail behmecpa@accunet.net.

BEAUTIFUL FOOD

Wilmette/March-May, September-November

Cooking instructor Charie MacDonald offers participation classes. Established in 1973. Maximum class/group size: 20. 20 programs per year. Facilities: 2,800-square-foot commercial kitchen of Beautiful Food, her catering and wholesale specialty food business. Also featured: Culinary tours.
Emphasis: Fresh foods, techniques, breads, pastas, pastries, soups, low-cholesterol foods.
Faculty: Charie MacDonald studied at Le Cordon Bleu, the École des Trois Gourmands (Provence), and with Simone Beck. A charter member of the IACP, she founded Beautiful Food in 1982.
Costs: Class fee is $45, payable in advance. The trip cost of $5,000 includes round-trip airfare, meals, lodging, and ground transport.

Location: Wilmette, northeast of Chicago near O'Hare International Airport.
Contact: Charie MacDonald, Beautiful Food, 2111 Beechwood Ave., Wilmette, IL 60091 US; 847-256-3979, Fax 847-853-0607.

CARLOS' RESTAURANT

Highland Park/Year-round

This restaurant offers luncheon demonstrations and A Day in the Kitchen individual participation classes. Established in 1993. Maximum class/group size: 25 demo.
Emphasis: Contemporary French cooking.
Faculty: Executive Chef Jacky Pluton.
Costs: Luncheon classes are $40; the full-day class is $140/person, which includes the class for one and dinner for two.
Location: Highland Park, 25 miles north of Chicago.
Contact: Carlos Nieto, Owner, Carlos' Restaurant, 429 Temple Ave., Highland Park, IL 60035 US; 847-432-0770, Fax 847-432-2047.

CHEZ MADELAINE COOKING SCHOOL & TOURS

Hinsdale, IL; Bellac, France/Year-round

Culinary professional Madelaine Bullwinkel, Cte. and Ctesse. Aucaigne de Sainte Croix offers 1- to 3-session hands-on classes, evening menu classes, summer weekend package. Established in 1977. Maximum class/group size: 6-12/class, 12/tour. 35-40 programs per year. Facilities: Tours: Restored 18th century Château de Sannat. Also featured: Three 7-day tours per year include hands-on classes, visits to markets, porcelain museum, cheese and foie gras artisans, meals in private homes, hunting for game and wild boar, mushroom picking.
Emphasis: Classes cover basics, techniques, ethnic cuisines, seasonal and holiday menus, preserving, soups, stocks, baking. Tours feature French regional cuisine.
Faculty: Madelaine Bullwinkel received the Diplome from L'Academie de Cuisine, is author of Gourmet Preserves, Chez Madelaine, and is a member of Les Dames d'Escoffier. Occasional guest instructors and cookbook authors.
Costs: Classes: $60-$75. Tours: $2,500 ($2,900) includes shared (single) lodging, most meals, ground transport. Summer weekend: $580 includes lodging and some meals.
Location: Classes: Hinsdale, 20 miles west of Chicago. Tours: Southeast of Bellac in the Limousin region of the Haute Vienne (3-1/2 hours south of Paris).
Contact: Madelaine Bullwinkel, Owner, Chez Madelaine Cooking School & Tours, 425 Woodside Ave., Hinsdale, IL 60521 US; 630-325-4177, Fax 630-655-0355, E-mail chezmb@aol.com, URL http://www.chezm.com.

COOKING CRAFT, INC.

St. Charles/September-May

This gourmet shop and deli offers 2 to 3 evening demonstrations per week and some participation courses. Established in 1982. Maximum class/group size: 24 demo. Also featured: Classes for children, private sessions.
Faculty: The main instructor has over 10 years of teaching experience. Other instructors specialize in ethnic topics.
Costs: $18-$22/class.
Location: Mid-size commuter town on the Fox River, 40 miles west of Chicago.
Contact: Anne Lorenz, Owner/Director, Cooking Craft, Inc., 1415 W. Main St., St. Charles, IL 60174 US; 630-377-1730, Fax 630-377-3665.

THE COOKING AND HOSPITALITY INSTITUTE OF CHICAGO

Chicago/Year-round *(See also page 41) (See display ad page 42)*

This private career school offers demonstration classes and series for nonprofessionals. Established in 1983. Maximum class/group size: 30. Facilities: 4 fully-equipped instructional kitchens, on-site restaurant.

Emphasis: Topics include seafood, herbs, pasta, charcuterie, pastry, techniques.
Faculty: School founder Linda Calafiore is a past state coordinator of vocational training programs. Instructors are ACF-certified and have demonstrated accomplishment in their field of expertise.
Costs: $40-$50/session, $200-$275/series.
Location: Downtown Chicago.
Contact: Jim Simpson, Director, The Cooking and Hospitality Institute of Chicago, 361 W. Chestnut, Chicago, IL 60610 US; 312-944-2725, Fax 312-944-8557, E-mail chic@chicnet.org, URL http://www.chic.edu.

THE COOKING SCHOOLS OF TREASURE ISLAND FOODS, INC.

Chicago, Lake Bluff/Year-round

This gourmet food and grocery market offers mostly demonstrations of seasonal and local food products, with emphasis on ability to duplicate recipes at home; classes for youngsters. Established in 1986. Maximum class/group size: 70 demo, 30 hands-on. 100-125 programs per year. Facilities: Loft overlooking store seats 70.
Emphasis: Current food trends, showcasing the market's foods and chefs.
Faculty: Anita J. Brown, Director, coordinates and implements programs and conducts classes. Guest chefs are from Chicago's top restaurants.
Costs: $8/session for demos, $18-$25 for hands-on classes.
Location: In Chicago, 10 minutes north of downtown; in Lake Bluff, 40 minutes north of Chicago.
Contact: Anita J. Brown, Director, The Cooking Schools of Treasure Island Foods, Inc., 3460 N. Broadway, Chicago, IL 60657 US; 773-327-4265, Fax 773-327-6337, URL http://www.tifoods.com.

CUISINE COOKING SCHOOL

Moline/September-June

Culinary professional Marysue Salmon offers hands-on classes. Established in 1979. Maximum class/group size: 10. 50 programs per year. Facilities: Remodeled large kitchen with 5 work stations and AGA range. Also featured: Food and wine pairing, classes for youngsters, private instruction, culinary tours.
Emphasis: French and Italian cuisines.
Faculty: Owner/teacher Marysue Salmon studied at La Varenne and with Simone Beck and earned a BS degree in Food Science from Iowa State University.
Costs: $45/class. Hotels are nearby.
Location: 3 hours west of Chicago, 3 hours east of Des Moines.
Contact: Marysue Salmon, Cuisine Cooking School, 1100 - 23rd Ave., Moline, IL 61265 US; 309-797-8613, Fax 309-797-8641.

FRONTERA GRILL/FRONTERA INSTITUTE

Chicago/Year-round

This restaurant offers demonstrations and occasional 1-week culinary tours of Oaxaca, Mexico. Established in 1992. Maximum class/group size: 25 demo. Facilities: Frontera Institute demonstration kitchen.
Emphasis: Mexican cuisine.
Faculty: Rick Bayless hosted a PBS Mexican cooking series, author of Authentic Mexican and Rick Bayless's Mexican Kitchen (IACP Cookbook of the Year), established Frontera Grill and Topolobampo restaurants, 1995 recipient of the James Beard Award as Chef of the Year.
Costs: Vary according to program.
Location: River North section of Chicago; Oaxaca, Mexico.
Contact: Jennifer Fite, Frontera Grill/Frontera Institute, 445 N. Clark, Chicago, IL 60610 US; 312-661-1434, Fax 312-661-1830, E-mail rbayl10475@aol.com.

LA VENTURÉ

Skokie/November-May

Culinary professional Sandra Bisceglie offers 6-session participation courses. Established in 1980. Maximum class/group size: 12. 300+ programs per year. Facilities: 600-square-foot professional-style kitchen. Also featured: Private classes, classes for children.

Emphasis: French, Italian, and Chinese cuisines; candy making and cake decorating; baking.

Faculty: Director-owner Sandra Bisceglie attended the French School Dumas Pere and Harrington Institute of Interior Design and has a certificate of completion from the National Institute for the Foodservice Industry.

Costs: $289 for six lessons, $359 for baking course.

Location: Skokie is adjacent to Chicago, 3 miles from O'Hare airport.

Contact: Sandra Bisceglie, La Venturé, 5100 West Jarlath, Skokie, IL 60077 US; 847-679-8845.

PRAIRIE KITCHENS COOKING SCHOOL

Chicago and suburbs/Year-round

This private school offers hands-on 1- to 5-session classes that feature updated classical techniques, certification program covering basic techniques of good cooking, seasonal, and ethnic foods. Established in 1989. 150+ programs per year. Facilities: Professional kitchens with work stations. Also featured: Children's classes, private instruction, visits to markets and food producers, sightseeing, dining in private homes, wine and food events for individuals and businesses, travel programs to Europe and Mexico.

Emphasis: Cooking for the home chef, women's foodservice apprenticeship program.

Faculty: Kristin James, professional chef and instructor, member of the IACP and Les Dames D'Escoffier; Carolynn Friedman, caterer and instructor, member of the IACP and AIWF.

Costs: ~$40/session.

Location: Chicago metropolitan area, including north, west, northwest, and central locations.

Contact: Carolynn Friedman, President, Prairie Kitchens Cooking School, PO Box 372, Morton Grove, IL 60053 US; 847-966-7574, Fax 847-966-7589. Additional contact: Kristin James, 773-883-2317, Fax 773-883-2394.

SAVEUR INTERNATIONAL COOKING SCHOOLS

Chicago and other major cities/Year-round

This manufacturer of copper cookware and professional cutlery offers demonstrations of basic techniques, equipment use, chef recipes with wine pairing and menu planning. Established in 1997. Maximum class/group size: 20. 40/market programs per year. Facilities: Restaurant kitchens. Also featured: Field trips to regional produce markets and farms.

Emphasis: Basic techniques and chef specialties prepared in Saveur International copper cookware.

Faculty: Restaurant chefs include Sarah Stegner (Ritz-Carlton, Chicago), Jimmy Schmidt (Rattlesnake Club, Detroit), Jeff Drew (Coyote Cafe, Santa Fe), Sandy D'Amato (Sanford's, Milwaukee), Hartmut Handke (Handke's Cuisine, Columbus, OH).

Costs: $50-$100/class.

Location: Includes Chicago, IL, Detroit, Ann Arbor and Rochester, MI, Minneapolis, MN, St. Louis, MO, Santa Fe, NM, Columbus and Cleveland, OH, Milwaukee and Madison, WI.

Contact: Georgia LaBomascus, Director of Operations, Saveur International, PO Box 654, Cary, IL 60013 US; 800-707-2117/847-516-0746, Fax 847-516-1539, E-mail enjoyfood@aol.com.

TASTINGS COOKING SCHOOL

Fairfield/Year-round

Health Information Services, a food and nutrition firm, offers demonstration and hands-on classes and 6-session courses that include Culinary Hearts Kitchen (from American Heart Assn), Cooking with Diabetes, Food and Menopause, basics for children and adults. Established in 1997. Maximum class/group size: 12 hands-on, 35 demo. 60+ programs per year. Facilities: Teaching classroom with adjoining kitchen. Also featured: Children's etiquette classes, private instruction.

Emphasis: Health-related food preparation, basic culinary skills.
Faculty: Owner Lisa K. Fieber, MS, RD, LD, a member of the American Dietetic Assn. Other instructors include Chef Wm. Connors of Washburne Trade School.
Costs: $25-$45 for a 2-hour class, $120 for an American Heart Association class.
Location: 100 miles southeast of St. Louis, MO, 70 miles west of Evansville, IN.
Contact: Lisa Fieber, Director, Tastings Cooking School, Rt. 15 East, Fairfield, IL 62837 US; 618-847-7025/800-664-6363, Fax 618-847-8148, E-mail fieberl@midwest.net.

TRUFFLES, INC.
O'Fallon/September-June

Caterer and educator Kathy Kneedler offers 2-session demonstration and participation courses. Established in 1980. Maximum class/group size: 8-20. Facilities: Professional home kitchen with grill and double ovens, overlooking a lake. Also featured: Tours of the Hill (Italian) in St. Louis.
Emphasis: Regional and international food, bread and pastry, party themes.
Faculty: Caterer and food consultant Kathy Kneedler, CCP, has taught cooking for 18+ years and is a past newsletter editor for the St. Louis Culinary Society and chef of The Mark Twain Bank.
Costs: $32-$40.
Location: About 45 minutes from a major airport and 20 minutes from St. Louis.
Contact: Kathy A. Kneedler, President and Director, Truffles, Inc., Cooking School, 910 Indian Springs Rd., O'Fallon, IL 62269 US; 618-632-9461, Fax 618-234-3701, E-mail KKneed1051@aol.com.

WHAT'S COOKING
Hinsdale/Year-round

Cookbook author Ruth Law offers demonstration and participation courses in Hinsdale and Far East tours that feature classes with professional chefs, gourmet dining, sightseeing, visits to food markets. Established in 1980. Maximum class/group size: 15 hands-on.
Emphasis: The cuisines of China, Thailand, Singapore, Malaysia, Indonesia, India, Korea, the Philippines, Japan, and Hawaii.
Faculty: Ruth Law is author of Julia Child Cookbook Award finalist Indian Light Cooking, Pacific Light Cooking, The Southeast Asia Cookbook, and Dim Sum-Fast and Festive Chinese Cooking. She studied in southeast Asian countries.
Location: Hinsdale, a Chicago suburb. Tours visit China, Hong Kong, Thailand, Singapore, Malaysia, Indonesia, India, Hawaii.
Contact: Ruth Law, What's Cooking, P.O. Box 323, Hinsdale, IL 60521 US; 630-986-1595, Fax 630-655-0912.

WILLIAMS-SONOMA
Chicago area/Year-round

This cookware store offers weekly classes on a variety of topics. Facilities: Full service kitchen with professional quality equipment.
Emphasis: Chef and cookbook author specialties.
Faculty: 3- and 4-star chefs, cookbook authors, in-house staff.
Costs: $30/class.
Location: Chicago, Oakbrook, Lake Forest, Woodfield.
Contact: Sue Ellen Flockencier, Cooking School Coordinator, Williams-Sonoma, 708-246-7669, Fax 708-246-5021, E-mail dfflock@aol.com.

WILTON SCHOOL OF CAKE DECORATING
(See also page 46) **Woodridge/February-November**

This private school offers 1-day to 2-week participation courses. Established in 1929. Maximum class/group size: 15-20. Facilities: The 2,200-square-foot school includes a classroom, teaching kitchen, student lounge, and retail store.
Emphasis: Cake decorating.

Faculty: Includes Sandra Folsom, Susan Matusiak, Nicholas Lodge, Wesley Wilton, and Elaine Gonzalez.
Costs: Range from $75 for 1 day to $650 for a 10-day course.
Location: A southwestern Chicago suburb, 25 miles from downtown.
Contact: School Secretary, Wilton School of Cake Decorating and Confectionery Art, 2240 W. 75th St., Woodridge, IL 60517 US; 630-963-7100 #211, Fax 630-963-7299.

INDIANA

COUNTRY KITCHEN, SWEETART, INC.
Fort Wayne/Year-round

This private school offers basic to advanced cake decorating courses, demonstration and participation classes on candies, desserts, and other topics. Established in 1964. Maximum class/group size: 35 hands-on/60 demo. 30-40 programs per year. Facilities: Classroom with tiered work and observation seating. Also featured: Classes for groups, children's parties.
Emphasis: Cake decorating, desserts, candies.
Faculty: More than 10 instructors.
Costs: Cake decorating courses range from $65-$70 each. Demonstrations range from $10-$40.
Location: Northeast Indiana.
Contact: Vi Whittington, Owner, Country Kitchen, SweetArt, Inc., 3225 Wells St., Fort Wayne, IN 46808 US; 219-482-4835, Fax 219-483-4091, E-mail ckfa@aol.com.

KITCHEN AFFAIRS
Evansville/January-November

This cookware store and school offers demonstration and participation classes. Established in 1987. Maximum class/group size: 12 hands-on/20 demo. 150+ programs per year. Facilities: 350-square-foot kitchen with 4 work stations. Also featured: Children's classes, private classesc.
Emphasis: Basic techniques, ethnic cuisines, menu planning.
Faculty: Restaurant chefs, professional instructors, cookbook authors, and school owners Shelly and Mike Sackett. Many instructors are IACP members.
Costs: $15-$60.
Location: Across from Evansville's largest shopping mall.
Contact: Shelly Sackett, Director, Kitchen Affairs, 4610 Vogel Rd., Evansville, IN 47715 US; 800-782-6762, URL http://www.kitchenaffairs.com.

THE OLSON ACADEMY
Norman/Year-round

This private school offers 1- to 3-day classes for advanced nonprofessionals. Established in 1994. Maximum class/group size: 15 demo, 6 hands-on. 10-15 programs per year. Facilities: 6 stations, dining/demo area. Also featured: Wine seminars, programs for youngsters.
Faculty: Joan Y. Olson, B.A., Indiana University; certificate from La Varenne; practical experience in France.
Costs: $50-$80/class.
Location: Hoosier Forest.
Contact: Joan Y. Olson, Director of Gastronomy, The Olson Academy, 10902 N. Co. Rd., 800 W, Norman, IN 47264 US; 812-497-3568, Fax 812-497-3020.

IOWA

COOKING WITH BONNIE
Urbandale/Year-round

Cooking instructor Bonnie Boal offers seasonally-themed classes, private instruction, dining in private homes. Established in 1995. Maximum class/group size: 50 demo, 25 hands-on. 100 pro-

grams per year. Facilities: The gourmet kitchen of a fire station, the tasting room of JT's Fine Wine & Spirits, other well-equipped local facilities, including private homes.
Emphasis: Fresh, seasonal recipes and wine pairings.
Faculty: Food writer and instructor Bonnie Boal, IACP member, BA in Journalism.
Location: Just west of Des Moines.
Contact: Bonnie Boal, Owner, Cooking with Bonnie, 9612 Valdez Dr., Urbandale, IA 50322 US; 515-276-0781, Fax 515-276-0778, E-mail BonnieBoal@aol.com.

COOKING WITH LIZ CLARK
Keokuk/Year-round

Culinary professional Liz Clark offers demonstration and participation classes. Established in 1977. Maximum class/group size: 12 hands-on/16 demo. 80 programs per year. Facilities: Hy Vee grocery store and Elizabeth Clark's renovated antebellum home, which also houses her restaurant. Also featured: Weekend intensives that offer continuing education units, culinary tours in the U.S. and abroad.
Emphasis: Seasonal and holiday menus and guest chef specialties.
Faculty: Liz Clark studied in Italy and France, received her diploma in the Cours Intensifs from La Varenne, studied at the Moulin de Mougins with Roger Vergé and at The Oriental in Bangkok. Other instructors have included Jill Van Cleve, Shirley Corriher, and Monique Hooker.
Costs: $39-$69/class. Bed & breakfast is located nearby.
Location: 50 miles south of Burlington airport.
Contact: Sandy Seabold, Coordinator, Southeastern Community College, 335 Messenger Rd., Box 6007, Keokuk, IA 52632-6007 US; 319-752-2731, Fax 319-524-8621, E-mail sseabold@secc.cc.ia.us, URL http://www.secc.ia.us/contedu/liz.

KANSAS

BARON'S SCHOOL OF PITMASTERS
Kansas City/Year-round

Barbecue expert Paul Kirk offers demonstration/participation classes and one-day intensives. A commercial barbecue school is planned. Established in 1991. Maximum class/group size: 15, divided into 2-person teams. Facilities: Outdoor cooking, indoor sauce and barbecue seasoning development and meat trimming. Also featured: Out-of-town courses.
Emphasis: Basics of traditional, backyard, and competition barbecue.
Faculty: 1 full-, 6 part-time instructors. Full-time instructor Paul Kirk, CWC, is 7-time World Barbecue Champion and winner of over 400 barbecue and cooking awards.
Costs: Basic course is $250. Lodging is available at local hotels and motels.
Location: Kansas City or any city.
Contact: Paul Kirk, CWC, Kansas City Baron of Barbecue, Baron's School of Pitmasters, 3625 W. 50th Terr., Shawnee Mission, KS 66205-1534 US; 913-262-6029, Fax 816-756-5860. Additional contact: Kansas City Barbecue Society 800-963-KCBS, 816-765-5891.

COOKING AT BONNIE'S PLACE
Wichita/September-May

Cooking instructor Bonnie Aeschliman offers demonstrations. Established in 1990. Maximum class/group size: 25. 40-50 programs per year. Facilities: Demonstration kitchen.
Emphasis: A variety of topics.
Faculty: Bonnie Aeschliman, CCP, has a master's degreee in food and nutrition; Dr. Phil Aeschliman is a member of the IACP.
Costs: $20-$30/class.
Location: One mile from Wichita.
Contact: Bonnie Aeschliman, Cooking at Bonnie's Place, 5900 E. 47th St., North, Wichita, KS 67220 US; 316-744-1981.

ELDERWOOD

Lawrence/Year-round

Culinary professional Debbie Elder offers monthly evening and weekend demonstration and participation classes. Established in 1996. Maximum class/group size: 6 hands-on/12 demo. Facilities: Home kitchen in a newly built residence in the Alvamar community in West Lawrence; facilities at an historic 1880's home. Also featured: Group classes, private instruction, culinary tours.

Emphasis: International and regional cuisines, guest chef specialties, home entertaining and wine pairing menus.

Faculty: Founder Debbie J. Elder has been associated with the culinary arts since 1980, planning meal functions, catered events, and group travel. Guest chefs from around the country also teach.

Costs: Ranges from $45-$95/class. Lodging can be arranged at the historic Eldridge Hotel in downtown Lawrence.

Location: Lawrence, home of the University of Kansas, is 45 minutes from Kansas City.

Contact: Ms. Debbie J. Elder, Founder, Elderwood, 1972 Carmel Dr., Lawrence, KS 66047 US; 785-842-5580, Fax 785-842-8917.

KENTUCKY

THE COOKBOOK COTTAGE

Louisville/Year-round

This cookbook store and school offers demonstration and participation classes. Established in 1986. Maximum class/group size: 10 hands-on/20 demo. 150 programs per year. Facilities: 1,200-square-foot classroom, which seats 30 and has overhead mirrors.

Emphasis: Herb and spice cookery, breads, international and regional cuisines, holiday and seasonal menus, guest chef specialties.

Faculty: Proprietor/instructor Stephen J. Lee earned a degree in Culinary Arts from the University of Kentucky, is food columnist for the Louisville Entertainer and a member of the IACP. Other faculty includes local cooking teachers and guest chefs.

Costs: $15-$30 for most classes, $48 for guest chefs.

Location: Louisville.

Contact: Stephen J. Lee, Proprietor, The Cookbook Cottage, 1279 Bardstown Rd., Louisville, KY 40204 US; 502-458-5227, Fax 502-473-7108, E-mail ckbkman@aol.com.

KROGER SCHOOL OF COOKING

Lexington/Year-round

This school in a food market offers 1- to 4-session demonstration and participation courses. Established in 1991. Maximum class/group size: 12 hands-on/16 demo. 150+ programs per year. Facilities: 180-square-foot kitchen with overhead mirror in Kroger market. Also featured: Private group classes, children's classes, birthday parties. TV series What's Cooking in Kroger's Kitchen.

Emphasis: Ethnic and regional cuisines, techniques, pastry, cake decorating, guest chef specialties.

Faculty: Instructors include IACP and ACF members, registered dietitians, and guest chefs Beatrice Ojakangas, Nathalie Dupree, Joanne Weir, and Jude Theriot.

Costs: $10-$28.

Contact: Dianne Holleran, Consumer Affairs Director, Kroger School of Cooking, 344 Romany Rd., Lexington, KY 40502 US; 606-269-1034, Fax 606-269-1034.

LOUISIANA

CHEF JOHN FOLSE CULINARY INSTITUTE

Thibodaux/Year-round *(See also page 55) (See display ad page 55)*

This university offers half-day to week-long demonstration/participation program. Established in 1994. Maximum class/group size: Varies with program. Facilities: 2 newly-equipped teaching kitchens, 2 demonstration classrooms. Also featured: Market visits, dining in fine restaurants, sightseeing, tours of food producers.

Emphasis: Cajun and Creole cuisine.

Faculty: Chef John Folse, CEC, AAC, executive chef and owner of Chef John Folse & Company, specializing in Cajun and Creole cuisine; and Institute faculty.

Costs: Varies.

Location: South Louisiana, 45 minutes from New Orleans, 75 minutes from Baton Rouge.

Contact: Dr. Jerald Chesser, Dean, Nicholls State University, P.O. 2099, Thibodaux, LA 70310 US; 504-449-7100, Fax 504-449-7089, E-mail jfci-jwc@nich-nsunet.nich.edu, URL http://server.nich.edu/~jfolse.

COOKIN' CAJUN COOKING SCHOOL

New Orleans/Year-round

Creole Delicacies, a company specializing in Cajun and Creole gourmet items, offers demonstration classes Monday through Sunday mornings. Established in 1988. Maximum class/group size: 1-75. Facilities: Theatre-style mirrored kitchen overlooking the Mississippi River. Also featured: Private classes, parties, fish classes for anglers.

Emphasis: Cajun and Creole cuisine.

Faculty: Susan Murphy and several other instructors.

Costs: $17/class.

Location: Riverwalk Marketplace near the New Orleans Convention Center.

Contact: Lissette Sutton, Owner, Cookin' Cajun Cooking School, #1 Poydras, Store #116, New Orleans, LA 70130 US; 504-523-6425, Fax 504-523-4787.

CREOLE COOKING DEMONSTRATION

New Orleans/October-May

The Hermann-Grima Historic House offers demonstrations of Creole cooking in New Orleans in the mid-ninteenth century. The program takes place in a restored 1831 kitchen. Established in 1985. Maximum class/group size: 20.

Emphasis: New Orleans food, culture, and history.

Costs: $5, includes tour of the Hermann-Grima Historic House. Group rates if booked in advance.

Location: The French Quarter.

Contact: Claude Stephens, Education Programs Mgr., The Hermann-Grima Historic House, 820 St. Louis St., New Orleans, LA 70112 US; 504-525-5661, Fax 504-568-9735, E-mail hggh@gnofn.org.

INTERNATIONAL DINING ADVENTURES (See page 350)

KAY EWING'S EVERYDAY GOURMET

Baton Rouge/Year-round

Culinary professional Kay Ewing offers participation classes. Established in 1985. Maximum class/group size: 8. 10-12 programs per year. Facilities: The Panhandler, a gourmet kitchen store. Also featured: Classes for youngsters held in the summer and during holiday.

Emphasis: International and cajun cuisines, full participation, menu classes.

Faculty: Kay Ewing is a member of the IACP and author of Kay Ewing's Cooking School Cookbook.

Costs: $30 for adults, $20 for children.

Contact: Kay Ewing, Owner, c/o The Panhandler, 3072-A College Dr., Baton Rouge, LA 70808 US; 504-927-4371, E-mail TimEwing@worldnet.att.net, URL http://Home.att.net/~TimEwing.

THE NEW ORLEANS SCHOOL OF COOKING

New Orleans/Year-round

This private school offers morning demonstrations Monday through Saturday. Established in 1980. Maximum class/group size: 20-200. Facilities: Large mirrored kitchen. Also featured: Classes for private groups.

Emphasis: Cajun and Creole cuisine.
Faculty: Kevin Belton is a self-taught cook and television personality.
Costs: $20/class.
Location: Historic (1830's) building in the French Quarter.
Contact: Kris Hicks, Director of Sales, The New Orleans School of Cooking, 524 St. Louis St., New Orleans, LA 70130 US; 504-525-2665, Fax 504-525-2922.

SPICE, INC.

New Orleans/Year-round

This retail food store and cooking school offers weekly classes on Tuesdays, Thursdays, and Saturdays. Established in 1997. Maximum class/group size: 45 demo, 18 hands-on. Facilities: High tech Viking-equipped classroom/kitchen with seating for 48 students. Also featured: Classes for youngsters.
Emphasis: New Orleans cooking, Bayona Restaurant specialties.
Faculty: Full-time chefs Susan Spicer, chef-owner of Bayona Restaurant, and Michelle Nugent, part-time chef Poppy Tooker, guest chefs from around the country.
Costs: $35-$75/class ($75 for hands-on classes).
Location: New Orleans Warehouse District, bottom of the Cotton Mill Apts.
Contact: Susan Spicer, Spice, Inc., 1051 Annunciation St., New Orleans, LA 70130 US; 504-558-9992, Fax 504-558-9993, E-mail spiceinc@gs.net.

WHAT'S COOKING

Baton Rouge/Year-round

The Culinary Shoppe cookware store and The Silver Spoon restaurant offer demonstration classes in international and regional cuisines. Established in 1993. Maximum class/group size: 40. 18 programs per year. Facilities: Fully equipped restaurant kitchen set up for demonstration classes. Also featured: Private instruction, classes for youngsters.
Faculty: Mike Mangham from The Silver Spoon, the staff of The Culinary Shoppe.
Costs: $45 and up.
Location: The Silver Spoon restaurant in Bocage Village.
Contact: Kathy Mangham, The Silver Spoon, 7731 Jefferson Hwy., Baton Rouge, LA 70806 US; 504-926-1172/504-927-4288, E-mail Silversp@aol.com.

MAINE

THE WHIP AND SPOON

Portland/March-May, September-November

This gourmet foods and cookware store offers demonstration classes. Established in 1980. Maximum class/group size: 25. 100+ programs per year. Facilities: Well-equipped teaching kitchen.
Emphasis: Ethnic and regional cuisines, healthful foods, guest chef specialties.
Faculty: Local cooks, chefs, caterers, culinary school graduates.
Costs: $15/class.
Location: The Old Port Exchange on Portland's waterfront.
Contact: Cindy Tubbs, Class Coordinator, The Whip and Spoon, 161 Commercial St., Portland, ME 04101 US; 800-937-9447, Fax 207-774-6261.

MARYLAND

THE CAKE COTTAGE, INC.

Baltimore/Year-round

This candy shop and school offers participation courses in basic and advanced cake decorating and demonstration and participation classes in candies, cake writing, puff pastry, petit fours, butter cream flowers, air brush decorating, and party foods. Established in 1977. Maximum class/group size: 30 hands-on/50 demo. Also featured: Children's courses.

Faculty: Carole studied at the Wilton School and has taught for over 20 years; Donna studied candymaking with chocolatiers in the U.S. and abroad for over 19 years.

Costs: Basic decorating course is $35, advanced course is $40, children's course is $25. Single sessions range from $8-$40.

Location: Baltimore's northeast section, 20 minutes from BWI airport.

Contact: Carole Eaves, The Cake Cottage, Inc., 8716 Belair Rd., Baltimore, MD 21236 US; 410-529-0200, Fax 410-529-6867.

THE CHINESE COOKERY, INC.

Silver Spring/Year-round

Culinary professional Joan Shih offers 8 levels of participation and demonstration courses in Chinese cuisine. Established in 1975. Maximum class/group size: 5. Facilities: Classroom/lab equipped for Chinese cooking, outdoor Chinese brick oven. Also featured: Japanese sushi class, classes for teenagers, private lessons for cooking professionals, market visits, restaurant kitchen tours, and culinary tours to the Far East.

Emphasis: Chinese cuisine: Basic, Advanced, Gourmet I, II, III, Szechuan, Hunan, vegetarian.

Faculty: Joan Shih, a chemist at the National Institutes of Health, received a certificate in Chinese cuisine in Taiwan and has taught Chinese and Japanese cooking on television and in schools.

Costs: Five-session courses are $150 and the sushi class is $40.

Location: North suburb of Washington, DC.

Contact: Joan Shih, President and Director, The Chinese Cookery, Inc., 14209 Sturtevant Rd., Silver Spring, MD 20905 US; 301-236-5311.

L'ACADEMIE DE CUISINE

(See also page 60) (See display ad page 61) **Bethesda/Year-round**

This proprietary avocational school offers 1- to 4-session demonstration and participation courses. Established in 1976. Maximum class/group size: 21-25. 400+ programs per year. Facilities: 21-station practice and pastry kitchen, 30-seat demonstration classroom. Also featured: Children's classes, private dinners, guest chef demos, 1-week culinary/culture trips to France that include trips to an Armagnac distillery, Bordeaux wineries, duck farm, Michelin-starred restaurants.

Emphasis: Techniques, international and regional cuisines, nutritional and low-fat foods, pastry, wine and food pairing, entertaining menus.

Faculty: 4 full- and 15 part-time. School President Francois Dionot, graduate of L'École Hoteliere de la Societe Suisse des Hoteliers and founder of the IACP, Chris Green, Patrice Dionot, Martha Bridgers.

Costs: $30-$75/session; trips are ~$2,500, which includes lodging, meals, planned excursions, ground transport.

Location: Bethesda campus, 3 miles northwest of D.C.

Contact: Chris Green, Managing Director, L'Academie de Cuisine, 5021 Wilson Lane, Bethesda, MD 20814 US; 301-986-9490, Fax 301-652-7970, E-mail LAcademie@aol.com, URL http://www.washingtonpost.com/yp/LAcademie.

PUDDING ON THE RITZ, INC.

Oxford/Year-round

This private school offers one-day programs. Established in 1996. Maximum class/group size: 60 demo, 16 hands-on. 50 programs per year. Facilities: Custom-designed professional facility. Also featured: Children's cooking and etiquette classes, dining in private homes.

Faculty: Professional chefs.

Costs: $40/class.

Location: 45 minutes east of Annapolis.

Contact: Pamela Meredith, Owner/Director, Pudding on the Ritz, Inc., 4776 Sailors Retreat Rd., Oxford, MD 21654 US; 410-820-4414, E-mail pudding@crosslink.net.

SOCIETY OF WINE EDUCATORS PROGRAMS

August-September *(See also page 377)*

This nonprofit organization for wine educators and consumers sponsors an annual 1-week conference featuring workshops from 60+ wineries; Overseas Study Tours to wine-producing regions worldwide; week-long Short Courses highlighting a specific region of North America. Established in 1977.

Emphasis: Wine education.

Faculty: Wine experts.

Contact: Geralyn Brostrom, Director, Society of Wine Educators, 8600 Foundry St., Mill Box 2044, Savage, MD 20763 US; 301-776-8569, Fax 301-776-8578, E-mail vintage@erols.com, URL http://www.wine.gurus.com.

MASSACHUSETTS

BOSTON UNIVERSITY CULINARY ARTS

Boston/Year-round *(See also page 61)*

This university offers 1- to 3-session demonstration and participation courses. Established in 1986. Maximum class/group size: 24-130. 20-25 programs per year. Facilities: Demonstration room. Also featured: Children's classes, market visits, food and wine pairing, domestic and foreign tours hosted by a culinary historian familiar with the region's food and wine.

Emphasis: Guest chef specialties.

Faculty: Jacques Pépin, Albert Kumin, Julia Child, Jasper White, Jody Adams, Franco Romagnoli, Daniel Bruce, Sandy Block, Chris Schlesinger, Nina Simonds, Julie Sahni.

Costs: Seminars range from $10-$125, full-day classes and 3-session courses range from $150-$300.

Contact: Rebecca Alssid, Director of Special Programs, Boston University Culinary Arts, 808 Commonwealth Ave., Boston, MA 02215 US; 617-353-9852, Fax 617-353-4130, E-mail ralssid@bu.com.

THE CAMBRIDGE SCHOOL OF CULINARY ARTS

Cambridge/Year-round *(See also page 62) (See display ad page 62)*

This proprietary school offers 1- to 5-session participation courses. Established in 1974. Maximum class/group size: 12. Facilities: 3 newly renovated kitchens and 3 demonstration classrooms; gas and electric commercial appliances. Also featured: Culinary tours.

Emphasis: Ethnic and regional cuisines, vegetarian meals, breads, pastries, event planning, appetizers, desserts.

Faculty: President and founder Roberta Avallone Dowling received diplomas from Julie Dannenbaum, Marcella Hazan, and Madeleine Kamman.

Costs: $65-$75/session.

Location: 5 miles from downtown Boston.

Contact: The Cambridge School of Culinary Arts, 2020 Massachusetts Ave., Cambridge, MA 02140 US; 617-354-2020, Fax 617-576-1963, URL http://www.cambridgeculinary.com.

THE CAPTAIN FREEMAN INN

Brewster, Cape Cod/January-March, November

This Victorian sea captain's mansion offers weekend courses that feature a hands-on Saturday class, wine tasting, and dinner. Established in 1993. Maximum class/group size: 12-18. 7 programs per year. Facilities: The inn's kitchen.

Emphasis: Northern, southern, and central Italy.

Costs: $380-$530, which includes inn lodging, breakfasts, and dinner.

Location: On Cape Cod.

Contact: Carol Edmondson, Innkeeper, The Captain Freeman Inn, 15 Breakwater Rd., Brewster, Cape Cod, MA 02631 US; 800-843-4664/508-896-7481, Fax 508-896-5618, E-mail visitus@capecod.net, URL http://www.captfreemaninn.com.

THE CHEF'S SHOP

Great Barrington, Holyoke/February-December

This cookware store offers classes, private instruction. Established in 1994. Maximum class/group size: 30 demo, 20 hands-on. 100 programs per year. Facilities: Cookware store's working kitchen. Also featured: Cookbook authors, food tasting, product demos.

Emphasis: Restaurant-quality dishes that can be created at home.

Faculty: Local and national restaurant chefs, cookbook authors, food industry experts.

Costs: $30/class.

Location: Downtown Great Barrington in the Berkshires, Holyoke Mall at Ingleside.

Contact: Joe Kopper, The Chef's Shop, 290 Main St., Great Barrington, MA 01230 US; 413-528-0135. 2nd location: Melissa Fregault, The Chef's Shop, Holyoke Mall, Holyoke, MA 01041; 413-538-7929.

THE KUSHI INSTITUTE COOKING SEMINARS

Becket/Year-round

Macrobiotic cooking for health, using natural, whole foods, founded by Michio and Aveline Kushi, offers 2 different weekend programs that each feature 4 cooking classes, 2 lecture classes, and morning exercise sessions. Established in 1978. Maximum class/group size: 16. Facilities: Former Franciscan Friar monastery. Also featured: Shiatsu massage, macrobiotic dietary consultation.

Emphasis: The Essentials of Macrobiotic Cooking, Naturally Gourmet Holiday Cooking.

Faculty: Wendy Esko, Diane Avoli, Carry Wolf, Mayumi Nishimura, Gail Jack, Warren Kramer, Ed Esko, Alex Jack, Charles Millman, John Kozinski.

Costs: Weekend seminars are $350, which includes meals and double occupancy country manor or dormitory lodging with shared bath. Private bath and single lodging are additional.

Location: The Berkshires, a 3-hour drive from Boston or New York City, near Lenox, Mass.

Contact: Program & Services Information, The Kushi Institute, P.O. Box 7, Becket, MA 01223 US; 800-975-8744, Fax 413-623-8827, E-mail kushi@macrobiotics.org, URL http://www.macrobiotics.org.

LA CUCINA COOKING SCHOOL

Edgartown, MA; Tuscany, Italy/March-December

The Tuscany Inn restaurant offers Friday-Sunday hands-on cooking vacations at the Tuscany Inn and 1-week culinary tours to Tuscany. Established in 1994. Maximum class/group size: 12 hands-on, 10 travel. 12 programs per year. Facilities: Commercial kitchen designed for teaching.

Emphasis: Tuscan cuisine.

Faculty: Tuscan native Laura Sbrana-Scheuer, co-owner, with husband Rusty Scheuer, of the Tuscany Inn.

Costs: $425 for the weekend, includes meals and lodging at the newly restored Inn (formerly the Captain Fisher House). Noncook guest $175 for room, $75 for meals.

Location: The Inn overlooks the Edgartown Harbor on Martha's Vineyard.

Contact: Laura Sbrana-Scheuer, La Cucina Cooking School, P.O. Box 2428, Edgartown, MA 02539 US; 508-627-5999, Fax 508-627-6605, E-mail 70652.3363@compuserve.com.

LE PETIT GOURMET COOKING SCHOOL

Wayland/Year-round

Culinary professional Fran Rosenheim offers 1- to 4-session demonstration/participation courses. Established in 1979. Maximum class/group size: 6-8. 80+ programs per year. Facilities: 330-square-foot kitchen with two workspaces. Also featured: Private classes.

Emphasis: Beginning to advanced French cuisine, wild game, specific subjects. Specialty is chocolate desserts and sauces.

Faculty: Fran Rosenheim studied with local chefs and at Le Cordon Bleu and La Varenne in Paris.

Costs: $60/session, $60/session for private lessons.

Location: 20 minutes from Boston.

Contact: Fran Rosenheim, Le Petit Gourmet Cooking School, 19 Charena Rd., Wayland, MA 01778 US; 508-358-4219, Fax 508-358-4291.

MARGE COHEN

Needham Heights/September-June

Culinary professional Marge Cohen offers 5-session demonstration and participation courses. Established in 1980. Maximum class/group size: 8. Facilities: Home kitchen. Also featured: Private classes, local culinary tours.

Emphasis: Basic to advanced Chinese, low fat, low salt.

Faculty: Marge Cohen has certificates from the Le Cordon Bleu and Weichuan Cooking School in Taiwan and has hosted a cable TV program.

Costs: Each 5-session course is $105, culinary tours are $35.

Location: Twenty minutes west of Boston.

Contact: Marge Cohen, P.O. Box 53, Needham Heights, MA 02194 US; 781-449-2688, Fax 781-449-7878, E-mail mscohen@gis.net.

TAMING OF THE STEW

Wellesley/Year-round

Culinary professional Sally Larhette offers 3 classes/week that feature pre- and post-cooking discussions, shopping and preparation tips, historical perspective, hands-on class of a full menu. Established in 1998. Maximum class/group size: 15 demo, 9 hands-on. 150 programs per year. Facilities: Fully-equipped professionally-styled home kitchen. Also featured: Private instruction, classes for youngsters, market and winery visits, dining in private homes.

Emphasis: Seasonal foods as the center of a healty diet with consideration given to cost and ease of preparation.

Faculty: Sally Larhette, CCP, member of the IACP, trained in French and Italian cuisines by Madeleine Kamman, studied in France and Italy, on Board of Directors of The Culinary Guild in New England.

Costs: $35-$50/class.

Location: 20 minutes from downtown Boston on MA Tpke. W, 1 block east of Wellesley College.

Contact: Sally Larhette, School Director, Taming of the Stew, 619 C Washington St., Wellesley, MA 02181 US; 781-235-1792, Fax 781-235-7714, E-mail larhette@tiac.net.

TERENCE JANERICCO COOKING CLASSES

Boston/September-June

Cookbook author Terence Janericco offers 1- and 6-session demonstration and participation courses. Established in 1966. Maximum class/group size: 6 hands-on/14 demo. 100+ programs per year. Facilities: Home kitchen. Also featured: Private classes.

Emphasis: Gourmet cooking, baking, ethnic and regional cuisines, specific subjects.

Faculty: Terence Janericco has operated a catering firm for over 25 years and teaches at education centers in Boston and at schools in New England and Michigan. Author of 12 books, including The Book of Great Hors d'Oeuvres, and The Book of Great Desserts.

Costs: Six-session courses are $360, single-sessions are $60.

Contact: Terence Janericco, Owner, Terence Janericco Cooking Classes, 42 Fayette St., Boston, MA 02116 US; 617-426-7458, Fax 617-426-7458.

MICHIGAN

CUISINE INTERNATIONAL ÉCOLE DE CUISINE

Michigan and France/Year-round

Culinary professional Chef Deborah Ward offers 1- to 5-day sessions that feature classes in French cuisine, pastry arts, wine studies, regional and international cuisines with cultural and sightseeing excursions, market visits, winery tours, visits to food producers. Established in 1997. Maximum class/group size: 6 hands-on, 12 travel. Facilities: Restaurant kitchen. Also featured: Restaurant training staff programs, specialty group programs.

Emphasis: French cuisine, pastry arts, wine studies.

Faculty: Chefs who have received international awards, the Premier Sommelier de France.
Location: Various locations in Michigan, the Pay Basque region in the south of France.
Contact: Chef Deborah Ward, Cuisine International École de Cuisine, 7045 W. Deer Rd., Box 249, Mears, MI 49436 US; 616-873-0274, Fax 616-873-0074/0274, E-mail cuisine@voyager.net, URL http://www.voyager.net/cuisineinternational.

KITCHEN GLAMOR...THE COOK'S WORLD

Five locations/September-May

This gourmet cookware store and cooking school offers demonstration and participation courses, pre-registration classes, and guest chef classes. Established in 1950. Maximum class/group size: 12-16 pre-reg. classes/25 hands-on/85 demo. 30-50 programs per year. Facilities: Kitchen auditorium has a 14-foot counter, 2 four-range burners, and overhead mirror. Also featured: National cookware catalog.
Emphasis: Techniques.
Faculty: Includes cake decorator Mary Ann Hollen, program director Toula Patsalis, and local chefs. Guest chefs have included Nicholas Malgieri and Carol Walters.
Costs: Demonstrations are $3 each, $30 for 12; pre-registration classes are $40 for local chefs, $25-$35 for others; guest chef demonstrations range up to $80/session.
Location: Redford Twsp, W. Bloomfield in Orchard Mall, Rochester in Great Oaks Mall, Walton at Livernols, Novi Town Ctr. Closest major city is Detroit.
Contact: Toula Patsalis, Program Director, Kitchen Glamor, 39049 Webb Ct., Westland, MI 48185-7606 US; 734-641-1244, Fax 734-641-1240. Redford 313-537-1300, Rochester 248-652-0402, W. Bloomfield 248-855-4466, Novi 248-380-8600.

NELL BENEDICT COOKING CLASSES

Birmingham/September-May

This adult community center offers evening demonstration classes. Established in 1970. Maximum class/group size: 40. 12 programs per year. Facilities: Teaching kitchen at The Community House.
Emphasis: Ethnic cuisines, breads, restaurant specialties.
Faculty: Nell Benedict studied at Le Cordon Bleu, La Varenne, and L'Arts Culinara and with James Beard, Jacques Pépin, and Roger Vergé. She has taught on television and is a Charter Member of the IACP.
Costs: $18/session.
Location: Approximately 18 miles north of downtown Detroit.
Contact: Nell Benedict, International Cuisine, Gourmet Cooking, 380 S. Bates St., Birmingham Community House, Birmingham, MI 48009 US; 248-664-5832.

SAVEUR INTERNATIONAL COOKING SCHOOLS (See page 212)

MINNESOTA

BYERLY'S SCHOOL OF CULINARY ARTS

St. Louis Park/Year-round

This school in an upscale market offers 1-session demonstration and participation classes. Established in 1980. Maximum class/group size: 14 hands-on/25 demo. 200+ programs per year. Facilities: Large teaching kitchen with overhead mirror. Also featured: Private and couple's classes, children's birthday classes.
Emphasis: Italian, Asian, and French cuisine; entertaining, seasonal, healthy cooking basic cooking series. Guest chefs.
Faculty: Culinary Services Manager Deidre Schipani has a diploma from L'Academie de Cuisine. Instructors include CIA graduate Carol Brown, columnist Mary Carroll, NPR host Lynne Rossetto Kasper, cookbook authors, IACP members, guest chefs.
Costs: Average $35/class.

Location: St. Louis Park, a suburb of Minneapolis, 15 minutes from downtown Minneapolis.
Contact: Deidre Schipani, Manager of Culinary Services, Byerly's School of Culinary Arts, 3777 Park Center Blvd., St. Louis Park, MN 55416 US; 612-929-2492, Fax 612-929-7756, URL http://www.Byerlys.com.

COOKS OF CROCUS HILL

St. Paul/Year-round

This gourmet retail store and cooking school offers 1-, 4-, and 5-session demonstration and participation courses. Established in 1976. Maximum class/group size: 12 hands-on/25 demo. 200+ programs per year. Also featured: Private group classes, corporate team-building classes.
Emphasis: Basics, ethnic cuisines, holiday and seasonal menus, single subjects, guest chef specialties.
Faculty: The 21-member faculty includes Jennifer Holloway, Andrew Zimmern, and Yvonne Moody. Guest chefs include Hugh Carpenter, Jim Dodge, and Joanne Weir.
Costs: The 4-session course is $250, the 5-session course is $300. Classes range from $40-$65.
Contact: Jennifer Holloway, Director, Cooks of Crocus Hill, 877 Grand Ave., St. Paul, MN 55105 US; 612-228-1333, Fax 612-228-9084.

FOOD FESTS

January, March-April, August-November

Publicist Gail Guggenheim sponsors cooking school get-away weekends that feature cooking classes, seminars, and tastings. Established in 1984. Maximum class/group size: Varies. Facilities: Complete kitchen with overhead mirrors set up on a stage; closed-circuit monitors. Also featured: Hotel amenities.
Emphasis: Menus for entertaining, new food trends, ethnic specialties, techniques, gourmet products.
Faculty: Area chefs and cooking teachers, usually certified members of the IACP.
Costs: $250-$279 per couple for two nights, including continental breakfasts.
Location: The Kahler Hotel in Rochester, MN; Lakeview Resort in Morgantown, WV; Olympia Park Hotel in Park City, UT; Sheraton San Marcos Resort in Phoenix, AZ.
Contact: Gail Guggenheim, Food Fests, 125 Country Lane, Highland Park, IL 60035 US; 847-831-4265, Fax 847-831-4266, E-mail Gailgugg@aol.com.

FOOD ON FILM

Minneapolis/April-May

The Twin Cities Chapter of Home Economists in Business and the IACP sponsor a 2-day intensive food styling seminar, held in even-numbered years, that features 16 rotating demonstration workshops. Established in 1982. Maximum class/group size: 50-75. Facilities: Conference rooms of the Minneapolis Hyatt Regency. Also featured: Keynote and luncheon speakers, cookbook gallery, social activities.
Emphasis: Food styling and photography.
Faculty: Changes each seminar. Includes food stylists and photographers, cooking instructors, chefs, food writers, and other industry professionals.
Costs: The 1996 program fee was $395 early bird, $425 regular, which includes breakfasts, lunches, and receptions. Lodging available at the Hyatt Regency Hotel.
Location: Downtown Minneapolis.
Contact: Nancy Iverson, Co-chair, Food on Film, 7227 West Fish Lake Rd., Maple Grove, MN 55311-2511 US; 612-420-4552, Fax 612-420-2469, E-mail niverson@juno.com.

SAVEUR INTERNATIONAL COOKING SCHOOLS (See page 212)

SCHOOL OF NATURAL COOKERY (See page 199)

MISSISSIPPI

THE EVERYDAY GOURMET

Jackson/January-November

These schools in two Jackson cookware stores offer demonstration and participation classes. Established in 1981. Maximum class/group size: 12 hands-on/36 demo. 100+ programs per year. Also featured: Guest chefs, lunch sessions, classes for children, tour groups.

Faculty: Includes Fran Ginn, Gayle Stone, Butchie Nations, Cheryl Welch, Julie Moore, school director Chan Patterson.

Costs: $20-$50; children's classes are $18.

Contact: Chan Patterson, Director, The Everyday Gourmet, Inc., 2905 Old Canton Rd., Jackson, MS 39216 US; 601-362-0723, Fax 601-981-3266, E-mail chanedg@aol.com.

MISSOURI

DIERBERGS SCHOOL OF COOKING

Four locations/Year-round

This cooking school with four locations offers 1-session demonstration and participation courses. Established in 1978. Maximum class/group size: 18. 800+ programs per year. Facilities: Soundproof enclosures in Dierbergs Supermarkets. Also featured: Classes for couples, children's classes, parent-child sessions.

Emphasis: Home cooking.

Faculty: In addition to the more than 30-member faculty of home economists and cooking instructors, guest teachers include Gerard Germain, executive chef, Tony's Restaurant; industry spokespersons, traveling chefs, and cookbook authors.

Costs: Adult classes range from $18-$22, guest classes are $20-$40.

Location: West Oak (Creve Coeur), Mid Rivers Center (St. Peters), Southroads (St. Louis County), Clarkson/Clayton Center (Ellisville).

Contact: Jeannie Rader, Home Economist, Dierbergs West Oak, 11481 Olive St. Blvd., Creve Coeur, MO 63141 US; 314-432-6505, Fax 314-432-2548. St. Peters 314-928-1117 (Loretta Evans), St. Louis 314-849-3698 (Lorene Greer), Ellisville 314-394-9504 (Nancy Lorenz).

JASPER'S

Kansas City/Year-round

This restaurant offers demonstration classes and holiday luncheon classes. Established in 1954. Maximum class/group size: 45 demo. 25 programs per year. Also featured: Private classes for groups of 20 or more, children's and couples' classes, wine classes.

Emphasis: Northern Italian cuisine.

Faculty: Executive Chef Jasper J. Mirabile, Jr., son of Jasper's founder, studied at La Varenne, and the Gritti Palace. He has cooked at the James Beard House. Jasper's received the Dirona Award, Travel/Holiday Award, the Mobil 4-Star award, the AAA 4-Diamond award.

Costs: $40/class.

Location: South Kansas City.

Contact: Jasper J. Mirabile, Jr., Jasper's, 1201 W. 103rd, Kansas City, MO 64114 US; 816-941-6600/800-35-JASPE, E-mail jmirablois@aol.com.

KEMPER CENTER COOKING SCHOOL

St. Louis/March-December

The Kemper Center of the Missouri Botanical Garden offers 1- to 3-session demonstration and participation courses. Maximum class/group size: 16-20. 60+ programs per year. Also featured: Gardening, arts & crafts, nature study.

Emphasis: A variety of topics.

Faculty: Includes caterers, nutritionists, cookbook authors, and cooking teachers.
Costs: $25-$35/session. Members receive a discount.
Contact: Glenn E. Kopp, Coordinator, Missouri Botanical Garden, Kemper Center Cooking School, Box 299, St. Louis, MO 63166-0299 US; 314-577-9441, Fax 314-577-9444.

KITCHEN CONSERVATORY

St. Louis/Year-round

This cooking school in a gourmet shop offers demonstration and participation classes. Established in 1984. Maximum class/group size: 18. 200+ programs per year. Facilities: Modern kitchen with front-row seats for all students. Also featured: Day trips to restaurants and shops, classes for children and teens.
Emphasis: Ethnic and regional cuisines, seasonal and entertaining menus, guest chef specialties.
Faculty: Local chefs, restaurateurs, caterers, and IACP members. Guest instructors have included Hugh Carpenter, Merle Ellis, Martin Yan, Joanne Weir, Paula Wolfert, and Perla Meyers.
Costs: $32-$75/class.
Location: Minutes from downtown.
Contact: Anne Schlafly, Owner, Kitchen Conservatory, 8021 Clayton Rd., St. Louis, MO 63117 US; 314-862-COOK, Fax 314-862-2110.

MISSOURI CULINARY INSTITUTE

Kansas City/January-November *(See also page 72)*

This private trade school offers 1- to 4-session demonstration and participation courses. Established in 1995. Maximum class/group size: 12. 12 programs per year. Facilities: Lab kitchen with work stations.
Emphasis: Gourmet food.
Faculty: 2 full-, 1 part-time. Includes Dorothy Kopp, who has over 50 years restaurant experience, and Terry Kopp, who owns The Missouri Gourmet restaurant and studied at Le Cordon Bleu.
Location: 35 miles from Kansas City.
Contact: Terry Kopp, Missouri Culinary Institute, Rte. 1, Box 224F, Lexington, MO 64067 US; 816-259-6464.

SAVEUR INTERNATIONAL COOKING SCHOOLS (See page 212)

SUZANNE CORBETT – CULINARY RESOURCES

St. Louis/September-May

Culinary professional Suzanne Corbett offers participation courses. Established in 1976. Maximum class/group size: 16. 14 programs per year. Facilities: Vocational schools, historic sites. Also featured: Wine instruction, Missouri wine country tours, summer workshops for groups.
Emphasis: Historic American foods, hearth-style baking, regional dishes, international cuisines.
Faculty: Suzanne Corbett, CCP, specializes in foods from the past, is a food and wine writer with 20 years experience, contributing editor to Rodale Press and Victoria Magazine.
Costs: $25-$50/class.
Contact: Suzanne Corbett, St. Louis Community College, 5850 Pebble Oak, St. Louis, MO 63128-1412 US; 314-487-5205, Fax 314-487-5335, E-mail pncor@aol.com.

TAKE PLEASURE IN COOKING!

Kansas City/September-May

Culinary professional Gloria Martin offers demonstration and participation classes. Established in 1988. Maximum class/group size: 6 hands-on/20 demo. 40-50 programs per year. Facilities: 280-square-foot home kitchen with overhead mirror. Also featured: Private classes and workshops.
Emphasis: Culinary arts for the home, herbs, yeast doughs, regional cuisines, creative techniques, ease of preparation.
Faculty: Gloria Martin, owner, is an IACP member and has taught cooking for 19+ years.
Costs: $35-$45/class.

Location: The Kansas City suburb of Raytown, 15 miles from downtown.
Contact: Gloria Martin, Owner/Teacher, Take Pleasure in Cooking!, 8612 E. 84th St., Kansas City, MO 64138 US; 816-353-6022, E-mail ggmartin2@juno.com.

NEVADA

NOTHING TO IT! CULINARY CENTER

Reno/Year-round

This cookware store and gourmet take-out cafe offers classes that cover techniques and a variety of cuisines. Established in 1995. Maximum class/group size: 55 demo, 16 hands-on. 200 programs per year. Facilities: Culinary center with retail cafe. Also featured: Classes for youngsters, private instruction.
Faculty: Two chefs on staff and nationally-known guest chefs.
Costs: $40-$125/class.
Location: South of Reno.
Contact: Jennifer Bushman, Owner/Academic Chef, Nothing To It! Culinary Center, 255 Crummer Ln., Reno, NV 89509 US; 702-826-2628, E-mail jennifer@nothingtoit.com, URL http://www.nothingtoit.com.

NEW HAMPSHIRE

LIYUEN CHINESE CULINARY SPECIALTIES

North Hampton/January-June, September-November

This Chinese restaurant and retail store offers 6 levels of hands-on instruction with 4 sessions/level. Established in 1981. Maximum class/group size: 12. Facilities: Professional demonstration kitchen.
Emphasis: Chinese cuisine.
Faculty: LiYuen Buesing, instructor since 1981.
Costs: $30/session, $90/level.
Location: 55 miles north of Boston, 5 miles south of Portsmouth, NH.
Contact: LiYuen Buesing, Instructor, LiYuen Chinese Culinary Specialties, 112 C. Lafayette Rd., North Hampton, NH 03862-2409 US; 603-964-8181, Fax 603-964-4279.

A TASTE OF THE MOUNTAINS COOKING SCHOOL

Glen/January-May, October-November

This private school founded at the Bernerhof Inn offers weekend courses for novices and intermediate cooks in April, May, October, and November; day courses every Wednesday and Thursday in January, February, March and April. Established in 1980. Maximum class/group size: 10 hands-on. Facilities: Bernerhof Inn restaurant kitchen. Also featured: Custom seminars for groups of 7 or more.
Emphasis: Basic techniques including knife handling, sauces, sauteeing, breads.
Faculty: Owner/Chef Mark Prine of Prine Place at the Bernerhof Inn, northern New England area guest chefs.
Costs: Weekend courses, which include shared lodging and meals, range from S439 (standard room) to $529 (suite); day rate is $299 for the weekend; class rate is $45-$55.
Location: Bernerhof Inn in the White Mountains near Conway, 2.5 hours north of Boston and 90 minutes west of Portland, ME.
Contact: Sharon Wroblewski, Owner, A Taste of the Mountains Cooking School, Box 240, Glen, NH 03838 US; 603-383-9132/800-548-8007, Fax 603-383-0809, E-mail stay@bernerhofinn.com, URL http://www.virtualcities.com/tastemt.htm. Other contact: Bernerhof Inn, http://www.bernerhofinn.com.

NEW JERSEY

ALFREDO'S CLASSICAL ITALIAN COOKING

Milltown/January-June *(See also page 313, Alimandi's Roman Culinary Classes)*

Chef Alfredo de Bonis offers 4-session courses that include a sit-down dinner with wine. Established in 1990. Maximum class/group size: 15. Facilities: Alfredo's home kitchen.

Emphasis: Italian regional and gourmet cookery.

Faculty: Alfredo de Bonis, a chef in his own restaurant for 25+ yrs, owns and operates Bravo Alfredo Catering and teaches at North Brunswick High School.

Costs: $150 for a 4-session course.

Location: Near New Brunswick, 35 miles from New York City.

Contact: Alfredo de Bonis, Chef Instructor, Alfredo's Classical Italian Cooking, 375 Tremont Ave., Milltown, NJ 08850-2013 US; 732-828-8460, E-mail nrwu55a@prodigy.com.

CHEF AND SOMMELIER FOR A DAY

Short Hills/Year-round

The Hilton at Short Hills offers Thursday programs for one or two persons, who plan and help prepare dinner, including wine selection, and dine on it that evening with up to 8 guests. Established in 1997. Maximum class/group size: 1-2. Facilities: Restaurant kitchen of the Hilton Hotel. Also featured: Wine tasting and discussion with sommelier, children's etiquette classes.

Emphasis: Chef specialties and wine tasting.

Faculty: Hilton Executive Chef Walter Leffler and Chef de Cuisine Paul Sale.

Costs: $150/person for guest chefs, $105/person for dining companions.

Location: 13 miles west of Newark International Airport.

Contact: George Staikos, Director of Restaurants, Chef and Sommelier for a Day, Hilton Hotel, 41 JFK Parkway, Short Hills, NJ 07078 US; 973-379-0100 #7838, Fax 973-379-1153.

CLASSIC THYME

Westfield/Year-round

This private school offers demonstration and hands-on classes on a variety of topics. Established in 1995. Maximum class/group size: 12-24. Facilities: Viking-equipped kitchen with overhead mirror and video monitoring. Also featured: Wine instruction, classes for youngsters, day trips, market visits, winery tours, fine dining, private instruction.

Emphasis: Basic techniques, special subjects.

Faculty: David P. Martone, CCP, Sheila Turteltaub.

Costs: $25-$85/session.

Location: 45 minutes from New York City.

Contact: David Martone, CCP and Sheila Turteltaub, Classic Thyme, 401 W. Broad St., Westfield, NJ 07090 US; 908-232-5445, Fax 908-789-4813, E-mail ClassicThyme@westfieldNJ.com, URL http://www.quintillion.com/westfield. Retail address: 161 E. Broad St., Westfield, NJ, 07090 US.

COOKINGSTUDIO

Four locations/Year-round

This cooking school in Kings Super Markets offers single and multi-session classes. Demonstration and participation courses cover a variety of subjects including techniques, single subject classes and full menu classes. Maximum class/group size: 22-28 demo, 12-15 hands-on. 150+ programs per year. Facilities: Fully-equipped enclosed kitchens, overhead mirrors for easy viewing. Also featured: Celebrity chef demonstrations, classes for couples and children.

Emphasis: Principles of Cooking, an 8-week diploma series, emphasizes a full range of cooking techniques. Additional series offered include Advanced Principles of Cooking & Mastery of Baking.

Faculty: More than 15-member resident faculty includes Carole Walter, Jean Yueh, and Kathleen Sanderson. Guest chefs include Nick Malgieri, David Rosengarten and Andre Soltner.

Costs: Principles of Cooking series is $55-$60/session. Individual classes range from $20-$50. Children's classes are $25-$35.
Location: Short Hills, Bedminster, Verona, Hillsdale.
Contact: Susan Loden, Manager, CooKINGStudio, 2 Dedrick Pl., West Caldwell, NJ 07006 US; 973-808-4277, Fax 973-575-3297.

COOKTIQUE
Tenafly/Year-round

This school and cook- and giftware store offers day and evening demonstration and participation sessions. Established in 1976. Maximum class/group size: 14 hands-on/25 demo. 150 programs per year. Facilities: 400-square-foot demonstration kitchen with overhead mirror. Also featured: Children's classes, birthday parties for children and adults.
Emphasis: Techniques, sauces, fish, guest chef specialties, dessert, pizza, pasta, menu classes.
Faculty: Culinary professionals and master chefs. Guest chefs have included Guiliano Bugialli, Marcella Hazan, Nicholas Malgieri, Lorenza de'Medici, Arthur Schwartz, and Jaques Pépin.
Costs: From $25; guest chef classes $50-$100.
Location: 16 miles from New York City.
Contact: Cathy McCauley, CCP, Director, Cooktique, 9 W. Railroad Ave., Tenafly, NJ 07670 US; 201-568-7990, Fax 201-568-6480, E-mail cooktique@email.msn.com, URL http://www.cooktique.com.

DOUBLE DOLPHIN COOKING ADVENTURES
Year-round

This cooking vacation sponsor offers 3- to 5-day cooking vacations to regions of the US that include hands-on classes, market visits, winery tours, sightseeing, and restaurant dinners. Established in 1996. Maximum class/group size: 10+. 10 programs per year.
Emphasis: Regional cuisines.
Contact: Susan Flythe, Double Dolphin Cooking Adventures, P.O. Box 399, Edgewater Park, NJ 08010 US; 609-387-4746, E-mail CookingAdv@aol.com.

EDIBLES...NATURALLY! COOKING SCHOOL
Princeton Junction/Year-round

This cafe and cooking school offers evening demonstration and participation classes. Established in 1994. Maximum class/group size: 12 hands-on/24 demo. 100+ programs per year. Facilities: Residential-style kitchen with Miele appliances, dinner theater seating for up to 36. Also featured: Children's, senior citizens', and singles' classes, field trips, tours, cruises, Cook with the Chef dinner parties.
Emphasis: American regional, international, ethnic, herbs and spices, vegetarian and low fat, macrobiotic, baking, basics.
Faculty: Guest chefs and cookbook authors, including Giuliano Hazan, Craig Shelton, Marc Browngold, Bobby Trigg.
Costs: $45-$60/class.
Location: Central New Jersey, an hour from Philadelphia and New York City.
Contact: Alice Miller, Director, Edibles...Naturally! Cooking School, 14 Washington Rd., Princeton Junction, NJ 08550 US; 609-936-8200, Fax 609-936-8855, E-mail ednatly@aol.com.

GINGERBERRY COOKING SCHOOL
Stewartsville/March-December

Culinary professional Nancy Wyant offers 1-session mostly demonstration classes on a variety of topics. Classes for adults and children. Established in 1988. Maximum class/group size: 11. Facilities: Home kitchen in English cottage-style home in rural setting. Also featured: Private and group classes, children's birthday parties.
Emphasis: Seasonal, regional, and holiday menus.
Faculty: Nancy Wyant.
Costs: $36/class.

Location: 90 minutes from Philadelphia, 1 hour from New York City, 30 minutes from Allentown, PA.

Contact: Nancy Wyant, Owner, Gingerberry Cooking School, 304 Rt. 173, Stewartsville, NJ 08886 US; 908-479-6062, E-mail gingercreek@hotmail.com.

GLORIA ROSE GOURMET LONG LIFE COOKING SCHOOLS

Springfield and Wayne/Year-round

This nutritional cooking school offers 3-session demonstration courses for individuals concerned with weight control, sound nutrition, and/or medical conditions requiring a modified diet. Established in 1984. Maximum class/group size: 8. Facilities: 300-square-foot teaching kitchen.

Emphasis: Nutritional counseling; gourmet cooking without added salt, sugar, fats, or oils; vegetarian diets, cultural and ethnic foods; analysis of product labels.

Faculty: Gloria Rose is author of Low Fat Cooking for Good Health, which is included in the program materials. Instructors are gourmet cooks, registered dietitians or nurses affiliated with New Jersey hospitals.

Costs: $285 cost/course includes textbook and private nutritional counseling.

Location: The Springfield school (central NJ) is 8 miles from Newark; the Wayne school (northern NJ) is 30 minutes from Manhattan.

Contact: Stuart Ehrlich, Director, Gloria Rose Gourmet Long Life Cooking Schools, 7 Pinecrest Terr., Wayne, NJ 07470 US; 973-831-0471, Fax 973-831-0471, E-mail gourmet1@bellatlantic.net, URL http://www.healthygourmet.com. Also: Gloria Rose, 48 Norwood Road, Springfield, NJ; 201-376-0942.

SEASHORE GOURMET GETAWAYS/ATLANTIC COMMUNITY COLLEGE

Mays Landing/July-August *(See also page 76) (See display ad page 76)*

This two-year college offers 1- to 4-day themed programs based on the food and philosophy of a celebrity chef who has authored a cookbook. Three days of hands-on instruction are followed by a demo/lecture by the chef. Established in 1997. Maximum class/group size: 30 demo, 20 hands-on. 4 programs per year. Facilities: Academy of Culinary Arts instructional kitchens. Also featured: Kids College summer cooking program for children.

Emphasis: Specialties of featured cookbook authors.

Faculty: Chef cookbook authors and Academy of Culinary Arts faculty.

Costs: $445 for 4-day program including signed cookbook and lunches, $79 for 1-day demo/lecture and lunch.

Location: Mays Landing, 15 miles from Atlantic City, 45 miles from Philadelphia.

Contact: Trish Krevetski, Asst. Dir. Workshops and Training, Academy of Culinary Arts, Atlantic Community College, 5100 Black Horse Pike, Mays Landing, NJ 08330 US; 609-343-4829, Fax 609-343-4823, E-mail krevetsk@atlantic.edu, URL http://www.atlantic.edu.

NEW MEXICO

CHEFS AT WORK

Santa Fe/September-June

Culinary professional Vikki Nulman offers 1-day and 1- or 2-week hands-on classic French cooking programs. Established in 1997. Maximum class/group size: 8. 20 programs per year. Facilities: 600-square-foot new professional kitchen with 5 ovens, Sub Zero appliances, individual work areas, mountain views. Also featured: Private instruction, market visits.

Emphasis: Classic French cooking.

Faculty: Chef Vikki Nulman is a graduate of the French Culinary Institute, a professional caterer, and studied in Venice with Marcella Hazan.

Costs: $70/day, $200 for 3 classes/week.

Contact: Chef Vikki Nulman, Chefs at Work, 4 High Ridge Rd., Santa Fe, NM 87501 US; 505-820-0377, Fax 505-820-0389.

GILDA LATZKY
Albuquerque/Year-round

Culinary professional Gilda Latzky (formerly of New York City) offers participation courses. Established in 1977. Maximum class/group size: 5. Also featured: Seasonal and private classes.

Emphasis: Italian, French, and American cuisine; baking; recipe modification and menu planning.

Faculty: Gilda Latzky has taught French and Italian cooking and baking for more than 27 years in New York City. She studied in France and Italy and at the CIA.

Costs: $35/session.

Location: Colorado, New Mexico, Texas, Arizona.

Contact: Gilda Latzky, 9900 Spain NE #V2120, Albuquerque, NM 87111 US; 505-296-1439.

JANE BUTEL'S SOUTHWESTERN COOKING SCHOOL
Albuquerque/Year-round

Culinary professional Jane Butel offers 5-day and weekend participation courses as well as demonstration sessions. Established in 1983. Maximum class/group size: 18 hands-on. 15-20 programs per year. Facilities: 2,000-square-foot kitchen with 6 work stations, demo area, overhead mirror. Also featured: Private group lessons, sightseeing, ballooning, visits to markets and wineries.

Emphasis: Traditional, innovative, and low-fat, New Mexican and Southwestern cuisine.

Faculty: Jane Butel, author of 16 cookbooks, including Jane Butel's Southwestern Kitchen, Southwestern Grill, Chili Madness, Hotter Than Hell, and Fiesta, founder of the Pecos Valley Spice Co., guest chefs and speakers.

Costs: The $1,695 ($895) 5-day (weekend) fee includes most meals and double occupancy lodging at La Posada de Albuquerque. Demonstrations average $40-$50 each.

Location: Old Town Albuquerque.

Contact: Jane Butel, Owner, Jane Butel's Southwestern Cooking School, c/o La Posada de Albuquerque, 125 Second NW, Albuquerque, NM 87102 US; 800-472-8229/505-243-2622, Fax 505-243-8297, E-mail cookie@abq.com, URL http://www.janebutel.com.

SANTA FE SCHOOL OF COOKING
Santa Fe/Year-round

This school, food market, and mail order catalog offers demonstration classes and smaller hands-on classes that include shopping at the Farmer's Market. Established in 1989. Maximum class/group size: 15 hands-on/44 demo. Facilities: Santa Fe-style kitchen with overhead mirrors. Also featured: Private classes, shopping trips to the Farmer's Market in August and September, culinary tours of northern New Mexico.

Emphasis: New Mexican and contemporary Southwestern cuisines, vegetarian, Mexican light cooking, and cuisines of Mexico classes.

Faculty: Includes owner/director Susan Curtis, author of Santa Fe School of Cooking Cookbook; cookbook authors, chefs from Santa Fe's top restaurants, guest celebrities.

Costs: Classes range from $30-$70; tours are ~ $800-$1,600, including some meals, field trips, lodging.

Location: The historic downtown district, 50 miles from Albuquerque International Airport.

Contact: Susan Curtis, Owner/Director, Santa Fe School of Cooking, 116 W. San Francisco St., Santa Fe, NM 87501 US; 505-983-4511, Fax 505-983-7540, E-mail cookin@nets.com, URL http://www.nets.com/cookin. Other contact: Nicole Curtis, Manager.

SAVEUR INTERNATIONAL COOKING SCHOOLS (See page 212)

NEW YORK

ALICE ROSS HEARTH STUDIOS
Smithtown/Year-round

Culinary historian Alice Ross offers hands-on classes in culinary history and traditional methods including hearth techniques, Native American and Civil War cookery, game butchery and preparation, baking using hearth and brick ovens. Facilities: Converted carriage house equipped with an open hearth, Victorian wood stove, outdoor wood-fired brick oven, water pump, smoke house, antique utensils.

Emphasis: Basic hearth cooking, early American cookery, seasonal preserving, game cookery, basic American cookery and its ethnic origins.

Faculty: IACP member Dr. Alice Ross is co-founder of Culinary Historians of New York, wrote her doctoral dissertation in food history, teaches at CCNY and other colleges, served as consultant to Colonial Williamsburg and The Lowell National Historic Park (MA).

Location: Long Island.

Contact: Alice Ross, Alice Ross Hearth Studios, 15 Prospect St., Smithtown, NY 11787 US; 516-265-9335, E-mail aross@li.net, URL http://www.li.net/~aross.

ANNA TERESA CALLEN ITALIAN COOKING SCHOOL

New York/September-June

Culinary professional Anna Teresa Callen offers 5-session participation and demonstration courses. Established in 1978. Maximum class/group size: 6. 8 programs per year. Facilities: Efficient home kitchen. Also featured: Culinary tours to Italy.

Emphasis: Italian regional cooking.

Faculty: IACP-member Anna Teresa Callen is author of The Wonderful World of Pizzas, Quiches and Savory Pies, Anna Teresa Callen's Menus for Pasta, and Italian Classics in One Pot. She also teaches at Peter Kump's New York Cooking School and NYU.

Costs: $625/course.

Location: Downtown Manhattan.

Contact: Anna Teresa Callen, Anna Teresa Callen Italian Cooking School, 59 W. 12th St., New York, NY 10011 US; 212-929-5640.

CAKES UNIQUELY YOURS SCHOOL OF CONFECTIONARY ARTS

New York/Year-round

Cake designer Ajike Williams offers 1- to 9-session courses in Wilton Method, marzipan, gumpaste, fondant, designer cookies, baking, and wedding cakes. Established in 1996. Maximum class/group size: 8-15. Facilities: The professional kitchen at Urban Horizons. Also featured: Wedding cake consultation.

Emphasis: Wilton Method courses, wedding cakes, gumpaste flowers.

Faculty: Cake designer Ajike Williams.

Costs: From $45-$50 for a single session to $500 for a 9-session baking course.

Location: 50 E. 168th St.

Contact: Ajike Williams, Cake Designer, Instructor, Cakes Uniquely Yours School of Confectionary Arts, 1258 Fteley Ave., Bronx, NY 10472 US; 718-617-4538.

CAROL'S CUISINE, INC.

Staten Island/Year-round

Culinary professional Carol Frazzetta offers 1- to 6-session demonstration and participation courses. Established in 1972. Maximum class/group size: 18 hands-on and demo. 130 programs per year. Facilities: Fully-equipped professional teaching kitchen with overhead mirror. Also featured: Private lessons, wine classes.

Emphasis: Techniques, baking, cake decorating, international and Italian cuisine.

Faculty: Owner/Director Carol Frazzetta, IACP accredited, advanced certificate from Le Cordon Bleu, studied at CIA, Wilton School of Cake Decorating, Marcella Hazan's School, L'Academie de Cuisine. Leonard Pickell, wine consultant.

Costs: $45-$55/session.

Location: Central Staten Island, 1 hour from Manhattan by ferry or bus.

Contact: Carol Frazzetta, Owner, Chef, Cooking Teacher, Carol's Cuisine, Inc., 1571 Richmond Rd., Staten Island, NY 10304 US; 718-979-5600.

COOKHAMPTON – SILVIA LEHRER

Water Mill/June-February

This private school offers 1- to 3-session demonstration and participation courses. Established in 1988. Maximum class/group size: 15 demo/6 hands-on. Facilities: 400-square-foot teaching kitchen. Also featured: Culinary tours.

Emphasis: A variety of topics, including do-ahead seasonal and entertaining menus, flavor of Provence, regional Italian cuisine, chef specialties.

Faculty: IACP-certified instructor Silvia Lehrer, former owner of Cooktique in Tenafly, NJ, is author of Cooking at Cooktique. She studied with James Beard, Simca Beck, and Giuliano Bugialli. Guest chefs of the Hamptons.

Costs: $70/session, $60 for guest chefs.

Location: The summer resort area of Southampton township.

Contact: Silvia Lehrer, Cookhampton, P.O. Box 765, Water Mill, NY 11976 US; 516-537-7831, Fax 516-537-3330.

COOKING BY THE BOOK, INC.

New York/September-July

Suzen and Brian O'Rourke offer evening participation classes that focus on a selected cookbook or chef menu. Established in 1989. Maximum class/group size: 20. 40+ programs per year. Facilities: Fully-equipped 500-square-foot kitchen with 6 work stations. Also featured: Private parties and cooking instruction, corporate events, children's programs, wine instruction, customized classes.

Emphasis: Hands-on approach to cooking.

Faculty: Suzen and Brian O'Rourke; authors regularly present.

Costs: $75/class.

Location: Downtown Manhattan, 25 minutes from Newark and La Guardia airports.

Contact: Suzen O'Rourke, Cooking by the Book, Inc., 13 Worth St., New York, NY 10013 US; 212-966-9799, Fax 212-925-1074, E-mail CBTBINC@aol.com, URL http://www.cookingbythebook.com.

COOKING CLASSES IN THE HAMPTONS

East Hampton/June-September

Culinary professionals Kristi Hood and Katherine Hartnett offers 8 bi-weekly classes that focus on fundamentals. Established in 1996. Maximum class/group size: 12. 1 programs per year. Facilities: Professional kitchen of The Mill House Inn, an 8-guestroom bed & breakfast rated 3-Crowns by the American B&B Assn.

Emphasis: Cooking fundamentals and fun.

Faculty: Kristi Hood, CPC, caterer, instructor at Peter Kump's New York Cooking School and New School; Katherine Hartnett, New York Restaurant School graduate and bed & breakfast owner.

Costs: $80/class, $40 for children.

Location: 2 hours east of New York City.

Contact: Kristi Hood, Cooking Classes in the Hamptons, 10 Woodcock Ln., East Hampton, NY 11937 US; 516-324-2201, Fax 516-324-2201, E-mail millhouseinn@worldnet.att.net, URL http://www.millhouseinn.com.

CORNELL'S ADULT UNIVERSITY

Ithaca/July-August

This university continuing education division offers an annual on-campus 4-week summer program consisting of 1-week workshops and courses in subjects that include cooking, history, current events, ecology, music, literature, architecture, and art. Established in 1968. Maximum class/group size: 12-20. Also featured: Supervised youth program offers activities geared to 5 age groups.

Emphasis: The yearly culinary workshop focuses on menu planning and techniques for creating appetizing and nutritionally sound meals.

Faculty: Cornell University faculty and staff.

Costs: Approximately $805/week for adults and $320-$470 for children, which includes tuition,

double occupancy dormitory lodging, meals, and planned activities. A $30 materials fee is extra.
Location: The 13,000-acre campus is in New York's Finger Lakes region.
Contact: Cornell's Adult University, 626 Thurston Ave., Ithaca, NY 14850-2490 US; 607-255-6260, Fax 607-254-4482, E-mail cauinfo@cornell.edu, URL http://www.sce.cornell/sce/cau/.

CUISINE GROUP, INC.

New York, France/Year-round

This special events producer has a model recipe-testing kitchen for television and photo sessions, guest chef demonstrations, food and wine promotion events, seminars. Established in 1985. Facilities: Throughout tri-state area. Also featured: Culinary trips and French immersion courses in southern France, yacht excursions.
Emphasis: Producers of food and/or wine related events.
Faculty: Founder-president Michèle Lyster trained at La Varenne and her family's restaurant; food stylist and consultant; on program committee of James Beard Foundation. Guest chefs have included Martin Yan, Anna Teresa Callen, Steven Schmidt, Jack Ubaldi.
Costs: Trips range from $2,000 to $3,000, including meals, lodging, and excursions.
Location: Various sites in New York City; trips to south of France.
Contact: Michèle Lyster, Director, Cuisine Group, Inc., 244 Madison Ave., New York, NY 10016-2815 US; 212-557-5702, Fax 212-681-8058, E-mail mcuisine@aol.com. In France: 129 Promenade des Anglais, 06200, Nice; (33) 493 374627, Fax (33) 493 440147.

THE CULINARY INSTITUTE OF AMERICA

Hyde Park/Year-round *(See also pages 13 and 80) (See display ad page 81)*

The CIA's Continuing Education Dept. Travel Programs offers a variety of adult education demonstration and hands-on courses, wine instruction, classes for youngsters. Facilities: Well-known hotels and restaurants. Also featured: Travel programs that include cooking demos, dining in fine restaurants and private homes, sightseeing, visits to food producers, markets, wineries.
Emphasis: Travel programs feature the cuisines of France, Italy, Spain, Switzerland.
Faculty: The CIA's chefs and instructors.
Costs: Costs of on-campus courses vary. Trips are approximately $3,500, which includes airfare, 5-star hotel lodging, some meals, planned activities.
Location: Adult education courses are held at the Hyde Park campus. Trip destinations include France, Italy, Spain, Switzerland.
Contact: Continuing Education Dept., The Culinary Institute of America, 433 Albany Post Rd., Hyde Park, NY 12538-1499 US; 800-888-7850, URL http://www.ciachef.edu. Travel Programs: Douglas White, Marketing Services Specialist, 914-451-1455, E-mail d_white@culinary.edu.

THE CULINARY LOFT

New York/Year-round

This private school offers optional hands-on vegetarian classes, wine tastings, and gourmet dinners with live music. Established in 1995. Maximum class/group size: 50 demo, 10 hands-on. Facilities: 2,000-square-foot SoHo loft with double kitchen. Also featured: Classes for youngsters, private instruction, loft available for private parties.
Emphasis: Mostly vegetarian with attention to health.
Faculty: Cookbook authors, chefs, and other culinary professionals.
Costs: $50-$250 all-inclusive. Lodging can be arranged for out-of-towners.
Location: Between Spring and Broome streets in SoHo.
Contact: Corinne Colen, Owner, The Culinary Loft, 515 Broadway #5A, New York, NY 10012 US; 212-431-7425, Fax 212-431-7816, E-mail CULINARYNY@aol.com.

DE GUSTIBUS AT MACY'S

New York/February-June, September-November

This independent school in Macy's department store operated by Arlene Feltman Sailhac offers demonstrations and on-location classes. Established in 1980. Maximum class/group size: 70 demo.

40 programs per year. Facilities: Professionally-equipped teaching kitchen. Also featured: Wine seminars, a recipe-writing course, and a media skills class.

Emphasis: Regional American cuisine, French and Italian cuisines, wine selection, menus for entertaining, guest chef specialties.

Faculty: Guest chefs and cookbook authors include David Bouley, Daniel Boulud, Bobby Flay, Anne Rosenzweig, and Alain Sailhac.

Costs: From $70/session to $260 for a series of 4.

Location: Macy's Herald Square store at 34th Street and 7th Avenue.

Contact: Arlene Feltman Sailhac, Owner/Director, De Gustibus at Macy's, 343 E. 74th Street, Apt. 9G, New York, NY 10021 US; 212-439-1714, Fax 212-439-1716, E-mail grtcooks@aol.com, URL http://starchefs.com/DeGustibus/.

THE FRENCH CULINARY INSTITUTE

(See also page 82) (See display ads pages 83, 85, 87) **New York/Year-round**

This proprietary institution for aspiring professionals and serious amateur cooks offers La Technique, a 22-session Saturday course that provides an overview of classical French cuisine; Essentials of Pastry, a 20-session Saturday course that provides an introduction to the philosophy of pastry making. Established in 1984. Maximum class/group size: 22. 6 programs per year. Facilities: 15,000-square-feet of kitchens, newly-equipped pastry and bread kitchens, demonstration amphitheater, L'École open-to-the-public restaurant.

Emphasis: French technique, pastry.

Faculty: 20 full-time, 9 part-time. Includes Dean of Special Programs Jacques Pépin, Dean of Culinary Studies Alain Sailhac, Dean of Pastry Arts Jacques Torres, Master Chef/Senior Lecturer Andre Soltner, Dean of Bread Baking Daniel Leader.

Costs: Tuition, which includes professional knife set, kitchen tools, and uniform, is $4,290 for La Technique and $3,650 for Essentials of Pastry.

Location: New York's historic SoHo district, adjacent to Chinatown and Little Italy.

Contact: Cynthia C. Marchese, Director of Admissions, The French Culinary Institute, 462 Broadway, New York, NY 10013-2618 US; 888-FCI-CHEF/212-219-8890, Fax 212-431-3054, E-mail cmarchese@frenchculinary.com, URL http://www.frenchculinary.com.

GRANDMA'S SECRETS

New York/Year-round

This baking company offers classes in basic and advanced pie baking and other desserts. Established in 1998. Maximum class/group size: 10. Facilities: Professional kitchen at Urban Horizons.

Emphasis: Pie baking, cakes, brownies.

Faculty: Regina McRae, founder of Grandma's Secrets, a home-baked goods to go service.

Costs: $25/class.

Location: 50 E. 168th St.

Contact: Regina McRae, Grandma's Secrets, 15 Sylvan Terr., New York, NY 10032 US; 212-740-3177/888-557-3273, Fax 212-740-3177.

HAY DAY COOKING SCHOOL (See page 201)

JULIE SAHNI'S SCHOOL OF INDIAN COOKING

(See also page 310) **New York /Year-round**

Cooking school owner and cookbook author Julie Sahni offers two 3-day hands-on courses that include shopping in an Indian/spice market: Indian Cooking and Understanding Spices and Herbs. Established in 1973. Maximum class/group size: 3. Facilities: Specially-designed teaching kitchen in Ms. Sahni's studio. Also featured: Each course can be taught anywhere in US as a 1-day introduction for 6 students.

Emphasis: Techniques, ingredients, healthful meal planning, historical background, social ideology.

Faculty: Julie Sahni is author of Classic Indian Cooking, Indian Vegetarian Cooking, and

Savoring Spices and Herbs. She is a member of the IACP and Les Dames D'Escoffier and has degrees in Architecture and Classical Dance.

Costs: $985 per person.

Location: Near the Brooklyn entrance to the Brooklyn Bridge.

Contact: Julie Sahni, Director, Julie Sahni's School of Indian Cooking, 101 Clark St., #13A, Brooklyn Heights, NY 11201 US; 718-625-3958, Fax 718-625-3456, E-mail jsahni@worldnet.att.net.

KAREN LEE IMAGINATIVE COOKING CLASSES & CATERING

New York, Amagansett/Year-round

Culinary professional Karen Lee offers participation classes that include two 5-day courses for out-of-towners, three 4-session courses that meet once weekly, and 3-day weekend seminars. 3- and 5-day courses meet in October and May, 4-session courses meet year-round. Established in 1972. Maximum class/group size: 9 hands-on.

Emphasis: Fusion and traditional Chinese cuisine, Italian cuisine, basic technique, vegetarian, entertaining menus.

Faculty: Owner and caterer Karen Lee apprenticed with Madame Grace Zia Chu and is author of Nouvelle Chinese Cuisine, Soup, Chinese Cooking Secrets, and The Occasional Vegetarian.

Costs: Tuition, which includes a copy of Ms. Lee's latest book, The Occasional Vegetarian $655 for the 5-day course, $440 for the 4-session course ($125/single class), and $380 for the weekend course.

Location: September-June: New York City's Lincoln Center area. July-Aug: Long Island.

Contact: Karen Lee Imaginative Cooking Classes & Catering, 142 West End Ave., #30V, New York, NY 10023 US; 212-787-2227, Fax 212-496-8178. In July and Aug: PO Box 1998, Amagansett, NY 11930; 516-267-3653, Fax 516-267-3114.

LA CUCINA DI CORINNE

Akron/Year-round

Culinary professional Corinne Derry offers hands-on classes in northern and regional Italian cuisine, history and culture of Italy. Established in 1997. Maximum class/group size: 20 demo, 10 hands-on, 20 travel. 80+ programs per year. Facilities: Professionally-equipped Italian-style home kitchen. Also featured: Programs for youngsters, private classes, market visits, visits to food producers, dining in private homes.

Emphasis: Northern Italian and regional Italian cuisine.

Faculty: Corinne Derry and guest chefs.

Costs: $35-$50.

Location: 20 miles east of Buffalo, 40 miles west of Rochester, 100 miles south of Toronto.

Contact: Corinne Derry, Owner, La Cucina di Corinne, 22 Colony Pl., Akron, NY 14001 US; 716-542-1525, Fax 716-542-1525, E-mail corinne@fcs-net.com.

LA CUISINE SANS PEUR

New York/October-August

Chef-de-Cuisine Henri-Etienne Lévy offers 5- and 6-session demonstration courses, including the 6-session basic course, and the 5-session intermediate, advanced, and baking courses. Specialty classes include desserts, fish, and game. Established in 1978. Maximum class/group size: 4. 20+ programs per year. Facilities: Traditional French well-equipped kitchen. Also featured: 1-week culinary vacations in Provence in September.

Emphasis: The regional cooking of France with emphasis on Alsace and Provence, cooking without recipes, basic to advanced courses.

Faculty: Chef and proprietor Henri-Etienne Lévy trained and worked in restaurant kitchens in France and Germany for 15 years.

Costs: $450/course.

Location: On Manhattan's Upper West Side, 10 minutes from Lincoln Center.

Contact: Henri-Etienne Lévy, chef/proprietor, La Cuisine Sans Peur, 216 W. 89 St., New York, NY 10024 US; 212-362-0638.

LAUREN GROVEMAN'S KITCHEN

Larchmont/Year-round

Culinary professional Lauren Groveman offers 5-session participation courses and individual classes on specific subjects. Classes for young people. Established in 1990. Maximum class/group size: 6 hands-on.

Emphasis: Techniques and preparation of comfort foods, breads, appetizers, edible gifts.

Faculty: Cookbook author and columnist Lauren Groveman.

Costs: Tuition is $450 for the 5-session course and $100 for a specialty class.

Location: About 30 minutes from New York City.

Contact: Lauren Groveman, CCP, President, Lauren Groveman's Kitchen, Inc., 55 Prospect Ave., Larchmont, NY 10538 US; 914-834-1372, Fax 914-834-3802, E-mail l.groveman@worldnet.att.net, URL http://www.laurengroveman.com.

LOOK WHO'S COOKING, INC.

Oyster Bay/Year-round

Culinary professional Barbara Sheridan offers 1- to 4-session demonstration and participation courses. Established in 1994. Maximum class/group size: 20 demo/10 hands-on. 200+ programs per year. Facilities: 800-square-foot well-equipped kitchen with 10 workspaces.

Emphasis: Gourmet cooking for the everyday cook, low fat cooking, baking, fundamentals and techniques, entertaining menus.

Faculty: Barbara Sheridan, graduate of N.Y. Institute of Technology Culinary Arts Program, attended Peter Kump's New York Cooking School and Le Cordon Bleu.

Costs: $50/session.

Location: Long Island, 20 miles from New York City.

Contact: Barbara M. Sheridan, Look Who's Cooking, Inc., 7 West Main St., Oyster Bay, NY 11771 US; 516-922-2400, Fax 516-379-6067.

MARY BETH CLARK

(See also page 323) (See display ad page 323) **New York/November-March, July-August**

Private school established by chef and author Mary Beth Clark offers 3-hour custom-designed, full-participation private lessons in Italian cuisine. Student brings home 3-4 dishes from the class. Established in 1977. Maximum class/group size: 1-2.

Emphasis: Italian cuisine.

Faculty: Cookbook author Mary Beth Clark operates the International Cooking School Of Italian Food And Wine week-long cooking courses in Bologna, Italy (May-July, September-October).

Costs: $500/session plus ingredients, payable in advance, refundable a week prior.

Location: Manhattan.

Contact: Mary Beth Clark, 201 E. 28th St., #15B, New York, NY 10016-8538 US; 212-779-1921, Fax 212-779-3248, E-mail MaryBethClark@worldnet.att.net, URL http://www.shawguides.com/cook/mbc.

MIETTE

New York/September-June

Tartine restaurant offers hands-on courses for youngsters, ages 8-15, and adults. Sessions meet Monday afternoons and evenings in the restaurant kitchen and conclude with a student-prepared dinner/graduation. Established in 1995. Maximum class/group size: 14. Facilities: The restaurant and kitchen of Tartine. Also featured: Corporate cook-ins, birthday parties, theme cooking, wine tastings.

Emphasis: French and other cuisines, etiquette, table setting, French language, healthful recipes.

Faculty: Chef Paul Vandewoude and his assistant, Mariette Bermowitz.

Costs: $30 ($35)/class, $100 ($120)/4 classes for youngsters (adults).

Location: West Greenwich Village in New York City.

Contact: Mariette Bermowitz, Miette, 253 W. 11th St., New York, NY 10014 US; 212-229-2611, Fax 212-229-1972, E-mail msmiette@aol.com.

THE NATURAL GOURMET INSTITUTE FOR FOOD AND HEALTH

New York/Year-round *(See also page 86) (See display ad page 86)*

This private trade school devoted to healthy cooking offers demonstration and participation programs: evening and weekend classes and series. Intensives include Basic Techniques, Pastry Arts, Food & Healing. Established in 1977. Maximum class/group size: 8-25. 120+ programs per year. Facilities: Include 2 kitchens, classroom, and bookstore. Also featured: Classes for adults and youngsters; beginning, advanced cooking and theory.

Emphasis: Vegetarian cooking, low/no fat, recipe adaptation, medicinal cooking, tofu, tempeh, seitan, fish, organic poultry, food and healing, international cuisines, whole grains, natural sweeteners.

Faculty: 10 faculty, includes founder Annemarie Colbin, MA, Certified Health Education Specialist, author of Food and Healing, The Book of Whole Meals, The Natural Gourmet; Co-Presidents/Directors Diane Carlson and Jenny Matthau, graduates of the school.

Costs: $697/week for summer intensives, classes and series are $40-$75/session. Lodging at local hotels, hostels, bed & breakfasts.

Location: Between 5th and 6th Avenues in Manhattan.

Contact: Susan Kaufman, Registration Manager, The Natural Gourmet Institute for Food and Health, 48 W. 21st St., 2nd Floor, New York, NY 10010 US; 212-645-5170 #106, Fax 212-989-1493, URL http://www.naturalgourmetschool.com. Other contact: Diane Carlson.

NEW SCHOOL CULINARY ARTS

New York/Year-round

The New School for Social Research offers 1- to 8-session demonstration and participation courses and weekend workshops. Established in 1919. Maximum class/group size: 10-20. 300+ programs per year. Facilities: Restored landmark townhouse with indoor and outdoor dining areas and a fully-equipped instructional kitchen. Also featured: On-site restaurant chef demonstrations, lectures on culture and cuisine, classes for youngsters, wine courses.

Emphasis: Culinary techniques, ethnic and regional cuisines, holiday menus, home entertaining, baking, light-style cooking.

Faculty: 50+ faculty headed by Gary Goldberg, co-founder Martin Johner, includes Miriam Brickman, Richard Glavin, Arlyn Hackett, Micheal Krondl, Harriet Lembeck, Lisa Montenegro, Robert Posch, Dan Rosati, Stephen Schmidt, Marie Simmons, Karen Snyder, Carole Walter.

Costs: $65-$80/session, $40 for youngsters, $15 for lectures, $325 for weekend workshops, $45-$65 for wine classes.

Location: Manhattan's Greenwich Village.

Contact: Gary A. Goldberg, Executive Director, New School Culinary Arts, 100 Greenwich Ave., New York, NY 10011 US; 212-255-4141/800-544-1978, Fax 212-229-5648, E-mail NSCulArts@aol.com, URL http://www.newschool.edu.

NEW YORK CAKE & BAKING DISTRIBUTORS, INC.

New York/Year-round

This private school and equipment store specializing in cake decoration and baking offers classes, courses, and intensives in cake decorating and baking. Established in 1978.

Emphasis: Cake decorating and baking.

Costs: From $35 per class to $250 for a 2-day intensive.

Contact: New York Cake & Baking Distributors, Inc., 56 W. 22nd St., New York, NY 10010 US; 212-675-2253.

NEW YORK UNIVERSITY

New York/Year-round *(See also page 90) (See display ad page 90)*

The Dept. of Nutrition and Food Studies, NYU's School of Education offers a variety of lecture and demonstration courses for food professionals and career changers. Established in 1986. Maximum class/group size: 10-35. 60 programs per year. Facilities: New teaching kitchen and library, computer, academic resources.

Emphasis: Food business and management, food history and culture, nutrition, food writing, food marketing.
Faculty: Foodservice and industry professionals, historians, authors.
Costs: Approximately $50-$200/session.
Location: Washington Square, Greenwich Village.
Contact: Carol Guber, Director of Food Programs, New York University, 35 W. 4th St., 10th Fl., New York, NY 10012-1172 US; 212-998-5588, Fax 212-995-4194, E-mail csg1@is2.nyu.edu, URL http://www.nyu.edu/education/nutrition/.

NORMAN WEINSTEIN'S BROOKLYN COOKING SCHOOL

Brooklyn/Year-round

Norman Weinstein, CCP, offers private classes. Established in 1974. Maximum class/group size: 6. Facilities: Hot Wok Catering's 150-square-foot kitchen.
Emphasis: Asian cuisines, authentic Western barbecue, knife skills workshops, basic techniques, Asian tasting dinners, Chinatown walking tours.
Faculty: Norman Weinstein, CCP, is author of 2 cookbooks, a founder of NYACT, has operated Hot Wok Catering for 18+ yrs, and is a spokesperson for Wusthof-Trident.
Costs: $40-$50/session.
Location: Brooklyn's Kensington area, accessible via subway from Manhattan.
Contact: Norman Weinstein, CCP, Director, Norman Weinstein's Brooklyn Cooking School, 412 E. 2nd St., Brooklyn, NY 11218 US; 718-438-0577, Fax 718-435-5595, E-mail hotwok@interport.net.

PETER KUMP'S NEW YORK COOKING SCHOOL

(See also page 92) (See display ads pages 93, 239, 240) **New York/Year-round**

This private school offers hands-on courses and workshops that include the 5-session, 25-hour Techniques of Fine Cooking series, offered over 70 times a year, frequently on a Monday-Friday schedule. Established in 1974. Maximum class/group size: 12 hands-on, 30+ demo. 800+ programs per year. Facilities: 27,000-square-foot facility (opens 12/98) includes 9 kitchens, wine studies center, confectionery lab, rooftop herb garden. Also featured: Other Techniques series: spa cuisine, Italian cooking, pastry and baking, cake decorating. Other courses: ethnic and regional cuisines, business topics, wines, cheesemaking, knife skills, walking tours.
Emphasis: Hands-on cooking instruction.
Faculty: 15 staff chef-instructors and guest chefs, including Julia Child, Bobby Flay, David Rosengarten, Madeleine Kamman, and Jean-George Vongerichten.
Costs: Hands-on classes range from $35 -$485. Housing and restaurant suggestions are provided to out-of-towners.
Location: Year-round classes at 50 W. 23rd St. and 307 E. 92nd St. in New York City.
Contact: Peter Kump's New York Cooking School, 50 W. 23rd St., New York, NY 10128 US; 800-522-4610/212-242-2882, Fax 212-242-0127, E-mail user700766@aol.com, URL http://www.pkcookschool.com.

SAPORE DI MARE

Wainscott/September-June

This regional Italian restaurant offers participation classes up to three times weekly. Established in 1990. Maximum class/group size: 15. Facilities: Restaurant kitchen.

Emphasis: Appetizers, pasta, main courses, pizza, breads, grilling, desserts.

Faculty: Restaurateur Pino Luongo, author of A Tuscan in the Kitchen. Instructors include the chefs from the restaurants.

Costs: $150/class.

Location: Long Island, 90 miles from New York City.

Contact: Kristen Dell'Aguzzo, Director, Sapore di Mare Summer Cooking Classes, P.O. Box 1357, Wainscott, NY 11975 US; 516-537-2764, Fax 516-537-1828.

SAVORY SOJOURNS, LTD.

New York/Year-round

This culinary tour provider offers one-day, weekend, and week-long culinary tours of New York City that include hands-on classes at Peter Kump's New York Cooking School, food events, dining at fine restaurants, market visits, winery tours, visits to food producers, walking tours. Established in 1997. Maximum class/group size: 8-15. Facilities: Peter Kump's New York Cooking School and other teaching facilities. Also featured: Customized tours for groups of 6 or more; spousal programs, corporate events.

Emphasis: Insider tours of the New York City food world.

Costs: Day tours start at $110, weekend tours start at $1,115, 5-day tours start at $1,530, all-inclusive.

Location: New York City.

Contact: Addie Tomei or Brian Patterson, President/Associate, Savory Sojourns, Ltd., 155 W. 13th St., New York, NY 10011 US; 212-691-7314/888-9-SAVORY, Fax 212-367-0984, E-mail Addie@savorysojourns.com, URL http://www.savorysojourns.com.

SCHOOL FOR CREATIVE CUISINE

Buffalo/Year-round

Culinary professional Maryann Bolles offers classes, private instruction, market visits. Established in 1988. Maximum class/group size: 20 demo, 12 hands-on. Facilities: Kitchen in private home.

Emphasis: Gourmet natural foods/vegetarian.

Faculty: Maryann Bolles, graduate of Natural Gourmet Cookery School.

Location: North Buffalo.

Contact: Maryann Bolles, Director, School for Creative Cuisine, 48 Rugby Rd., Buffalo, NY 14216 US; 716-875-8571, E-mail mabolles@aol.com.

THE SEASONAL KITCHEN

Pittsford/Year-round

This private school offers morning and evening demonstrations. Established in 1980. Facilities: Well-equipped country kitchen. Also featured: Classes for men, couples, and groups.

Emphasis: Easy-to-prepare and seasonal recipes.
Faculty: Ginger and Dick Howell are members of the IACP.
Costs: $33-$36.
Location: A suburb of Rochester.
Contact: The Seasonal Kitchen, 610 W. Bloomfield Rd., Pittsford, NY 14534 US; 716-624-3242, E-mail dickhcook@aol.com.

TO GRANDMOTHER'S HOUSE WE GO

New York/September-June

Culinary professional Susan Baldassano offers classes taught by older men and women who cook dishes that have been passed on. Established in 1995. Maximum class/group size: 8. 8-10 programs per year. Facilities: The instructor's home kitchen. Also featured: Trips to Mexico and Italy.
Emphasis: Home cooking using traditional recipes and preparation methods.
Faculty: Susan Baldassano studied at The Natural Gourmet Cookery School and Peter Kump's NY Cooking School and was head chef of Angelica's Kitchen, a vegetarian macrobiotic organic restaurant.
Costs: $40-$55 for classes, $1,400-$1,800 for 7- to 10-day trips.
Location: New York City, Long Island, New Jersey.
Contact: Susan Baldassano, To Grandmother's House We Go, 342 Ninth St., #2, Brooklyn, NY 11215 US; 718-768-6197.

VINTAGE HUDSON VALLEY COOKING VACATIONS & SEMINARS

Hudson Valley/Year-round

These ten country inns offers inn to inn cooking vacations (9 sessions at different inns on 3 consecutive days), seminars that focus on summer dishes and market visits, special occasion recipes, heritage cookery, celebrity chef and food arts seminars, classes for brides and grooms. Established in 1994. Maximum class/group size: 12. 12 programs per year. Facilities: Kitchens of the country inns and historic homes. Also featured: Dining in country inns, visits to food producers and wineries, summer theater packages.
Emphasis: Inn specialties.
Faculty: The inn's CIA-trained chefs, including Melissa Kelly of Old Chatham Sheepherding Company Inn.
Costs: $75-$125.
Location: 30 minutes to 2 hours north of Manhattan.
Contact: Maren Rudolph, President, Vintage Hudson Valley Cooking Vacations & Seminars, P.O. Box 288, Irvington, NY 10533 US; 914-591-4503, Fax 914-591-4510, E-mail vintagehudsn@earthlink.com, URL http://www.vintagehudsonvalley.com.

NORTH CAROLINA

COOKS & CONNOISSEURS COOKING SCHOOL

New Bern/March-April, October-November

This cookware store and school offers demonstration and participation classes. Established in 1983. Maximum class/group size: 10-13. 16 programs per year. Facilities: 400-square-foot space in a 3,600-square-foot building.
Emphasis: Regional southern, French, Italian, and local chef specialties. Wine tastings.
Faculty: Proprietor Bill and Bev Haight and staff.
Costs: One-day classes range from $15-$22, 4-week series, $60-$80.
Location: Near Tryon Palace Restoration and 30 miles from beaches.
Contact: Bill and Bev Haight, Owners, Cooks & Connoisseurs Cooking School, 1907 S. Glenburnie Rd., New Bern, NC 28562 US; 919-633-2665, Fax 919-634-9888, E-mail wrhaight@coastalnet.com, URL http://www.cooksandconns.com.

THE GRANDE GOURMET COOKING SCHOOL

Wilmington/Year-round

Culinary professional Robin Hackney offers demonstration and participation classes. Established in 1989. Maximum class/group size: 25 demo/10 hands-on. 50+ programs per year. Also featured: Children's and private classes.

Emphasis: Italian, Chinese, and French cuisines, techniques, low-fat cookery, cake decorating, entertaining menus, wine tastings.

Faculty: Owner Robin Hackney studied with Nick Malgieri, Karen Lee, and others; occasional guest instructors.

Costs: $30-$45/session.

Location: The southeastern beach resort of Wilmington, 2 hours from Raleigh.

Contact: Robin Hackney, Owner, The Grande Gourmet Cooking School, 1605 Country Club Rd., Wilmington, NC 28403 US; 910-763-1764, Fax 910-763-0658.

JOHN C. CAMPBELL FOLK SCHOOL

Brasstown/Year-round

This folk school offers 3- to 6-day courses relating to herbs, food folklore, and whole foods cookery. Established in 1925. Maximum class/group size: 3-12 hands-on. 6 programs per year. Facilities: 380-acre campus with 27 buildings. Designated an Historical District by the National Register of Historic Places. Also featured: Hundreds of art, craft, music, folk dance courses in a wide variety of disciplines.

Emphasis: Herbs, natural foods.

Faculty: Herbalist Carol Blackburn and wild foods expert Ila Hatter.

Costs: Tuition $244-$295; room and board plans available include dormitories, shared rooms, campground.

Location: 7 miles east of Murphy, NC, off US 64 in a mountain valley. Closest airports (2-1/2 hour drive) are Chattanooga, Knoxville, Asheville, Atlanta.

Contact: John C. Campbell Folk School, One Folk School Rd., Brasstown, NC 28902-9603 US; 800-FOLK-SCH/828-837-2775, Fax 828-837-8637, URL http://www.grove.net/~jccfs/.

ROOSTERS GOURMET MARKET & GIFTS

Greensboro/January-November

This cookware store offers demonstration classes. Established in 1983. Maximum class/group size: 24. 150+ programs per year. Facilities: Fully-equipped teaching kitchen with overhead mirror. Also featured: Wine tastings, classes for private groups.

Faculty: Mary James Lawrence, CCP, and Lucy Hamilton, who holds a Cordon Bleu certificate; guest chefs.

Costs: $12-$60.

Location: A restored area of Greensboro.

Contact: Mary James Lawrence, Owner, Roosters Gourmet Market & Gifts, 401 State St., Greensboro, NC 27405 US; 336-272-2665, Fax 336-272-3465, E-mail mjml@aol.com.

THE STOCKED POT & CO.

Winston-Salem/Year-round

This school and gourmet kitchen store offers classes. Established in 1980. Maximum class/group size: 35-50. 100+ programs per year. Facilities: Fully-equipped teaching kitchen with overhead mirror, outdoor grilling facilities. Also featured: Bridal shower cooking classes, Lunch and Learn sessions, junior chef classes, wine tastings.

Emphasis: Ethnic and regional cuisines, seasonal dishes, nutritional foods, wine seminars.

Faculty: Marietta Pagani, Vickie Nielsen, Tom Peters, Garren Brannon, Donald McMillan. Guest teachers include Hugh Carpenter, Giuliano Hazan, Michelle Braden, Tom Young.

Costs: $15-$75/class, Lunch & Learn and children's classes $15, celebrity chef classes $45-$50.

Location: Winston-Salem's Reynolda Village, the former Richard Joshua Reynolds estate, ~20

miles from Piedmont Triad Int'l. Airport, adjacent to Wake Forest University.
Contact: Cindy Wilson, Pres., GBB, Inc., The Stocked Pot & Co., 111-B Reynolda Village, Winston-Salem, NC 27106 US; 910-722-3663, Fax 910-725-5034.

OHIO

BUEHLER'S FOOD MARKETS

Five locations/Year-round

This food market offers demonstration and participation classes at its 5 locations. Established in 1983. Maximum class/group size: 30 demo/12 hands-on. Facilities: Teaching areas with theater-type seating and overhead mirrors. Also featured: Child, teen, and parent-child classes.
Emphasis: Seasonal menus, nutrition, children's birthday parties.
Faculty: Staff home economists and guest instructors.
Costs: $6-$20/session.
Location: Delaware, Dover, Medina, Wadsworth, and Wooster, Ohio.
Contact: Mary McMillen, Director of Consumer Affairs, Buehler's Food Markets, P.O. Box 196, 1401 Old Mansfield Rd., Wooster, OH 44691 US; 330-264-4355 #256, Fax 330-264-0874, E-mail mmcmillen@buehlers.com.

THE CLEVELAND RESTAURANT COOKING SCHOOL

(See also page 99) **Cleveland/Year-round**

This restaurant offers classes for hobbyists and 1-week seminars. 10-week program, which includes 8 weeks hands-on, 2 weeks externship. Established in 1986. Maximum class/group size: 25 demo, 6 hands-on. Facilities: Include a teaching kitchen, demonstration area, a restaurant kitchen.
Emphasis: Seasonal ingredients from local farms.
Faculty: Chef Parker Bosley, owner of Parker's Restaurant and Catering, and the restaurant staff.
Location: 5 minutes from downtown Cleveland.
Contact: The Cleveland Restaurant Cooking School, 2801 Bridge Ave., Cleveland, OH 44113 US; 216-771-7130, Fax 216-771-8130.

COLUMBUS STATE CULINARY ACADEMY

(See also page 99) (See display ad page 244) **Columbus/Year-round**

This division of Columbus State Community College offers demonstration and participation classes. Established in 1995. Maximum class/group size: 18-30. 160 programs per year. Also featured: Wine instruction, classes for youngsters. Facilities: Two professionally-equipped kitchens.
Emphasis: International cuisines, theme classes, guest chefs, vegetarian and healthy cooking.
Costs: $45 for demonstrations, $65 for hands-on classes.
Contact: Carolyn Claycomb, Program Coordinator, Columbus State Culinary Academy, 550 E. Spring St., Box 1609, Columbus, OH 43216-1609 US; 614-227-3655/2665, Fax 614-227-5973, E-mail CClaycom@cscc.edu.

COOKS'WARES CULINARY CLASSES

Cincinnati/September-July

This kitchenware store offers demonstration and participation classes. Established in 1992. Maximum class/group size: 24 demo/12 hands-on. 100 programs per year. Facilities: 300-square-foot teaching kitchen with overhead mirror. Also featured: Wine tasting, private and children's classes.
Emphasis: Basics, ethnic and regional cuisines, specific subjects.
Faculty: Includes Jude Theriot, Marilyn Harris, and Chefs Paul Teal, Juan Robert de Cavel, David Cooke, George Geary, Anita Hirsch, Scott Anderson, John Kinsella, Suneeta Vaswani.
Costs: Range from $25-$50, $15 for youngsters.
Location: Approximately 20 miles from central Cincinnati.
Contact: Amy Tobin and Nancy Pigg, Directors, Cooks'Wares Culinary Classes, 11344 Montgomery Rd., Cincinnati, OH 45249 US; 513-489-6400, Fax 513-489-1211.

DOROTHY LANE MARKET SCHOOL OF COOKING

Dayton/September-June

This supermarket offers demonstration and participation classes. Established in 1984. Maximum class/group size: 20 demo/10 hands-on. 60-80 programs per year. Facilities: 400-square-foot teaching kitchen with overhead mirror and new appliances. Also featured: Children's classes, wine instruction, market visits, private classes.

Emphasis: Includes basic techniques, ethnic and regional cuisines, entertaining menus, guest chef specialties, specific subjects.

Faculty: Includes professional chefs, caterers, cookbook authors, home economists, and dietitians. Guest chefs have included Giuliano Bugialli, Hugh Carpenter, Perla Meyers, Susan Purdy, George Geary, Betty Rosbottom, Claire Criscuolo, Martin Yan.

Costs: Adult classes range from $35 to $48 for celebrity chefs. Children's classes are $25.

Contact: Deb Lackey, School of Cooking Director, Dorothy Lane Market School of Cooking, 2710 Far Hills Ave., Dayton, OH 45419 US; 937-299-3561, Fax 937-299-3568, E-mail cooking@dorothylane.com, URL http://www.dorothylane.com.

GOURMET CURIOSITIES, ETC.

Sylvania/March-June, September-November

The Creative Cooking School of Gourmet Curiosities offers demonstration and participation courses. Established in 1974. Maximum class/group size: 36 demo/16 hands-on. 75 programs per year. Facilities: 1,000-square-foot kitchen with overhead mirror. Also featured: Children's classes, wine tastings, private classes, market visits, demonstrations at off-site functions.

Emphasis: Latest cooking techniques by certified chefs.

Faculty: Geneva and Bruce Williams attended La Varenne, are members of the ACF, have served

on consumer panels, and had recipes accepted for publication. Other instructors include area chefs.
Costs: $20-$30/session.
Location: 11 miles from Toledo and 60 miles from Detroit.
Contact: James F. Wilson, Gourmet Curiosities, Etc., 5700 Monroe St., Sylvania, OH 43560 US; 419-882-2323/800-465-8843, Fax 419-885-3011, E-mail francewil@aol.com, URL http://www.gourmetcuriosities.com.

HANDKE'S CUISINE COOKING CLASS

Columbus/September-June

This restaurant offers demonstrations and private classes year-round for groups of 20 or more. Established in 1991. Maximum class/group size: 32.
Emphasis: American and European cuisine.
Faculty: Hartmut Handke, CMC.
Costs: $39.
Contact: Margot Handke, Handke's Cuisine Cooking Class, 520 S. Front St., Columbus, OH 43215 US; 614-621-2500, Fax 614-621-2626.

THE LORETTA PAGANINI SCHOOL OF COOKING

(See also page 100) **Chesterland/Year-round**

This private school affiliated with Lakeland Community College offers 1- to 4-session demonstration and participation courses. Established in 1981. Maximum class/group size: 28 demo/12 hands-on. 400+ programs per year. Facilities: Large professional kitchen with overhead mirror, professional equipment. Also featured: Couples classes, a young gourmet series, 7- to 10-day gastronomic tours and cruises to Italy, local trips to food-related sites.
Emphasis: Professional techniques.
Faculty: Owner/director Loretta Paganini, born and schooled in Italy, culinary consultant to area restaurants, food writer, cookbook author and guest chef on local TV. Guest faculty includes local chefs, teachers, and visiting professionals from many countries.
Costs: $25-$45/session. Tour and cruise prices range from $2,450-$4,850, including airfare, lodging, meals, planned excursions.
Location: 25 miles east of Cleveland.
Contact: Loretta Paganini, Owner/Director, The Loretta Paganini School of Cooking, 8613 Mayfield Rd., Chesterland, OH 44026 US; 440-729-1110, Fax 440-729-6459, E-mail lpscinc@msn.com.

THE PROVINCIAL KITCHEN

Huron/April-September

Certified Culinary Professional Lynn Nelson offers demonstration classes. Established in 1996. Maximum class/group size: 8. 60 programs per year. Facilities: 400-square-foot home teaching kitchen with demonstration mirror, wood-fired outdoor oven, culinary library.
Emphasis: Country French and Italian menus using seasonal ingredients, featuring sauces, pasta, cheese, savories, sweets, country breads, pizzas, flatbreads.
Faculty: Food editor and stylist Lynn Nelson, CCP, has a certificate in cuisine and pastry from Le Cordon Bleu-Paris, has taught cooking since 1979, and develops recipes and menus for promotional literature. Guest chefs and teachers are also featured.
Costs: $45/class, 3 classes for $120. Guest instructors $50-$100.
Location: On the south shore of Lake Erie in the Isle St. George wine growing region, 1 hour from Cleveland and Toledo, 2 hours from Columbus and Detroit.
Contact: Lynn Nelson, Proprietor, The Provincial Kitchen, PO Box 330, Huron, OH 44839 US; 419-433-5668, Fax 419-433-5668, E-mail ProKitchen@aol.com, URL http://www.provincialkitchen.com.

SAVEUR INTERNATIONAL COOKING SCHOOLS (See page 212)

UNIVERSITY HOSPITALS SYNERGY CULINARY SCHOOL

Mayfield Heights/Year-round

University Hospitals of Cleveland offers demonstration classes. Established in 1993. Maximum class/group size: 10-25. 140 programs per year. Facilities: Newly-equipped facility, video monitors

Emphasis: Healthful, high flavor, low fat cuisine.

Faculty: Culinary Director is Michelle M. Gavin, CCP. Registered dietitians are present at each class.

Costs: $30/class or 2 for $50.

Location: 20 minutes east of downtown Cleveland.

Contact: Michelle Gavin, Director, University Hospitals Synergy Culinary School, 5850 Landerbrook Dr., #110, Mayfield Heights, OH 44124 US; 216-646-2300, Fax 216-646-2322.

WESTERN RESERVE SCHOOL OF COOKING

Hudson/Year-round

This cookware store (formerly Zona Spray Cooking School) offers multi-session demonstration and hands-on courses on a variety of topics, hands-on professional techniques courses. Established in 1972. Maximum class/group size: 38 demo, 10-25 hands-on. Facilities: 500-square-foot kitchen with 3 ovens and 3 work stations. Also featured: Classes for youngsters, private instruction, market visits, sessions on catering and food writing.

Emphasis: Techniques, basics, pastry, low-fat, Japanese, herbs and spices, cake decorating, breads.

Faculty: Professional chefs, caterers, and cookbook authors.

Costs: Demos $28 ($45 guest chef), professional courses $495, gastronomic tours $850-$3,000.

Location: 20 miles south of Cleveland, 15 miles east of Akron.

Contact: Carole M. Ferguson, Owner, Western Reserve School of Cooking, 140 N. Main St., Hudson, OH 44236 US; 330-650-1665, Fax 330-650-6920.

OKLAHOMA

COOKING SCHOOL OF TULSA

Tulsa/Year-round

This cookware shop offers demonstration and participation classes. Established in 1991. Maximum class/group size: 20 demo/8 hands-on. 130+ programs per year. Facilities: Fully-equipped teaching kitchen with overhead mirror. Also featured: Wine tastings.

Emphasis: Simple techniques, menus for entertaining, ethnic cuisines, guest chef specialties.

Faculty: Proprietor Keith Lindenberg, local chefs, visiting guest chefs, nutritionists.

Costs: $35-$45/class.

Contact: Keith Lindenberg, President, Cooking School of Tulsa, 8264 S. Lewis, Tulsa, OK 74137 US; 918-298-7110, Fax 918-298-7117.

GOURMET GADGETRE, LTD.

Lawton/October-June

This cookware store offers cooking demonstrations. Established in 1980. Maximum class/group size: 20.

Emphasis: Breads, candies, ethnic cuisines.

Faculty: June Harris.

Costs: $15/class.

Contact: June Harris, President, Gourmet Gadgetre, Ltd., 1105 Ferris, Lawton, OK 73507 US; 580-248-1837.

OREGON

CAPRIAL'S BISTRO & WINE

Portland/Year-round

This restaurant offers cooking demonstrations. Established in 1992. Maximum class/group size: 24. 40 programs per year.

Faculty: Caprial and John Pence.
Costs: $45/class.
Location: 10 minutes southeast of downtown.
Contact: Caprial or John Pence, Owners, Caprial's Bistro & Wine, 7015 S.E. Milwaukie, Portland, OR 97202 US; 503-236-6457, Fax 503-233-4382, E-mail cbistro@mci2000.com.

CARL'S CUISINE

Salem/Year-round

Culinary professional Carl Meisel offers demonstration classes. Established in 1978. Maximum class/group size: 12.
Emphasis: Ethnic and regional cuisines, seasonal menus, specific subjects.
Faculty: Proprietor Carl Meisel has traveled and studied in Europe, Thailand, and regions of the U.S. He is a consultant on menu planning, travel, and kitchen design.
Costs: $24/class.
Location: Downtown Salem.
Contact: Carl's Cuisine, 333 Chemeketa St. NE, Salem, OR 97301 US; 503-363-1612, Fax 503-363-5014, E-mail cookin@teleport.com, URL http://www.carlscuisine.com.

COOKS, POTS & TABLETOPS

Eugene/Year-round

This cookware store offers demonstration and hands-on classes in a variety of cuisines, including ethnic and American regional foods, techniques, winemaking, food and wine pairing. Established in 1979. Maximum class/group size: 12. 130-150 programs per year. Facilities: Semi-circular cooking area that seats 12, professional equipment. Also featured: Off-premise classes at scenic locales, camp for cooks.
Emphasis: Techniques, chef specialties, food and wine pairing.
Faculty: Local chefs and winemakers.
Costs: $20-$50/class.
Location: 110 miles south of Portland in central Oregon.
Contact: Kathy Campbell-Ellis, Co-owner, Cooks, Pots & Tabletops, 2807 Oak St., Eugene, OR 97405 US; 541-338-4339, Fax 541-338-4339.

ROBERT REYNOLDS, NORTHWEST FORUM

(See also page 104) **Portland, OR; Seattle, WA; Niort, France/Year-round**

This private school offers a 1-week master class for those seeking advanced culinary study. Established in 1997. Maximum class/group size: 6. Facilities: Restaurant kitchen in Portland, winery kitchen in the countryside, cooking school classroom in Seattle, restaurant kitchen in France. Also featured: Visits to producers, growers, wineries, those involved in the arts of the table, some restaurant dining.
Emphasis: Theory, methods, practice, techniques, physics, chemistry, history, culture: advanced study leading to cook's independence.
Faculty: Robert Reynolds operated Le Trou Restaurant in San Francisco for 15 yrs; operated aux Gastronomes cooking school in France; created and directed professional program in Colorado; co-directs Provence 3D cooking program in France.
Costs: $1,250 in Portland and Seattle includes meals and local transport. $2,450 in Niort, France, includes meals, 1st class hotel lodging and local transport.
Location: Portland, OR (spring and summer), Seattle, WA (summer), Niort, France (fall).
Contact: Robert Reynolds, Northwest Forum for Advanced Culinary Study, 222 SE 18th Ave., Portland, OR 97214 US; 888-733-3391, Fax 503-233-1934, E-mail rowbear@ibm.net, URL http://www.RobertReynoldsCooks.com.

PENNSYLVANIA

CHARLOTTE-ANN ALBERTSON'S COOKING SCHOOL

Philadelphia/Year-round

Culinary professional Charlotte-Ann Albertson offers 1- to 4-session demonstration and some participation courses. Established in 1973. Maximum class/group size: 25 demo/15 hands-on/40 travel. 75 programs per year. Also featured: Market tours, children's classes, wine seminars and dinners, European culinary vacations, winter classes in private homes in Florida.

Emphasis: Ethnic cuisines, holiday menus, grilling, wine, kitchen design, food science and marketing.

Faculty: Charlotte-Ann Albertson, IACP member and certified teacher, studied at La Varenne and Le Cordon Bleu. Others include CIA-trained Philadelphia chefs, caterers and experts such as Shirley Corriher and Barbara Tropp.

Costs: Classes: $35-$45/session. Trips: $75-$3,200 includes bus transport or airfare from Philadelphia, hotel lodging, most meals, planned excursions.

Location: Madsen Design Center, 8 miles from downtown Philadelphia, suburban commercial kitchens.

Contact: Charlotte-Ann Albertson, Owner/Director, Charlotte-Ann Albertson's Cooking School, PO Box 27, Wynnewood, PA 19096-0027 US; 610-649-9290, Fax 610-649-9290, E-mail caacooking@aol.com, URL http://www.usnetway.com/cookingschool.

CLASS COOKING

Bryn Mawr/October-May

Culinary professional Susan Winokur offers hands-on and demonstration classes that can be taken individually or as a series. Established in 1986. Maximum class/group size: 8 hands-on, 12 demo. 30 programs per year. Facilities: 14-foot counter, standard cooking appliances, overhead mirror. Also featured: Private classes, classes for special groups.

Emphasis: French, Italian, and American cuisines, menu planning, basics and more advanced cooking techniques.

Faculty: IACP-member Susan Winokur earned a BS degree from Cornell University, MS in Education from the University of Pennsylvania. Guest chefs.

Costs: $50-$65/class. Profits go to charity.

Location: 20 minutes from Philadelphia.

Contact: Susan Winokur, Owner/teacher, Class Cooking, PO Box 751, Bryn Mawr, PA 19010 US; 610-527-1338, Fax 610-527-6069.

COOKING WITH CLASS

Shohola/March-August

Culinary professional Sheelah Kaye-Stepkin offers hands-on classes and series. Established in 1993. 100+ programs per year. Facilities: Commercial kitchen studio and cookbook library. Also featured: Private instruction available.

Emphasis: Basics, ethnic and regional cuisines, special occasion dishes, cake decorating, sugar artistry and design, guest chef specialties, specific subjects.

Faculty: Founder Sheelah Kaye-Stepkin and guest chefs from New York City and surrounding area.

Costs: $50-$150/session.

Location: On Twin Lakes, 90 minutes from Manhattan and 5 minutes from Milford.

Contact: Sheelah Kaye-Stepkin, Cooking with Class, 301 Main St., Hawley, PA 18428 US; 800-226-6540/717-226-8200, Fax 717-251-7922/717-226-8201, E-mail sheelah@ptd.net.

THE COOKING COTTAGE AT CEDAR SPRING FARM

Sellersville/September-June

Culinary professional Winnie McClennen offers demonstration classes and specialized series. Established in 1992. Maximum class/group size: 12. 90 programs per year. Facilities:

Demonstration kitchen with overhead mirror. Also featured: Private group classes and market trips, trips to France, Italy, Ireland, and Austria.
Emphasis: Various types of cooking.
Faculty: Winnie McClennen and her daughter, Peggi Clauhs, guest chefs.
Costs: $35-$45.
Location: Rural Upper Bucks County, between Allentown and Philadelphia.
Contact: Peggi Clauhs, Co-owner, The Cooking Cottage at Cedar Spring Farm, 1731 B Old Bethlehem Pike, Sellersville, PA 18960 US; 215-453-1828, Fax 215-257-6177, E-mail TheCCottag@aol.com.

COOKING AT TURTLE POND

Quakertown/Year-round

Corporate chef Una Maderson offers single and 3-session demonstration and hands-on classes, weekend vacation courses, wine appreciation. Maximum class/group size: 4-6. Facilities: Country home kitchen with cookbook library at Turtle Pond, a 23-acre wildlife retreat with a 2-acre pond.
Emphasis: Professional techniques, international, vegetarian, entertaining. Focus on flavor and healthy eating.
Faculty: Una Maderson is a professional chef with a degree in hotel and restaurant management. She's also a registered nurse.
Costs: $40-$55/session, includes full-course meal. Weekend courses by arrangement. Payment with reservation.
Location: Upper Bucks County, 90 miles from New York City, 30 miles north of Philadelphia.
Contact: Una Maderson, Turtle Pond, 210 Axehandle Rd., Quakertown, PA 18951-4904 US; 215-538-2564, E-mail unaturtlepnd@earthlink.net.

THE COOK'S CORNER, INC.

Yardley/Year-round

This school in a cookware store offers 1 demonstration class/week. Established in 1987. Maximum class/group size: 35. Facilities: Full demonstration kitchen with overhead mirror.
Emphasis: Ethnic and regional cuisines, desserts, pastas, breads, salads, guest chef specialties, wine appreciation.
Faculty: Owner Catherine Rowan and guest cookbook authors, professional instructors, and restaurant chefs.
Costs: $40-$45.
Location: Yardley is 30 miles north of Philadelphia and 1 mile from Trenton, N.J.
Contact: The Cook's Corner, Inc., 90 W. Afton Ave., Yardley, PA 19067 US; 215-493-9093.

CRATE

Pittsburgh/March-December

This retail kitchenware store offers day and evening demonstration and participation courses. Established in 1978. Maximum class/group size: 40 demo/12 hands-on. 100-120+ programs per year. Facilities: New demonstration kitchen with 8 burners, regular, convection, and microwave ovens, overhead mirror, new professional kitchen available late 1997. Also featured: January and February food-related seminars.
Emphasis: Italian, Chinese, Mediterranean, and French cuisines, vegetarian, bread, herbs, biscotti, filo, cookies, cakes, guest specialties.
Faculty: Includes chefs and owners of top local restaurants, owners of culinary businesses, and professional caterers. Guests have included Mary Beth Clark, Joanne Weir, Alice Medrich, Perla Meyers, Marlene Sorosky, and Giuliano Bugialli.
Costs: Average class cost is $30-$40; guests are higher.
Location: In Pittsburgh's South Hills-Scott Township.
Contact: Linda Wernikoff, Owner, Crate, Greentree Road Shopping Ctr, Pittsburgh, PA 15220 US; 412-341-5700, Fax 412-341-6321.

JANE CITRON COOKING CLASSES

Pittsburgh/October-January, April-May

Cooking teacher Jane Citron offers demonstration classes; culinary trips to France, Italy, Napa Valley. Established in 1978. Maximum class/group size: 12. 20 classes, annual trip programs per year. Facilities: Well-equipped home kitchen. Also featured: Private classes, market visits.

Emphasis: French, Italian and contemporary.

Faculty: Jane Citron studied with Marcella Hazan, Madeleine Kamman, Jacques Pépin, and Roger Vergé. She is food editor and writes a column for Pittsburgh Magazine.

Costs: $55/class.

Location: The Murdoch Farms section of Pittsburgh.

Contact: Jane Citron, Jane Citron Cooking Classes, 1314 Squirrel Hill Ave., Pittsburgh, PA 15217 US; 412-621-0311, Fax 412-765-2511, E-mail janecooks@aol.com.

KATHY D'ADDARIO'S COOKING TECHNIQUES

Ambler/January-June, September-November

Culinary professional Kathy D'Addario offers 10-session participation courses. Established in 1994. Maximum class/group size: 6-8. 50 programs per year. Facilities: Large home kitchen. Also featured: Wine instruction, market visits, private classes, small group lessons.

Emphasis: Traditional techniques, low-fat meals, seasonal menus, holiday entertaining.

Faculty: Kathy D'Addario is a graduate of The Restaurant School and studied at Le Cordon Bleu, Giuliano Bugialli's in Florence, and Peter Kump's New York Cooking School.

Costs: $35-$50/session.

Location: A 30-minute drive from Philadelphia.

Contact: Kathy D'Addario's Cooking Techniques, 858 Tennis Ave., Ambler, PA 19002 US; 215-643-5883, Fax 215-257-6681, E-mail kathcook@aol.com.

THE KITCHEN SHOPPE AND COOKING SCHOOL

Carlisle/September-May

This kitchenware, gift and gourmet store offers demonstration and hands-on classes. Established in 1974. Maximum class/group size: 36 demo/15 hands-on. 150 programs per year. Facilities: Well-equipped professional kitchen with 6 work stations, overhead mirrors, audio system. Also featured: Classes for youngsters and private groups.

Emphasis: Afternoon tea, breads and pastries, grilling and smoking, ethnic and vegetarian cuisines, guest chef specialties.

Faculty: Proprietor Suzanne Hoffman, IACP, instructors Sherry Ball, Helen Davenport, Jim Lupia, Diana Povis. Guest chefs include Hugh Carpenter, George Geary, Joanne Weir.

Costs: $30-$50/session; children's classes $18-$25.

Location: 20 minutes west of Harrisburg, 2-1/2 hours west of Philadelphia and northwest of Washington DC and Baltimore.

Contact: Suzanne Hoffman, The Kitchen Shoppe and Cooking School, 101 Shady Lane, Carlisle, PA 17013 US; 800-391-COOK/717-243-0906, Fax 717-245-0606, E-mail kshoppe@pa.net, URL http://www.kitchenshoppe.com.

THE LEARNING STUDIO

Malvern/Year-round

This adult education provider offers 1- to 3-session hands-on courses on a variety of topics. Established in 1998. Maximum class/group size: 16. Facilities: Gas-equipped cooking classroom. Also featured: Private instruction, market visits.

Emphasis: To enable the student to understand the basics of good and creative cooking and be able to begin the process on their own.

Faculty: Area restaurant chefs, most graduates of The CIA.

Costs: $59-$99 plus $5-$15 materials fee/class.

Location: 30 miles from Center City Philadelphia.

Contact: Miriam Hughes, Vice-President, The Learning Studio, 412 East King St., Malvern, PA 19355 US; 610-578-0600, Fax 610-578-0680, E-mail LSMalvern@AOL.com.

RANIA'S COOKING SCHOOL

Pittsburgh/March-May, September-November

This restaurant offers over 20 demonstrations each season. Established in 1984. Maximum class/group size: 24. Also featured: Children's classes, wine instruction, private classes.

Emphasis: Ethnic and regional cuisines, holiday foods, appetizers to desserts.

Faculty: Proprietor Rania Harris and chefs Michael Barbato (Westin Wm. Penn), Joe Nolan (Cafe Allegro), Bill Fuller (Casbah), Stuart Marks (Rania's Catering); Sharryn Campbell, wine tasting.

Costs: $35/class, children's class is $20.

Location: 7 minutes from downtown Pittsburgh.

Contact: Rania's Cooking School, 100 Central Sq., Pittsburgh, PA 15228 US; 412-531-2222, Fax 412-531-7242, E-mail ranias@usaor.net, URL http://www.wtaetv.com.

THE RESTAURANT SCHOOL

(See also page 114) (See display ad page 115) **Philadelphia/October-June**

This proprietary career school offers hands-on and demonstration classes in a variety of culinary arts disciplines. Established in 1974. Maximum class/group size: 24-85. 200+ programs per year. Facilities: Include 4 classroom kitchens, two 75-seat demonstration kitchens, pastry shop, and restaurant. Also featured: Culinary weekends.

Emphasis: Holiday menus, wines, guest chef and cookbook author series.

Faculty: The 12-member professional faculty, nationally known guest chefs, guest cookbook authors.

Costs: $40-$60/class.

Location: Restored mansion in University City.

Contact: Michele Gambino, Community Education Administrator, The Restaurant School, 4207 Walnut St., Philadelphia, PA 19104 US; 215-222-4200 #3046, Fax 215-222-4219, URL http://www.therestaurantschool.com.

SCHOOL OF CULINARY ARTS

Sewickley/Year-round

This private school and culinary professional Gaynor Grant offer hands-on courses that include Techniques of Fine Cooking, Pastry and Baking, Italian, international and other topics. Established in 1995. Maximum class/group size: 8. 30 programs per year. Facilities: Professional kitchen. Also featured: Private instuction.

Emphasis: French cuisine.

Faculty: Gaynor Grant is a Certified Culinary Professional and member of the IACP.

Location: Near Pittsburgh.

Contact: Gaynor Grant, Director, School of Culinary Arts, 515 Broad St., Sewickley, PA 15143 US; 412-741-8671, Fax 412-741-8671.

WINNER INSTITUTE OF ARTS & SCIENCES

Transfer/Year-round

This private school offers 4-session evening hands-on courses on a variety of topics. Established in 1997. Maximum class/group size: 15. Facilities: High-tech kitchen, banquet room, 4 classrooms, fully-equipped computer lab and library.

Faculty: 3 full-time ACF-certified chef instructors, 1 part-time chef instructor, 1 part-time academic instructor.

Costs: $149 per course.

Location: 90 minutes from Pittsburgh, Cleveland, and Erie.

Contact: John Matsis, Executive Director, Winner Institute of Arts & Sciences, One Winner Place, Transfer, PA 16154 US; 888-414-CHEF/724-646-CHEF, Fax 724-646-0218, E-mail wias@infonline.net, URL http://www.by1.com/pa/winnerchefs.

RHODE ISLAND

ACCADEMIA ITALIANA DI CUCINA

Providence/January-March

Walter Potenza offers 5-day programs that include Italian regional cooking, Italian Jewish cooking, Neopolitan pastries, terracotta cookery, children's etiquette and cooking programs. Established in 1988. Maximum class/group size: 12 hands-on, 20 demo, 15 travel. 10 programs per year. Facilities: Renovated palazzo with restaurant facilities and professional kitchen with modern equipment. Also featured: Wine and food tours of Italy.

Emphasis: Renaissance culture through cooking, Italian food resources, food in history and geography, restaurant operation, food production.

Faculty: Chefs Walter Potenza, Luca Regoli, Stefano di Sano, Giancarlo Rosedorne, Giovanni Cozza.

Costs: $1,560, includes lodging in 5-star European-style inn, breakfasts and dinners. 50% due with reservation, cancellation 1 week prior for full refund.

Location: Central Providence, a 5-minute walk from downtown, 50 minutes from Boston, 3 hours from NY.

Contact: Walter Potenza, Director, Accademia Italiana di Cucina, 265 Atwells Ave., Providence, RI 02903 US; 401-273-2652, Fax 401-273-6879, E-mail feedback@chefwalter.com, URL http://www.chefwalter.com.

RHODE ISLAND SCHOOL OF DESIGN

Providence/Year-round

This university offers 1- to 3-session classes and travel programs. Classes include celebrity chef and cooking series, excursions to nearby culinary locales, international programs that combine art and cooking and include market visits, winery tours, sightseeing. Established in 1877. Maximum class/group size: 38 demo, 15 hands-on, 20 travel. 75 programs per year. Facilities: Residential and commercial fine appliances. Also featured: Classes that relate food to culture, such as Food & the Movies and Art & Food.

Emphasis: Small classes taught by local, national, and international chefs.

Faculty: Includes Madeleine Kamman, Barbara Kafka, Emily Luchetti, Chris Schlesinger, Jasper White, Lynn Rosetto Kasper, Johanne Killeen and George Germon, Barbara Lynch, Michael Schlow, Wayne Gibson, Steve Johnson, Bruce Tillinghast, Jules Ramos.

Costs: $55-$65/class, ~$2,500 for trips.

Contact: Janet S. Egan, Culinary Arts Coordinator, Rhode Island School of Design, 2 College St., Providence, RI 02903 US; 401-454-6205, Fax 401-454-6363.

SAKONNET MASTER CHEFS SERIES

Little Compton/September-June

Sakonnet Vineyards offers full-day demonstration and participation classes. Established in 1980. Maximum class/group size: 12. 10 programs per year. Facilities: Large main work table and counter, which serves as individual work areas.

Emphasis: Guest chef specialties; wine selection and food pairing.

Faculty: Has included Johanne Killeen and George Germon (Al Forno); Maureen Pothier (Bluepoint Oyster Bar); Jasper White (Jasper's); Todd English (Olives); Michael Schlow (Cafe Louis); Wayne Gibson (Castle Hill Inn); Casey Riley (Agora).

Costs: $80-$100. Accommodations can be arranged.

Location: 30 minutes from New Bedford and Fall River, Mass., 40 minutes from Providence, 75 minutes from Boston.

Contact: Sakonnet Vineyards, P.O. Box 197, Little Compton, RI 02837 US; 401-635-8486, Fax 401-635-2101, E-mail SakonnetRI@aol.com.

SOUTH CAROLINA

BOBBI COOKS II

Hilton Head Island/September-June

Culinary professional Bobbi Leavitt offers demonstration and participation classes. Established in 1993. Maximum class/group size: 10 hands-on/15 demo. 35 programs per year. Facilities: Large fully-equipped home kitchen. Also featured: Classes for youngsters, wine and food pairing, couples' classes, private instruction.

Emphasis: Smoking and grilling, ethnic, entertaining, heart healthy, techniques.

Faculty: Bobbi Leavitt studied at Johnson & Wales; Master Chefs Institute; Michael James French Chefs School. Past president NYACT. IACP, AIWF, and James Beard Foundation member.

Costs: $28-$40/class. Resort facilities, tennis, golf, ocean are nearby.

Location: 35 miles north of Savannah airport, 10 minutes from Hilton Head airport.

Contact: Bobbi Leavitt, Owner, Bobbi Cooks II, 9 Baynard Pk., Hilton Head Island, SC 29928 US; 843-671-5902, Fax 843-671-5902, E-mail BobbiCooksatHHI@juno.com. Additional contact: Lindy Russell, Registrar, 843-842-9855, Fax 843-689-5440.

HOPPIN' JOHN'S, INC.

Charleston/Year-round

Cookbook store owner/author John Martin Taylor offers twice weekly classes. Established in 1997. Maximum class/group size: 20.

Emphasis: Southern cuisine.

Faculty: John Martin Taylor is author of four books. Other instructors include cookbook authors and noted chefs.

Costs: $30/session.

Location: In Historic Charleston, 2 blocks from the City Market, adjoining Hoppin' John's cookbook store.

Contact: John Martin Taylor, Hoppin' John's, Inc., 30 Pinckney St., Charleston, SC 29401 US; 800-828-4412/843-577-9429, Fax 843-577-6932, E-mail hoppinjon@aol.com.

IN GOOD TASTE

Charleston/January-November

This gourmet shop offers demonstration and participation classes. Individual theme classes, wine classes, generally 4 in a series. Established in 1983. Maximum class/group size: 8-14. Facilities: Well-equipped teaching kitchen. Also featured: Bed & breakfast tours.

Emphasis: Ethnic and regional cuisines, techniques, breads, wine appreciation.

Faculty: School owner Jacki Boyd, Roland Gilg, and Bonnie Caracciolo; local chefs.

Costs: $30-$60/session, includes full dinner and wine.

Location: West of the Ashley on the way to Middleton Gardens, Drayton Hall Plantation and Magnolia Plantation. 10 minutes from downtown.

Contact: Jacki Boyd, In Good Taste, 1901 Ashley River Rd., Charleston, SC 29407 US; 803-763-5597, Fax 803-763-5597, E-mail rgaryboyd@compuserve.com.

TENNESSEE

CLASSIC GOURMET COOKING SCHOOL

Nashville/Year-round

This cookware and gourmet foods store offers demonstration and participation courses ranging from individual classes to a 9-week semi-professional course for the home chef. Established in 1991. Maximum class/group size: 30 demo/25 hands-on. 120 programs per year. Facilities: Teaching kitchen with overhead mirror. Also featured: Wine and food classes, private instruction, classes for youngsters.

Emphasis: Semi-professional for home cooks or entry-level career in the hospitality industry.
Faculty: Restaurant chefs, caterers, culinary instructors, cookbook authors, visiting guest chefs.
Costs: From $35 for a demonstration to $70 for a hands-on class.
Location: Green Hills, west of downtown Nashville, with Opryland, BelleMeade Mansion, Cheekwood, The Hermitage within a few miles drive.
Contact: Hilda Pope, Classic Gourmet Cooking School, 3900 Hillsboro Rd., Nashville, TN 37215; 615-383-8700, Fax 615-383-8788, URL http://www.citysearch.com/nas/classicgourmet.

RAJI RESTAURANT

Memphis/February, November

This restaurant offers full-day workshops. Maximum class/group size: 15. Facilities: Raji Restaurant kitchen.
Emphasis: Fusion (French Indian) cooking.
Faculty: Chef Raji Jallepalli.
Costs: $250.
Contact: Raji Jallepalli, Raji Restaurant, 712 W. Brookhaven Circle, Memphis, TN 38117 US; 901-685-8723, Fax 901-767-2226.

UT COMMUNITY PROGRAMS

Knoxville/Year-round

University of Tennessee's community programs offers weekly lectures, demonstration and hands-on classes. Established in 1970. Maximum class/group size: 20. Also featured: Wine instruction, market visits.
Emphasis: Multi-ethnic cuisines, vegetarian, low-fat cooking.
Faculty: Culinary and wine professionals.
Costs: Ranges from $25-$39/session.
Contact: Program Manager, UT Community Programs, 600 Henley St., Knoxville, TN 37996-4110 US; 423-974-0150, Fax 423-974-0264, E-mail utcommunity@gateway.ce.utk.edu, URL http://www.ce.utk.edu/communityprograms.

TEXAS

BBQ WORLD HEADQUARTERS

Austin/Year-round

This restaurant offers monthly 2-hour classes in Smoking 101 (beef, pork, chicken). All of the beef is Certified Angus. Established in 1997. Maximum class/group size: 12. Facilities: Full kitchen. Also featured: Classes for youngsters, private instruction, market visits.
Emphasis: Barbecued beef, pork, chicken.
Faculty: ACF and WACS member.
Costs: $20/class.
Location: North Austin.
Contact: Duke Bischoff, Owner, BBQ World Headquarters, 6701 Burnet Rd., Austin, TX 78757 US; 512-323-9112, Fax 512-323-9115, E-mail bbqw@web-brothers.com, URL http://www.web-brothers.com/bbq/.

BLAIR HOUSE COOKING SCHOOL

Wimberley/Year-round

This country inn offers 3-day hands-on cooking vacations that focus on a specific theme, such as seafood, vegetables, or international cuisines. Established in 1992. Maximum class/group size: 20.
Emphasis: Technique, chemistry, presentation.
Faculty: Innkeeper/Chef Jonnie Stansbury trained for 10 years under Ken Wolfe, the French and Viennese-trained Executive Chef of Ernie's in San Francisco. Blair House was named one of the Top 10 Places to Stay in Texas by The Dallas Morning News.

Costs: $425 includes lodging for two nights, meals, and wine.

Location: On 85 acres in Texas Hill Country, 45 minutes from San Antonio International Airport or Robert Mueller Airport in Austin.

Contact: Mary Schneider, Director/Public Relations, Blair House Cooking School, 100 Spoke Hill Ln., Wimberley, TX 78676 US; 512-847-1111, Fax 512-847-8820, E-mail info@blairhouseinn.com, URL http://www.blairhouseinn.com.

CENTRAL MARKET COOKING SCHOOL

Austin, San Antonio/Year-round

This food market offers classes on a variety of topics including international cuisines and guest chef specialties, children's classes. Established in 1994. Maximum class/group size: 15 hands-on, 35 demo. 350+ programs per year. Facilities: New high-tech kitchen and classroom. Also featured: Health awareness programs, community educational out-reach, market visits.

Emphasis: Techniques and culinary expertise for the home cook.

Faculty: Well-known guest chefs and cookbook authors.

Costs: $25-$50/class.

Contact: Cathy Cochran-Lewis, Cooking School Manager, Central Market Cooking School, 4001 N. Lamar Blvd., Austin, TX 78756 US; 512-458-3068, Fax 512-206-1010, E-mail s06180@heb.com, URL http://www.centralmarket.com. In San Antonio: Kathy Gottsacker, Manager, Central Market Cooking School, 4821 Broadway, San Antonio, TX 78209; 210-368-8617, Fax 210-826-3253.

CREATING CULINARY OPPORTUNITIES

Houston/Year-round

Culinary professional Ann Iverson offers 1- and 2-day participation courses in a variety of cuisines. Established in 1993. Maximum class/group size: 12. Facilities: 340-square-foot private kitchen with 12 work areas.

Emphasis: Northern Italian and Mediterranean cuisines; other cuisines are also offered.

Faculty: Guest chefs and authors and Ann Iverson, who studied with Giuliano Bugialli, Mary Beth Clark, Marcella and Victor Hazan, and Lorenza di Medici.

Costs: Range from $125-$250/session.

Location: Galleria area.

Contact: Ann Iverson, Owner, Creating Culinary Opportunities, 2902 West Lane Dr., Unit E, Houston, TX 77027 US; 713-622-6936, Fax 713-622-2924, E-mail annci@aol.com.

CREATIVE CUISINE & CATERING

(See also page 255)

Austin/Year-round

This private culinary arts school offers beginners' programs, baking/pastry, specialty courses, party planning, holiday menus, Texas wines, regional/international cuisine. Established in 1998. Maximum class/group size: 10-25. Facilities: Well-equipped commercial kitchen. Also featured: Visits to markets and food producers, private instruction, culinary tours, summer classes for hospitality educators.

Faculty: Professional instructors, guest chefs.

Costs: $30-$85/class.

Location: Two-story building in Central Austin, shared with a private, for-profit catering company, Private Affairs Catering. Near Texas hill country.

Contact: Glenn Mack, Director of Education, Creative Cuisine & Catering, 2823 Hancock Dr., Austin, TX 78731 US; 512-451-5743, Fax 512-467-9120, E-mail chefs@texas.net, URL http://chefs.home.texas.net.

CUISINE CONCEPTS

Fort Worth/Year-round

Author, food stylist, and food and wine writer Renie Steves offers private wine and cooking instruction designed to the student's requests. May include entertaining and tablesetting. Established in 1979. Maximum class/group size: 1-2. Facilities: Kitchen in a private home. Also fea-

tured: Group classes for up to 26 persons, hands-on classes for up to 9.
Emphasis: A variety of topics, including stocks, sauces, low-fat cooking, menu planning, techniques.
Faculty: Owner Renie Steves, CCP, chair of the IACP Foundation studied with Madeleine Kamman, James Beard, Julia Child, Nick Malgieri, and the Hazans.
Costs: $75/hour for one student, $100 for two, plus a $7/hour assistant fee and minimal marketing expense. Each class is a 4-hour minimum.
Location: Ft. Worth's west side, near the Kimbell Museum, 35 minutes west of DFW Airport.
Contact: Renie Steves, Cuisine Concepts, 1406 Thomas Pl., Ft. Worth, TX 76107-2432 US; 817-732-4758, Fax 817-732-3247, E-mail RenieSteves@msn.com.

CULINARY INSTITUTE ALAIN & MARIE LENÔTRE

Houston/Year-round *(See also page 121) (See display ads pages 95 and 121)*

This private French export corporation offers Level II intermediate Chef de Parti, Level III advanced Sous Chef, Level IV Master Chef: A) classic and modern cooking; B) classic and modern baking (pastry and bread), chocolate and ice cream. Classes are 90% hands-on. Established in 1998. Maximum class/group size: 10-15 hands-on. Facilities: 4 classrooms, meeting room, student cafeteria in a 28,000-square-foot free-standing modern new building. Also featured: Saturday, evening, and summer sessions, programs for youngsters, private instruction, travel programs, French language instruction.
Emphasis: French cooking and baking, chocolate, ice cream, sorbet, wedding cake, international cuisines, a la carte cooking, catering cuisine, food and wine pairing, French language.
Faculty: Technical Director Partick LeNôtre, elected one of the best 30 chefs by the French culinary critic press.
Costs: $950/week.
Location: Central Houston, 30 minutes from NASA Space Center, 1 hour from Galveston Island on the Gulf of Mexico.
Contact: Alain LeNôtre, President, Culinary Institute Alain & Marie LeNôtre, 7070 Allensby, Houston, TX 77022-4322 US; 888-LeNotre/713-692-0077, Fax 713-692-7399, E-mail lenotre@wt.net, URL http://www.lenotre-alain-marie.com.

DESIGNER EVENTS COOKING SCHOOL

Bryan/Year-round

Culinary professional Merrill Bonarrigo offers monthly participation classes. Food and wine pairing seminars are provided by Messina Hof Wine Cellars. Established in 1992. Maximum class/group size: 20. 12 programs per year.
Emphasis: Menus prepared by Texas chefs.
Costs: $35-$75/session. Lodging is available at bed & breakfast at Vineyard.
Location: The Messina Hof Winery estate in the Brazos Valley, 100 miles from Houston and 90 miles east of Austin.
Contact: Merrill Bonarrigo, Owner, Designer Events Cooking School, 4545 Old Reliance Rd., Bryan, TX 77808 US; 409-778-9463, Fax 409-778-1729, URL http://www.messinahof.com.

DOLORES SNYDER GOURMET COOKERY SCHOOL

Irving/February-May, September-November

Culinary professional Dolores Snyder offers classes. Established in 1976. Maximum class/group size: 10 hands-on/20 demo. 16 programs per year. Facilities: Kitchen with 6 work stations. Also featured: Wine classes, private instruction, dining in fine restaurants.
Emphasis: Entertaining with English tea, French, Asian.
Faculty: Dolores Snyder, CCP.
Costs: $40-$50/class.
Location: Near Dallas.
Contact: Dolores Snyder, Owner, Dolores Snyder Gourmet Cookery School, P.O. Box 140071, Irving, TX 75014-0071 US; 972-717-4189, Fax 972-717-1063, E-mail RHS629@aol.com.

THE HEALTHY GOURMET

San Antonio/Year-round

This private school offers cooking for prevention and remediation of illness. Low fat, no dairy or sugar. Program directed towards developing simple home cooking skills for fast, fresh meals. Established in 1996. Maximum class/group size: 7. Facilities: Large kitchen that seats 10 comfortably, adoining herb garden.

Emphasis: Health supportive cooking. Whole foods, vegetarian, macrobiotic, Chinese dietary therapy.

Faculty: Founder Mary Martha McNeel has been teaching and practicing for 15+ yrs. She has sponsored workshops in San Antonio with Annmarie Colbin, Herman and Cornelia Aihara, and Ed Ware, a local herb horticulturalist.

Costs: $225 covers 12 hours of instruction over 4 days.

Location: Near Quarry shopping center on McCullough.

Contact: Mary Martha McNeel, Founder/owner/manager/instructor, The Healthy Gourmet, 7062 McCullough Ave., San Antonio, TX 78216 US; 210-826-4591, E-mail pleas@salsa.net.

HUDSON'S ON THE BEND COOKING SCHOOL

Austin/Year-round

This restaurant offers one demonstration/participation class the third Sunday of each month at the restaurant and the first Thursday of each month at the home of Chef Blank. Established in 1993. Maximum class/group size: 20-25. 12 programs per year. Facilities: Hudson's on the Bend restaurant kitchen, Chef Blank's home kitchen. Also featured: Sightseeing.

Emphasis: Wild game, seafood, smoking, sauces, chef specialties.

Faculty: Executive Chef Jay Moore, a CIA graduate, and Owner/Chef Jeff Blank, creators of Hudson's on the Bend Gourmet Sauces.

Costs: $90/session.

Location: West of the city in a restored rock ranch house.

Contact: Shanny Lott, Hudson's on the Bend, 4304 Hudson Bend Rd., Austin, TX 78734 US; 512-266-1369/800-996-7655, Fax 512-266-1399, URL http://austin.citysearch.com/E/2/austex/0003/09/81. Also: 4304 Hudson Bend Rd., Austin, TX 78734.

THE KITCHEN SHOP AT THE GREEN BEANERY

Beaumont/Year-round

This cafe and cookware store offers demonstration and limited participation classes. Established in 1992. Maximum class/group size: 30 demo/15 hands-on. 75 programs per year. Facilities: 20-seat demonstration kitchen area. Also featured: Culinary tours.

Emphasis: Basics, ethnic and regional cuisines, pastries, specific subjects.

Faculty: Glenn Watz, chef/owner of the Green Beanery Cafe for 19 years; local and visiting instructors.

Costs: $20-$35/class.

Location: Beaumont is 90 miles east of Houston.

Contact: Carolyn Wood, Owner, The Kitchen Shop at the Green Beanery, 2121 McFaddin Ave., Beaumont, TX 77701 US; 409-833-5913, Fax 409-832-9738.

LAKE AUSTIN SPA RESORT

Austin/June-September

Lake Austin Resort offers 3-, 4-, and 7-night vacation packages that include spa treatments and cooking classes. Established in 1997. Maximum class/group size: 65. Facilities: Demonstration kitchen overlooking the hill country. Also featured: Hiking, yoga, tennis, biking, sculling, kayaking, canoeing.

Emphasis: American regional, Mediterranean, Mexican, Thai.

Faculty: Terry Conlan, food expert and author of the Lean Star Cuisine cookbook, guest instructors from Cooking Light and Eating Well magazines.

Costs: 3-day packages start at $1,000, including meals, lodging, and spa treatments.

Location: 25 minutes from downtown Austin, on the shores of Lake Austin.

Contact: Lake Austin Spa Resort, 1705 S. Quinlan Park Rd., Austin, TX 78732 US; 512-372-7300, Fax 512-266-1572, E-mail info@lakeaustin.com, URL http://www.lakeaustin.com.

LE PANIER

Houston/Year-round

This private cooking school offers demonstration and participation classes. Established in 1980. Maximum class/group size: 45 demo/15 hands-on. 200 programs per year. Facilities: Well-equipped teaching area that offers theater seating, a large overhead mirror, and several cooking and work spaces. Also featured: Classes for youngsters, basic techniques series, catering courses.

Emphasis: Ethnic cuisines, breads, entertaining menus, cooking for health, main course dishes, pastries and desserts.

Faculty: Owner/Director LaVerl Daily teaches basic techniques. Most other classes are taught by guest chefs, teachers, and cookbook authors, including Giuliano Bugialli, Giuliano Hazan, Nicholas Malgieri, Hugh Carpenter, and Shirley Corriher.

Costs: Range from $35-$60 per session, children's classes are $20.

Location: Near intersection of Kirby and Holcombe Streets.

Contact: LaVerl Daily, Director, Le Panier, 7275 Brompton Rd., Houston, TX 77025 US; 713-664-9848, Fax 713-666-2037.

THE MANSION ON TURTLE CREEK

Dallas/Year-round

This Mobil 5-star, AAA 5-diamond hotel and restaurant offers a demonstration class, special dinner, or both each month. Established in 1994. Maximum class/group size: 10-150. 5 programs per year. Facilities: Function rooms: Pavilion, Promenade, Sheppard King and FDR Suites. Also featured: Chef for a Day, Cooking Class & Dining Etiquette for Children.

Emphasis: Specific cuisines.

Faculty: Chef Dean Fearing co-hosts the classes. Guest chefs have included Wolfgang Puck, Julia Child, Jacques Pépin, and Emeril Lagasse.

Costs: Classes range from $145-$225, including tax. Special room rates begin at $215/night, excluding tax.

Location: Turtle Creek residential area, 10 minutes from downtown Dallas.

Contact: Patty Sullivan, Director of Marketing, The Mansion on Turtle Creek, 2821 Turtle Creek Blvd., Dallas, TX 75219; 214-559-2100 x136, Fax 214-526-5345, URL http://www.rosewood-hotels.com.

THE NATURAL EPICUREAN ACADEMY OF CULINARY ARTS

Austin/Year-round *(See also page 124)*

This private school offers natural whole foods cooking classes and seminars with vegetarian/vegan emphasis. Established in 1994. Maximum class/group size: 20. 88 programs per year. Facilities: Professionally equipped home-style kitchen. Also featured: Classes for youngsters, private instruction, visits to markets and food producers, sightseeing.

Emphasis: Natural whole foods cooking, vegetarian/vegan emphasis, natural remedies.

Faculty: 3 full- and 3 part-time instructors who have completed a qualifying chef program.

Costs: $205/level, $195/seminar, $62/class.

Location: Central Austin.

Contact: Elizabeth Foster, CEO, The Natural Epicurean Academy of Culinary Arts, 902 Norwalk Ave., Austin, TX 78705 US; 512-476-2276, Fax 512-476-2298.

NATURAL FOODS COOKING SCHOOL

Houston/September-July

This private school offers demonstration classes. Established in 1989. Maximum class/group size: 30. 100 programs per year. Facilities: Houston locations. Also featured: Classes for youngsters, market tours, private classes.

Emphasis: Natural foods, vegetarian, vegan, macrobiotic cooking.

Faculty: Nutritional counselor Marian Bell has taught for more than 23 years.

Costs: $20-$30/class.

Contact: Natural Foods Cooking School, 4418 Woodvalley, Houston, TX 77096 US; 713-723-8868, E-mail mbell@accesscomm.net.

RICE EPICUREAN MARKETS COOKING SCHOOL

Houston/Year-round

This market offers demonstration and participation classes. Established in 1990. Maximum class/group size: 45 demo/16 hands-on. 300 programs per year. Facilities: 1,100-square-foot classroom with overhead mirror. Also featured: Private classes, children's classes and cooking camp, market tours.

Faculty: Local and out-of-town chefs and teachers, including Emeril Lagasse, Michael Chiarello, Anne Willan, Robert Del Grande.

Costs: $20-$75/session.

Location: On a major road in a residential area.

Contact: Peg Lee, Director, Rice Epicurean Markets Cooking School, 6425 San Felipe, Houston, TX 77057 US; 713-954-2152, Fax 713-789-9853.

STAR CANYON COOKING SCHOOL

Dallas/Year-round

This restaurant offers demonstration courses. Established in 1994. Maximum class/group size: 50. 40-50 programs per year. Facilities: Demonstration kitchen, classroom, closed-circuit TV monitors.

Emphasis: New Texas cuisine.

Faculty: Chef Stephan Pyles, a founder of Southwestern cuisine.

Costs: Demonstration and tasting is $60-$75.

Location: Minutes from Love Field, 5 minutes from downtown.

Contact: Carla Waites, Cooking Class Coordinator, Star Canyon, 3102 Oak Lawn Ave., No. 144, Dallas, TX 75219 US; 214-520-8111, Fax 214-520-2667, URL http://www.starcanyon.com.

TINA WASSERMAN'S COOKING & MORE...

Dallas/September-May

This private school offers hands-on classes on a variety of subjects with focus on techniques and understanding the process. Established in 1982. Maximum class/group size: 14 hands-on, 50 demo. 35-40 programs per year. Facilities: 500-square-foot home kitchen. Also featured: Programs for youngsters, private classes.

Emphasis: Science of cooking, baking, traditional Jewish, ethnic cuisines.

Faculty: Tina Wasserman has 29 years experience teaching cooking. She has a BS in Home Economics from Syracuse University and an MA in Merchandising and Foods from NYU.

Costs: $40/class.

Location: North Dallas area, a 15-minute drive from downtown.

Contact: Tina Wasserman, Owner/Instructor, Tina Wasserman's Cooking & More..., 7153 Lavendale, Dallas, TX 75230 US; 214-369-6269, Fax 214-369-4307, E-mail magicook@aol.com.

UTAH

FOOD FESTS (See page 224)

VERMONT

CULINARY MAGIC COOKING SEMINARS

Ludlow/June-October

The Mobil 4-Star Governor's Inn offers 3-day weekend hands-on cooking vacations. Established in 1992. Maximum class/group size: 16. 5 programs per year. Facilities: Vermont Country Inn. Also featured: Visit to an antique cooperative.

Emphasis: The Inn's healthy gourmet specialties, presentation.

Faculty: Deedy Marble, innkeeper and chef since 1982, studied with Madeleine Kamman, Lorenza de Medici, and Roger Vergé. She and her husband, Charlie, have received 21 national culinary awards and placed fifth in the World Chef Competition.
Costs: Cost is $370-$428 double, $540-$600 single occupancy, which includes Inn lodging, tax and gratuities, most meals, beverages, planned activities. The Inn has 8 guest rooms, one suite.
Location: The Okemo Valley, 132 miles from Hartford Airport, 135 miles from Boston, and 230 miles from New York.
Contact: Chef Deedy Marble, Culinary Magic Cooking Seminars, Governor's Inn, 86 Main St., Ludlow, VT 05149 US; 800-468-3766.

NEW ENGLAND CULINARY INSTITUTE

Essex/January-April, July, September-November *(See also page 126) (See display ad page 127)*

This private career institution offers culinary vacation weekends on various topics. Experience the Magic of Baking is held each October. Established in 1979. Maximum class/group size: 112. 8 programs per year. Facilities: The Inn at Essex, a 97-room country hotel. Also featured: Dining in fine restaurants, day trips, sightseeing.
Emphasis: Baking and other topics.
Faculty: Noted chefs from NECI.
Costs: Cost is $400, which includes double occupancy lodging at the AAA 4-Diamond Inn at Essex; single supplement $75.
Location: 15 minutes outside Burlington.
Contact: Amy Trubek, Director of Continuing Education, New England Culinary Institute, 250 Main St., Dept. S, Montpelier, VT 05602 US; 802-863-5231, Fax 802-223-0634, URL http://www.neculinary.com.

THE VERMONT WINE AND FOOD FESTIVAL

Stratton/July

American Express and Craftproducers sponsor an annual 3-day event in mid-July that features classes, seminars, discussions, demonstrations, dinners and dancing, wine auction. Established in 1997. Maximum class/group size: ~2,000. 1 programs per year. Facilities: Stratton Mountain Resort and The Equinox Resort and Spa.
Emphasis: Foods and wines presented by noted chefs and winemakers.
Faculty: Noted chefs, winemakers, sommeliers, authors of books on food and wine.
Costs: $295 core festival pass for 3 days includes seminars, tastings, brunch. Day rate is $150. Special events (auction, dinners) range from $60-$100.
Location: Near Manchester, Vermont.
Contact: Craftproducers, P.O. Box 300, Charlotte, VT 05445 US; 802-425-3399, Fax 802-425-3711, E-mail craftpro@together.net, URL http://www.craftproducers.com/wine1.htm.

THE WOODSTOCK INN & RESORT

Woodstock/Year-round

This Mobil 4-star resort, AAA 4-diamond dining room offers a Chef for a Day program on Thursdays, Fridays, and Saturdays that includes hands-on preparation of the evening's specials and a tour of the kitchen. Established in 1997. Maximum class/group size: 2. Facilities: Kitchen of The Woodstock Inn.
Emphasis: Inn specialties.
Faculty: The Resort's chefs.
Costs: $50 in addition to the lodging rate ($159-$325 double/night). Reservations required 48 hours prior.
Location: East central Vermont, 2-1/2 hours from Boston.
Contact: Tom List, Inn Manager, The Woodstock Inn & Resort, 14 The Green, Woodstock, VT 05091-1298 US; 802-457-1100, Fax 802-457-6699, E-mail email@woodstockinn.com, URL http://www.woodstockinn.com.

VIRGINIA

HELEN WORTH'S CULINARY INSTRUCTION
Charlottesville (Ivy)/Year-round

This private school founded in Cleveland 1940, New York City 1947, Virginia 1980 offers one-on-one lessons. Established in 1940. Maximum class/group size: 1. Facilities: Modern kitchen with new equipment.

Emphasis: Essential skills, cooking equipment, kitchen efficiency, food purchasing, aesthetics, table refinements, wine appreciation.

Faculty: Helen Worth, Les Dames d'Escoffier member and author of Cooking Without Recipes and Hostess Without Help, initiated and taught a food and wine appreciation course at Columbia University and Charlottesville's University of Virginia.

Costs: $75/hour.

Location: 68 miles west of Richmond, 118 miles from Washington, D.C.

Contact: Helen Worth, Director, Helen Worth's Culinary Instruction, 1701 Owensville Rd., Charlottesville (Ivy), VA 22901-8825 US; 804-296-4380.

JUDY HARRIS' COOKING SCHOOL
Alexandria/September-June

Culinary professional Judy Harris offers participation and demonstration classes. Established in 1978. Maximum class/group size: 12 hands-on/20 demo. 65 programs per year. Facilities: Large, well-equipped kitchen, culinary herb and vegetable gardens. Also featured: Private group classes, culinary tours, restaurant trips.

Emphasis: International and American regional cuisines, basic techniques, healthy cooking, baking, holiday and entertaining menus.

Faculty: Judy Harris studied at La Varenne in Paris. Well known guest chefs, teachers, and cookbook authors also teach, including Hugh Carpenter, Jacques Blanc, and Jacques Haeringer.

Costs: $42-$60/session.

Location: Ten miles from Washington, D.C., five miles from Old Town, Alexandria, and Washington National Airport.

Contact: Judy Harris, Judy Harris' Cooking School, 2402 Nordok Place, Alexandria, VA 22306 US; 703-768-3767, E-mail sselwyn@tidalwave.net, URL http://www.tidalwave.net/~sselwyn.

THE PATRICIAN SCHOOL OF COOKING
Sterling/March-December

This private school offers 1- to 4-session courses that include the hands-on Fine Cookery Series of 4-session basic, intermediate, and advanced cookery and basic baking and pastry courses. Established in 1991. Maximum class/group size: 15 demo, 5 hands-on, 18 travel. 60+ programs per year. Facilities: Well-equipped residential kitchen with central work station. Also featured: Wine instruction, programs for youngsters, day trips, private instruction. Travel programs include market visits, winery tours, visits to food producers, fine dining.

Emphasis: Techniques of cooking, menu planning and working without recipes. Regional American, Irish, and Eastern European cuisines.

Faculty: Mary Kinkelaar, instructor and proprietor. Guest chefs/instructors teach specific topics.

Costs: $295 for each 4-session course in the Fine Cookery Series. Single topic classes range from $25-$60.

Location: ~20 miles northwest of Washington, DC, along the border of Fairfax and Loudoun Counties.

Contact: Mary Kinkelaar, Instructor and Proprietor, The Patrician School of Cooking, 20366 Marguritte Sq., Sterling, VA 20165 US; 703-444-0317, E-mail mp_kinkelaar@prodigy.com.

A PINCH OF THYME COOKING SCHOOL

Alexandria/October-June

This private school offers series and single sessions on nutritional cookery, ethnic and international cuisines, kitchen skills, and other topics. Established in 1989. Maximum class/group size: 20 hands-on, 32 demo. 30+ programs per year. Facilities: Demonstration and hands-on facilities. Also featured: Private instruction, visits to markets and food producers, dining in private homes.

Emphasis: Low fat cookery.

Faculty: Robyn Webb, principal instructor and author of Robyn Webb's Memorable Menus Made Easy, and guest instructors.

Costs: $165 for a series, $45 for a single class.

Contact: Robyn Webb, A Pinch of Thyme Cooking School, 308 S. Payne St., Alexandria, VA 22314 US; 703-683-5034, Fax 703-683-3097, E-mail robynwebb@aol.com, URL http://www.robynwebb.com.

WILLIAMSBURG INN CLASSIC AFFAIRS WEEKENDS

Williamsburg/January-March, August

Colonial Williamsburg Foundation offers 2-day theme weekend programs that include lectures and demonstrations by noted chefs and wine experts. Established in 1992. Maximum class/group size: 100. Also featured: Reception, wine tastings, guest ticket to the Historic Area, tour of the Williamsburg Winery, fitness center, golf.

Emphasis: Theme programs devoted to a specific culture or cuisine.

Faculty: Has included Andre Soltner, retired chef of NYC's Lutece restaurant; Gerard Pangaud of Gerard's Place in Washington, DC, and Jimmy Sneed of The Frog and the Redneck in Richmond, VA.

Costs: Ranges from $190-$285, which includes double occupancy lodging, continental breakfasts, gourmet dinner, and planned activities. Lodging: The Mobil 5-star Williamsburg Inn, Providence Hall, Colonial Houses and Taverns, and Williamsburg Lodge.

Location: Colonial Williamsburg.

Contact: Trudy Moyles, Concierge, Williamsburg Inn Classic Affairs Weekends, P.O. Box 1776, Williamsburg, VA 23187-1776 US; 757-220-7979, Fax 757-220-7096.

WASHINGTON

BON VIVANT SCHOOL OF COOKING

Seattle/Year-round

This private school founded by Louise Hasson offers 4- and 9-session certificate courses, demonstration classes. Established in 1977. Maximum class/group size: 20 demo. 150+ programs per year. Facilities: Home kitchens. Also featured: Assistant program for graduates of certificate courses.

Emphasis: Basic techniques, foundations of fine cuisine, breads, pastry, seasonal specialties, regional and international cuisines.

Faculty: Louise Hasson has a BA in education, 20 years of teaching and catering experience, and is a certified member of the IACP. She studied at the Cordon Bleu, Badia a Coltibuono and Regalaeli. Other instructors include Northwest chefs and teachers.

Costs: $295 for 12 classes, $275 for an additional 12 classes, $469 for 20 classes.

Location: Seattle and suburban areas.

Contact: Louise Hasson, Director, Bon Vivant School of Cooking, 4925 NE 86th, Seattle, WA 98115 US; 206-525-7537, Fax 206-523-2992, E-mail louise@bon-vivant.com, URL http://www.bon-vivant.com.

COOK'S WORLD COOKING SCHOOL

Seattle/Year-round

This cookware store offers demonstration and participation courses. Established in 1990. Maximum class/group size: 20 demo/12 hands-on. 200+ programs per year. Facilities: 400-square-foot professionally-designed instruction kitchen with overhead mirrors, large teaching classroom.

Also featured: Wine instruction, private classes.
Emphasis: French, Italian, Indian, Pacific Northwest, gourmet vegetarian, baking, Thai, basics and intermediate series.
Faculty: Nancie Brecher, IACP member, who studied at the CIA, Peter Kump's, and La Varenne; local chefs and professional food experts.
Costs: $29-$36/class.
Location: North of downtown Seattle.
Contact: Nancie Brecher, Director, Cook's World, 2900 NE Blakeley St., Seattle, WA 98105 US; 206-528-8192, Fax 206-782-8239, E-mail cooksworld@aol.com.

COOKING WITH CHRISTIE
Olympia-Lacey/Year-round

Culinary professional and caterer Christie O'Loughlin offers classes on a variety of topics including seasonal specialties, seafood, Northwest regional cuisine, and Thai, Italian, Greek, and French cuisines. Established in 1991. Maximum class/group size: 12 demo, 8 hands-on. 8-12 programs per year. Facilities: Stools surround cooktop and prep areas. Also featured: Visits to markets, food producers, dining in private homes.
Emphasis: Ethnic and Northwest regional cooking using fresh, local ingredients.
Faculty: Christie O'Loughlin has studied with local chefs for 20 yrs.
Location: Lacey, 50 miles south of SeaTac Airport, 60 miles south of Seattle on Interstate 5.
Contact: Christie O'Loughlin, Cooking with Christie, PO Box 232, Olympia, WA 98507-0232 US; 360-459-0862, E-mail cologhlin@juno.com.

EVERYDAY GOURMET COOKING SCHOOL/CULINARY TOURS
Seattle, Bellevue/September-June

Culinary professional Beverly Gruber offers 10- and 18-session hands-on certificate courses and 1- to 6-session participation and demonstration courses. Established in 1988. Maximum class/group size: 12 hands-on certificate/30 demo courses. Facilities: Large, multi-station work island in Larry's Gourmet Markets. Also featured: Apprentice/assistant program, custom classes, culinary travel.
Emphasis: Basic techniques, pastry, kitchen survival skills, ethnic specialties.
Faculty: Beverly Gruber is a cum laude graduate of Madeleine Kamman's 2-year professional cooking school, an IACP-Certified Teacher, member of Italian Culinary Institute, and Les Dames d'Escoffier, and has taught professionally for more than 15 years.
Costs: $35-$40/class.
Location: Larry's Gourmet Market in Bellevue, Seattle area.
Contact: Beverly Gruber, Director, Everyday Gourmet Cooking School/Culinary Tours, 677 120th Ave. NE, #155, Bellevue, WA 98005 US; 206-363-1602/425-451-2080, Fax 206-363-1602/425-451-2080, E-mail gormayschl@aol.com, URL http://members.aol.com/gormaytrvl. Additional contact: Sabine Bradshaw, Barbara Gardener.

THE HERBFARM
Fall City/Year-round

This restaurant and herb nursery offers demonstration classes and events. Established in 1974. Maximum class/group size: 28. 300+ programs per year. Facilities: Open kitchen of The Herbfarm Restaurant, top-rated in the Northwest by the Zagat Guide. Also featured: Classes in horticulture, basketry, herbal crafts, herbal medicine, wines of the Pacific Northwest, a Father's Day weekend microbrewery festival, and the Northwest Wine Festival in August.
Emphasis: Pacific Northwest cuisine, herbs.
Faculty: Jerry Traunfeld, The Herbfarm Restaurant chef and co-author of Seasonal Favorites from The Herbfarm, and local guest chefs.
Costs: $27-$45/class.
Location: Situated on 13 rural acres 30 minutes east of Seattle, The Herbfarm has 17 public gardens, an organic garden, and a gift shop.

Contact: BJ Duft, Cooking School Director, The Herbfarm, 32804 Issaquah-Fall City Rd., Fall City, WA 98024 US; 206-784-2222, Fax 206-789-2279, E-mail herborder@aol.com, URL http://www.theherbfarm.com. Other contact: Carrie Van Dyck.

INTERNATIONAL DINING ADVENTURES (See page 350)

LE GOURMAND RESTAURANT

Seattle/Year-round

This restaurant offers a demonstration class the last Sunday and Monday of each month. Established in 1986. Maximum class/group size: 20.

Emphasis: French and Northwest regional cuisine.

Faculty: Le Gourmand Chef Bruce Naftaly, a founder of the Northwest cuisine movement.

Costs: $35.

Location: Near Seattle's Ballard District, 25 minutes from Sea-Tac airport.

Contact: Bruce Naftaly, Chef/Owner, Le Gourmand Restaurant, 425 N.W. Market St., Seattle, WA 98107 US; 206-784-3463.

ROBERT REYNOLDS, NORTHWEST FORUM (See page 247)

SCHOOL OF NATURAL COOKERY (See page 199)

SUR LA TABLE

Kirkland/Year-round *(See also page 194))*

This cookware store offers demonstrations and hands-on classes covering a variety of cuisines, the basics, special subjects, chef specialties. Established in 1996. Maximum class/group size: 16 hands-on, 36-40 demo. 100-150 programs per year. Facilities: Full demonstration kitchen with TV monitors and overhead mirror over teaching island, hands-on tables. Also featured: Professional classes, corporate team-building classes, culinary walking tours, market visits, programs for youngsters.

Emphasis: Basic cooking and baking, topics of current interest featuring popular restaurant chefs and authors.

Faculty: Local and nationally known chefs, restaurateurs, and cookbook authors.

Costs: $35-$100/class, includes a 10% merchandise discount coupon.

Location: 10 miles northeast of Seattle on Lake Washington.

Contact: Martha Aitken, Cooking Program Coordinator, 90 Central Way, Kirkland, WA 98033 US; 425-827-1311(store), 206-682-7175 (office).

WEST VIRGINIA

FOOD FESTS (See page 224)

THE GREENBRIER GOURMET COOKING CLASSES

White Sulphur Springs/May-October

The Greenbrier resort offers daily hands-on classes for resort guests. Children's classes each Wednesday and Saturday afternoon (June-Aug). Established in 1995. Maximum class/group size: 12. Facilities: Large demonstration and participation kitchen with overhead mirror. Also featured: Resort amenities.

Emphasis: Preparation for a single menu from appetizer to dessert.

Faculty: Greenbrier culinary staff.

Costs: $100/adult session, $35/child session.

Location: The Mobil 5-Star, AAA 5-Diamond resort is in the Allegheny mountains, 15 minutes from the Greenbrier Valley Airport in Lewisburg.

Contact: Riki Senn, Cooking School Coordinator, The Greenbrier Gourmet Cooking Classes, 300 West Main St., White Sulphur Springs, WV 24986 US; 800-228-5049, Fax 304-536-7893, URL http://www.greenbrier.com. Also: Townley Aide, Mgr. of Public Relations.

LA VARENNE AT THE GREENBRIER

(See also page 301) (See display ad page 301) **White Sulphur Springs/March-April**

The Greenbrier and La Varenne offer 5-day cooking vacations that feature daily morning demonstration classes and optional hands-on instruction in the afternoons. Established in 1977. Maximum class/group size: 60. 8 programs per year. Facilities: Large demonstration kitchen with overhead mirror. Also featured: Receptions, dinners, resort amenities.

Emphasis: Contemporary American and International cuisine, culinary technique.

Faculty: Anne Willan, founder and director of École de Cuisine La Varenne, food columnist, TV show food host, and author of more than a dozen cookbooks; Greenbrier chefs; guest food personalities.

Costs: $2,000 includes lodging and meals. Hands-on class is $100 extra. Resort amenities include golf, tennis, horseback riding, skeet and trap, hiking, spa, swimming, concerts, live music, dancing.

Location: The Mobil 5-Star, AAA 5-Diamond resort is in the Allegheny mountains, 15 minutes from the Greenbrier Valley Airport in Lewisburg.

Contact: Cindy McCutcheon, Cooking School Coordinator, La Varenne at the Greenbrier, 300 West Main St., White Sulphur Springs, WV 24986 US; 800-228-5049, Fax 304-536-7893, E-mail The_Greenbrier@csx.com, URL http://www.greenbrier.com. Additional contact: Townley Aide, Manager of Public Relations.

WISCONSIN

CREATIVE CUISINE COOKING SCHOOL

Milwaukee/Year-round

Cookbook author, culinary instructor Karen Maihofer offers demonstration classes. Established in 1977. Maximum class/group size: 20. 100+ programs per year. Also featured: Private demonstrations, programs for groups.

Emphasis: Ethnic cuisines, use of food processor, pasta, baking, vegetarian, heart-healthy, entertaining menus, and holiday foods.

Faculty: IACP-Certified Karen Maihofer studied with Julia Child, James Beard, and Guiliano Bugialli. Her books include Foods For Entertaining, Holiday Cuisine, Salads & Muffins, Pasta For All Seasons, Summertime Cooking, and Fast & Fabulous Entertaining.

Costs: $30-$35/class.

Location: Twenty minutes north of downtown Milwaukee; Naples, on Florida's Gulf Coast. May-Nov in Milwaukee; Dec-Apr in Florida.

Contact: Karen Maihofer, Owner, Creative Cuisine Cooking School, 9458 N. Regent Ct., PO Box 17664, Milwaukee, WI 53217 US; 414-352-0975/941-947-9879. Also: 20610 Rivers Ford, Estero, FL 33928, Tel# 941-947-9879 (Nov-Apr).

ÉCOLE DE CUISINE

Kohler/Year-round

This school of professional cooking for the home chef offers intensive weekend participation courses, 5-day classic cuisine courses, demonstration classes. Private tours to food producers, restaurants, special events, customized programs for groups. Established in 1988. Maximum class/group size: 10 hands-on/48 demo. 50 programs per year. Facilities: 2,500-square-foot professionally-equipped facility. Also featured: Annual 8-day Food Tour of Paris escorted by Jill Prescott features cooking class, chocolate seminar, market visits, dining in a Michelin-starred restaurant, tours of bread, pastry, and kitchenware shops.

Emphasis: Professional cooking techniques for home chefs, classic French cuisine, bistro cooking, Italian cuisine, bread baking, pastry, regional cooking, wine course, sauces, stocks, soups.

Faculty: Jill L. Prescott, professionally trained in Paris at École LeNôtre, École de Cuisine La Varenne, and Ritz-Escoffier, and host of a PBS cooking series, Professional Cooking for the Home Chef (1999). Guest chefs.

Costs: Participation courses $185-$895, single demonstration classes start at $35. Lodging at resorts and hotels with restaurants, golfing, fishing, skiing, biking, spas, and shopping mall.

Location: A resort area about 50 miles from Milwaukee Airport, 100 miles from Chicago's O'Hare Airport.

Contact: Jill L. Prescott, École de Cuisine, 765 H Woodlake Dr., Kohler, WI 53044 US; 920-451-9151, Fax 920-451-9152, E-mail Ecole@execpc.com, URL http://www.execpc.com/~ecole/.

ORANGE TREE IMPORTS COOKING SCHOOL

Madison/January-June, August-October

This cookware store offers single session evening classes taught by area chefs covering a variety of topics. Established in 1981. Maximum class/group size: 10 demo and hands-on. Facilities: Upstairs classroom kitchen in gourmet store.

Emphasis: Small, informal classes with emphasis on learning new skills.

Faculty: Experienced staff of local chefs and cooking experts.

Costs: $25/class.

Location: Near the UW football stadium in Madison.

Contact: KT Ellenbecker, Cooking School Director, Orange Tree Imports Cooking School, 1721 Monroe St., Madison, WI 53711 US; 608-255-8211, Fax 608-255-8404, E-mail info@orangetreeimports.com, URL http://www.orangetreeimports.com.

SAVEUR INTERNATIONAL COOKING SCHOOLS (See page 212)

WISCONSIN SCHOOL OF COOKERY

Cascade/Year-round

Culinary professional Richard Baumann offers demonstrations. Established in 1991. Maximum class/group size: Varies.

Faculty: Richard Baumann, member of FWA, writes for magazines, local newspapers, is author of Wisecrackers, and produces and hosts TV cooking shows.

Location: 50 miles from Milwaukee.

Contact: Wisconsin School of Cookery, W6248 Lake Ellen Dr., Cascade, WI 53011-1322 US; 920-528-8015, Fax 920-528-8811, E-mail cbaumann@excel.net.

AFRICA

CHRISTINA MARTIN SCHOOL OF FOOD AND WINE

Durban/February-November

This private school offers weekly courses focusing on a specific theme. Established in 1988. Facilities: Auditorium, delicatessen, 60-seat restaurant, 80-seat conference venue, garden restaurant.

Emphasis: Theme courses.

Faculty: Christina Martin is a Maitre Chef de Cuisine & Commandeur Associé de la Commanderie des Cordons Bleus de France. Instructors include vice-principal Michelle Barry, Chef de Cuisine Gerhard Van Rensburg, and Andrew White.

Costs: R90-R170/course.

Location: 10 minutes from city center.

Contact: Christina Martin, Principal and Owner, Christina Martin School of Food and Wine, PO Box 4601, Durban, 4000 South Africa; (27) 31-3032111, Fax (27) 31-233342, E-mail chrismar@iafrica.com.

FPT SPECIAL INTEREST TOURS

Kenya, India, Belgium/February, April-May, July

This tour operator offers 7- to 17-day trips that include A Tea Lover's Tour of India, A Coffee Lover's Tour of Kenya, and A Chocolate Lover's Tour of Belgium. Maximum class/group size: 30. Also featured: Visits to farms, research centers, and food producers, cooking demonstrations, cultural excursions.

Emphasis: Tea, coffee, chocolate.

Costs: Ranges from $2,250-$4,900, includes round-trip airfare, shared deluxe lodging, ground

transport, planned activities.
Location: India, Kenya, Belgium.
Contact: Diana Altman, Director, FPT Special Interest Tours, 186 Alewife Brook Pkwy., Cambridge, MA 02138 US; 800-645-0001/617-476-1142, Fax 617-661-3354, E-mail dma@fpt.com, URL http://www.fpt.com.

SILWOOD KITCHEN CORDONS BLEUS COOKERY SCHOOL
Rondebosch Cape/Year-round

This career school offers guest chef demonstrations. Established in 1964. Maximum class/group size: 4 groups of 10 students each. 6 programs per year. Facilities: A 200-year-old coach-house converted into a demonstration and experimental kitchen, 3 additional kitchens, demonstration hall, and a library. Also featured: 1-week hands-on classes for children twice yearly.
Faculty: 11-member faculty. Includes school principal Alicia Wilkinson, nutrition instructors Jeanette Rietmann, menu reading instructor Alisa Smith, and practical supervisors Louise Faull, Peggy Loebenberg, Carianne Wilkinson, Pauline Copus and Shirley Henderson.
Location: Cape Town.
Contact: Mrs. Alicia Wilkinson, Silwood Kitchen, Silwood Rd., Rondebosch Cape, South Africa; (27) 21-686-4894, Fax (27) 21-686-5795.

AUSTRALIA

AGL COOKING SCHOOL
Chatswood, Bondi/February-November

This gas company offers evening and day courses. Maximum class/group size: 45. Facilities: Well-equipped gas kitchen.
Emphasis: Basic techniques, guest chef specialties, specific topics.
Faculty: Head of School Lyn Sykes, author of magazine and newspaper columns with 25 years of food experience; qualified cooking instructors; guest chefs from Australia and abroad.
Costs: Ranges from A$35-A$150.
Contact: Sally Muller, Head of School, AGL Cooking School, 31 Newland St., 1st Flr., Bondi Junction, NSW, 2022 Australia; (61) 2-9389-8934, Fax (61) 2-9389-2675.

ARTIS – ARTIS NOOSA COOKING SCHOOL
Noosa Heads/February-October

This restaurant and private cooking school offers a variety of hands-on 3-day events. Established in 1995. Maximum class/group size: 20 hands-on/40 demo. Facilities: Modern demonstration kitchen and licensed restaurant. Also featured: Tours to local food producers and growers, wine tastings and lectures, tours of So. Australian wineries.
Emphasis: Regional and international cuisines.
Faculty: The restaurant's executive chef and guest chefs.
Costs: Begins at A$85/day. A center for tourism, Noosa offers a range of hotels and motels.
Location: On Queenslands Sunshine Coast, about 100 miles north of Brisbane.
Contact: David. Horton, ARTIS, 8 Noosa Dr., Noosa Heads, QLD, 4567 Australia; (61) 74 472 300, Fax (61) 74 472 383.

ACCOUTREMENT COOKING SCHOOL
Sydney/April-October

This private school offers culinary tours and classes. Established in 1976. Maximum class/group size: 35. 100 programs per year.
Emphasis: Thai, Japanese, Italian, French, Middle Eastern, and Indian cuisines, seafood, salads, desserts, guest chef specialties.
Faculty: Proprietor Susan Jenkins, who trained at École LeNôtre and worked with many chefs; Australian guest chefs; chefs from abroad.

Costs: $60/course.

Contact: Susan A. Jenkins, Accoutrement Cooking School, 611 Military Road, Mosman, Sydney, NSW, 2088 Australia; (61) 2-969-7929, Fax (61) 2-969-7929.

AMANO

Perth/Year-round

This cookware store offers 2- to 3-session demonstration and participation courses. Established in 1982. Maximum class/group size: 16 hands-on/36 demo. 80-90 programs per year. Facilities: Teaching kitchen with overhead mirror and individual work areas. Also featured: Culinary tours to France, Bali, and Italy.

Faculty: School director is IACP-member Beverly Sprague. Instruction is given by prominent Australian and international culinary professionals.

Costs: Range from A$40-A$70 per session; series range from A$90-A$180.

Location: The beach-side suburb of Perth, Western Australia's capitol.

Contact: Beverly Sprague, Director, Amano, 12 Station St., Cottesloe, Perth, 6011 Australia; (61) 8-(0)9384-0378, Fax (61) 8-(0)9385-0379.

BEVERLEY SUTHERLAND SMITH COOKING SCHOOL

Mt. Waverley/Year-round

Cookbook author Beverley Sutherland Smith offers 1- to 3-session demonstration courses. Established in 1967. Maximum class/group size: 15. 30+ programs per year. Facilities: Mirrored teaching kitchen that overlooks an herb garden.

Emphasis: Instructor and guest chef specialties, ethnic and regional dishes.

Faculty: Beverley Sutherland Smith, Vice Conseillere Culinaire Chaine des Rotisseurs, contributed to Epicurean and Gourmet, authored 25 books and won the Australian Gold Book award, food writer for The Sun Herald newspaper and Enjoy magazine.

Costs: Starts at A$49.

Location: Mt. Waverley, a Melbourne suburb, is about 12 miles from city center.

Contact: Beverley Sutherland Smith, Beverley Sutherland Smith Cooking School, PO Box 2134, Mt. Waverley, Victoria, 3149 Australia; (61) 3-9802-5544, Fax (61) 39-802-7683.

THE COOK, THE ARTIST COOKERY SCHOOL

Brisbane/Year-round

Culinary professional Roz MacAllan offers 3-session demonstration and participation courses. Established in 1992. Maximum class/group size: 7 hands-on/12 demo. Facilities: Professional kitchen in private home. Also featured: Escorted overseas cooking tours.

Emphasis: Italian, Thai, Asian cuisines, entertaining in warm climates.

Faculty: Director Roz MacAllan.

Costs: A$55/session, A$155/course, payable in advance.

Location: About 3 miles from city center.

Contact: Roz MacAllan, MS, The Cook, The Artist Cookery School, P.O. Box 152, Brisbane Market, QLD, 4106 Australia; (61) 7-3870-4101, Fax (61) 7-3870-3760, E-mail macallan@gil.com.au.

COUNCIL OF ADULT EDUCATION

Melbourne/Year-round

This educational organization offers 2- to 6-session cooking and catering courses and full-day classes. 70 programs per year. Also featured: Workshops and travel programs on a wide range of topics, including art, crafts, photography, performing arts, recreation, personal development, history, writing, languages, literature, and nature.

Emphasis: International cuisines, microwave, breads, vegetarian cookery, catering, wine appreciation.

Faculty: Cooking instructors and guest chefs.

Costs: Nonrefundable tuition ranges from A$20-A$40 per session for multi-session courses and from A$60-A$75 for a full-day class. Discounts for seniors and pensioners. Credit cards accepted.

Contact: Dianne Berlin, Asst. Director, Council of Adult Education, 256 Flinders Street, Melbourne, Victoria, 3000 Australia; (61) 3-9652-0611, Fax (61) 3-9652-0793.

ELISE PASCOE COOKING SCHOOL

Sydney/March-December

Culinary professional Elise Pascoe offers 1- to 2-session demonstration and participation courses and weekend workshops. Established in 1975. Maximum class/group size: 6 hands-on/20 demo. 70 programs per year. Facilities: Home kitchen with 3 work stations and overhead mirror. Also featured: Private and men only classes, culinary tours of Italy.

Emphasis: Technique and theory, Mediterranean cuisine, Italian, Thai, and modern Australian cooking.

Faculty: Elise Pascoe is a free-lance food writer, television presenter, and author of 5 cook books. She trained at Le Cordon Bleu and La Varenne in Paris and with Roger Vergé in France and Angelo Paracucchi in Italy.

Costs: A$170 per 3-session course, single session A$70, workshop A$135.

Location: Sydney's eastern suburbs, Wagga Wagga, Condobolin, Moree, Walcha, North Star and Southern Highlands NSW; 20 minutes from Sydney airport.

Contact: Elise Pascoe, Managing Director, Elise Pascoe Cooking School, 1/44 Darling Point Road, Darling Point, NSW, 2027 Australia; (61) 2-9363-0406, Fax (61) 2-9363-3122, E-mail elisep@ozemail.com.au.

THE FOODLOVERS WORKSHOP

Box Hill (Melbourne)/Year-round

Culinary professional Sherry Clewlow offers range from 1-session classes to 16-hour 2-day courses. Instruction is hands-on. Established in 1993. Maximum class/group size: 6-8. Facilities: Home-based business, semi-commercial kitchen. Also featured: Optional full day gourmet tour of Melbourne plus lunch and local wines.

Emphasis: New Australian cuisine.

Faculty: Owner Sherry Clewlow, Executive Boardroom chef who trained in Canada and Australia.

Costs: Ranges from A$45/class to A$250 for a course. Optional gourmet tour of Melbourne $120.

Contact: Sherry Clewlow, Instructor Chef/Tour Operator, The Foodlovers Workshop, 6 Sewell St., Box Hill, Victoria, 3129 Australia; (61) 3-9899-9292, Fax (61) 3-9899-9292, E-mail flw@ozemail.com.au.

H.T.A. COOKERY SCHOOL WITH BARBARA HARMAN

Brisbane/Year-round

The Hospitality Training Association sponsors 1- to 2-session demonstration and 3- to 4-session participation courses. Established in 1996. Maximum class/group size: 16 hands-on/35 demo. Facilities: Auditorium with overhead mirror and professional kitchen. Also featured: Recipe advice, local food tours.

Emphasis: Australian and Asian cuisines, vegetarian recipes, regional foods, entertaining.

Faculty: Barbara Harman, cookery school head, has 18 year experience and was previously head of the Boral Gas Cookery Service.

Costs: Ranges from A$35-A$60 per session, A$60 for visiting chef classes. Payment due 1 week prior.

Location: Five minutes from Brisbane in Fortitude Valley. Near Chinatown and Bistro district.

Contact: Barbara Harman, H.T.A. Cookery School with Barbara Harman, 269 Wickham St., Fortitude Valley, Brisbane, QLD, 4006 Australia; (61) 7-3257-0377.

HARRY'S CHINESE COOKING CLASSES

Sydney/February-December

Culinary professional Harry Quay offers three 8-session demonstration courses on a rotating basis. Established in 1977. Maximum class/group size: 20. Facilities: Rented halls and schools with kitchen facilities. Also featured: Children's classes, private classes, 4-week courses.

Emphasis: Basic Chinese, advanced Chinese, and Thai cuisines.
Faculty: A third generation chef, Harry Quay has more than 30 years experience.
Costs: A$20/session, includes full meal, payable at class.
Contact: Harry Quay, Harry's Chinese Cooking Classes, 47 Bruce Street, Brighton-le-Sands, NSW, 2216 Australia; (61) 2-9567-6353, Fax (61) 2-9567-3653.

HOWQUA-DALE GOURMET RETREAT

Mansfield/March-November

Howqua-Dale country house-hotel resort offers 4-day and weekend participation courses and 6-day gourmet cycling tours of Australia's wine regions. Established in 1977. Maximum class/group size: 12. 12 programs per year. Facilities: Horse-shoe shaped pavilion with specialized equipment. Also featured: Fishing, skiing, swimming, horseback riding, bird-watching.
Emphasis: Fresh local foods and modern Australian cuisine; wine appreciation.
Faculty: Co-owner Marieke Brugman, a noted food writer and cooking demonstrator, conducts the classes. Her partner, Sarah Stegley, acts as hostess and instructs students in wine selection.
Costs: All-inclusive fee is approximately A$700 for the weekend course, A$1,200 for the 4-day course, and A$2,200 for the tour, excluding transport.
Location: A 40-acre estate on the Howqua River 18 miles from Mansfield, a country town 128 miles northeast of Melbourne.
Contact: Marieke Brugman, Howqua-Dale Gourmet Retreat, P.O. Box 379, Mansfield, Victoria, 3722 Australia; (61) 35-777-3503, Fax (61) 35-777-3896.

JOHANNA MINOGUE COOKING SCHOOL

Paddington, Sydney/Year-round

Culinary professional Johanna Minogue offers 3-session hands-on courses and 1-week spring and fall cooking vacations to Positano, Italy. Established in 1991. Maximum class/group size: 8 hands-on, 12 travel. 8-10 programs per year. Facilities: Private home. Also featured: 3- and 4-day school holiday cooking classes for teenagers.
Emphasis: Modern Australian cuisine, including Mediterranean and Asian.
Faculty: Johanna Minogue, food commentator for a Sydney radio station, 20+ yrs catering experience, instructor in Sydney evening colleges. Italian Chef Tanina Vanacore Attanasio teaches on Italy trips.
Costs: 3-session courses A$195, teenage 4-day course A$200, Italy trip A$5,000, including airfare and lodging.
Location: 2 miles from Sydney.
Contact: Johanna Minogue, Proprietor, Johanna Minogue Cooking School, 25 Prospect St., Paddington, 2021 Australia; (61) 2-9331-5175, Fax (61) 2-9331-4352.

LE CORDON BLEU – SYDNEY

Sydney/Year-round *(See also pages 154 and 301) (See display ad page 149)*

This private school with other locations in Paris, London, Tokyo, and Ottawa offers half-day to 10-month hands-on and demonstration courses: daily demonstrations, gourmet sessions, bread baking, wine appreciation, floral art, guest chefs, 10-week cuisine and pastry courses, catering, table decoration, children's courses. Established in 1996. Maximum class/group size: 10. Facilities: Classrooms designed to resemble professional working kitchens. Individual workspaces with refrigerated marble table tops, demonstration kitchen.
Emphasis: The Classic Cycle.
Faculty: French and Australian master chefs, international staff.
Costs: $100 for a full-day course to $5,250 for 10 weeks.
Location: ~7 miles from city center.
Contact: Geoff Montgomery, Director, Le Cordon Bleu Sydney, 250 Blaxland Rd., Ryde, NSW, 2112 Australia; (61) 2-9808-8307, Fax (61) 2-9807-6541, E-mail Geoff.Montgomery@tafensw.edu.au, URL http://www.cordonbleu.net. In U.S. and Canada: 800-457-CHEF.

MARCEA WEBER'S COOKING SCHOOL – FOOD AS MEDICINE

Faulconbridge/Year-round

Culinary professional Marcea Weber offers demonstration and hands-on classes. Established in 1980. Maximum class/group size: 12-14. Also featured: Women's workshops including healing ailments.

Emphasis: Macrobiotic and Chinese-herbal cuisine.

Costs: A$30-$35/class.

Location: Inner city of Sydney and Melbourne, an hour from Sydney.

Contact: Marcea Weber, Marcea Weber's Cooking School, 56 St. George's Crescent, Faulconbridge, NSW, 2776 Australia; (61) (02) 4751-1680, Fax (61) (02) 4751-1680.

MARGARET RIVER WINE & FOOD FESTIVAL

Margaret River/November

This annual festival offers an annual festival featuring a 3-day master class sponsored by Ma Cuisine Cooking School at a wine estate. 1 programs per year.

Contact: Pauline McLeod, Margaret River Wine & Food Festival, Bussell Highway, Margaret River, 6285 Australia; (61) 97-572911, Fax (61) 97-573287.

MELBOURNE FOOD & WINE FESTIVAL

Melbourne/March-April

This four-week festival 4-week festival consisting of 30 events that include master classes, dinners with chefs and vintners, market and vineyard tours, quality produce, award-winning wines, fine restaurants, bistros, and cafes. Established in 1993. 1 program per year. Facilities: Events range from A$20-A$185 each.

Contact: Coralie Stupart, Administration Mgr., Melbourne Food & Wine Festival, Level 1, 477 Collins St., Melbourne, VIC, 3000 Australia; (61) 3-9628-5008, Fax (61) 3-9628-5007, E-mail foodfest@enternet.com.au.

NATURAL FOODS VEGETARIAN COOKING SCHOOL

Sydney/May, November

Culinary professional Myrna Fenn offers 5-session demonstration and participation courses. Special requests honored. Established in 1987. Maximum class/group size: 4 hands-on/36 demo. 2 programs per year. Facilities: Private kitchen with 4 work stations.

Emphasis: Nutrition, fruits, legumes, wholegrains, vegetables.

Faculty: Certified cooking demonstrator Myrna Fenn.

Costs: A$150.

Location: Lower North shore, 10 minutes from city centre.

Contact: Myrna Fenn, Natural Foods Vegetarian Cooking School, 20/21 Rangers Road, Cremorne, Sydney, NSW, 2090 Australia; (61) (02) 9953-7175.

PARIS INTERNATIONAL COOKING SCHOOL

Sydney/Year-round

Culinary professional Laurent Villoing offers 9-session courses that include French, Italian, Asian, Mediterranean, Health Conscious, European, Basics, Vegetarian, Vegan. Day workshops include Bush Tucker, Pastry & Cakes, Bread, Cajun, Chocolate, Stocks & Sauces, Cake Decorating, Food Carving, Sushi. Established in 1994. Maximum class/group size: 24 demo, 16 hands-on. Facilities: Fully-equipped high school kitchens and community centers in Sydney. Also featured: Classes for youngsters, wine/beer/cocktail classes, art of the table, floral art.

Emphasis: Practical classes where students learn to cook independently under close supervision.

Faculty: French native teacher and coordinator Laurent Villoing trained at École LeNôtre, has 20 years experience as hotel chef, lectured at London catering colleges and in Sydney, appears on Australian and international TV, radio, press. Other specialist teachers.

Costs: Ranges from A$29 for a 1-day workshop to A$100 for a 9-session course.

Location: includes Chatswood, Vaucluse, Maroubra, Matraville, North and South Sydney.

Contact: Laurent Villoing, Coordinator/Manager, Paris International Cooking School, 21/1 Mosman St., Mosman, NSW, 2088 Australia; (61) 2-9969-4687, Fax (61) 2-9969-4687.

THE SCHOOL HOUSE

Port Douglas/May-September

This private school offers once monthly 1-week vacation programs that include five hands-on half-day classes. Established in 1993. Maximum class/group size: 8. 12 programs per year. Facilities: The School House's air-conditioned restaurant kitchen. Also featured: Day trip and picnic in Far North Queensland, swimming in Mossman Gorge, market visits, sightseeing, tours of food producers.

Emphasis: Specialties of noted Australian chefs.

Faculty: Noted Australian restaurant chefs conduct each session.

Costs: A$2,350, which includes shared lodging, most meals and wines, planned activities. Lodging is at The School House, which has 6 guest bedrooms, 3 with ensuite.

Location: 50 minutes north of Cairns, in Far North Queensland.

Contact: Michael Edwards, The School House, P.O. Box 275, Port Douglas, QLD, 4871 Australia; (61) 2-9251-9475, Fax (61) 2-9251-9552. Also: 61 Lower Fort St., Millers Point 2000, Australia.

SUNNYBRAE COUNTRY RESTAURANT AND COOKING SCHOOL

Victoria/March-December

This restaurant offers 1-day courses covering all aspects of cooking techniques. Established in 1991. Maximum class/group size: 14. 100 programs per year. Facilities: Fully equipped professional teaching kitchen with garden views, large vegetable garden. Also featured: Market visits, winery tours, visits to food producers, dining in private homes, sightseeing. Private instruction.

Emphasis: Modern Australian cuisine using fresh local produce.

Faculty: Chef George Biron, international and Australian guest chefs.

Costs: A$65/class.

Location: 90 minutes from Melbourne, near the Great Ocean Rd. and Lorne, near township of Birregurra near Colac.

Contact: George Biron, Chef, Sunnybrae Country Restaurant and Cooking School, Crn Cape Otway Rd. and Lorne Rd. Birregurra, Victoria, 3242 Australia; (61) 3-52362276, Fax (61) 3-52362276, E-mail sunnybrae@ne.com.au.

SYDNEY SEAFOOD SCHOOL

Pyrmont, Sydney/Year-round

Sydney Fish Market Pty. Ltd. offers hands-on courses in seafood preparation. Established in 1989. Maximum class/group size: 48. 200+ programs per year. Facilities: Practical kitchen and 66-seat demonstration auditorium with tiered seating and overhead mirror. Also featured: Children's classes, trade program for commercial cooks.

Emphasis: Seafood cookery, guest chef specialties, advanced techniques, seafood buying and handling, sushi and sashimi.

Faculty: Qualified home economists who are seafood specialists, guest chefs from Sydney's top restaurants.

Costs: Nonrefundable tuition, payable in advance, ranges from A$45-A$130 per course.

Location: The Sydney Fish Market, which has a fish auction hall, fish retail outlets, gourmet deli, fruit and vegetable store, seafood cafe, and a sushi bar.

Contact: Roberta Muir, Manager, Sydney Seafood School, Locked Bag 247, Pyrmont, NSW, 2009 Australia; (61) (02) 9660-1611, Fax (61) (02) 9552-1661, E-mail sss@sydneyfishmarket.com.au, URL http://www.sydneyfishmarket.com.au. Also: P.O. Box 247, Pyrmont, NSW 2009, Australia.

TAMARA'S KITCHEN

Melbourne/February-November

Culinary professional Tamara Milstein offers 3- to 5-session demonstration and participation courses. Established in 1989. Maximum class/group size: 12. 40 programs per year. Facilities: Large shopfront with 12 workspaces at 490 Tooronga Rd. Also featured: Market visits, private classes, children's classes, gourmet tours.

Emphasis: Includes breadmaking, pasta, risotto, modern Jewish menus, fish, European desserts, Asian flavors, Indian cuisine, Mediterranean, 30-minute entrees, high flavor low fat.

Faculty: Two full-time teachers, including Tamara Milstein CCP, who trained in Europe and the U.S.

Costs: A$120-A$220/course.

Location: Inner eastern suburb of Melbourne, 15 minutes from the city.

Contact: Tamara Milstein, Owner, Tamara's Kitchen, 2 Garden St., Hawthorn East, VI VIC 3123 Australia; 613-988-24906, Fax 613-988-23436, E-mail tamara@iaccess.com.au, URL http://www.tamaraskitchen.com.au.

THORN PARK COOKING SCHOOL

Clare Valley/January-March, June-July, September, November

This country inn offers 2- and 3-day weekend and midweek programs. Established in 1992. Maximum class/group size: 8-10. Facilities: 2 kitchens, small hotel/inn, commercial. Also featured: Wine tours and art classes.

Emphasis: A variety of topics, including seasonal and entertaining menus, wines, and specific subjects.

Costs: Approximately A$400-A$550 includes accommodations for 2 nights.

Location: South Australia.

Contact: David Hay, Co-owner, Thorn Park Country House, College Rd., Sevenhill via Clare, 5453 Australia; (61) 8-8843-4304, Fax (61) 8-8843-4296, E-mail thornpk@capri.net.au.

VICTORIA'S KITCHEN OF CREATIVE COOKING

Mt. Hawthorn/February-September

Culinary professional Victoria Blackadder offers 1- and 2-session demonstration courses. Established in 1981. 12-15 programs per year.

Emphasis: International cuisines, entertaining menus, vegetarian cookery, microwave, beef, poultry and seafood preparation, purchasing, storage, economy.

Faculty: Food consultant Victoria Blackadder was named Australian Women's Weekly Best Cook in Australia in 1981 and established a catering business in 1986. She is author of Victoria's Kitchen.

Costs: Range from A$14-A$20 per session, payable in advance.

Location: The Home Base Exhibition Centre Auditorium, 10 minutes from Perth city center.

Contact: Victoria Blackadder, Victoria's Kitchen of Creative Cooking, P.O. Box 278, Mt. Hawthorn, 6016 Australia; (61) 9-443-2266.

WHOLEFOOD VEGETARIAN & JAPANESE

Melbourne/Year-round

This private school offers hands-on classes in wholefood organic cuisine, macrobiotic style, dairy-, wheat- and sugar-free, low fat; traditional Japanese. Established in 1986. Maximum class/group size: 5-15. 10-15 programs per year. Also featured: Private instruction, classes for youngsters.

Emphasis: Health-supportive wholefood organic cuisine and traditional Japanese cuisine.

Location: Prahran, Middle Park, Caulfield.

Contact: Sandra Dubs, Natural Food Consultant, Wholefood Vegetarian & Japanese, PO Box 523, Malvern, Melbourne, VIC, 3144 Australia; (61) 15-360323, E-mail dubs@netspace.net.au.

YALUMBA WINERY COOKING SCHOOL

Barossa Valley/Year-round

This winery offers classes and 1-day weekend workshops. Established in 1985. Maximum class/group size: 60. Facilities: Kitchen, cooking school, dining room, tasting rooms, cellars, petanque pitch. Also featured: Winery tours, specialized tastings.

Emphasis: Wine and food combinations.

Costs: Local bed & breakfast details available.

Location: About 40 miles from Adelaide.

Contact: Jane Ferrari, Yalumba Winery Cooking School, P.O. Box 10, Angaston, SA, 5353 Australia; (61) 8-561-3200, Fax (61) 8-561-3393, E-mail jferrari@samsmith.com.au, URL http://ww.samsmith.com.au/yalumba/index.html.

AUSTRIA

HERZERL TOURS "A TASTE OF VIENNA"

Vienna/March-May, October-December

The City of Vienna, Austria, sponsors 1-week trips that include 4 classes at Vienna's Am Judenplatz cooking school. Established in 1994. Maximum class/group size: 18. 3+ programs per year. Also featured: Market visit, winery tour, trip to tableware displays in the Imperial Palace, sightseeing, concert.

Emphasis: Viennese cuisine.

Costs: $2,395, includes airfare from New York City, lodging, some meals, planned activities.

Contact: Susanne Servin, Herzerl Tours, 127 W. 26th St., New York, NY 10001 US; 800-684-8488/212-366-4245, Fax 212-366-4195, E-mail sms@herzerltours.com, URL http://www.herzerltours.com.

BELGIUM

FPT SPECIAL INTEREST TOURS (See page 266)

BRAZIL

ACADEMY OF COOKING & OTHER PLEASURES, BRAZIL

Ouro Preto and Paraty/Year-round

Brazilian cuisine expert Yara Castro Roberts sponsors 1-week vacation programs that feature hands-on classes in Brazilian cuisine, including the regional dishes of Amazon, Bahia, and Minas Gerais. Established in 1996. Maximum class/group size: 10-16. 10 programs per year. Facilities: Hands-on classes are held in professional kitchens of 5-star hotels; lectures and videos are presented in conference rooms. Also featured: Day trips to islands, sugar cane distillery tour, coffee plantation, underground gold mine, fine dining, market visits, cultural activities, dancing, craft workshops, sports.

Emphasis: Brazilian culinary arts and its relationship with the various aspects of Brazilian culture.

Faculty: Academy founder Yara Castro Roberts is Emmy Award nominee host for PBS-WGBH cooking show series and graduate of the Boston U. Culinary Arts Program. Dr. Moacyr Laterza is a Brazilian history professor. Local chefs and food artisans teach their specialties.

Costs: $2,650, which includes double occupancy lodging, breakfasts, other meals in fine restaurants, planned activities. Lower rate for non-participant guest. Rooms are individually decorated and have private baths.

Location: Ouro Preto, in the mountains of Minas Gerais, an hour from Rio by plane, an hour from Belo Horizonte by car.

Contact: Yara Castro Roberts, Director/Instructor, Academy of Cooking & Other Pleasures, Brazil, 256 Marlborough St., Boston, MA 02116 US; 617-262-8455, Fax 617-267-0786, E-mail rroberts@tiac.net.

SANDY'S BRAZILIAN & CONTINENTAL CUISINE (See page 202)

CANADA

ART OF FOOD COOKING SCHOOL

Toronto/Year-round

Culinary professional Merla McMenomy offers 3 weekly demonstration classes. Established in 1991. Maximum class/group size: 10. Also featured: Spring culinary tours to Provence, private instruction.

Emphasis: International cuisines and special occasion menus.
Faculty: Merla McMenomy, IACP member, who studied at Peter Kump's New York Cooking School and La Varenne; local guest chefs.
Costs: $68, payable with registration.
Location: Mid-town Toronto, 45 miles from Toronto International Airport.
Contact: Merla McMenomy, Art of Food Cooking School, 98 Walker Ave., Toronto, ON, M4V 1G2 Canada; 416-975-5088, Fax 416-960-9337, E-mail henley@interlog.com.

BIRTHE MARIE'S COOKING SCHOOL

Brampton/September-June

Culinary professional Birthe Macdonald offers 1 or 2 demonstrations per week. Established in 1977. Maximum class/group size: 12. Facilities: Home kitchen.
Faculty: IACP-member Birthe Macdonald learned to cook in Denmark.
Costs: Approximately C$50 per session.
Location: Toronto.
Contact: Birthe Macdonald, Birthe Marie's Cooking School, 88 Hillside Dr., Brampton, ON, L6S 1A6 Canada; 905-453-6647.

BONNIE STERN SCHOOL OF COOKING

Toronto/Year-round

This private school offers demonstration and participation classes and series. Established in 1973. Maximum class/group size: 30 demo. 100+ programs per year. Facilities: Interchangeable demonstration/participation area with overhead mirror. Also featured: Private group classes.
Emphasis: Basic techniques, ethnic and regional cuisines, low fat cookery, holiday menus.
Faculty: Bonnie Stern, a George Brown College graduate, studied with Simone Beck and Marcella Hazan and authored 6 cookbooks, including Simply Heartsmart Cooking and Bonnie Stern's Appetizers; Linda Stephen, a George Brown College graduate; and guest instructors.
Costs: C$70/session, C$260/6-week course.
Contact: Bonnie Stern, Founder, Bonnie Stern School of Cooking, 6 Erskine Ave., Toronto, ON, M4P 1Y2 Canada; 416-484-4810, Fax 416-484-4820, E-mail bonnie@bonniestern.com, URL http://www.bonniestern.com.

CANADORE COLLEGE ARTSPERIENCE

(See also page 140) **North Bay/July**

This Summer School of the Arts offers a 1-month arts program featuring weekly 5-day hands-on classes in Italian, Greek, Tex-Mex, and Chinese cuisines. Other workshops include crafts, visual and media arts, creative writing, dance, youth programs. Established in 1978. Facilities: Canadore College facilities. Also featured: Performances, exhibits, art talks, readings, discussions, social activities.
Emphasis: Culinary experience weeks focusing on different cultures.
Faculty: Chefs and kitchen technicians from Canadore College's School of Hospitality & Tourism.
Costs: C$102 + materials fee. College residence lodging C$96/week.
Location: The 700-acre campus is situated on a wooded escarpment overlooking Lake Nipissing and the city of North Bay.
Contact: Joan Annandale, Canadore College Artsperience, 100 College Dr., P.O. Box 5001, North Bay, ON, P1B 8K9 Canada; 705-474-7600 x5401, Fax 705-494-7462, E-mail annandaj@cdrive.canadorec.on.ca, URL http://www.canadorec.on.ca.

CAREN'S COOKING SCHOOL

Vancouver/September-May

This private school offers 3- to 4-session evening demonstration courses. Established in 1978. Maximum class/group size: 32. 60 programs per year. Facilities: Overhead mirror, butcher block demonstration table, 8 gas burners. Also featured: Wine classes, classes for children, culinary tours to Europe.

Emphasis: Italian, French, and Continental Asian cuisine.
Faculty: Owner Caren McSherry-Valagao, CCP, trained at the Cordon Bleu, the CIA, and The Oriental in Bangkok. She studied with Julia Child, Jacques Pépin, and Paul Prudhomme.
Costs: Range from C$40-C$50 per class. Cancellations with 1 week notice receive credit.
Location: Ten minutes from downtown, near Victoria and Hastings St.
Contact: Caren McSherry-Valagao, Caren's Cooking School, 1856 Pandora St., Vancouver, BC, V5L 1M5 Canada; 604-255-5119, Fax 604-253-1331.

CHEZ SOLEIL
Stratford/Year-round

This bed & breakfast offers demonstration and hands-on cooking weekends and classes. Established in 1996. Maximum class/group size: 6. Facilities: Commercial kitchen. Also featured: Wine instruction, classes for youngsters, day trips, market visits, private instruction, cooking library, discussions on cooking resources.
Emphasis: The guests' choice. Can include vegetarian, Tuscan, East Indian, food styling, food economy, appetizers, breads, entertaining.
Faculty: Liz Mountain, 13 years chef experience, graduate of George Brown College; Janet Sinclair, graduate and valedictorian of Stratford Chefs School class of 1995.
Costs: C$250/person for a weekend, which includes 2 nights lodging at an English Tudor cottage bed & breakfast with 3 private rooms and baths, breakfasts, and dinner originating from the day's class.
Location: Near Stratford Shakespeare Theatre, 3 minutes from downtown, 90 minutes from the Toronto airport.
Contact: Janet Sinclair, Chez Soleil Cooking School, 120 Brunswick St., Stratford, ON, N5A 3M1 Canada; 519-271-7404, Fax 519-271-7404. Additional contact: Liz Mountain.

COOKING LIGHT
Montreal/September-May

Culinary professional Marilyn Flaherty offers demonstration and participation classes. Established in 1989. Facilities: Small, well-equipped kitchen.
Emphasis: Healthful foods.
Faculty: Marilyn Flaherty earned a bachelor's degree in Home Economics.
Costs: A 3-class series is C$85.
Location: Pointe Claire, a Montreal suburb.
Contact: Marilyn Calder Flaherty, Cooking Light, Fish, Beans, Grains & Lentils, 29 Cedar Ave., Pointe Claire, QB, H9S 4X9 Canada; 514-695-4117.

COOKING AT THE PRUNE
Stratford/April-May

This restaurant offers hands-on weekend menu courses. Established in 1990. Maximum class/group size: 12. 5-6 programs per year. Facilities: Professional restaurant kitchen. Also featured: Market visits, visits to food producers, dining in private homes.
Emphasis: Restaurant specialties.
Faculty: 4-star Chef Bryan Steele, culinary instructor at Stratford Chefs School.
Costs: C$70-C$85/class.
Contact: Bryan Steele, Cooking at the Prune, 151 Albert St., Stratford, ON, N5A 3K5 Canada; 519-271-5052, Fax 519-271-4157, E-mail ddprune@cyg.net, URL http://www.cyg.net/~oldprune/.

COOKING STUDIO
Winnipeg/September-June

Culinary professional Marisa Curatolo offers 4- and 6-session demonstration and participation courses. Established in 1994. Maximum class/group size: 14. Facilities: 1,000-foot commercial kitchen. Also featured: Children's birthday parties, private dinners, Saturday workshops.
Emphasis: Ethnic and regional cuisines, nutrition, guest chef specialties.
Faculty: Owner Marisa Curatolo earned a degree in Foods and Nutrition from the University of

Manitoba, completed Chef Training at the Dubrulle French Culinary School, and studied at Peter Kump's New York School, École LeNôtre and Le Cordon Bleu.

Costs: The 4-session course is $149.80, 4-session course is $150; Saturday workshops range from $40-$45 each.

Contact: Marisa Curatolo, Cooking Studio, 3200 Roblin Blvd., Winnipeg, MB, R3R OC3 Canada; 204-896-5174, Fax 204-888-0628.

COOKING WITH SUSAN LEE

London/January-March, May-July, September-November

Culinary professional Susan Lee offers 3-hour participation classes. Established in 1988. Maximum class/group size: 12. Facilities: Large family kitchen. Also featured: Private lessons.

Emphasis: Seasonal foods, ethnic and regional cuisines, entertaining menus.

Faculty: Home economist and food writer Susan Lee and local chefs.

Costs: Classes are C$45 each.

Contact: Susan Lee, Cooking with Susan Lee, 1011 Wellington St., London, ON, N6A 3T5 Canada; 519-439-1423.

THE COOKING WORKSHOP

Toronto/January-April, October-November

This private school offers weekend hands-on workshops, evening workshops/demonstrations. Established in 1985. Maximum class/group size: 8-12, private classes up to 24. Facilities: Industrial kitchen of Dufflet Pastries, a cafe/bakery equipped with skylights and double ovens; a private home kitchen, and large cooking labs. Also featured: Wine-tastings, private classes for groups (including large, corporate events), culinary tours to Italy and Toronto.

Emphasis: Italian cuisine, breads, pastry, foundations (soups, stocks, sauces, dressings).

Faculty: Maria Pace, author of The Little Italy Cookbook, studied at La Varenne and with Marcella Hazan. Baker Paula Bambrick trained at George Brown College. Doris Eisen creates bread and pastry recipes. Chef Steve Jukic, European trained, leads culinary events.

Costs: Range from C$75-C$90. Group rates available.

Location: Three blocks west of Bathurst Street in Toronto's Little Italy; Dufflet Pastries (787 Queen St. West). The Columbus Centre or Humber College.

Contact: Maria Pace, Owner, The Cooking Workshop, 10 Beaconsfield Ave., Ste. 2, Toronto, ON, M6J 3H9 Canada; 416-588-1954, Fax 416-588-1954, E-mail marypace@enoreo.on.ca.

COOKSCHOOL AT THE COOKSHOP

Vancouver/Year-round

This cookware store offers demonstration and participation classes, 2 classes/day, 6 days/week. Established in 1992. Maximum class/group size: 8-20. Facilities: 1,000-square-foot area with overhead mirror. Also featured: Private lessons/functions, wine pairing, nutrition counseling, cookbook author signings, guided tours of wineries and other facilities.

Emphasis: Fresh and healthy.

Faculty: School director, restaurateur, and teacher Nathan Hyam; 60 guest chefs from local hotels and restaurants.

Costs: Range from C$19-C$199 per class. Payment in advance. Good hotels within 5 minute walk.

Location: Downtown Vancouver.

Contact: Peter Haseltine, COOKSCHOOL at the COOKSHOP, 3-555 W. 12th Ave., Vancouver, BC, V5Z 3X7 Canada; 604-873-5683, Fax 604-876-4391.

DUBRULLE INTERNATIONAL CULINARY & HOTEL INSTITUTE

(See also page 141) **Vancouver/Year-round**

This private career school offers demonstration and hands-on evening, weekend and multi-week techniques courses. Established in 1982. Facilities: The 6,000-square-foot facility has classrooms, 3 teaching kitchens with fully-equipped working stations, student dining areas. Also featured: Wine instruction, classes for youngsters, custom classes.

Emphasis: Basic and advanced techniques, regional Italian cuisine, Thai cooking, Provencal French cuisine.
Faculty: Classically-trained chefs.
Costs: C$55-C$785.
Location: A block from Broadway and Granville, 3 hours from Seattle.
Contact: Robert Sung, Director of Admissions, Dubrulle French Culinary School, 1522 W. 8th Ave., Vancouver, BC, V6J 4R8 Canada; 604-738-3155/800-667-7288, Fax 604-738-3205, E-mail cooking@dubrulle.com, URL http://www.dubrulle.com.

EMPIRE COOKING SCHOOL
Woodstock

Cookbook author Charlotte Empringham offers demonstration classes. Established in 1983.
Emphasis: Microwave cookery.
Faculty: Home economist and cookbook author Charlotte Empringham.
Costs: C$25 per hour for private classes, C$185 for a group, plus food and mileage.
Contact: Charlotte Empringham, Empire Cooking School, 124 John Davies Dr., Woodstock, ON, N4T 1N2 Canada; 519-421-2837.

GREAT COOKS
Toronto/September-June

This cooking school offers afternoon and evening demonstration and hands-on classes. Established in 1989. Maximum class/group size: 24. 60+ programs per year. Facilities: 600-square-foot kitchen with overhead mirrors, similar to a chef's table. Also featured: Wine tastings, group classes, culinary trips, tours of Toronto markets, dining in top restaurants.
Emphasis: International and regional cuisines, menus for entertaining, vegetarian meals, local and out of town guest chef specialties, wine appreciation.
Faculty: More than 30 Toronto chefs, including Mark McEwan, Arpi Magyar, Jean Pierre Challet, Martin Kouprie and Dufflet Rosenberg, owner of Dufflet Pastries.
Costs: Range from C$70-C$100/class includes dinner and wine.
Location: Downtown Toronto.
Contact: Esther Rosenberg, Proprietor/Director, Great Cooks, 787 Queen Street West, Toronto, ON, M6J 1G1 Canada; 416-703-0388, Fax 416-703-9832, E-mail esther@ciphermedia.com, URL http://www.ciphermedia.com/greatcooks.

HEALTHY GOURMET INDIAN COOKING
Oakville, Ontario/September-July

This private school Established in 1993. Maximum class/group size: 8. Facilities: Kitchen and dining area.
Emphasis: Simple, healthy, nutritious and economical home-style Indian cooking that isn't overspiced, greasy, or labor-intensive.
Faculty: Arvinda Chauhan has been teaching since 1993. She has made guest appearances on TV and conducted demos at the annual Good Food Festival.
Costs: $30, $35 and $40 per class. $135 for 6-week Indian beginners class.
Location: 30 minutes west of Toronto.
Contact: Arvinda Chauhan, Cooking Instructor, Healthy Gourmet Indian Cooking, 1334 Creekside Dr., Oakville, ON, L6H 4Y2 Canada; 905-842-3215, E-mail hgic@interlog.com.

HOLLYHOCK FARM
Cortes Island/November-May

This center for healing arts offers workshops in the practical, creative, spiritual, and healing arts, including cooking. Established in 1983. 70 programs per year. Also featured: Yoga, meditation, birdwalks, body work, star talks, drawing, painting, dancing, ceramics, photography, writing, kayaking.

Emphasis: Open-pit style cooking of seafood and fresh produce.
Faculty: Includes cook, teacher, and author James Barber.
Costs: C$900, which includes dormitory or semi-private lodging and meals.
Location: About 100 miles north of Vancouver, Canada.
Contact: Oriane Lee Johnston, Program Director, Hollyhock Farm, Box 127, Manson's Landing, Cortes Island, BC, V0P 1K0 Canada; 800-933-6339, Fax 604-935-6424, E-mail hollyhock@oberon.ark.com, URL http://www.go-interface.com/hollyhock.

THE INN AT BAY FORTUNE

Bay Fortune/May-October

This country inn rated three stars by Where to Eat in Canada offers 1-day participation sessions that focus on enhancing tasting ability, understanding ingredients, and a behind the scenes look at a working kitchen. Established in 1992. Maximum class/group size: 2-20. Facilities: Fully-equipped professional kitchen, library, herb gardens. Also featured: Wild mushroom (seasonal) picking, herb and vegetable gardening, field trips, Chef's Tasting Menu.
Emphasis: Contemporary creative cuisine in a country inn setting.
Faculty: New York Chef Michael Smith, CEC, a CIA graduate.
Costs: Tuition is C$150, which includes meals and kitchen gear. Double occupancy country inn rooms are C$130 to C$185 with full breakfast.
Location: 45 minutes from the nearest airport in Charlottetown.
Contact: Chef Michael Smith, The Inn At Bay Fortune, Bay Fortune, PEI, C0A 2B0 Canada; 902-687-3745, Fax 902-687-3540, E-mail innatbayft@auracom.com, URL http://www.innatbayfortune.com.

MAMMA WANDA COOKING SCHOOL

Kirkland/May-August

Culinary professional Wanda Calcagni offers demonstrations and hands-on classes. Maximum class/group size: 8.
Emphasis: Northern Italian cuisine.
Location: 10 minutes from Montreal City.
Contact: Wanda Calcagni, Mamma Wanda Cooking School, 1 Viger St., Kirkland, QB, H9J 2E4 Canada; 514-695-0864.

MANOR CUISINE'S CREATIVE COOKING

Pointe Claire/September-June

Culinary professional Ausma Groskaufmanis offers 6- to 8-session beginner to advanced courses (7 levels). Established in 1984. 10 programs per year.
Emphasis: Techniques, creative presentation, boning and butterflying, garnishing, napkin folding, cake decorating, contemporary low fat and low cholesterol cuisine, and fruit, vegetable, and ice sculpture.
Faculty: Ausma Groskaufmanis, B.Sc.
Costs: C$23 to C$25 per session.
Location: Montreal.
Contact: Ausma Groskaufmanis, Manor Cuisine's Creative Cooking, 6 Manor Crescent, Pointe Claire, QB, H9R 4S9 Canada; 514-697-7015, Fax 514-697-7015, E-mail ogga@axess.com.

McCALL'S SCHOOL OF CAKE DECORATION, INC.

(See also page 144) **Etobicoke**

This private school offers all-day hands-on workshops and 1- to 4-session cake decorating, baking, chocolate, and specialty courses. Established in 1976. Facilities: 1,000 sq ft of teaching space with overhead mirrors and two 20-seat classrooms.
Faculty: Includes school director Nick McCall, and Kay Wong.
Costs: C$50-C$100/session. Lodging available at area hotels.
Location: A western subdivision of Toronto.

Contact: Nick McCall, President, McCall's School of Cake Decoration, Inc., 3810 Bloor St. West, Etobicoke, ON, M9B 6C2 Canada; 416-231-8040, Fax 416-231-9956, E-mail decorate@mccalls-cakes, URL http://www.mccalls-cakes.com.

---

*

MY PLACE FOR DINNER

Toronto, Ontario/Year-round

Culinary professional Debbie Diament offers hands-on cooking workshops three times weekly (evenings and Saturdays). Established in 1996. Maximum class/group size: 6. Facilities: Kitchen and dining room. Also featured: Private group workshops.

Emphasis: Ethnic cuisines, low-fat, entertaining and holiday menus.

Faculty: Debbie Diament, cookbook writer, recipe tester.

Costs: $60/workshop.

Location: Toronto's Riverdale neighborhood, 1 block north of The Danforth, 1 block east of Chester subway.

Contact: Debbie Diament, My Place for Dinner, 56 Arundel Ave., Toronto, ON, M4K 3A4 Canada; 416-465-7112, Fax 416-465-7112, E-mail mpfd@idrect.com.

NATURAL FOODS COOKING SCHOOL

Montreal/January-May, October-November

Culinary professional Bonnie Tees offers The Basics of Healthy and Inter-natural Cooking program, which consists of 6-session participation classes and Saturday workshops. Established in 1988. Maximum class/group size: 8. Facilities; Private kitchen with large gas stove and oven, working island. Also featured: Individual instruction and group classes on specific themes.

Emphasis: Healthy and low-fat foods, food history, ethnic cuisine, cutting techniques.

Faculty: Founder Bonnie Tees has 10 years' experience in natural cooking. She was head cook at the Macrobiotic Institute of Switzerland and studied at The Natural Gourmet Cooking School in New York.

Costs: C$175 for 5 sessions.

Contact: Bonnie Tees, Natural Foods Cooking School, 4865 Harvard, #6, Montreal, QB, H3X 3P1 Canada; 514-482-1508.

NEELAM KUMAR'S NORTH INDIAN CUISINE

Kirkland/January, April, September

Culinary professional Neelam Kumar offers 6-session participation courses. Established in 1984. Maximum class/group size: 7. 4 programs per year. Facilities: 140-square-foot kitchen.

Emphasis: Vegetarian and non-vegetarian Indian cuisine.

Faculty: Neelam Kumar has a university degree in Home Science.

Costs: C$100.

Location: Western Montreal.

Contact: Neelam Kumar, Neelam Kumar's North Indian Cuisine, 6 Daudelin St., Kirkland, QB, H9J 1L8 Canada; 514-697-4029, Fax 514-737-2342.

RUNDLES RESTAURANT COOKING SCHOOL

Stratford/March-May

This restaurant offers hands-on weekend courses that feature four menus. Established in 1987. Maximum class/group size: 14. 7 programs per year. Facilities: Commercial restaurant kitchen.

Emphasis: Restaurant menus.

Faculty: Rundles Chef Neil Baxter, 20 yrs experience, European trained, head of cookery at Stratford Chefs School.

Costs: C$295, which includes uniform.

Location: Newar Stratford Shakespeare Theatre, 90 minutes from Toronto.

Contact: Neil Baxter, Rundles Restaurant Cooking School, 9 Cobourg St., Stratford, ON, N5A 3E4 Canada; 519-271-6442, Fax 519-273-6603.

SHERWOOD INN COUNTRY COOKING WEEKENDS

Port Carling/January, Marcy, May, November-December

This country inn offers hands-on theme weekends. Established in 1992. Also featured: Wine discussion and restaurant kitchen tour.

Faculty: The chefs of Sherwood Inn.

Costs: Cost ranges from C$338 to C$427, depending on accommodation, and includes lodging at the CAA/AAA 4-Diamon Sherwood Inn, most meals, health club, mountain biking, and cross country skiing equipment in the winter.

Location: 2 hours north of Toronto.

Contact: Arthur Lambert, Assistant General Manager, Sherwood Inn Country Cooking Weekends, P.O. Box 400, Port Carling, ON, P0B 1J0 Canada; 705-765-3131, Fax 705-765-6668, E-mail sherwood@muskoka.com, URL http://www.countryinns.com.

CHINA

CHOPSTICKS COOKING CENTRE

(See also page 146) **Kowloon/September-January, April-June**

This private trade school offers customized gourmet tours of varying duration that include a demonstration lesson, dry and wet market visits, restaurant kitchen visits, and dining in local restaurants. Maximum class/group size: 20 maximum. Facilities: Professional kitchen with facilities for participation sessions. Also featured: 1- to 4-hour demonstration classes, half-day tourist classes, 1-day courses, 1- and 4-week intensive course.

Emphasis: Understanding of Chinese culinary culture.

Faculty: Cecilia J. Au-Yang, proprietor, with 30+ yrs experience; Caroline Au-Yeung, graduate of hotel management school, with 10+ yrs experience; other professionals.

Costs: Tour cost begins at $900, which includes most meals, ground transport. Group classes begin at $90 for a 2-hour demonstration, $500-$900 for an individually selected 1-day course. Local lodging averages $650/week.

Location: Kowloon, in the city center.

Contact: Caroline Au-yeung, Director, Chopsticks Cooking Centre, 108 Boundary St., G/Fl., Kowloon, Hong Kong, China; (852) 2336-8433, Fax (852) 2338-1462, E-mail cauyeung@netvigator.com.

THE HOME MANAGEMENT CENTRE

Year-round

The Public Affairs Dept. of The Hongkong Electric Co., Ltd. sponsors 1- to 8-session courses in northern and southern Chinese cookery, European and southeast Asian cuisines, including dim sum, sushi, Italian pasta, English cookies. Established in 1977. Maximum class/group size: Up to 40. 400+ programs per year. Facilities: 6,500-square-foot facility with two electric cooking theatres and two special interest lecture rooms. Also featured: Programs for youngsters, private instruction; special interest classes in sewing, flower arrangement, knitting, pando arts.

Emphasis: No-flame cooking.

Costs: HK$85-HK$450/course.

Contact: Teresa Tang, Manager, The Home Management Centre, 10/F, Electric Centre, 28 City Garden Rd., North Point, Hong Kong, China; (852) 2510-2828, Fax (852) 2510-7870.

HONG KONG FOOD FESTIVAL

March

The Hong Kong Tourist Association sponsors an annual 2-week event featuring approximately 20 Cooking with Great Chefs demonstration classes that are held in the kitchens of participating hotels and restaurants. Maximum class/group size: 15. 1 programs per year. Also featured: T'ai chi lessons, theme parties and banquets, and tours of tea companies, wedding cake bakeries, and Sai Kung villages.

Emphasis: Thai, Cantonese, Japanese, and European specialties.
Contact: Hong Kong Tourist Assn., 18 Whitfield Rd., North Point, Hong Kong, China; (852) 2807-6543, URL http://www.hkta.org. In the U.S.: 590 Fifth Ave., 5th Flr., New York, NY 10036; 212-869-5008/9, Fax 212-730-2605.

TOWNGAS COOKING CENTRE

Year-round

The Hong Kong and China Gas Company Limited sponsors 1- to 4-session demonstration and practical courses on Chinese, Southeast Asian, Western cuisines, including baking and dim sum. Established in 1977. Maximum class/group size: 100 demo, 24 hands-on. Facilities: Theatre-style demonstration room that accommodates 100 persons, practical room that accommodates 24. Also featured: Programs for adults, youngsters, tourists.
Emphasis: Chinese cuisine and dim sum.
Faculty: Professional home economists and freelance tutors.
Costs: HK$75-HK$900.
Contact: Annie Wong, Business Development Mgr., Towngas Cooking Centre, Basement, Leighton Centre, 77 Leighton Rd., Causeway Bay, Hong Kong, China; (852) 2576-1535, Fax (852) 2894-8058, E-mail cooking_centre@hkcg.com, URL http://www.hkcg.com.

ENGLAND

ACORN ACTIVITIES

Herefordshire/April, June, September, November

This activity holiday provider offers a 2-day gourmet cooking course. Established in 1989. Maximum class/group size: 12. Facilities: Well-equipped kitchen with individual work areas and cookers. Also featured: Study tours, courses, and programs relating to air sports, water sports, ball sports, horseback riding, hunting, shooting, falconry, fishing, motor sports, arts & crafts, music, and languages.
Emphasis: Low-fat gourmet cooking.
Faculty: A professional chef who has appeared on the BBC2 Food and Drink program.
Costs: Tuition is £100. Accommodations, including breakfast, range from farmhouses and cottages at £20 per night to luxury hotels at £95 per night.
Location: Herefordshire, near the Welsh border, 120 miles (a 3-hour train ride) west of London.
Contact: Charles Cordle, Managing Director, Acorn Activities, P.O. Box 120, Hereford, HR4 8YB England; (44) (0)1432-830083, Fax (44) (0)1432-830110, E-mail acornactivities@btinternet.com, URL http://www.acornactivities.co.uk`.

AGA WORKSHOP

Buckinghamshire/February-July, September-November

Culinary professional Mary Berry offers 1- and 2-day Aga demonstration workshops. Established in 1990. Maximum class/group size: 20. 50 programs per year. Facilities: Watercroft, Mary Berry's home. Also featured: Half AGA cooking, half gardening days.
Emphasis: Using the Aga cooker for grilling and frying, saving fuel, fresh herbs and vegetables, entertaining and holiday cookery, specialized equipment.
Faculty: Mary Berry studied at the Paris Cordon Bleu and the Bath College of Home Economics and has a City and Guilds teaching qualification. Author of 20 cookery books, she was cookery editor of Ideal Home Magazine and contributes to Radio 2.
Costs: One-day (two-day) workshop is £89.30 (£173.90), which includes lunch and VAT. Group bookings of 4 or more for 1 day are £84.60 per person. A list of nearby bed & breakfasts is provided.
Location: Watercroft, situated on 3 acres of informal garden, is 30 miles from London and Heathrow airport, accessible by railway from London Marylebone.
Contact: Lucy Young, Assistant to Mary Berry, Aga Workshop, Watercroft, Church Rd., Penn, Buckinghamshire, HP10 8NX England; (44) (0)1494-816535, Fax (44) (0)1494-816535.

THE BATH SCHOOL OF COOKERY

Bath, Somerset/Year-round

Bassett House, a listed grade II 18th-century country house, offers 1-day, weekend, 4-day, and 4-week demonstration and participation courses. Established in 1988. Maximum class/group size: 10. 30+ programs per year. Facilities: Large kitchen equipped with Aga cookers, microwaves, and modern appliances and a smaller kitchen for vegetable/salad preparation, pastry, and chocolate. Also featured: Tennis, cycling, croquet, walking.

Emphasis: Everyday French, Italian, basic to advanced food preparation and technique, creative cuisine, seasonal, holiday, and special occasion dishes, ethnic menus.

Faculty: Sallie Caldwell studied in England, France and the Far East, operated a catering firm and restaurant in Bath, and was principal teacher of a cookery school.

Costs: Resident (non-resident) tuition, which includes meals, is £450 (£390) for the 4-day courses, £1,770 (£1,570) for the 4-week Master Course; day demonstrations are £65 each, weekends are £280 (£220).

Location: 10 minutes from Bath, 70 minutes from London, in the Limpley Stoke Valley on 6 acres of herb, fruit, and vegetable gardens, with access to the River Avon.

Contact: D.K.S. Caldwell, Director, The Bath School of Cookery, Bassett House, Claverton, Bath, Somerset, BA2 7BL England; (44) (0)1225-722498, Fax (44) (0)1225-722980.

BOOKS FOR COOKS COOKING SCHOOL

London/May-July, September-March

This specialty bookstore (established 15 years) offers workshops that include The Basics, which guides students through the techniques of a specific area of cookery. Established in 1995. Maximum class/group size: 8 hands-on/22 demo. 100+ programs per year. Facilities: Purpose-built demonstration kitchen with overhead mirror located above the bookstore. Also featured: Children's breadmaking classes, 3-day cookery course.

Emphasis: Techniques, verdura, canapés, flavored breads, stir-fries, anitpasti, French bistro, vegetarian Christmas, Moroccan.

Faculty: Cookbook authors Eric Treuillé, Ursula Ferrigno, Victoria Blashford Snell, Celia Brooks Brown.

Costs: Range from £10 (for children's classes) and £20 for workshops to £60 for 3-day courses.

Location: Just off Portobello Road in London's Notting Hill area.

Contact: Eric Treuillé, Workshops, Books for Cooks Cooking School, 4 Blenheim Crescent, London, W11 1NN England; (44) (0)171-221-1992, Fax (44) (0)171-221-1517, E-mail info@booksforcooks.com.

BUTLERS WHARF CHEF SCHOOL

(See also page 147)

London/Year-round

This private vocational school offers a range of demonstration classes and weekend courses. Established in 1995. Maximum class/group size: ~15. Facilities: Training/production kitchen, specialized demonstration theatre, restaurant, study areas.

Faculty: The 11-member faculty includes Director John Roberts, and chefs Gary Witchalls, Nicky Hopkins, Denzil Newton, Perry Reeves. Guest faculty includes chefs, food and wine experts, and restaurateurs.

Location: Central London.

Contact: Mr. John Roberts, Director, Butlers Wharf Chef School, Cardamom Bldg., 31 Shad Thames, London, SE1 2YR England; (44) (0)171-357-8842, Fax (44) (0)171-403-2638.

CAROLINE HOLMES – GOURMET GARDENING

Paris/Year-round

Culinary professional Caroline Holmes offers 1-day demonstration courses and 5-day tours that visit local food producers and herb gardens in United Kingdom and France, weekends in conjunction with Hilton Hotels. Established in 1983. Maximum class/group size: Minimum 10. Also featured: Tailored itineraries for groups.

Emphasis: Growing, maintaining, and using herbs: East Anglian specialties.

Faculty: Caroline Holmes holds a Certificate in Gourmet Cookery and City and Guilds Horticulture. She works with the Museum of Garden History, Hintlesham Hall, Hilton Hotels, BBC Good Food magazine and was chairman of the Herb Society.

Costs: One-day courses begin at £40; 5-day tours, U.K.-based, from £430; French-based from £330; UK weekends from £150. Lodging is simple to 4-star.

Location: London, Ipswich, Portsmouth and West Country in England; Paris, France.

Contact: Mrs. Susan Standley, Caroline Holmes-Gourmet Gardening, Denham, Bury St. Edmunds, Suffolk, IP29 5EE England; (44) (0)1284-810653, Fax (44) (0)1284-810653.

CONFIDENT COOKING

Wiltshire/February-December

This private school offers monthly demonstrations and residential weekend hands-on courses that include market visits, visits to food producers, mushroom forays, sightseeing, and dining in a country home. Established in 1996. Maximum class/group size: 10 hands-on, 25 demo. Facilities: High-tech professional kitchen with overhead mirror and video monitor. Also featured: Children's classes, private instruction.

Emphasis: Vegetarian cuisine, breads.

Faculty: Cook, author, and food consultant Caroline Yates and guest chefs.

Costs: £30-£35/class, £250-£275 for weekend courses, which includes meals, lodging, and planned activities.

Location: Wiltshire, 10 minutes from Marlborough, 30 minutes from Salisbury, 30 minutes from Bath.

Contact: Caroline Yates, The Manor Farmhouse, Little Cheverell, Devizes, Wiltshire, SN10 4JP England; (44) (0)1380-840396. Phone # after 12//98: (44) (0)1380-812844.

CONSTANCE SPRY COOKERY & FLOWER ARRANGING

Farnham Surrey/September-July

This private school offers demonstrations and hands-on workshops devoted to specialties of the season.

Emphasis: Seasonal specialties.

Faculty: Includes Francesca Fenwick, Annie Grubb, and Elizabeth Bowyer.

Costs: £35 for demonstrations, £55 for workshops.

Contact: Martine Frost, Constance Spry, Ltd., Moor Park House, Moor Park Lane, Farnham Surrey, GU9 8EN England; (44) 1252-734477, Fax (44) 1252-712011.

COOKERY AT THE GRANGE

Frome/Year-round *(See also page 147)*

This private school offers weekend and 5-day hands-on courses in addition to its certificate program. Established in 1981. Maximum class/group size: 14-20. Facilities: Main kitchen, cold kitchen, herb garden. Also featured: Wine instruction.

Emphasis: European and international cuisines, herbs, wine tasting.

Faculty: Jane and William Averill (Grange trained) and teaching staff.

Costs: All-inclusive rates range from £320 for the weekend to £590 for the 5-day program.

Location: 90 minutes from London by train.

Contact: Jane and William Averill, Cookery at The Grange, Whatley, Frome, Somerset, BA11 3JU England; (44) (0)1373-836579, Fax (44) (0)1373-836579, E-mail cookery.grange@clara.net, URL http://www.hi-media.co.uk/grange-cookery.

THE CORDON VERT COOKERY SCHOOL

Altrincham/Year-round *(See also page 148)*

The Vegetarian Society UK offers weekend and day courses on a variety of vegetarian topics in addition to the diploma course. Established in 1984. 40 programs per year.

Faculty: Sarah Brown began the courses in 1982, based on her BBC-TV series, Vegetarian

Kitchen. Tutors include Rachel Markham, Lyn Weller, Deborah Clarke, Chico Francisco, John Williams.

Costs: Weekend residential (non-residential) courses range from £190 (£160); day courses are £45, which includes lunch. Local hotel/motel lodging.

Location: Ten miles south of Manchester.

Contact: Lyn Weller, Cordon Vert Cookery School, Parkdale, Dunham Road, Altrincham, Cheshire, WA14 4QG England; (44) (0) 161-928-0793, Fax (44) (0) 161-926-9182, E-mail vegsoc@vegsoc.demon.co.uk, URL http://www.veg.org/veg/orgs/vegsocuk/.

DIVERTIMENTI COOKERY DEMONSTRATIONS

London/February-July, September-November

This specialist kitchen and tableware store offers informal cookery demonstrations and wine tasting. Established in 1994. Maximum class/group size: 25. Facilities: Cookery theatre in the Divertimenti flagship store.

Emphasis: New World cooking, Mediterranean and Italian cooking. Basics such as pastry, sauces, game.

Faculty: Lyn Hall organizes the demos. Demonstrators include chefs and food writers Alastair Little, Brian Turner, Gordon Ramsay, and Jean-Christophe Novelli.

Costs: £30/session.

Location: Central London, South Kensington tube station.

Contact: Dannielle Gibb, Assistant Manager, Divertimenti Cookery Demonstrations, 139-141 Fulham Rd., London, SW3 6SD England; (44) (0)171-581-8065, Fax (44) (0)171-823-9429, E-mail al.schnei@zetnet.co.uk.

EARNLEY CONCOURSE

Sussex, near Chichester/Year-round

This resident center for courses and conferences founded by the Earnley Trust, Ltd. educational charity offers weekend demonstration and participation courses. Established in 1975. Maximum class/group size: 12. 15 programs per year. Facilities: Fully-equipped kitchen workshop, which has demonstration and dining areas.

Emphasis: Japanese and Chinese cookery, cooking for health, vegetarian dishes, advanced techniques, special occasion dishes, baking, making chocolates.

Faculty: Includes Deh-Ta Hsiung, Steven Page, Mary Whiting, and Lucy Shaw-Baker.

Costs: Each course is priced from £142, which includes lodging and meals. Nonresident tuition is £98, which includes lunch. Cost of ingredients is additional. Amenities include art and craft studios, computer room, heated pool, gardens.

Location: A rural setting in West Sussex, 6 miles south of Chichester.

Contact: Owain Roberts, Earnley Concourse, Earnley, Chichester, West Sussex, PO20 7JL England; (44) (0)1243-670392, Fax (44) (0)1243-670832, E-mail earnley@interalpha.co.uk.

FRANCES KITCHIN COOKING COURSES

Somerset/March-November

Culinary professional Frances Kitchin offers 1-day demonstration courses. Established in 1987. Maximum class/group size: 10. 8 programs per year. Facilities: Large kitchen in a country house.

Emphasis: English, French, Italian, and Indian cuisine.

Faculty: Frances Kitchin, a qualified home economist and chef who has lectured at Strode College for 21 years, is a freelance writer and author of 2 cookbooks, and has a weekly cookery spot on radio, (BBC).

Costs: Day courses are £20, including lunch. B&B within walking distance.

Location: Stoney Mead, a country house, is 12 miles from Taunton, on the railway line from Paddington (London).

Contact: Frances Kitchin, Owner, Frances Kitchin Cooking Courses, Stoney Mead, Curry Rivel, Langport, Somerset, TA10 0HW England; (44) (0)1458-251203, Fax (44) (0)1458-251203.

HAZLEWOOD CASTLE

North Yorkshire/Year-round

This restored Norman castle offers demonstrations of a variety of topics.

Emphasis: Seasonal dishes, individual topics.

Faculty: Guest chefs.

Costs: £18.50-£35/class.

Contact: Hazlewood Castle, Paradise Lane, Hazlewood, North Yorkshire, LS24 9NJ England; (44) (0)1937-535353, Fax (44) (0)1937 530630.

HINTLESHAM HALL

Suffolk/Year-round

This country hotel offers demonstration classes. Maximum class/group size: 12. 20 programs per year. Also featured: Interior design day, flower arranging.

Emphasis: Vegetarian and fish dishes, seasonal menus, herb cookery, sugarcraft, specific topics.

Faculty: Hintlesham's Chef Alan Ford and guest instructors.

Costs: Each class is £48. Nightly lodging, which includes continental breakfast and VAT, begins at £85 single, £115 double. Hintlesham Hall, on 175 acres, offers golf, tennis, snooker, trout fishing, clay pigeon shooting, horseback riding, pool, spa.

Location: About five miles west of Ipswich, Suffolk, an hour drive from Stansted Airport, and an hour train ride from London.

Contact: Claire Hills, Hintlesham Hall, Hintlesham, Ipswich, IP8 3NS England; (44) (0)1473-652268, Fax (44) (0)1473-652463.

ITALIAN SECRETS

Beaconsfield, Bucks/Year-round

This private school offers demonstration and hands-on 1- to 5-day programs. Titles include Basic Italian, Effortless Entertaining, Vegetarian Italian Style, Effortless Entertaining Plus, Italian Cuisine, A Taste of Italy. Established in 1995. Maximum class/group size: 18 demo, 12 hands-on. 40 programs per year. Facilities: High tech kitchen with island for 12 people to work together. Also featured: Private instruction.

Emphasis: Classic Italian cuisine.

Faculty: Anna Venturi, born and raised in Milan, Italy, learned how to cook from grandmother and family cook.

Costs: £46 to £159.

Location: Beaconsfield, small town 25 miles northwest of London, easily accessible by train and motorway.

Contact: Anna Venturi, Owner, Italian Secrets, 13 The Broadway, Penn Rd., Beaconsfield, Bucks, HP9 2PD England; (44) 1494 676136, Fax (44) 1494 714599, E-mail anna@italiansecrets.co.uk, URL http://www.italiansecrets.co.uk.

LA CUISINE IMAGINAIRE VEGETARIAN COOKERY SCHOOL

Hertfordshire/Year-round

This private school offers full- and half-day demonstration courses and 4-day practical certificate courses. Established in 1989. Maximum class/group size: 6 hands-on/12 demo. Facilities: Well-equipped demonstration and practical kitchen, lounge area. Also featured: Private instruction.

Emphasis: Vegetarian cuisine for dinner parties, breadmaking, international dishes using unusual ingredients.

Faculty: Director and instructor Roselyne Masselin is a food writer, cookbook author, and runs a large catering company, Catering Imaginaire.

Costs: Half-day Bushey course is £45, four-day course is £435.

Location: Bushey, Hertfordshire, 30 minutes from Heathrow airport and central London.

Contact: Roselyne Masselin, La Cuisine Imaginaire Vegetarian Cookery School, PO Box 70, Bushey, Hertfordshire, WD2 2NQ England; (44) (0)1923-250099, Fax (44) (0)1923-250030.

LA PETITE CUISINE

London/Year-round

Culinary professional Lyn Hall, former chef and restaurateur offers 6-session master classes. Established in 1977. Maximum class/group size: 1 or 6. Facilities: Bulthaup's kitchen studio. Also featured: Private hands-on 5-lesson courses in Lyn Hall's home teaching kitchen.

Emphasis: Modern gourmet cuisine, French, Italian, Chinese, Italian.

Faculty: Owner Lyn Hall, BA, MFCA holds certificates in wine, bread, cake decorating, and butchery, and is an Olympic Culinary Gold Medalist.

Costs: Master class series is £550, private courses are £800 for five lessons.

Location: South Kensington in London, Bulthaup on Wigmore St., West End.

Contact: Lyn Hall, La Petite Cuisine, 21 Queen's Gate Terrace, London, SW7 5PR England; (44) 171-584-6841, Fax (44) 171-225-0169.

LE CORDON BLEU – LONDON

(See also pages 154 and 301) (See display ad page 149) **London/Year-round**

This private school acquired by Le Cordon Bleu-Paris in 1990 offers daily half-day demonstrations, 1-, 3- and 5-day gourmet hands-on sessions, evening classes, 1-month courses. Established in 1933. Maximum class/group size: 8. 50 programs per year. Facilities: Specialized equipment; classrooms resemble professional kitchens; individual workspaces with refrigerated marble tops; demonstration rooms with video. Also featured: Children's workshops, wine classes, cheese courses, guest chef demonstrtions, culinary excursions.

Emphasis: Intensive and comprehensive culinary cuisine and pastry courses.

Faculty: All staff full time. Chefs all professionally qualified with experience in Michelin-starred and fine quality culinary establishments.

Costs: Range from £15-£415.

Location: London's West End, close to Oxford and Bond Streets.

Contact: Anne-Laure Trehorel, Enrollment Officer, Le Cordon Bleu, 114 Marylebone Lane, London, W1M 6HH England; (44) (0)171-935-3503, Fax (44) (0)171-935-7621, E-mail information@cordonbleu.co.uk, URL http://www.cordonbleu.net. Toll free in the U.S. and Canada: 800-457-CHEF.

LEITH'S SCHOOL OF FOOD AND WINE

(See also page 148) **London/Year-round**

This private school offers 1- and 4-week holiday courses; 1-week fish, healthy eating, dinner party, and Christmas cooking courses; 10-session beginner and advanced evening courses, Saturday demonstrations, 5-session Certificate in Wine course, special wine and food classes. Established in 1975. Maximum class/group size: 48 demos, 16 hands-on. Facilities: Demonstration theatre, 3 kitchens, preparatory kitchen, library, changing room. Also featured: Private instruction.

Faculty: 13 full-, 2 part-time. School founder and cookbook author Prue Leith is former Veuve Cliquot Business Woman of the Year. Principal is Caroline Waldegrave, vice-principals is A. Cavaliero.

Costs: Basic Certificate course is £1,510, 10-session courses are £400, demonstrations are £45, Wine Certificate course is £200. Housing list provided.

Location: A refurbished Victorian building in Kensington, the center of London.

Contact: Judy Van DerSande, Registrar, Leith's School of Food and Wine, 21 St. Alban's Grove, London, W85 5BP England; (44) (0)171-229-0177, Fax (44) (0)171-937-5257, E-mail info@leiths.com, URL http://www.leiths.com.

THE MANOR SCHOOL OF FINE CUISINE

(See also page 150) **Widmerpool/Year-round**

This private school offers a 5-day Foundation course, 4-day Entertaining course, theme weekends, and day and evening courses. Also featured: Water sports, clay pigeon shooting, horseback riding, golf.

Emphasis: Vegetarian, Thai, and Indian cooking and wine courses. Focus on healthy eating, holiday cookery, seasonal menus, Aga cookery, specific topics.

Costs: Inclusive of VAT, tuition is £390 resident (£295 nonresident) for Foundation course, £330 (£285) for Entertaining course, £130 for weekend courses. Lodging at the Manor.

Contact: The Manor School of Fine Cuisine, Old Melton Road, Widmerpool, Nottinghamshire, NG12 5QL England; (44) (0)1949-81371.

THE MOSIMANN ACADEMY

London/Year-round *(See also page 150)*

This private school offers half- and one-day demonstrations and courses that focus on such themes as Italian, Thai, Spanish, Indian, chocolate, pastry, and wines. Established in 1996. Maximum class/group size: 60. 50 programs per year. Facilities: New seminar and demonstration theater, library of Anton Mosimann's 6,000 cookery books. Also featured: Tailor-made corporate courses.

Emphasis: The Anton Mosimann philosophy (Cuisine Naturelle), which eschews the use of fats and alcohol.

Faculty: 3 full- and 3 part-time, including Anton Mosimann, author of 9 cookery books, Shaun Hill, John Burton-Race, Brian Turner, Jean-Christophe Novelli.

Costs: £85 for demonstration and lunch at Mosimann's Dining Club in Belgravia, £35 demonstration only.

Location: Centrally located in Battersea, London.

Contact: The Mosimann Academy, 5 William Blake House, The Lanterns, Bridge Lane, London, SW11 3AD England; (44) 171-924-1111, Fax (44) 171-924-7187, E-mail academy@mosimann.com, URL http://www.mosimann.com.

THE MURRAY SCHOOL OF COOKERY

Farnham, Surrey/Year-round

This private school, formerly the Winkfield Place cookery school, offers demonstration theme classes, hands-on master classes covering specific dishes, 4-week courses covering a variety of topics, cookery holidays in France and Italy. Established in 1989. Facilities: The wing of a large country house.

Emphasis: Theme classes, master classes, cookery holidays.

Costs: Day demos £33, master classes £67.50, 4-week courses £900 (£300/week), trips to the Dordogne £350 include full board, Sicily £650 (£870) includes airfare, shared (single) lodging, and half board.

Location: 35 miles from London. Cookery holidays are scheduled to The Dordogne, France, and Sicily, Italy.

Contact: Paulette M. Murray, The Murray School of Cookery, Glenbervie House, Holt Pound, Farnham, Surrey, GU10 4LE England; (44) 1420-23049.

PAUL HEATHCOTE'S SCHOOL OF EXCELLENCE

Manchester/Year-round *(See also page 151)*

This private school offers demonstrations and hands-on classes on a variety of topics. Established in 1997. Maximum class/group size: 12 hands-on, 40 demo. Facilities: Professionally-equipped kitchen with 6 individual work spaces, demonstration auditorium with projector and screen.

Faculty: Paul Heathcote, chef and owner of four restaurants, and his staff of instructors.

Location: Central Manchester, off Deansgate.

Contact: Administration Office, Paul Heathcote's School of Excellence, Jackson Row, Deansgate, Manchester, M2 5WD England; (44) (0)161-839-5898, Fax (44) (0)161-839-5897, E-mail cookeryschool@heathcotes.co.uk, URL http://www.heathcotes.co.uk.

RAYMOND BLANC'S LE PETIT BLANC ÉCOLE DE CUISINE

Oxford/October-April *(See also page 290. Cuisine International)*

Le Manoir country inn offers 5-day cooking vacations that feature hands-on stage 1, stage 2, and stage 3 classes. Established in 1991. Maximum class/group size: 8. Facilities: Individual work areas in the restaurant kitchen.

Emphasis: Contemporary French cuisine, including appetizers, fish and vegetables, meat and vegetables, pastries.
Faculty: Chef Raymond Blanc owns and operates Le Manoir. Sessions are taught by head chef Clive Fretwell.
Costs: Course cost of £1,150 includes all meals, service, VAT, and lodging at Le Manoir aux Quat' Saisons, which has the highest classification of Relais & Châteaux. Lodging free for non-cooking guests.
Location: Seven miles from Oxford and 40 miles from London.
Contact: Raymond Blanc, Le Manoir aux Quat' Saisons, Great Milton, Oxford, NO OX9 7PD England; (44) (0)1-844-278-881. In the U.S.: Judy Ebrey, Cuisine International, P.O. Box 25228, Dallas, TX 75225; 214-373-1161, Fax 214-373-1162, E-mail CuisineInt@aol.com, URL http://www.cuisineinternational.com.

SQUIRES KITCHEN INTERNATIONAL SCHOOL

(See also page 151) **Farnham, Surrey/Year-round**

This private school offers 1-hour to 3-day demonstration and participation courses. Established in 1987. Maximum class/group size: 12 hands-on/35 demo. Facilities: A kitchen with specialized equipment and materials. Also featured: Classes for youngsters, private instruction.
Faculty: 17 full- and part-time. Members of the British Sugarcraft Guild. Guest tutors include Eddie Spence, Alan Dunn, Toribi Peck.
Location: Period building in a suburban area, a short walk to train station, 45 minutes from London.
Contact: Course Coordinator, Squires Kitchen, Int'l School of Sugarcraft & Cake Decorating, 3 Waverley Lane, Farnham, Surrey, GU9 8BB England; (44) (0)1252-711749, Fax (44) (0)1252-714714, E-mail school@squires-group.co.uk, URL http://www.squires-group.co.uk.

TANTE MARIE SCHOOL OF COOKERY

(See also page 151) **Surrey/Year-round**

This private school offers certificate courses thrice yearly, 3- to 5-day hands-on courses, 1-day theme demonstrations. Established in 1954. Maximum class/group size: 4 groups of 12 students each. Facilities: Include 5 modern teaching kitchens, a mirrored demonstration theatre, and a lecture room. Also featured: Wine instruction, private group demonstrations.
Emphasis: Basic and advanced skills, wine appreciation, specific subjects.
Faculty: 12 full- and part-time. Qualified to work in state schools and many have held catering positions. All undergo teacher training. Well-known TV cookery demonstrators, a noted wine expert, and local tradesmen also present.
Location: A turn-of-the-century country mansion near the center of Woking, a small country town approximately 25 minutes by train from London.
Contact: Margaret A. Stubbington, Registrar, Tante Marie School of Cookery, Woodham House, Carlton Rd., Woking, Surrey, GU21 4HF England; (44) (0)1483-726957, Fax (44) (0)1483-724173, E-mail info@tantemarie.co.uk, URL http://www.tantemarie.co.uk.

EUROPE

THE ART OF LIVING – CULINARY TOURS WITH SARA MONICK

(See display ad page 290) **Tuscany/March-May, September-November**

Culinary and travel professional Sara Monick offers 5- to 11-day culinary tours that include hands-on and demonstration classes. Established in 1986. Maximum class/group size: 8-14. Also featured: Visits to markets, food producers, wineries, private homes and gardens; dining at fine restaurants; sightseeing; language classes; custom tours for private groups.
Emphasis: French, Italian, Spanish, and Moroccan cuisines.
Faculty: Tour escort Sara Monick, a cooking instructor since 1977 and Certified Member of the IACP, owns The Cookery in Minneapolis. She studied with Madeleine Kamman, Jacques Pépin, Nicholas Malgieri, and Giuliano Bugialli. Local chefs.

Costs: $2,300-$4,000, which includes double occupancy lodging, most meals, and planned excursions. Lodging in first-class hotels or private homes.
Location: France (Provence, Paris), Italy (Tuscany, Amalfi Coast), Spain (Barcelona), Morocco.
Contact: Sara Monick, Hilliard & Olander Ltd., 608 Second Ave. So., Minneapolis, MN 55402 US; 612-333-1440, 800-229-8407, Fax 612-333-3554, E-mail DHilli608@aol.com, URL http://www.HilliardOlander.com.

CHARLOTTE-ANN ALBERTSON'S COOKING SCHOOL (See page 248)

CUISINE ECLAIRÉE
Marrick, North Yorkshire/Year-round
Culinary professional Elaine Lemm offers 1-day to 1-week vacations that include daily hands-on cooking classes. Established in 1994. Maximum class/group size: 8-10. Facilities: Hotel and restaurant kitchens, private homes, castles. Also featured: Wine tastings, vineyard visits, table decoration.
Emphasis: Vegetarian, appetizers, seasonal and holiday specialties, classical French cuisine, Tuscan cooking using fresh, local ingredients.
Faculty: Elaine Lemm, Chef de Cuisine and owner, studied at the Ritz Escoffier in Paris.
Costs: Range from £50-£850, which includes meals and shared lodging.
Location: Marrick, North Yorkshire, England; Central Stockholm, Sweden; Lucca, Italy; Provence, France.
Contact: Elaine M. Lemm, Owner, Cuisine Eclairee, 5, the Poplars, Newton-on-Ouse, York, Y06 2BL England; (44) (0)1347-848557, Fax (44) (0)1347-848557, E-mail cuisine.eclairee@compuserve.com, URL http://www.diningpages.com. Sweden contact: Lena Nilsson Erleman, (46) 8-716-1584.

CUISINE INTERNATIONAL
Italy, France, England, Spain, Ireland, Brazil/Year-round *(See display ad page 291)*
This tour operator specializing in culinary vacations offers cooking schools and culinary tours in Italy, France, England, Spain, Ireland, Brazil. Established in 1988. Maximum class/group size: 8-20, depending on program. Facilities: Hotel and restaurant kitchens, private homes, castles, monasteries. Also featured: Excursions to food-related and historical sites; visits to wineries, tastings of local products, shopping for local products.
Emphasis: Culinary tours and cooking classes in regional cuisines.
Faculty: Owner Judy Ebrey, CCP.
Costs: Vary from $1,850 to $4,700 per week including lodging, meals, planned activities, ground transport.
Location: Italy: Tuscany, Amalfi, Puglia, Rome, Sicily, Venice, Umbria, Lucca. France: Provence, Paris, Gascony. Spain: Barcelona. Ireland: County Cork. Brazil: Ouro Prato.
Contact: Judy Ebrey, Owner, Cuisine International, P.O. Box 25228, Dallas, TX 75225 US; 214-373-1161, Fax 214-373-1162, E-mail CuisineInt@aol.com, URL http://www.cuisineinternational.com.

CULINARY INSTITUTE OF AMERICA (See page 234)

CULTURAL HOMESTAY INTERNATIONAL

France and Italy/April-May, August-October

This travel company offers 2-week hands-on programs that feature a homestay with a chef, sightseeing, wine classes and winery tours, visits to markets and food producers, dining in fine restaurants and private homes. Established in 1980. Maximum class/group size: 6-12. 2-3 programs per year. Facilities: Home of the chef or a local restaurant. Also featured: Ponza program includes boating, snorkeling, swimming, 4-day stay in Rome, optional visit to Pompeii.

Emphasis: French and Italian regional cuisine.

Faculty: Noted chef of the region.

Costs: $2,895 Aix, $3,325 Florence, $2,875 Ponza, includes airfare from/to U.S., lodging, most meals, planned activities.

Location: Aix-en-Provence, France; Florence, Island of Ponza, Italy.

Contact: Gayle Peebles, Outbound Manager, Cultural Homestay International, 2455 Bennett Valley Rd., #210B, Santa Rosa, CA 95404 US; 800-395-2726/707-579-1813, Fax 707-523-3704, E-mail chigaylep@msn.comec, URL http://www.chinet.org/outbound.html.

FRIENDS & FOOD INTERNATIONAL, INC.

France and Italy/March-November

This private school offers 9- to 11-day programs that include cooking classes in local homes and trips to markets, farms, gardens, art history sites. Established in 1994. Maximum class/group size: 8-12. 6-10 programs per year. Facilities: Historic restored country estates with private facilities. Some professional restaurant kitchens and local farm houses.

Emphasis: Local ingredients, developing a refined palate.
Faculty: Professional chefs, art historians, native language speakers fluent in English, local experts on food, wine, culture.
Costs: $2,300 includes meals, lodging, ground transport.
Location: Tuscany (near Lucca), Veneto (near Venice), Provence (near Avignon).
Contact: Mark Haskell, President, Friends & Food International, Inc., 1707 Taylor St., NW, Washington, DC 20011 US; 202-726-4616, Fax 202-726-4616, E-mail MkHaskell@AOL.com.

INLAND SERVICES, INC. (See La Cuisine de Provence, p. 301, and L'Amore di Cucina Italiana, p. 327)

(See display ad above)

THE INTERNATIONAL KITCHEN, INC.

France and Italy/Year-round *(See display ad page 293)*

This travel company specializing in culinary tours offers week-long and shorter programs that include hands-on instruction in the cuisine of the region, market visits, winery tours, visits to food producers, dining in private homes and fine restaurants, sightseeing, cultural activities. Established in 1995. Maximum class/group size: 8-12. Facilities: Include farmhouse, home, and restaurant kitchens. Also featured: Some programs include spa treatments, walking, bicycling, barging, a truffle hunt, shopping at designer factory stores, music festivals.
Emphasis: French and Italian cuisine.
Faculty: Restaurant chefs.
Costs: France: $1,475-$2,500, which includes B&B to deluxe lodging, meals, planned activities. Italy: $1,900-$3,600, which includes farmhouse to deluxe lodging, meals, planned activities.
Location: France: Provence, Burgundy, Bordeaux. St. Emilion, Paris. Italy: Tuscany, Umbria, Portofino, Amalfi Coast, Sicily.
Contact: Karen Herbst, The International Kitchen, Inc., 1209 N. Astor St., #11N, Chicago, IL 60610 US; 800-945-8606, Fax 847-295-0945/312-654-8446, E-mail info@intl-kitchen.com, URL http://www.intl-kitchen.com. Additional contact: Catherine Merrill.

JC FOOD & WINE TOURS

Italy and France/Year-round

Special interest tour operator Joyce Capece offers customized itineraries to Italy, France, Napa Valley. Occasional trips to other countries in the UK, Europe, southeast Asia. Established in 1992. Maximum class/group size: 6-18. Also featured: Small group tours to visit wineries, private villas, agritourismos, markets, food/wine producers; dining at fine and local restaurants, truffle hunts, sightseeing, meeting locals. Private tours offered.
Emphasis: Food and wine travel focusing on specialties of the regions.
Faculty: Winemakers, food producers, chefs, restaurateurs, culinary consultants. Joyce Capece, a 30-year travel professional, specializes in food and wine tours and special interest travel.
Costs: Rates are based on countries, regions, seasons, and length of program, usually 7-12 days.
Contact: Joyce Capece, CTC, DS, JC Food & Wine Tours, 480 Madera Ave., #2, Sunnyvale, CA 94086 US; 408-732-0891, Fax 408-749-9479.

MURRAY SCHOOL OF COOKERY (see page 288)

RHODE ISLAND SCHOOL OF DESIGN (See page 252)

THE WANDERING SPOON

Greece and Portugal/March-May, September-November

Cookbook author Lucille Haley Schechter offers 1-week hands-on culinary programs for private groups of 6 to 8. Established in 1983. Maximum class/group size: 10. Also featured: Dining in fine restaurants, visits to markets, vineyards, and cultural centers.

Emphasis: Mediterranean and international cooking techniques.

Faculty: Former Harper's Bazaar magazine editor Lucille Haley Schechter is co-author of The International Menu Diabetic Cookbook and is Professional Member of the James Beard Fdn.

Costs: One-week session, excluding airfare, is $2,150, which includes double occupancy lodging in deluxe Mediterranean hotels.

Location: Greece and Portugal, including Athens, Crete, Corfu, Santorini, and the Algarve.

Contact: Lucille H. Schechter, Wandering Spoon, 340 East 57th St., New York, NY 10022 US; 212-751-4532, Fax 212-753-1714, E-mail bilucille@aol.com, URL http://www.wanderingspoon.com.

FINLAND

HELSINKI CULINARY INSTITUTE

Helsinki/Year-round

The owners/chefs of restaurant Kanavaranta offer tailor-made 1-day hands-on theme workshops for 8-12, 2-3 hour demos for 15-25, Shop and Cook lunch and dinner courses for 8-14. Instruction in Finnish with simultaneous English translation. Established in 1994. Maximum class/group size: 10-25. Facilities: Teaching kitchen situated above the restaurant Kanavaranta.

Emphasis: New and special flavors of Finland, Scandinavian cookery.

Faculty: Chef de Cuisines Eero Makela and Gero Hottinger, Maria Planting, visiting chefs from Finland and Europe.

Costs: $50/person for demos, $200-$400 for workshops (incl 5-course meal), 500 FIM-600 FIM for Shop & Cook courses.

Location: Katajanokka, in the center of Helsinki.

Contact: Maria Planting, Helsinki Culinary Institute, Kanavanta 3 F, Helsinki, 00160 Finland; (358) 9-6222633, Fax (358) 9-6222616, E-mail maria.planting@mplanti.pp.fi.

FRANCE

A LA BONNE COCOTTE EN PROVENCE

Provence/April-June, August-October

Culinary professional Lydie Marshall offers 5-day cooking vacations that include classes and sightseeing. Established in 1971. Maximum class/group size: 6-8. 6-8 programs per year. Facilities:

Large country kitchen, outdoor terrace for dining.
Emphasis: French cuisine, from simple regional Provence recipes to haute cuisine.
Faculty: Lydie Marshall, author of Cooking with Lydie Marshall, A Passion for Potatoes, and Chez Nous.
Costs: $1,600, which includes lodging in Lydie Marshall's small château.
Location: Nyons, about 45 miles northeast of Avignon.
Contact: Lydie P. Marshall, A La Bonne Cocotte en Provence, Château Feodal, Nyons, 26110 France; (33) 475-26-45-31, Fax (33) 475-26-09-31, E-mail ciboulette@juno.com.

ANDRE DAGUIN HOTEL DE FRANCE

Gascogne/October-April

This hotel offers 3-day to 2-week courses. Established in 1985. Maximum class/group size: 6. Facilities: The hotel's restaurant kitchen. Also featured: A tour of the Armagnac region with visits to wine cellars, a foie gras duck farm, and a farmer's market.
Emphasis: The cuisine of Gascony; foie gras, confit de canard, and other duck preparations.
Faculty: Under supervision of Chef Andre Daguin, proprietor.
Costs: Fee, which includes meals and hotel lodging, is approximately 2,850FF for 3 days.
Contact: Andre Daguin, Andre Daguin Hotel de France, 2, place de la Liberation, Auch en Gascogne, 32003 France; (33) (0)562-61-71-71, Fax (33) (0)562-61-71-81, E-mail hotelfrance@relaischateaux.fr, URL http://www.integra.fr/relaischateaux/hotelfrance.

ART OF LIVING – CULINARY TOURS WITH SARA MONICK (See page 289)

THE ASSOCIATION CUISINE AND TRADITION (ACT)

Arles/March-July, September-November

This private school offers evening, full day, weekend, half- and full-week hands-on programs that feature seasonal ingredients, discussion of wine and food pairing. Established in 1996. Maximum class/group size: 4-8. Facilities: Traditional Provencal kitchen outfitted with professional equipment for teaching. Also featured: Half- and full-week programs include such excursions as market visits, winery tours, herb collecting, a jeep ride to the Camargue, visit to an olive oil mill, sightseeing.
Emphasis: Provencal cuisine, dishes selected from the instructor's 1,000+ recipes.
Faculty: Author of l'Archeologie de la Cuisine.
Costs: $90/eve class, $130 full day, $180 weekend, $390 half week, $900 full week. 1- to 4-star hotels nearby begin at $40/night.
Location: Central Arles, 5 minutes walk from train station, 30 minutes drive from Avignon, 1 hour from Marseilles International Airport.
Contact: Madeleine Vedel, Course Coordinator, The Association Cuisine and Tradition (ACT), 30, rue Pierre Euzeby, Arles, 13200 France; 33-(0)4-90-49-69-20, Fax 33-(0)4-90-49-69-20, E-mail actvedel@provnet.fr.

AT HOME WITH PATRICIA WELLS: COOKING IN PROVENCE

Vaison-La-Romaine/February, May-June, August-September

American journalist and author Patricia Wells offers 5-day cooking vacations that include 4 hours of daily hands-on instruction. Established in 1995. Maximum class/group size: 6. 7 programs per year. Facilities: Ms. Wells' 18th-century farmhouse kitchen, which has a wood-fired bread oven. Also featured: Visits to local markets, wine tastings, dinner at a fine restaurant; February truffle workshop.
Emphasis: Provencal cuisine.
Faculty: Patricia Wells has lived in France since 1980, is restaurant critic of The International Herald Tribune, and authored 6 books, including Bistro Cooking and Simply French.
Costs: $3,000, including meals and planned activities. A list of recommended lodging is supplied.
Location: Ms. Wells' hilltop home is outside Vaison-la-Romaine, ~ 30 miles northeast of Avignon.
Contact: Judith Jones, Program Coordinator, Cooking in Provence, 708 Sandown Pl., Raleigh, NC 27615 US; 919-870-5955, Fax 919-846-2081, E-mail jj708@mindspring.com.

BARGE INN FRANCE

(See also page 292, International Kitchen) **Burgundy, Southern France/May-October**

This private school offers a six-night cruise aboard the luxury barge Occitane with hands-on cooking lessons and market visits. Maximum class/group size: 6.

Emphasis: French regional cuisine.

Faculty: Hazel Young, professionally-trained French chef.

Costs: $2,500 includes lodging, meals, excursions.

Location: Burgundy Canal and Canal du Midi.

Contact: Hazel Young, Barge Inn France, 6 Quai du Canal, St. Usage, 21170 France; (33) (0)3-80-29-19-61. In the U.S.: The International Kitchen, Inc., 1209 N. Astor St., #11N, Chicago, IL 60610; 800-945-8606, Fax 847-295-0945/312-654-8446, E-mail info@intl-kitchen.com, http://www.intl-kitchen.com.

BICYCLING & COOKING IN BORDEAUX

(See also page 292, International Kitchen) **Bordeaux/Year-round**

This private school offers 7-night cooking vacations that include fully-supported bicycle rides. Maximum class/group size: 8. Facilities: 3-star family-run B&B with pool and private bathrooms. Also featured: Visits to markets and food producers, winery tours, sightseeing.

Emphasis: Bordeaux cuisine.

Faculty: Local chef.

Costs: $2,500 includes lodging and meals, use of bicycle, planned excursions.

Location: Near city of Bordeaux.

Contact: Florent Maillot, Owner, Carpe Diem in France, 204 Rue Mouveyra, Bordeaux, 33000 France; (33) (0)5-56986726, Fax (33) (0)5-56986523. In the U.S.: The International Kitchen, Inc., 1209 N. Astor St., #11N, Chicago, IL 60610; 800-945-8606, Fax 847-295-0945/312-654-8446, E-mail info@intl-kitchen.com, http://www.intl-kitchen.com.

CHÂTEAU COUNTRY COOKING COURSE

Montbazon-en-Touraine/March-May, October-December

Denise Olivereau-Capron and her son, Xavier offer 6-day participation courses. Established in 1986. Maximum class/group size: 8-15. Facilities: The château's renovated kitchen. Also featured: Dining at fine restaurants and visits to the Tours flower market, the châteaux of the region, a goat cheese farm, the Chinon markets, and the caves of Vouvray.

Emphasis: French regional cuisine.

Faculty: Chef Edouard Wehrlin.

Costs: Course fee of 12,500FF single, 11,500FF double, includes château lodging, meals, wine tastings, planned excursions. Complimentary stay for 8 paying guests.

Location: Le Domaine de la Tortiniere, a 19th century manor-house château in the Loire Valley, in châteaux country 6 miles south of Tours.

Contact: Denise Olivereau-Capron, Château Country Cooking Course, Montbazon en Touraine, 37250 France; (33) (0)247-34-35-00, Fax (33) (0)247-65-95-70, E-mail domaine.tortiniere@wanadoo.fr. Additional contact: Xavier Olivereau, Mona Augis.

CHEZ MADELAINE COOKING SCHOOL & TOURS (SEE PAGE 210)

COOKERY LESSONS AND TOURAINE VISIT

Brehemont/October-May

Maxime and Eliane Rochereau in the 18th-century Le Castel de Bray et Monts offers 1-week hands-on vacation programs that include daily instruction. Established in 1983. Maximum class/group size: 15. 6 programs per year. Facilities: The manor's restaurant kitchen. Also featured: Visits to châteaux, wineries, pastry shops, sightseeing, shopping.

Emphasis: French cuisine.

Faculty: Chef Maxime Rochereau, who was chef at the Ritz Hotel in Paris.

Costs: Cost is 6,900FF (5% discount for a couple), which includes lodging at the manor, some

meals, wine, and planned activities. 1,500FF deposit is required; refund with 30 days notice.
Location: A vineyard village on the Loire River in the château region, 16 miles from Tours and a one-hour (155 miles) train ride from Paris.
Contact: Maxime Rochereau, Cookery Lessons and Touraine Visit, Brehemont, Langeais, 37130 France; (33) (0)247-96-70-47, Fax (33) (0)247-96-57-36.

COOKING AT THE ABBEY

Salon-de-Provence/March-June, September-December

The Hostellerie Abbaye de Sainte Croix resort sponsors 3-, 4-, and 7-day vacation participation courses; afternoon classes (minimum 6 students) on request. Established in 1987. Maximum class/group size: 6-12. Facilities: Restaurant kitchen, which has 4 ovens and 12 work stations. Also featured: Local sightseeing and vineyard visits.
Emphasis: Provencal cuisine.
Faculty: Chef P. Morel of the Abbey's Michelin 1-star restaurant.
Costs: 4,100FF (420FF) for 3 days to 10,900FF (1,200FF) for 7 days, which includes most meals, double (single supplement) lodging in the 12th-century Abbey's Roman-style rooms, planned activities. Additional afternoon class is 620FF.
Location: The Abbey, a member of Relais & Châteaux, is 2 miles from Salon, 20 miles northeast of the nearest airport, 20 miles west of Aix-en-Provence.
Contact: Catherine Bossard, Director, Cooking at the Abbey, Abbaye de Sainte Croix, Route Val-de Cuech, Salon-de-Provence, 13300 France; (33) (0)490-56-24-55, Fax (33) (0)490-56-31-12, E-mail saintecroix@relaischateaux.fr, URL http://www.relaischateaux.fr/saintecroix. U.S. Contact: Michael Giammarella, EMI Int., Box 640713, Oakland Gardens, NY 11364-0713; 800-484-1235, #0096; 718-631-0096; Fax 718-631-0316.

COOKING WITH FRIENDS IN FRANCE

Châteauneuf de Grasse *(See display ad above)*

This vacation school on the property once shared by Julia Child and Simone Beck offers 6-day participation courses. Established in 1993. Maximum class/group size: 8. 28-30 programs per year. Also featured: Visits to the Forville Market, a butcher shop, cheese ripener, cutlery shop, and Michelin 2-star restaurant kitchens; demonstration by a French chef.
Emphasis: French cuisine, including techniques, tricks, menu-planning, lighter dishes.
Faculty: Proprietor/instructor Kathie Alex apprenticed at Roger Vergé's Le Moulin de Mougins, assisted well-known chefs at the Robert Mondavi Winery, studied catering at École LeNôtre, and studied with and assisted Simone Beck at her school.
Costs: $1,850, which includes shared lodging (some private baths), breakfasts, lunches, wine tastings, planned excursions (car required). Lodging at La Pitchoune or La Campanette, private homes formerly owned by Julia Child and Simone Beck.
Location: On the Cote d'Azur, about 9 miles from Cannes, 4 miles from Grasse, and 20 miles from Nice International Airport. Golf and horseback riding are nearby.

Contact: Kathie Alex, Cooking with Friends in France, La Pitchoune, Domaine de Bramafam, Châteauneuf de Grasse, 06740 France; (33) (0)493-60-10-56, Fax (33) (0)493-60-05-56. U.S. Contact: Jackson & Co., 29 Commonwealth Ave., Boston, MA 02116; (617) 350-3837, Fax (617) 247-6149, E-mail mail@jackson-co.com.

COOKING WITH THE MASTERS

Three locations/June, September

Chef Michel Bouit offers 1- to 2-week hands-on programs (1-week at Le Vieux Moulin, 1- to 2-weeks at Tecomah and Ecully). Established in 1990. Maximum class/group size: 10-20. 2 programs per year. Facilities: Full service restaurant/hotel kitchens. Also featured: Excursions to local artisans, markets, and vineyards; sightseeing and shopping in Paris.

Emphasis: Classical and regional French cuisine, culinary terms and service, industry-related visits, customized tours.

Faculty: Jean-Pierre Silva, chef-owner of Le Vieux Moulin; Gerard Marquoin, director of Tecomah; 1989 ACF National Chef of the Year Michel Bouit, CEC, AAC, president of MBI Inc., which specializes in culinary tours, competitions, consulting, public relations.

Costs: 2-week student course at Tecomah (dorm lodging) ~$3,000; 1-week professional course at Le Vieux Moulin or other luxury country Auberge lodging $2,550. Fees include airfare, lodging, ground transport, most meals, excursions.

Location: Le Vieux Moulin, Bouilland, 9 miles from Beaune; Tecomah, Jouy-en-josas, a suburb of Paris near Versailles; Ecully, 10 miles from Lyon.

Contact: Michel Bouit, President, Cooking with the Masters, P.O. Box 1801, Chicago, IL 60690 US; 312-663-5701, Fax 312-663-5702, E-mail mbi@worldofmbi.com, URL http://www.worldofmbi.com. Additional contact: Elizabeth Bergin, Vice President.

COOKING IN PROVENCE

Crillon le Brave/November

Hostellerie de Crillon le Brave offers 6-day hands-on cooking vacations. Established in 1992. Maximum class/group size: 8. 4 programs per year. Facilities: The hotel's restaurant kitchen. Also featured: Market visits, winery tour, truffle hunting, visits to Avignon and other Provence sites, dining at the hotel and fine restaurants, golf, tennis, cycling, hiking.

Emphasis: Provencal and Mediterranean cuisine.

Faculty: Chef de Cuisine Philippe Monti, a native of Provence, trained at Pic, l'Esperance, Auberge de l'Ill, and Taillevent.

Costs: Cost is $2,500 ($2,900), which includes meals, double (single) occupancy lodging at Hostellerie de Crillon le Brave, a member of Relais & Châteaux, planned excursions; supplement for non-cooking partners is $850.

Location: 25 miles from Avignon.

Contact: Craig Miller, Cooking in Provence, Place de l'Eglise, Crillon le Brave, 84410 France; (33) (0)490-65-61-61, Fax (33) (0)490-65-62-86, E-mail crillonbrave@relaischateaux.fr, URL http://www.crillonlebrave.com.

CUISINE ECLAIRÉE (See page 290)

CUISINE GROUP, INC. (See page 234)

CUISINE INTERNATIONAL ÉCOLE DE CUISINE (See page 222)

CUISINE EN PROVENCE

(See also pages 290 and 292, Cuisine Intl. and Intl. Kitchen) **Provence/Year-round**

This private school offers 4- and 7-day cooking vacations that feature daily hands-on classes, tastings at 3 wineries, visits to markets and food producers, dining at local restaurants. Established in 1996. Maximum class/group size: 8-12. 10-15 programs per year. Facilities: Domaine de la Reparade traditional kitchen at le Mas des Graviers and Château de Vins. Also featured: Tennis, cycling, swimming.

Emphasis: Light Mediterranean and Provencal cuisine, pastry and sauces.
Faculty: Jean-Marie Carret, born in Provence, trained in France, is a freelance chef in London. His partner is Charleric Gensollen.
Costs: From $1,850 for the 1-week program includes shared lodging, meals, planned activities.
Location: 3 Provence locations: a 16th-century castle 40 minutes from St-Tropez, an 18th-century mas 20 minutes from Marseilles airport, an 18th-century farmhouse.
Contact: Jean-Marie Carret, Owner, Cuisine en Provence, 21 Ellerslie Rd.-Garden Flat, London, W12 7BN England; (44) (0)181-740-9193, Fax (44) (0)181-740-9193, E-mail jean-marie@cuisine-en-provence.com, URL http://www.cuisine-en-provence.com. In the U.S.: Karen Herbst, International Kitchen, 800-945-8606, Fax 847-295-0945; E-mail info@intl-kitchen.com. Judy Ebrey, Cuisine Int., Box 25228, Dallas, TX 75225; E-mail CuisineInt@aol.com.

ÉCOLE DES ARTS CULINAIRES ET DE L'HOTELLERIE, EACH

Lyon/Ecully/Year-round *(See also page 153)*

This private career school offers 1-week introductory courses for amateurs of French gastronomic tradition. Established in 1990. Facilities: 13 seminar rooms, 2 computer labs, 8 teaching kitchens, pastry/pantry facilities, video-equipped amphitheatre, sensory analysis lab, restaurants. Also featured: A la carte sessions taught in various languages can be organized (minimum 6 students).
Emphasis: French cuisine and pastry, menu composition, restaurant-quality desserts and decorative accompaniments, wines.
Faculty: In addition to the Board of Trustees, headed by Paul Bocuse, the permanent teaching staff includes 2 Meilleurs Ouvriers de France, Pastry Chef Alain Berne, and Restaurant Chef Alain Le Cossec.
Costs: 7,500FF fee includes lodging and meals.
Location: The restored 19th century Château du Vivier, in a 17-acre wooded park in the Lyon-Ecully University-Research Zone, 10 minutes from downtown Lyon.
Contact: Eleonore Vial, Angeline Phan, International Programs, École des Arts Culinaires et de l'Hotellerie, EACH Lyon, Château du Vivier, B.P. 25, Ecully, Cedex, 69131 France; (33) (0)478-43-36-10, Fax (33) (0)478-43-33-51, E-mail information@each-lyon.com, URL http://www.each-lyon.com.

ÉCOLE DES TROIS PONTS

Roanne/May-October *(See display ad page 299)*

This French cookery and French language institute in a château offers 1-week French Provincial Cookery courses in English courses, 1-week Cookery and French courses that feature 4 afternoon hands-on cooking classes, 2 market visits, guided excursion, and an optional wine course. Established in 1991. Maximum class/group size: 6-8 cookery, 6 language. 7 programs per year. Facilities: Classes and lodging are at the Château. Also featured: A cooking-only option (instruction in English), courses in general and intensive French, private instruction.
Emphasis: Provincial French cookery and French language courses.
Faculty: Professional chef for the cookery courses, professional native French teachers for the French language courses.
Costs: 5,900FF-7,700FF for Cookery course, 6,500FF-8,300FF for Cookery and French course, which includes lodging at the Château, most meals, planned excursions. Pool, tennis, and equestrian centers are nearby. Bikes are available.
Location: The 17th-century Château de Matel is on 32 acres of park and forest, 5 minutes from the center of Roanne in the Burgundy/Beaujolais area, 1 hour from Lyons.
Contact: Mrs. Margaret O'Loan Mr. René Dorel, Directors, École des Trois Ponts, Château de Matel, Roanne, 42300 France; 33-4-77-71-53-00, Fax 33-4-77-70-80-01, E-mail info@3ponts.edu, URL http://www.3ponts.edu. U.S. Contact: Michael Giammarella, EMI Int., Box 640713, Oakland Gardens, NY 11364-0713; 800-484-1235, #0096; 718-631-0096; Fax 718-631-0316, E-mail mgiamma@cuny.campus.mci.net.

ETOILE BLEU MARINE

La Rochelle/March-May, September-November

Maybelle Iribe offers 1-week cooking vacations (12 hours of hands-on instruction). Established in 1991. Maximum class/group size: 8. 6 programs per year. Facilities: 200-square-foot kitchen with garden. Also featured: Visits to cognac distilleries, wineries, oyster beds, food producers, market tour, sightseeing.

Emphasis: Regional French cooking and seafood.

Faculty: Emi Taya, a graduate of the University of Tokyo; restaurateurs Fred Nillson and Mirko Bettini, oenologist George Caviste.

Costs: Cost is $2,000, which includes meals, lodging (6-twin and double rooms with bath in Maybelle's Bed & Breakfast), and planned activities.

Location: La Rochelle is 3 hours from Paris by TGV train and one hour by airline TAT direct.

Contact: Maybelle Iribe, Charente Maritime, Etoile Bleu Marine, 33, rue Thiers, La Rochelle, 17000 France; (33) (0)546-41-62-63, Fax (33) (0)546-41-10-76, E-mail 106510.2775@compuserve.com. U.S. Contact: Ms. M.J. Drinkwater, Town & Country Travel, Sacramento, CA; 916-483-4621.

FOODSEARCH PLUS, INC. (See page 200)

FRANCE AUTHENTIQUE

Domfront, Normandy/Year-round

France on Your Plate offers 1-week customized vacation programs that feature cooking classes in restaurant kitchens. Established in 1994. Maximum class/group size: 8-10. Also featured: Visits to local food producers, the copper-making center of Villedieu-les-Poeles, and the World War II landing beaches, cycling, fishing, horseback riding, swimming, tennis, and golf.

Emphasis: Cuisine of Normandy.

Faculty: Prominent local chefs.

Costs: $1,200-$2,000 per week, which includes most meals, planned activities, and lodging at La Maison de la Resistance.

Location: The Normandy region, 3 hours from Paris.

Contact: William T. Fleming, Jr., France Authentique, 835 S. Lucerne Blvd., #308, Los Angeles, CA 90005 US; 213-857-1805/800-966-0266, Fax 213-857-1728, E-mail fleming@helix.net, URL http://www.helix.net/~fleming.

FRENCH KITCHEN COOKING SCHOOL AT CAMONT

Gascony/Year-round

Kate Ratliffe, cookbook author, chef, and canal barge captain, offers programs that include the 1-week Country Kitchens of Gascony gastronomic cook's tour of fine dining, touring and cooking, and the 5-day hands-on In the Kitchen... workshop based on seasonal subjects. Established in 1987. 12+ programs per year. Facilities: 18th-century stone kitchen of Camont, a canal-side relais featuring original brick fireplace and potager garden. Also featured: Visits to farmer's markets in medieval villages, armagnac cellars, foie gras farms, fine restaurants, bakeries, wineries, escorted

sightseeing. Custom classes for groups.
Emphasis: Regional country cooking of southwest France and Gascony.
Faculty: Kate Ratliffe, owner/chef and author of the cookbook/travelogue A Culinary Journey in Gascony.
Costs: $2,650 for the 1-week Country Kitchens of Gascony, $1,895 for the 5-day In the Kitchen...., includes lodging, meals, planned activities.
Location: Gascony, southwest France near Agen, 4 hours by train from Paris, 90 minutes from Bordeaux and Toulouse.
Contact: Kate Ratliffe, The French Kitchen Cooking School at Camont, 5 Ledgewood Way, #6, Peabody, MA 01960 US; 800-852-2625/978-535-5738, Fax 978-535-5738, E-mail KateRatliffe@compuserve.com.

HOLIDAYS IN THE SUN IN THE SOUTH OF FRANCE
Gordes/March-July, September-October, December

Les Megalithes school in a private country home offers 1-week participation courses with instruction in English, French, and German. Special programs available in Feb - truffle specialties; Jul - summery cooking. Established in 1980. Maximum class/group size: 6. Facilities: Indoor and outdoor home kitchen. Also featured: Market visits, horseback and bicycle riding, handicraft shopping, visits to museums and historic sites.
Emphasis: Cuisine of Provence, including appetizers, main courses, breads, jellies, holiday dishes.
Faculty: Sylvie Lallemand, president/founder of the Association des amis de la cuisine et des traditions provencales, learned to cook from her mother and grandmother and studied with Roger Vergé. Author of Enchanted Provence, cookbook and travelogue.
Costs: 3,400FF, which includes private room and bath at Les Megalithes, pool, and meals. A non-refundable 200FF deposit is required.
Location: Gordes, near Avignon, 3-½ hours from Nice.
Contact: Sylvie Lallemand, Les Megalithes, Gordes, 84220 France; (33) (0)490-72-23-41.

THE JAMES BEARD FOUNDATION AND B & V ASSOCIATES
Burgundy/April-May, October

This organization and travel provider offers 6-day barge river cruises through the French countryside that feature an itinerary designed by the guest chef and one food-related event each day. Established in 1998. Maximum class/group size: 12-51. 3 programs per year. Facilities: The barges are the 51-passenger Anacoluthe and the 12-passenger La Belle Epoque. Also featured: Market visits, cooking demonstrations, tastings of foie gras, wine, and cheese.
Emphasis: Cuisine of James Beard Award-winning chefs.
Faculty: Includes Todd English, Michael Ginor, Jean-Louis Palladin, Thierry Rautureau.
Costs: $2,500-$3,400 includes meals, wines, ground transport from Paris. Business-class airfare is $2,000. Proceeds partially benefit The James Beard Foundation.
Location: Burgundy.
Contact: Michele Seligmann, B & V Associates, 914-939-2309, Fax 914-939-4255, E-mail betelgeuse@aol.com, URL http://www.theverybest.com. Add'l contact: Naomi Kabak, Continental Waterways, 800-217-4447, Fax 212-688-3778.

LA CUISINE DE MARIE-BLANCHE
Paris/September-July

Culinary professional Marie-Blanche de Broglie (formerly Princess Ere 2001) offers 1- to 4-week participation courses with instruction in French, English, or Spanish. Diplomas: Grand Diplome Princesse Marie-Blanche de Broglie, Le Diplome de La Cuisine de Marie-Blanche. Established in 1975. Maximum class/group size: 6-8. 10 programs per year. Facilities: 50-sq-meter kitchen, 40 sq-meter lecture room for slides. Also featured: Discovering French Style for groups; French Cheeses, French Pastry & Tasting. A Gastronomic Tour of France: French Wines, Table Setting, candlelight dinner, visit to a Parisian market.

Emphasis: The art of entertaining at home.
Faculty: Marie-Blanche de Broglie, founder and director, is author of The Cuisine of Normandy in English, A La Table des Rois in French.
Costs: One class 760FF, 5 classes 3,010FF, 10 classes 5,510FF, 3 pastry classes 1,760FF, The Little Pastry Chef 150FF, 1-week course in (Cooking) l'Art de Vivre (5,500FF) 4,500FF, 1 class 1,500FF, Grand Diplome 19,500FF, Le Diplome 45,000FF.
Location: Near the Eiffel Tower subway station, École Militaire.
Contact: Marie-Blanche de Broglie, Manager, La Cuisine de Marie-Blanche, 18 Ave. de la Motte-Picquet, Paris, 75007 France; (33) (0)145-51-36-34, Fax (33) (0)145-51-90-19, E-mail infocmb@CuisineMB.com, URL http://www.CuisineMB.com/index.htm.

LA CUISINE DE PROVENCE

(See display ad page 292) **L'Isle sur la Sorgue/May, June, October, November**

This travel company offers 1-week hands-on culinary vacations. Established in 1997. Facilities: Kitchen of the Domaine de la Fontaine. Also featured: Shopping expeditions; visits to outdoor markets, wineries, olive oil and chocolate producers; a corkscrew museum; a cultural evening in Avignon; dining at fine restaurants.
Emphasis: Provencal and regional French cuisine.
Faculty: Jean-Claude Aubertin, the winner of Mouton Cadet and Poele d'Or competitions, and a member of Academie Culinaire de France and Les Diciples D'Escoffier.
Costs: $2,195 ($1,995 for non-cook guest) includes meals, shared lodging ($300 single supplement) at the Domaine de la Fontaine, planned activities.
Location: L'Isle sur la Sorgue, about a half hour drive from Avignon.
Contact: Ralph P. Slone, La Cuisine de Provence, Inland Services, Inc., 360 Lexington Ave., New York, NY 10017 US; 212-687-9898, E-mail incook@earthlink.net, URL http://home.earthlink.net/~incook/.

LA CUISINE SANS PEUR (SEE PAGE 236)

LA VARENNE

(See also pages 154 and 265) (See display ad above) **Burgundy/June-July, September-October**

This private school offers a Master Class program with five day-long sessions (hands-on classes, 1 demonstration, wine tasting). Maximum class/group size: 15. Facilities: The school is in the 17th-century Château du Feÿ, a registered historic monument owned by founder Anne Willan. Also featured: visits to Joigny and Chablis, market place and vineyard tours, dinner at a fine country restaurant.
Emphasis: Classic, contemporary, and regional French cuisine, bistro cooking, pastry, wine appreciation, guest chef specialties.
Faculty: Visitor Series features different chef-instructors each week.
Costs: $2,950, which includes transport from Paris, full board, shared twin lodging at Château du Feÿ, and planned activities. Single supplement $400.
Location: 90 minutes south of Paris. Amenities include a tennis court and an outdoor swimming pool in season.

Contact: La Varenne, P.O. Box 25574, Washington, DC 20007 US; 800-537-6486, Fax 703-823-5438, E-mail LaVarenne@compuserve.com, URL http://www.lavarenne.com.

L'ACADEMIE DE CUISINE (See pages 60 and 219)

LE CORDON BLEU – PARIS

Paris/Year-round *(See also pages 154, 270, 287,338) (See display ad page 149)*

This private school offers half-day to one-month courses, including daily half-day demonstrations, 3- to 4-day gourmet sessions, 1-month initiation courses in cuisine and pastry, market tours, demonstrations, diploma and certificate program. Established in 1895. Maximum class/group size: 15+ demo, 8-12 hands-on,. 50+ programs per year. Facilities: 2 demonstration rooms, 4 practical kitchens professionally equipped, individual workspaces. Also featured: Workshops for children, floral art, evening wine classes and excursions, cheese courses, guest chef demonstrations, Saturday classes and workshops.

Emphasis: Intensive and comprehensive culinary cuisine and pastry courses focusing on French culinary techniques.

Faculty: Full-time professional chefs from Michelin-starred restaurants including 2 Meilleur Ouvrier de France.

Costs: Ranges from 220FF/half-day to 750FF/full-day hands-on workshop to 4,950FF/4-day session.

Location: Southwest Paris near the Eiffel Tower and Montparnasse.

Contact: Sabine Bailly, Director of Admissions, Le Cordon Bleu, 8, rue Leon Delhomme, Paris, 75015 France; (33) (0)1-53-68-22-50, Fax (33) (0)1-48-56-03-96, E-mail info@cordonbleu.net, URL http://www.cordonbleu.net. Toll free in the U.S. and Canada: 800-457-CHEF.

LE MARMITON – COOKING IN PROVENCE

Avignon/September-June

This hotel-restaurant offers 1- and 5-day participation courses. Established in 1994. Maximum class/group size: 12. Facilities: Restored 19th century kitchen with its original wood-fired cast-iron stove and restored counters. Also featured: Visit to wineries,markets, sightseeing, special itineraries can be arranged through concierge.

Emphasis: Provencal cuisine.

Faculty: Christian Etienne, Robert Brunel, Jean-Claude Aubertin, Daniel Hebet, Roger Hennequin, Frederique Feraud.

Costs: 1,345FF-1,915FF daily (7,670FF-10,300FF 6 days) double occupancy, nonresident 600FF (2,500FF), 250FF daily for non-cook guest.

Location: La Mirande, a period town house, is about 65 miles from Marseille international airport and 6 miles from Avignon's domestic airport.

Contact: Martin Stein, Artist Director, Le Marmiton-Cooking in Provence, 4, place de La Mirande, Avignon, 84000 France; (33) (0)490-85-93-93, Fax (33) (0)490-86-26-85, E-mail mirande@worldnet.net, URL http://www.la-mirande.fr.

L'ÉCOLE DES CHEFS

Paris and other locales/Year-round *(See also page 155) (See display ad page 303)*

Culinary professional Annie Jacquet-Bentley offers 5 to 6 days one-on-one with 2- and 3-star Michelin chefs in their restaurants. Established in 1998. Maximum class/group size: 1. Facilities: French 2 and 3- star restaurant kitchens. Also featured: Visits to food market and kitchen equipment store.

Faculty: 17 Michelin-starred chefs, including Troisgros, Georges Blanc and Alain Passard. Pastry program at Laduree with Pierre Herme, catering program at Potel & Chabot with Jean-Pierre Biffi.

Costs: $1,950 ($2,450) for 2-star (3-star) restaurant includes meals with staff.

Location: Paris, Lyon and French countryside (Provence, Brittany, Champagne, Alps and Burgundy).

Contact: Annie Jacquet-Bentley, President, L'École des Chefs, PO Box 183, Birchrunville, PA 19421 US; 610-469-2500, Fax 610-469-0272, E-mail info@leschefs.com, URL http://www.leschefs.com.

L'ÉCOLE DE CUISINE DU DOMAINE D'ESPERANCE

(See also page 290, Cuisine Intl.) **La Bastide d'Armagnac/October-January, April, June**

This 18th-century country house offers 1-week hands-on vacations that include theoretical instruction in the mornings and preparation of the evening meal in the afternoons. Established in 1993. Maximum class/group size: 9. 6 programs per year. Facilities: Large country kitchen with 8 work areas. Also featured: Market trip and visits to nearby wine cellars; weekend courses for groups of 4 or more.

Emphasis: Seasonal French cuisine.

Faculty: Natalia Arizmendi, recipient of the Cordon Bleu Grand Diplome, has taught cooking and pastry for more than 10 years.She is tri-lingual in French, English, and Spanish.

Costs: 8,500FF, which includes double occupancy lodging (private bath) and meals at the Domaine. Amenities include outdoor swimming pool and tennis court.

Location: Gascony, in southwest France, 90 minutes from Bordeaux.

Contact: Claire de Montesquiou, L'École de Cuisine du Domaine d'Esperance, Mauvezin d'Armagnac, La Bastide d'Armagnac, 40240 France; (33) (0) 558-44-68-33, Fax (33) (0) 558-49-85-93. In the U.S.: Judy Ebrey, Cuisine International, P.O. Box 25228, Dallas, TX 75225; 214-373-1161, Fax 214-373-1162, E-mail CuisineInt@aol.com, URL http://www.cuisineinternational.com.

LES CASSEROLES DU MIDI

Avignon/September-June

Culinary professional Olga Manguin offers 1-week courses that cover 20-30 recipes. Established in 1993. Maximum class/group size: 4. Also featured: Shopping for ingredients, visits to farms, wineries, and food producers, sightseeing in Provence, swimming and tennis. Relative or friend can travel with you and pay as traveling guest.

Emphasis: Provencal, Italian, North African cuisine.

Faculty: Italian-born Olga Manguin was owner/chef of Le Cafe des Nattes in Avignon for 15 years and speaks English, Italian, and German.

Costs: Daily rate of 1,000FF includes meals, single or double lodging, ground transport, excursions. The Manguin home has 4 bedrooms with ensuite bath.

Location: About 70 miles from Marseille, 140 miles from Nice, 450 miles from Paris.

Contact: Olga Manguin, Les Casseroles du Midi, Ile de la Barthelasse, Avignon, 84000 France; (33) (0)490-85-55-94, Fax (33) (0)490-82-59-40.

LES LIAISONS DELICIEUSES

Six regions/September-May

This culinary tour company offers one-week vacations in different provinces that include 12 hours of hands-on cooking instruction. Established in 1994. Maximum class/group size: 8-10. 6-8 programs per year. Facilities: Hotel and restaurant kitchens with individual workspaces. Also featured: Visits to wineries, food producers, markets, and restaurants, sightseeing, hiking and biking. Custom trips available to various regions of France.

Emphasis: French regional cuisine.
Faculty: Founder Patti Ravenscroft is tour director and translator. Classes are taught by Michelin-star restaurant chefs and proprietors.
Costs: $1,950-$2,850, includes lodging, meals. Lodging: Hotel Les Pyrenees (Basque), Hostellerie du Vieux Moulin (Burgundy), L'Auberge de la Truffe (Dordogne), Auberge La Feniere (Provence), Hotel Jean Paul Jeunet (Jura), Château St. Paterne (Normandy).
Location: Includes Burgundy, Brittany, Normandy, Dordogne, Jura, Loire, Auvergne, Provence, the Pays-Basque.
Contact: Patti R. Ravenscroft, Les Liaisons Delicieuses, 4710 - 30th St. N.W., Washington, DC 20008 US; 202-966-4091, Fax 202-966-4091, E-mail cookfrance@cookfrance.com, URL http://www.cookfrance.com.

LES SAVEURS DE PROVENCE
Merindol-les-Oliviers/Year-round

This gastronomic association offers 3- to 6-day hands-on programs, including grape harvest weeks in autumn. Established in 1995. Maximum class/group size: 4-6. 12 programs per year. Facilities: Custom-built fully-equipped kitchen, library of wine and cook books. Also featured: Winery tours and instruction, market visits, dining in private homes and local restaurants, sightseeing, visits to food producers.
Emphasis: French Mediterranean cooking, wine and food pairing.
Faculty: Elizabeth Miller, graduate of City & Guilds of London in Hotel & Restaurant Catering and L'Academie du Vin in Paris; Isabelle Bachelard, former director of L'Academie du Vin, French wine journalist.
Costs: 1,500FF/day, which includes bed & breakfast lodging in private country houses, all meals and local wines, ground transport, planned activities. Wine tutorial is 650FF/half-day session including wine.
Location: In Provence, 40 miles from Avignon.
Contact: Elizabeth Miller, President, L'Association Les Saveurs de Provence, 26170 Merindol-les-Oliviers, Drome Provencale, France; (33) (0)475-28-78-12, Fax (33) (0)475-28-90-11.

LEVERNOIS CULINARY SCHOOL
Burgundy/January-March, August-September *(See also page 292, International Kitchen)*

Hostellerie de Levernois offers 4-day optional hands-on courses with preparation of a complete menu. Maximum class/group size: 10. Facilities: Large, well-equipped professional kitchen of the Restaurant Hostellerie de Levernois. Also featured: Wine instruction, private classes, afternoon excursions to cultural and historic sites, market visits, winery tours, goat cheese farm tours, fine dining, shopping, golf and tennis.
Emphasis: Classical and traditional French cooking, seafood, fois gras, escargot, pastry and desserts, cheese, wine appreciation.
Faculty: The Crotet family of chefs includes father Jean and sons Christophe and Guillaume. Christophe, who conducts the classes, was at Troisgros, Bocuse, Girardet, and now at Hostellerie de Levernois.
Costs: $1,684 includes meals, excursions, and shared lodging in the Michelin 2-star rated Relais and Châteaux of Hostellerie de Levernois. Single supplement $350.
Location: A 10-acre estate in the village of Levernois, 3 miles from Beaune, 2 hours southeast of Paris by train.
Contact: Gaby Crotet, Levernois Culinary School, Hostellerie de Levernois, Beaune, NO 21200 France; (33) (0)380-24-73-58, Fax (33) (0)380-22-78-00, URL http://www.integra.fr/relais-chateaux/levernois. In U.S.: The International Kitchen, 1209 N. Astor St., Chicago, IL 60610; 800-945-8606, Fax 847-295-0945, info@intl-kitchen.com, http://www.intl-kitchen.com. Lynn Parks 305-673-0310, Fax 305-673-0046.

MAS DE CORNUD

(See display ad above) **St. Remy-de-Provence/February-December**

David and Nitockrees Carpita's cooking school in their 18th-century Provencal farmhouse/country inn offers a one-week course, La Cuisine Provencale et Méditerranéenne, offers 5 days of participation cooking classes, market shopping, and visits to artisans producing regional products. Participants plan menus, learn table settings and napkin folding. Established in 1993. Maximum class/group size: 8. Facilities: The school has a newly installed air-conditioned professional teaching kitchen, a wood-burning oven, and an herb and vegetable garden. Also featured: Wine tastings, a walk in the footsteps of Vincent Van Gogh, boules.

Emphasis: Provencal and Mediterranean cuisine.

Faculty: Nitockrees Tadros Carpita, CCP, a French-Egyptian, is a member of the IACP and received culinary training in France. She has taught in Europe, the Middle East, and the U.S.

Costs: $2,300 includes shared lodging ($2,100 non-participant) at Mas de Cornud, a 4-star country inn with 6 guest rooms with private baths, meals, ground transport, and outings. Amenities include a swimming pool and boules court.

Location: 30 minutes to Avignon train station.

Contact: David Carpita, School Administrator, Mas de Cornud, Route de Mas Blanc, St. Remy-de-Provence, 13210 France; (33) (0)490-92-39-32, Fax (33) (0)4-90-92-55-99, E-mail mascornud@compuserve.com.

PASSPORT TO PROVENCE

(See also page 292, International Kitchen) **Provence/September-April, June**

Hostellerie Berard offers 5-night programs that include 4 hands-on cooking lessons, visits to markets and food producers, winery tours, sightseeing. Maximum class/group size: 10. Facilities: Restaurant of the hotel.

Emphasis: Provencal cuisine.

Faculty: Rene Berard, who studied at LeNôtre and with Roger Vergé.

Costs: $1,475 includes lodging at the family-run Hostellerie Berard, breakfasts and lunches, excursions. Hotel amenities include en-suite bathrooms and a swimming pool.

Location: Near Marseille, the Bandol region.

Contact: Sandra Berard, Passport to Provence, Rue Gabriel Peri, La Cadiere d'Azur, 83740 France; (33) (0)-94-983740. In the U.S.: The International Kitchen, Inc., 1209 N. Astor St., #11N, Chicago, IL 60610; 800-945-8606, Fax 847-295-0945/312-654-8446, E-mail info@intl-kitchen.com, http://www.intl-kitchen.com.

PROMENADES GOURMANDES

(See also page 292, International Kitchen) **Paris/Year-round**

This private school offers half-day to one-week hands-on cooking programs that feature food-related expeditions to markets, bakeries, butchers, and wine shops. Established in 1995. Maximum class/group size: 2-8. Facilities: Paule Caillat's renovated apartment kitchen, which has granite countertops and the latest appliances. Also featured: Accompanied dining at popular bistros.

Emphasis: A contemporary version of French cuisine bourgeoise featuring menus that are easy to recreate at home.
Faculty: Paule Caillat, born and raised in Paris and college-educated in the U.S. She apprenticed to a Cordon Bleu chef and is an experienced teacher, caterer, and food consultant.
Costs: From $200 for a half-day class to $560 for two full days plus two walking tours.
Location: On the Right Bank, in Paris' Marais district.
Contact: Paule Caillat, Owner, Promenades Gourmandes, 187, rue du Temple, Paris, 75003 France; (33) (0)1-48045684, Fax (33) (0)1-42785977. In the U.S.: The International Kitchen, Inc., 1209 N. Astor St., #11N, Chicago, IL 60610; 800-945-8606, Fax 847-295-0945/312-654-8446, E-mail info@intl-kitchen.com, http://www.intl-kitchen.com.

PROVENCAL GETAWAY VACATIONS

Curnier/May-June, September-November

Cooking instructor Eileen Dwillies offers 6-day participation vacations in a restored 16th century house, 12-day cooking/touring vacations that include Provence, Venice, and Tuscany. Established in 1994. Maximum class/group size: 4. 7 programs per year. Facilities: Home-style kitchen. Also featured: Daily tours of outdoor markets, vintners caves, artists' workshops, and olive mills.
Emphasis: Provenáal and Mediterranean cuisine.
Faculty: Eileen Dwillies, author of 9 cookbooks, has taught cooking for 20 years and is a former food editor and TV show host.
Costs: $1,000, which includes meals, tours, and twin lodging with shared bath for the 6-day program, $2,900 for the 12-day program.
Location: 75 minutes north of Avignon, 2-3 hours south of Lyon.
Contact: Eileen Dwillies, Provencal Getaway Vacations, #222, 525 Wheelhouse Square, Vancouver, BC, V5Z 4L8 Canada; 604-876-8722, Fax 604-876-1497, E-mail eileen@bc.sympatico.ca.

PROVENCE IN 3 DIMENSIONS

Carpentras/Year-round *(See also page 292, Robert Reynolds)*

This private school offers weekly demonstration and hands-on sessions. Established in 1996. Maximum class/group size: 6-20. Facilities: Professional kitchen in a château, plus dining room. Also featured: Winery tours and instruction, day trips, visits to markets and food producers, master classes for culinary professionals.
Emphasis: Authenticity of products and artisans, Provencal cuisine.
Faculty: Michel Depardon, trained at Lausanne, owner/chef of Sous les Micocouliers in Eygalieres; Robert Reynolds, French-trained chef and former restaurateur in San Francisco, academic director of Cooking School of the Rockies.
Costs: $2,250 for a 5-day session, which includes single occupancy lodging, meals, and ground transport. Lodging is at local first class hotels. Tuition is reduced for those making their own lodging arrangements.
Location: Downtown Carpentras, which has a population of 40,000 and is 30 minutes from Avignon.
Contact: Michel Depardon, President, Provence in 3 Dimensions, Allee des Tilleuls, Carpentras, 84000 France; (33) (0)490-67-02-90, Fax (33) (0)490-67-02-91, E-mail mdepar@avignon.pacwan.com. In the U.S.: Robert Reynolds, 222 SE 18th Ave., Portland, OR 97214; 888-733-3391, Fax 503-233-1934, E-mail rowbear@ibm.net.

PROVENCE ON YOUR PLATE

Provence/May-June, September-October

Certified Culinary Professional Connie Barney offers 1-week villa-based culinary tour programs. Three hands-on and demonstration classes per week. Established in 1993. Maximum class/group size: 8 maximum. Facilities: Well-equipped home kitchen. Also featured: Day trips, market visits, winery tours, dining in fine restaurants and private homes, visits to food producers, sightseeing.
Emphasis: Regional Provencal cookery. Unpretentious home cooking using fresh herbs, fish,

olive oil, fruit, vegetables.

Faculty: Connie Barney, CCP, holder of Grand Diplome from La Varenne, former director of Roger Vergé's cooking school in Mougins, recipe consultant for Markets of Provence.

Costs: $1,950/week, includes lodging, meals, ground transport, entry fees. La Mole: 6 bedrooms, 6 baths in restored 18th-century farmhouse on 20 wooded acres. Near Roussillon: 7 bedrooms, 7 baths in traditional home set in lavender fields. Both have swimming pools.

Location: 2 locations including La Mole, 10 miles southwest of St. Tropez; near Roussillon, 25 miles east of Avignon.

Contact: Connie Barney, Owner, Provence on Your Plate, 915 E. Blithedale, #7, Mill Valley, CA 94941 US; 800-449-2111/415-281-5644, Fax 415-389-0736, E-mail conbarn@aol.com, URL http://www.provenceonyourplate.com.

RESIDENTIAL SUMMER COOKERY COURSES IN BURGUNDY

Grancey le Château/June-July, September

This tour organizer offers one-week hands-on courses that focus on a specific subject, such as fish and sauces, meat and accompaniments, hors d'oeuvres and desserts. Includes visits to the market in Dijon and Burgundy vineyards. Maximum class/group size: 6-10. Facilities: Classes and lodging at a 200-year-old farmhouse. Also featured: Gourmet Weekends for private groups of 6-10.

Emphasis: Cuisine of France's Burgundy region.

Faculty: Michel Robolin, an English-speaking French chef who has taught in the UK, France, and the U.S.

Costs: £665 includes lodging, meals, and all planned activities.

Location: 5 hours from Calais, near Dijon.

Contact: Penny Easton, Encounters, Ltd., Garden House, Orchard Ct., Chillenden, Canterbury, Kent, CT3 1YA England; (44) 1304-841136, Fax (44) 1304-841136.

RESTAURANT HOTEL LA COTE SAINT JACQUES

Joigny, Burgundy/March

This Michelin 3-star restaurant offers 3-day demonstration programs that feature French cuisine, market visit, wine instruction, dining in the restaurant. Established in 1996. Maximum class/group size: 8-15. 10 programs per year. Facilities: New teaching kitchen. Also featured: Boat rides, golf, vineyard visits.

Emphasis: French cuisine.

Faculty: Jean Michel Lorain, chef of La Cote Saint Jacques.

Costs: 3,500FF-3,800FF (4,200FF-4,800FF) includes standard (superior) lodging at Hotel La Cote Saint Jacques, a member of Relais & Châteaux, most meals; 810FF (950FF) for second person in same room; 250FF class only.

Location: On the Yonne River in Joigny, about an hour southeast of Paris.

Contact: Jean-Michel Lorain, Restaurant Hotel La Cote Saint Jacques, 14, Faubourg de Paris, Joigny, 89300 France; (33) (0)3-86-62-09-70, Fax (33) (0)3-86-91-49-70, E-mail lorain@relais-chateaux.fr, URL http://www.integra.fr/lorain.

THE RHODE SCHOOL OF CUISINE

(See also page 332) (See display ad page 308) **Theoule-sur-Mer/March-November**

This French villa offers weekly 7-day hands-on vacation programs. Established in 1993. Maximum class/group size: 10. Facilities: Custom built teaching kitchen. Also featured: Cultural excursions, dining at Michelin-star restaurants, market visits, activities for non-cooking guests.

Emphasis: French cuisine, sauces, pastry, wine and cheese.

Faculty: Chef Frederic Riviere, a graduate of Les Sorbets, who has cooked at Michelin-star restaurants.

Costs: $2,295 ($2,595), which includes shared (single) lodging, meals, and ground transport. Non-cook rate is $1,795 ($1,995). Le Mas des Oliviers, a modernized 5-bedroom/5-bath villa, has a 3,500-square-foot terrace, pool, and Jacuzzi.

Location: On the coast in the Esterels, overlooking the Mediterranean.
Contact: Tim Stone, The Rhode School of Cuisine, 800-447-1311, Fax 415-388-4658, URL http://www.to-gastronomy.com.

RITZ-ESCOFFIER ÉCOLE DE GASTRONOMIE FRANCAISE

Paris/Year-round *(See also page 155) (See display ad page 157)*

This private school in the Hotel Ritz offers 1-week hands-on theme and pastry courses, half- and full-day workshops, demonstrations Monday, Tuesday, and Thursday afternoons. Established in 1988. Maximum class/group size: 8-10 hands-on/40 demo. Facilities: 2,000-square-foot custom-designed facility includes a main kitchen, pastry kitchen, conference room/library, changing rooms. Also featured: Floral arranging, wine tastings, children's courses, custom-designed programs for groups.
Emphasis: Themes include brasserie and bistro cooking, fish, Provencal specialties, sauces, wild game, chocolate, wine.
Faculty: 7 full-time, 3 part-time. School Director Jean-Philippe Zahm, Chef de Cuisine Blaise Volckaert, Christophe Bellet, and Chef Patissier Gilles Maisonneuve.
Costs: Demonstrations are 275FF each (6 for the price of 5), themed courses start at 5,750FF. Lodging packages are available on a limited basis.
Location: Central Paris, near 3 main subway entrances and most major department stores and gourmet stores.
Contact: M. Jean-Philippe Zahm, Director, Ritz-Escoffier École de Gastronomie Francaise, 15, Place Vendome, Paris Cedex 01, 75041 France; (33) (0)143-16-30-50, Fax (33) (0)1-43-16-31-50, E-mail ecole@ritzparis.com.

ROBERT REYNOLDS, NORTHWEST FORUM (See page 247)

ROGER VERGÉ COOKING SCHOOL

Mougins/September-July

Restaurant l'Amandier offers 2-hour demonstrations from Tuesday through Saturday. Established in 1984. Maximum class/group size: 20. 12 programs per year. Facilities: Restaurant kitchen.
Emphasis: Seasonal menus, Provencal cuisine.
Faculty: Michel Duhamel and other chefs from the Moulin de Mougins restaurant.
Costs: 300FF per class, 1,350FF for 5 classes. Booking is desired 48 hours in advance.
Location: Near Cannes.
Contact: Sylvie Charbit, Manager, Roger Vergé Cooking School, Restaurant l'Amandier, Mougins Village, 06250 France; (33) (0)493-75-35-70, Fax (33) (0)493-90-18-55, E-mail mougins@relais-chateaux.fr, URL http://www.relaischateaux.fr/mougins.

THE SAVOUR OF FRANCE

Burgundy and Provence/May-October

This specialty travel company offers 6-day cooking and wine-tasting tours that include château lodging, private winery tours, dining at Michelin 3-star restaurants, visits to local producers. Established in 1995. Maximum class/group size: 12. 6-7 programs per year. Facilities: Professional château kitchen in Burgundy, home-style kitchen in Provence. Also featured: Day cruise on the Mediterranean, barging the Burgundy Canal, balloon excursion.
Emphasis: Classic regional French cuisine paired with fine wines.
Faculty: Burgundy: local French chefs, including Chef Vignaud of the Michelin-star Hostellerie des Clos. Provence: Veronique Marget, professionally trained in France. Tour leader Darrin Anderson is bilingual.
Costs: $2,295 includes lodging, meals, wines, ground transport, planned activities except ballooning.
Location: Burgundy: Charny, 90 minutes southeast of Paris; Ecutigny, 20 minutes from Beaune; Provence, near Gordes, 30 minutes from Avignon.
Contact: David Geen, Director, Villas and Voyages, 2450 Iroquois Ave., Detroit, MI 48214 US; 800-827-4635, Fax 313-331-1915.

A TASTE OF PROVENCE

Le Bar sur Loup/May-June, September-October

This private school offers week-long hands-on culinary vacations in a private farmhouse in Provence. Includes a 1-day class at Roger Vergé cooking school, market visits, winery tours, sight-seeing, visits to food producers and private homes. Established in 1983. Maximum class/group size: 8. 14 programs per year. Facilities: Large Provencal kitchen.
Emphasis: Provencal cooking.
Faculty: Tricia Robinson, owner; faculty at Roger Vergé cooking school.
Costs: $1,800, includes lodging, most meals, planned activities.
Location: Between Grasse and Vence, 40 minutes by car from Nice.
Contact: Tricia Robinson, Owner, A Taste of Provence, 925 Vernal Ave., Mill Valley, CA 94941 US; 415-383-9439, Fax 415-383-6186, E-mail info@tasteofprovence.com, URL http://www.tasteof-provence.com.

THE VACATIONING TABLE, INC.

(See also page 200, Foodsearch Plus) **Gascony, Provence, Paris/May-October**

This private school offers weekly culinary trips that feature hands-on classes combined with culinary cultural excursions, including market visits, winery tours, visits to food producers, and sight-seeing. Established in 1994. Maximum class/group size: 12-14. Facilities: Fully equipped rustic and hotel kitchens.
Faculty: Karen Hanson, culinary educator, author, food consultant.
Costs: $3,000-$3,800 includes meals, small luxury hotel lodging, ground transport, planned activities.
Location: Regions of France.
Contact: Karen Hanson, 258 Florida Rd., Ridgefield, CT 06877 US; 203-438-0422.

A WEEK IN BORDEAUX

Bordeaux/June-September

Jean-Pierre and Denise Moulle offer 6-day culinary participation vacations each summer. Established in 1988. Maximum class/group size: 8. 4 programs per year. Facilities: Professional kitchen of Château La Louviere and a small farmhouse kitchen with a grilling fireplace. Also featured: Visits to châteaux, wine estates, markets, cheese shops, medieval villages, oyster beds, regional inns, and Michelin-starred restaurants. Also available: sight-seeing and shopping.

Emphasis: French classic cuisine, regional cuisine of Gascony, wine appreciation.

Faculty: Jean-Pierre Moulle graduated from École Hoteliere in Toulouse, served as chef at Chez Panisse, and consults for restaurants. Bordeaux native Denise Moulle opened wine shops in California's Bay Area and markets her family's French château wines in the U.S.

Costs: Cost of $2,800 (single supplement $200) includes lodging at the Château Mouchac, meals, planned excursions, and ground transport.

Location: France's Bordeaux region.

Contact: Denise Lurton-Moulle, A Week in Bordeaux, P.O. Box 8191, Berkeley, CA 94707-8191 US; 510-848-8741, Fax 510-845-3100.

A WEEK IN PROVENCE

Gordes/February-June, September-December

Sarah and Michael Brown offer week-long vacation programs, scheduled every other week, that include daily cooking demonstrations with as much hands-on participation as desired. Established in 1995. Maximum class/group size: 6. Also featured: History, architecture, and art tours of the region, visits to food markets and wineries.

Emphasis: Provence cuisine and culture.

Faculty: Sarah Brown spent her childhood in France, has a Ph.D. in Art History, and studied cooking since she was 7; gallery owner Michael Brown represented agricultural and food interests as a lobbyist in Washington, DC, and exports wines to the U.S.

Costs: Cost of $1,500 includes lodging in the Brown's converted village farmhouse (2 double and 4 single bedrooms, 4 baths), breakfasts and dinners, excursions. Amenities include swimming pool, library, hiking, riding, biking, golfing.

Location: 45 minutes from Marseilles-Provence airport, 30 minutes from Avignon train station.

Contact: Sarah and Michael Brown, A Week in Provence, Les Martins, Gordes, 84220 France; (33) (0)490-72-26-56, Fax (33) (0)490-72-23-83.

INDIA

FPT SPECIAL INTEREST TOURS (See page 266)

JULIE SAHNI'S SCHOOL OF INDIAN COOKING

India/February and October *(See also page 235)*

Cooking school owner and cookbook author Julie Sahni offers 16-day cultural and culinary tours that include visits to farm kitchens, spice and tea plantations, markets and bazaars, and private receptions. Established in 1973. 2 programs per year.

Emphasis: The cuisine and culture of India

Faculty: Julie Sahni is author of Classic Indian Cooking, Indian Vegetarian Cooking, and Savoring Spices and Herbs. She is a member of the IACP and Les Dames D'Escoffier and has degrees in architecture and classical dance.

Costs: ~$4,975 includes shared lodging (single supplement $1,475), most meals, ground transport, planned activities. Lodging in deluxe hotels or best available.

Location: Western, southern, and northern India.

Contact: Julie Sahni, Director, Julie Sahni's School of Indian Cooking, 101 Clark St., #13A, Brooklyn Heights, NY 11201 US; 718-625-3958, Fax 718-625-3456, E-mail jsahni@worldnet.att.net.

INDONESIA

THE SERAI COOKING SCHOOL

Bali/February-July, September-December

This resort offers 2- and 5-day programs that feature a half of each day in hands-on cooking classes, the other half exploring Bali's culinary and cultural aspects. Established in 1997. Maximum class/group size: 10. 20-25 programs per year. Also featured: Market tours, visits to food producers, sightseeing, dinners, cultural events, concluding Balinese feast.

Emphasis: Indonesian cuisine with particular emphasis on Balinese food.

Faculty: The Serai's executive chef, Jonathan Heath, an Australian who trained and apprenticed in Australia and has lived in Indonesia for 10 yrs.

Costs: $250 ($650) for the 2- (5-) day course, which includes most meals. Nonparticipant guest fee is $35 ($125). Lodging at The Serai starts at $130/night plus 21% tax and service charge. $100 deposit.

Location: The Serai resort hotel is on a secluded beach in East Bali, ~50 miles from Denpasar Airport.

Contact: Paul Walters, General Manager, The Serai Cooking School, P.O. Box 13, Manggis, Karangasem, Bali, 80871 Indonesia; (62) 363-41011, Fax (62) 363-41015, E-mail serai@ghmhotels.com.

IRELAND

BALLYMALOE COOKERY SCHOOL

(See also page 290, Cuisine International) **Midleton, County Cork/Year-round**

This proprietary school offers 1- to 5-day hands-on vacation programs that include a 1-day Christmas cooking demonstration and a weekend Entertaining course. Established in 1983. Maximum class/group size: 44 hands-on/60 demo. 30 programs per year. Facilities: Include a specially-designed kitchen with gas and electric cookers, mirrored demonstration area with TV monitors, gardens that supply fresh produce. Also featured: Fishing and golf.

Emphasis: A variety of topics, including entertaining menus, seafood, vegetarian dishes.

Faculty: 4 full- and 4 part-time. Includes Principal Darina Allen, IACP-certified Teacher & Food Professional, her brother, Rory O'Connell, both trained in the Ballymaloe House restaurant kitchen, and her husband Tim Allen. Guest chefs are featured.

Costs: 1-day courses are IR£98, weekend courses are IR£235, 5-day courses are IR£375. Accommodation in self-catering cottages at school during short courses is IR£12/night shared, IR£14.50/night single.

Location: A village near the sea in southern Ireland, 1 hour from Cork Intl Airport, 4 hours from Dublin. Airport and train station pick-up is available.

Contact: Tim Allen, Ballymaloe Cookery School, Kinoith, Shanagarry, County Cork, Midleton, Ireland; (353) 21-646785, Fax (353) 21-646909, E-mail enquiries@ballymaloe-cookery-school.ie, URL http://www.ballymaloe-cookery-school.ie. In the U.S.: Judy Ebrey, Cuisine International, P.O. Box 25228, Dallas, TX 75225; 214-373-1161, Fax 214-373-1162, E-mail CuisineInt@aol.com, URL http://www.cuisineinternational.com.

BALTIMORE INTERNATIONAL COLLEGE

(See also page 58) **Lough Ramor/Year-round**

BIC's European campus offers programs that are open to nonprofessionals. Established in 1987. Also featured: Golf, tennis, fishing, boating.

Costs: Lodging at The Park Hotel-Deer Park Lodge.

Location: The 100-acre campus is 50 miles from Dublin.

Contact: Steven Solomon, Director of Communications, Baltimore International College, 17 Commerce St., Baltimore, MD 21202 US; 800-624-9926, Fax 410-752-3730.

COUNTRY HOUSE COOKERY – BERRY LODGE

Miltown Malbay, Co. Clare/January-September, November

Culinary professional Rita Meade offers a 2-day summer vacation program: A Taste of Irish Cookery. Established in 1994. Maximum class/group size: 10. Facilities: Modern traditional farmhouse kitchen, bed & breakfast, restaurant. Also featured: Day tour of local cultural and historical sites, dinners at local restaurants, golf, fishing, swimming, sightseeing.

Emphasis: Traditional and new Irish country cookery.

Faculty: Rita Meade learned to cook from her mother and qualified as a Home Economist. She studied in England, France and Italy and has taught cooking for nearly 30 years.

Costs: Summer program: 2-day/7-hour cookery tuition is IR£100. Dinner IR£20. Bed & Breakfast IR£22.50/night at Berry Lodge, a 19th century Victorian country house with modern amenities. Total IR£220. Non-participant partner IR£110. Single supplement IR£7/night.

Location: Rural western Ireland, 200 yards from the sea in West Clare, 160 miles from Dublin, and 38 miles from Shannon International Airport.

Contact: Rita Meade, Owner/Teacher, Country House Cookery-Berry Lodge, Annagh, Miltown Malbay, Co. Clare, Ireland; (353) (0)65-87022, Fax (353) (0)65-87022.

OLD RECTORY COUNTRY HOUSE AND RESTAURANT

Wicklow, Cty. Wicklow/March-April, October-December

This country inn offers one-day courses in vegetarian cooking entertaining and holiday menus, fish and shellfish. Maximum class/group size: 12.

Emphasis: Organic produce and edible flowers.

Faculty: Linda Saunders, owner/chef for 20+ yrs. The Old Rectory is in the Blue Book of Irish Country Houses and received the 1997 Egon Ronay Award for Best Healthy Eating in Ireland.

Costs: IR£50/class.

Location: 30 miles south of Dublin in eastern Ireland, on the Irish Sea.

Contact: Linda Saunders, Owner/chef, Old Rectory Country House and Restaurant, Wicklow, County Wicklow, Ireland; (353) 404-67048, Fax (353) 404-69181, E-mail mail@oldrectory.ie, URL http://www.oldrectory.ie.

ISRAEL

TNUVA, TRAINING CENTER FOR FOOD CULTURE

Tel Aviv/September-July

Tnuva, the largest food distributing company in Israel offers demonstration classes that teach adults and children how to prepare food inexpensively and efficiently. Established in 1973. Facilities: Amphitheater. Also featured: Sessions in English for groups of 25 or more.

Emphasis: Russian, Middle Eastern, Moroccan, Italian, and Chinese cuisines; microwave, vegetarian, and nutritional cookery; baking; fruit preparation and marzipan garnishing.

Faculty: School director Tova Aran and a teaching staff of food writers, home economics instructors, hotel/restaurant chefs, pastry chefs, and caterers.

Costs: Classes for English-speaking groups are about 65 Shekels/person.

Contact: Tova Aran, Tnuva, Training Center for Food Culture, 47 Ben-Gurion Blvd., Tel Aviv, Israel; (972) 3-5243-157, Fax (972) 3-5230-055.

ITALY

A TAVOLA CON LO CHEF

Rome/Year-round

This private school offers amateur and professional training courses include basic and advanced cooking, confectionery, pizza, wine tasing, bartender, monographic lessons on specific themes, 1-week hands-on or demonstration courses. Established in 1992. Maximum class/group size: 20 hands-on, 30

demo. Facilities: 3 well-equipped professional kitchens. Also featured: Children's programs, private instruction for groups, market visits. By request: winery tours, sightseeing, visits to food producers.
Emphasis: Traditional and modern Italian and regional cuisine.
Faculty: Includes Antonio Sciullo, Laura Ravaioli, Renzo Barchieri, Massimiliano Bacich, Gianfranco Calidonna, Giuseppe Cespi Polisiani, Alberto Ciarla, Leonardo Di Carlo, Nazzareno Lavini, Salvo Leanza, Mauro Lotti, Maurizio Maesiri, Agata l'arisella Caraccio.
Costs: $60/$120 for each 3/4-hour lesson.
Location: Rome's historic center, next to St. Peter Cathedral.
Contact: Maria Teresa Meloni, Fiorella D'Agnano, A Tavola con lo Chef, Via dei Gracchi 60, Rome, 00192 Italy; (39) 06-32 22 096, Fax (39) 06-320 34 02.

ACQUERELLO ADVENTURES

Year-round

This restaurant offers culinary and wine tours for individuals and groups. Established in 1994. Also featured: Market visits, winery tours, sightseeing, visits to food producers, dining in private homes.
Emphasis: Culinary and wine travel programs, personalized service.
Location: All regions of Italy.
Contact: Anne Paterlini, Travel Advisors, 619 E. Blithedale Ave., Mill Valley, CA 94941 US; 415-383-2323, Fax 415-383-8929.

ALIMANDI'S ROMAN CULINARY CLASSES

(See also page 228, Alfredo's Classical Italian Cooking) **Rome/January-March**

The Hotel Alimandi offers 6-day cooking vacations that include hands-on classes, visits to markets and wineries, sightseeing, and dinner every night at a different restaurant in and around Rome. Established in 1997. Maximum class/group size: 15. 12 weekly programs per year. Facilities: Hotel Alimandi's newly remodeled kitchen. Also featured: Tours of the Castelli Romani and a pasta factory.
Emphasis: Roman cuisine & gourmet cookery.
Faculty: Chef Alfredo de Bonis, owner/operator of Bravo Alfredo Catering and instructor in Italian Regional Cooking at North Brunswick Twp. High School, has been in the restaurant business for 25 years.
Costs: $1,495 includes shared lodging at Hotel Alimandi, meals, planned activities.
Location: Hotel Alimandi is next to the Vatican Museum and close to the Metro stop Ottaviano.
Contact: Alfredo de Bonis, Chef Instructor, Alfredo's Classical Italian Cooking, 375 Tremont Ave., Milltown, NJ 08850-2013 US; 732-828-8460, E-mail nrwu55a@prodigy.com.

APICIUS – THE ART OF COOKING AT LORENZO DE'MEDICI

(See also page 158) **Florence/Year-round**

This private school, a member of the Federation of European Schools, offers 10-day food/wine/culture programs, 1-month language/cooking/wine courses, wine and cooking classes, 1-year diploma program. Established in 1973. Maximum class/group size: 5-20. 15 programs per year. Facilities: 4-story building containing large kitchen with modern teaching equipment, cooking island, additional cooking facilities. Also featured: Market tours, wine tasting sessions, gastronomic walking tours.
Emphasis: Italian and historic Renaissance cuisine, wine appreciation.
Faculty: Chefs, restaurateurs, culinary professionals, includes founder/director Gabriella Ganugi, author of The Four Seasons of the Tuscan Table.
Costs: $1,900 for 10-day programs includes lodging and activities, $500-$690 tuition for 1-month programs, $3,100 tuition for semester programs, $70 for group classes, $35 for wine classes, $190 for private instruction.
Location: In the San Lorenzo district, between the Duomo and the central train station.
Contact: Dr. Gabriella Ganugi, Director, Lorenzo de/Medici Institute, Via Faenza, Florence, 50123 Italy; (39) 055-28-73-60, Fax (39) 055-239-89-20, E-mail LDM@dada.it, URL http://www.dada.it/ldm.

ART OF LIVING – CULINARY TOURS WITH SARA MONICK (See page 289)

AVIGNONESI WINE AND FOOD WORKSHOPS

Florence/April-June, September-November *(See also page 318, Culinary Arts, International)*

This wine-producing estate offers 1-week cooking vacations that include demos and hands-on classes with a chef, a pizza maker, and a Tuscan home cook; meals in a trattoria, restaurant, private home, and the villa of a Contessa; market visits. Established in 1994. Maximum class/group size: 12. 12 programs per year. Facilities: Professional kitchen. Also featured: Visit to a cheesemaker and the country's largest enoteca in Montepulciano, Pienza, and Siena. A day with author and wine expert Burton Anderson.

Emphasis: Tuscan cuisine, including such artisanal foods as olive oil, cheese, pasta, and wine.

Faculty: Culinary Director Rolando Beramendi, Program Director Pamela Sheldon Johns, Wine Program and Artisan Foods Director Burton Anderson, Chef Massimiliano Mariotti.

Costs: $2,450, which includes double occupancy lodging in a medieval village with fully restored rooms and private baths, all meals and wine, ground transportation, planned activities.

Location: Pick-up and drop-off in Florence. Private shuttle to accommodations near Pienza and classes near Montepulciano.

Contact: Pamela Sheldon Johns, Avignonesi Wine and Food Workshops, 1324 State St., #J-157, Santa Barbara, CA 93101 US; 805-963-7289, Fax 805-963-0230, E-mail CulinarArt@aol.com, URL http://www.avignonesi.it. Also: Palazzo Avignonesi, 53045 Montepulciano, Italy; (39) (0)578-757874, Fax (39) (0)578-757847.

BADIA A COLTIBUONO

Siena/May-July, September-October *(See also page 290, Cuisine International)*

Cookbook author Lorenza de'Medici offers 5-day vacation courses (Mon-Sat) that include demonstration and participation cooking classes. Established in 1985. Maximum class/group size: 15. 12 programs per year. Facilities: Large teaching kitchen. Also featured: Visits to food producers, wineries, private estates; dining as guests of Lorenza and at private homes, villas and castles; trip to the Palio horse race (July); dinner as guests at a Siena contrada.

Emphasis: Regional Italian cooking and wines.

Faculty: Lorenza de' Medici, author of several cookbooks and cooking manuals and a PBS TV series.

Costs: Land cost is $4,100 single, $3,500 double, which includes lodging at Badia a Coltibuono, an 11th century estate that produces wine, vinegar, honey, olive oil, and other products. Amenities include a cookbook library, swimming pool, and sauna.

Location: 20 miles north of Siena, 40 miles south of Florence.

Contact: Lorenza de' Medici, Badia a Coltibuono, Gaiole in Chianti, Siena, 53013 Italy; Fax (39) 0577-749235. In the U.S.: Judy Ebrey, Cuisine International, P.O. Box 25228, Dallas, TX 75225; 214-373-1161, Fax 214-373-1162, E-mail CuisineInt@aol.com, URL http://www.cuisineinternational.com.

BAROLO WINE COUNTRY COOKING

Piedmont region/September

These four restaurants offer 6-day programs that feature 4 cooking classes in 4 top restaurants, lunches/dinners in other fine restaurants, 7 wine tastings with owners, 4 winemakers' dinners, and visits to 2 wine museums, 1 castle, 9 towns, and a market. Established in 1995. Maximum class/group size: 7. 2 programs per year. Facilities: Restaurant kitchens.

Emphasis: Piedmontese antipasti, pasta, meat, desert menus with Barolo and Barbaresco area wines.

Faculty: Chefs of Real Castello, Gran Duca, La Cascata, La Contea restaurants. Tour leaders are Piedmontese Elio Sabena and Canadian Margaret Cowan.

Costs: C$2,695, which includes lodging, meals, planned activities.

Location: 9 towns in the Barolo and Barbaresco wine country hills of northwestern Italy, an hour's drive south of Turin.

Contact: Margaret Cowan, Barolo Wine Country Cooking, 310-1184 Denman St., Vancouver, BC, V6G 2M9 Canada; 800-557-0370, Fax 604-681-4909, E-mail mcowan@portal.ca, URL http://www.italycookingschools.com.

BED AND BREAKFAST IN TUSCANY

Tuscany/May, October-November

Culinary professional Lucia Luhan offers 1-week and mini-courses. Established in 1985. Maximum class/group size: 10. 3-4 programs per year. Facilities: Ms. Luhan's family farm/bed & breakfast has a farm kitchen with individual work areas. Also featured: Shopping and sightseeing.

Emphasis: Tuscan cuisine, including pastas and pizza.

Faculty: Restaurateur and caterer Lucia Ana Luhan completed master's degree studies in public relations from Boston University and studied in Europe and South America. She owns What's Cooking? in Newport Beach and Luciana's Ristorante in Dana Point, California.

Costs: $1,800 for one week includes lodging and most meals.

Location: Central Italy's wine country, a 5-minute drive from Montecatini Terme, less than 30 minutes from Florence, 90 minutes from Siena.

Contact: Food and Wine Appreciation Program, PO Box 3582, Dana Point, CA 92629 US; 714-488-7694, Fax 714-488-7694.

BELLA CUCINA

(See also page 200)

Florence and Venice

Culinary professional Carol Borelli offers travel programs to Italy that combine scholar-guided visits to art museums and historic sites, lectures on the history of art and cuisine, cooking instruction, and excursions. Established in 1996. Maximum class/group size: 8-20. Facilities: Professional kitchens.

Emphasis: Fine cooking in the seasonal traditions of Italy's regions.

Faculty: Carol Borelli, artist and educator, studied at Le Cordon Bleu and with chefs worldwide.

Costs: ~$3,400-$5,600 all inclusive.

Contact: Carol Borelli, Owner, Bella Cucina, P.O. Box 421, New Canaan, CT 06840 US; 203-966-4477, Fax 203-966-8781, E-mail cborelli@earthlink.net.

BIKE RIDERS TOURS

Umbria and Sicily/May, September-October

This travel company offers 1-week tours (3 in Umbria, 2 in Sicily) that feature hands-on classes and 15- to 35-miles cycling/day with chefs, market visits, wine classes, winery tours. Established in 1990. Maximum class/group size: 16 maximum. 7 programs per year. Facilities: Inn and restaurant kitchens. Also featured: Umbria trip concludes with overnight and dinner at Le Tre Vaselle, in Torgiano; restaurants with regional cuisine.

Emphasis: Regional Italian cuisine.

Faculty: Umbria: Ron Suhanosky (Galleria Italiana, Boston), Rick Moonen (Oceana, NYC), Alex Lee (Restaurant Daniel, NYC), Francesco Ricchi (Cesco and I Ricchi, DC).

Costs: $2,700 ($2,620) for Umbria (Sicily), includes support van, lodging (villas and 4- and 5-star inns in Umbria, villas and island retreats in Sicily), most meals, planned activities. Bike rental $150. Single supplement $385.

Location: Umbria: Spello, Assisi, Montefalco, Spoleto, Todi, Torgiano. Sicily: Taormina, villages at base of Mt. Aetna, islands of Lipari, Vulcano, Salina.

Contact: Eileen E. Holland, Director, Bike Riders Tours, P.O. Box 130254, Boston, MA 02113 US; 800-473-7040/617-723-2354, Fax 617-723-2355, E-mail info@bikeriderstours.com, URL http://www.bikeriderstours.com.

CAPEZZANA WINE & CULINARY CENTER

Carmignano, Florence/March-June, September-November

This private school and winery offers 5-day hands-on vacation programs designed for food professionals and others interested in Tuscan cuisine. Covers basic and advanced concepts and includes market visits, winery tours, visits to food producers, dining in private homes. Established

in 1994. Maximum class/group size: 14. 8-10 programs per year. Facilities: The family kitchen at Tenuta di Capezzana and restaurant kitchens. Also featured: One day is devoted to Tuscan wines.
Emphasis: Food and wine of Tuscany.
Faculty: The Capezzana Wine & Culinary Center family chef, directors, family members, visiting chefs. Wine program is taught by a Master of Wine.
Costs: $2,300-$3,000, includes lodging, meals, all activities.
Location: Via di Capezzana 100 in Carmignano, Florence.
Contact: Marlene Levinson, Capezzana Wine & Culinary Center, 55 Raycliff Terr., San Francisco, CA 94115 US; 415-928-7711, Fax 415-928-7789, E-mail mlcooker@pacbell.net.

CHIANTI IN TUSCANY – ITALIAN COOKERY AND WINE

Gaiole in Chianti/April-October

This private school offers continuous 1-week hands-on cooking courses and Italian language lessons. Established in 1986. Maximum class/group size: 10. Facilities: Podere Le Rose, the instructors' home, which has a well-equipped typical country kitchen. Also featured: Wine lesson, winery tour, Chianti tour.
Emphasis: Northern and southern Italian cookery, fresh ingredients, easy-to-make recipes; Italian language.
Faculty: The Bevilacqua de'Mari family assisted by a Tuscan chef.
Costs: 1,100,000 Lira for classes only. Lodging upon request at a 13th-century restored Italian farmhouse. Various packages offered.
Location: A 30-minute drive from Siena and an hour from Florence.
Contact: Countess Simonetta de'Mari di Altamura, Chianti in Tuscany, Poggio S. Polo 2, Lecchi, Gaiole (SI), 53010 Italy; (39) 055-294511, Fax (39) 055-2396887, E-mail centro.pontevecchio@dada.it, URL http://www.firenze.net/cpvchianti. Additional contact for Australia and New Zealand: Mark James, 87 Muston St., Mosman 2088, Australia; phone/fax (61) (0)2 99681120.

CICLISMO CLASSICO

Tuscany/June-July, September

This tour operator specializing in bicycle and walking vacations offers 1-week walking trips that include 5 cooking lessons, shopping at Mercato Centrale in Florence, dining at trattorias, visits with olive oil, wine, and cheese producers, 3- to 5-hour walks daily. Established in 1989. Maximum class/group size: 8-15. 2 programs per year. Facilities: The Cordon Bleu Cooking School in Florence. Also featured: Cycling and walking tours through Italy.
Emphasis: Tuscan cuisine.
Faculty: Gabriella Mari of the Cordon Bleu Cooking School in Florence.
Costs: $2,650 includes lodging, most meals, ground transport. Single supplement $250.
Location: From Florence to the Chianti region.
Contact: Lauren Hefferon, Director, Ciclismo Classico, 13 Marathon St., Arlington, MA 02174 US; 800-866-7314/781-646-3377, Fax 781-641-1512, E-mail info@ciclismoclassico.com, URL http://www.ciclismoclassico.com.

CLASSIC TUSCANY

Figline Valdamo, Florence/Year-round *(See also page 292, International Kitchen)*

Torre Guelfa restaurant offers six-night hands-on cooking vacations customized to the level of the students. Maximum class/group size: 4. Facilities: Restaurant kitchen. Also featured: Visits to markets and food producers, winery tours, sightseeing.
Emphasis: Tuscan cuisine.
Faculty: Claudio Piantini, head chef at Torre Guelfa.
Costs: $2,300 includes shared lodging, meals, excursions. Lodging at the Villa Casagrande in Figline and 4-star Hotel Lungarno in Florence.
Location: Four nights in Figline Valdarno, two nights in Florence.

Contact: George Firias, Classic Tuscany, Borgo La Croce, 30/r, Florence, 50121 Italy; (39) 055-2479880. In the U.S.: The International Kitchen, Inc., 1209 N. Astor St., #11N, Chicago, IL 60610; 800-945-8606, Fax 847-295-0945/312-654-8446, E-mail info@intl-kitchen.com, http://www.intl-kitchen.com.

CONTESSA PICCOLOMINI'S SIENA AND COUNTRYSIDE

Siena, Tuscany/October

This travel company offers a 1-week travel program that features hands-on classes, visits to markets and wineries, private hospitality at the countess's villa and other sites, sightseeing to Siena, San Gimignano, Churches of Monte Oliveto Maggiore and Sant'Antimo. Established in 1998. Maximum class/group size: 12. 1 programs per year. Facilities: Restaurants Il Poggio Antico in Montalcino and Locanda dell'Amoroso in Sinalunga.

Emphasis: Private hospitality emphasizing food, wines and history/customs of Contessa Piccolomini.

Faculty: Contessa Piccolomini's sister, a Tuscan cook, Chef Roberto Manetti, and the chef of the Lovanda dell'Amorosa hotel in Sinalunga.

Costs: $3,455 includes lodging in the apartments of the Borgo Licignanello Bandini, most meals, planned activities.

Location: Near Lucignano d'Arbia, about a 25 minutes drive southeast of Siena.

Contact: Margot Cushing, CTC, Principal, Vantaggio Tours c/o Linden Travel Bureau, 41 E. 57th St., New York, NY 10022 US; 800-808-6237/212-421-3320, Fax 212-421-2790, E-mail mcushing@lindentravel.com, URL http://www.vantaggio.com.

COUNTRY WALKERS

Tuscany/May-June, September-October

This travel company specializing in walking tours offers 1-week walking trips that include three days of cooking classes as well as market visits, wine tastings, and sightseeing. Established in 1980. Maximum class/group size: 18. 3-4 programs per year.

Emphasis: Regional cuisine of Tuscany.

Faculty: Tuscan chefs.

Costs: $2,600, which includes shared lodging, meals, and planned activities.

Location: Lucca, Ripoli di Lari, San Gimignano, Volterra, Colle val d'Elsa.

Contact: Country Walkers, P.O. Box 180, Waterbury, VT 05676 US; 800-464-9255/802-244-1387, Fax 802-244-5661, E-mail ctrywalk@AOL.com, URL http://www.countrywalkers.com.

CUCINA CASALINGA (See page 200)

CUCINA DEL SOLE

Sicily/Year-round

Amelia Tours travel company offers 8-day trips that include 5 half-day cooking classes, visits to a cheese farm, winery, pastry shops, and markets, and guided excursions to Siracusa, Piazza Armerina, and Mt. Etna. Maximum class/group size: 12 max. 2 programs per year. Also featured: 3-night independent vacations that include 3 half-day cooking lessons.

Emphasis: Sicilian cuisine.

Faculty: Gastronomic journalist and TV personality Eleonora Consoli.

Costs: $1,995 ($816) for the 8-day (3-night) trip includes lodging at the 4-star Villa Paradiso Dell'Etna and most meals.

Contact: Amelia Tours, 28 E. Old Country Rd., Hicksville, NY 11801 US; 800-742-4591, Fax 516-822-6220, E-mail ameliatours@worldnet.att.net.

CUCINA TOSCANA

Florence/Year-round

This culinary travel service offers customized 1-day to 1-month gastronomic excursions in Italy. Established in 1983. Maximum class/group size: 2-25. Facilities: Restaurant kitchens. Also featured: Antiquing, garden visits, walking tours of Florence, shopping, excursions to the Tuscan countryside.

Emphasis: Regional Italian cuisine.
Faculty: Proprietor Faith Heller Willinger is author of Red, White and Greens: The Italian Way with Vegetables and directs the Hotel Cipriani culinary program. Her assistant, Laura Kramer, has a degree in Medieval studies and leads tours to Tuscany.
Costs: From $275/day plus expenses for 1-3 persons. Group rates start at $150/day and include meals and transport. Rates are lower for longer trips.
Location: Regions of Tuscany.
Contact: Faith Heller Willinger, Cucina Toscana, via della Chiesa, 7, Florence, 50125 Italy; (39) 055-2337014, Fax (39) 055-2337014. In the U.S.: Vivian 847-432-1814, Fax 847-432-1889.

CUISINE ECLAIRÉE (See page 290)

CULINARY ARTS, INTERNATIONAL

Santa Barbara/Year-round *(See also page 314, Avignonesi Wine and Food Workshops)*

This cookware store and culinary travel provider offers programs that range from 3-hour classes to day trips to week-long excursions, including a fall truffle and risotto week in Piemonte and a spring week in Emilia Romagna with emphasis on the region's balsamic vinegar and Parmigiano-Reggiano cheese. Established in 1990. Maximum class/group size: 18 hands-on/30 demo. Facilities: Restaurants, cooking schools, and farms. Also featured: Visits to markets and artisan food producers, dining in fine restaurants and private homes.
Emphasis: Regional cuisines with a focus on artisanal foods.
Faculty: Program Director Pamela Sheldon Johns, author of Parmigiano! and Balsamico!.
Costs: Classes are $40, trips are $2,450-$2,650, which includes 3- or 4-star lodging, meals, and all planned activities.
Location: 90 minutes north of Los Angeles; Alba and Torino in Italy's Piemonte region; Modena, Parma, and Bologna in the Emilia Romagna region.
Contact: Pamela Sheldon Johns, Director, Culinary Arts, International, 1324 State St., J-157, Santa Barbara, CA 93101 US; 805-963-7289, Fax 805-963-0230, E-mail CulinarArt@aol.com.

DIANE SEED'S CULINARY EXPERIENCES

Rome, Puglia/April-June, September-November *(See also page 290, Cuisine International)*

Culinary professional Diane Seed offers demonstration and hands-on classes in Rome and 6-day hands-on cooking vacations in Puglia. Established in 1996. Maximum class/group size: 12 Rome/15 Puglia. Also featured: Rome: market visits, dining in fine restaurants, private instruction; Puglia: visits to a cheese maker, olive grove, and oil production plant.
Emphasis: Italian regional cuisine.
Faculty: Diane Seed, British cooking teacher and author, who has lived in Rome for 28 years.
Costs: Puglia: $2,500, which includes meals and lodging at Puglia at Il Melograno, a family owned 5-star Relais & Châteaux hotel. Amenities include a health center, heated indoor pool, solarium, gymnasium, swimming pool, tennis courts.
Location: Central Rome, near the Palazzo Piazza Venezia; Puglia, in the Truilly region on the Adriatic, south of Bari.
Contact: Diane Seed, Diane Seed's Roman Kitchen, Via del Plebiscito 112, Rome, 00186 Italy; (39) 06-6797-103, Fax (39) 06-6797-109, E-mail 100525.1613@compuserve.com. In the U.S.: Judy Ebrey, Cuisine International, P.O. Box 25228, Dallas, TX 75225; 214-373-1161, Fax 214-373-1162, E-mail CuisineInt@aol.com, URL http://www.cuisineinternational.com.

EDDA SERVI MACHLIN'S COOKING IN TUSCANY

Tuscany/May-June, September-October

Cookbook author Edda Servi Machlin offers 1-week hands-on programs that include 5 classes and tours tailored to participants' interest, including places of artistic, cultural, Jewish, and culinary significance, musical evening in Florence, lessons in basic Italian, visits to homes of Florentine Jews. Established in 1998. Maximum class/group size: 12. 4 programs per year. Facilities: The kitchen of the estate, which has tennis courts, swimming pool with solarium, vineyards, view of Florence. Also

featured: Separate kosher style and rabbinically-supervised kosher weeks are offered.
Emphasis: Italian Jewish recipes.
Faculty: Edda Servi Machlin, a native of Tuscany and author of The Classic Cuisine of the Italian Jews, Volumes I and II, and Child of the Ghetto, Coming of Age in Fascist Italy, 1926-1946, A Memoir, columnist for La Cucina Italiana.
Costs: $3,500 ($3,750) for kosher style (strictly kosher) weeks include meals, luxury lodging, planned activities.
Location: In Tuscany, 12 miles from Florence.
Contact: E.S. Machlin, Edda Servi Machlin's Cooking in Tuscany, PO Box 203, Croton-on-Hudson, NY 10520 US; 914-271-8924, Fax 914-271-6552, E-mail 74252.1014@compuserve.com, URL http://www.giropress.com.

ESPERIENZE ITALIANE

Tuscany, Veneto, Umbria/May, September-November

Lidia Bastianich, co-owner of Felidia Ristorante, offers 7- to 10-day trips that feature cooking demonstrations, dining in fine restaurants, art tours, meetings with wine producers, chefs, and contemporary Italian artists, visits to wine estates. Established in 1997. Maximum class/group size: 15-20. 4 programs per year. Also featured: Custom-designed trips.
Emphasis: Regional Italian cuisine and wines, Renaissance art.
Faculty: Lidia Bastianich, author of La Cucina di Lidia and star of PBS series; Burton Anderson, author of The Wine Atlas of Italy; art historians Tanya Bastianich and Shelly Burgess.
Costs: $3,800-$4,950, includes 4- and 5-star shared hotel lodging (including Chianti's Le Piazze, Florence's Helvetia and Bristol, Venice's Gritti Palace, Rome's Hotel Eden), most meals, planned excursions, ground transport.
Location: Tuscany: Chianti, Siena, San Gimignano, Florence. Veneto: Verona, Vicenza, Venice, Friuli. Umbria: Rome, Palo, Assisi, Perugia, Todi.
Contact: Shelly Burgess, Program Director, Esperienze Italiane Travel c/o Felidia Ristorante, 243 E. 58th St., New York, NY 10022 US; 212-758-1488, Fax 212-935-7687, E-mail shelly@lidiasitaly.com, URL http://www.lidiasitaly.com. Additional contact: Judy Ebrey, Cuisine International, Box 25228, Dallas, TX 75225; 214-373-1161, Fax 214-373-1162, Email CuisineInt@aol.com, URL http://www.cuisineinternational.com.

ETOILE CULINARY INSTITUTE OF THE ARTS

(See also page 158)

Venice/Year-round

This private school offers 3-day programs that include market visits, winery tours, sightseeing, visits to food producers, dining in private homes. Established in 1985. Facilities: Modern, well-equipped laboratories, classrooms, library.
Emphasis: Regional Italian themes.
Faculty: 40 chef faculty.
Costs: $875, includes lodging and meals.
Location: 30 miles from Venice, near Chioggia, on the Venice lagoon across from the Adriatic beach.
Contact: Diana Place, U.S. Representative, Essence of Italy, P.O. Box 956, Boca Raton, FL 33429 US; 888-213-5678, 561-361-0301, Fax 561-361-0301, E-mail essenceofitaly@worldnet.att.net, URL http://www.etoile.org.

FROM MARKET TO TABLE IN A ROMAN KITCHEN

Rome/Year-round

This private school offers 1-day and half-day hands-on and demonstration courses. All lessons conclude with a full meal and local wine. Established in 1995. Maximum class/group size: 1-5. 60 programs per year. Facilities: Professional kitchen. Also featured: Day trips, private instruction. Several program options include in-depth visits to markets, purveyors, producers, with special attention to seasonal foods and regional recipes.
Emphasis: Regional Italian cuisine.

Faculty: Carla Lionello, pastry chef and restaurant consultant; Jon Eldan, baker and food researcher.

Costs: $185/couple for half-day lesson, $325/couple for full-day.

Location: Historic center of Rome, near the Trevi Fountain.

Contact: Carla Lionello, From Market to Table in a Roman Kitchen, Via Due Macelli, 106, Rome, 00187 Italy; (39) 06-699-20435, Fax (39) 06-699-20435, E-mail md2063@mclink.it, URL http://www.wheninrome.com.

GABRIELE'S TRAVELS TO ITALY

Amalfi, Assisi/March-June, September-November

This special interest tour provider offers 1-week cooking vacations that include daily hands-on classes. Established in 1992. Maximum class/group size: 8-16. 20+ programs per year. Facilities: Modern kitchens. Also featured: Day trips, market visits, winery tours and instruction, dining in fine restaurants, sightseeing, visits to food producers, ceramics shopping.

Emphasis: Southern Italian, Mediterranean, and Umbrian cuisines, fresh herbs and vegetables, roasted meats, game, truffles.

Faculty: Maria Maurillo-Fabrizi; Giuseppe Liuccio, food historian and president of the Academy of Medieval Cooking in Italy; Ezio Falcone, food historian and gastronome; Chef Enrico Cosentino, member of the Italian Federation of Chefs.

Costs: $1,850-$2,160, including some meals, shared lodging and private bath, excursions. Lodging is at a country villa in Assisi and the 19th-century 50-room Hotel Cappuccini Convento in Amalfi.

Location: 2 miles east of Assisi on the slopes of Monte Subasio; the Amalfi coast.

Contact: Gabriele Dellanave, Owner, Gabriele's Travels to Italy, 3037 14th Ave. NW, Rochester, MN 55901 US; 507-287-8733, Fax 507-287-9890, E-mail gabriele@hps.com, URL http://www.cookinginitaly.com.

GIOVANNA PASSERI

Monticello (Lecco)/February-June, September-December

This culinary professional offers programs that vary according to the season. Established in 1983. Maximum class/group size: 14. 12 programs per year. Also featured: Programs for youngsters, wine instruction, day trips, market visits, winery tours, dining in fine restaurants and private homes, private classes.

Emphasis: Classical cooking, traditional Italian pastries, local specialties, fish, historic recipes, Pugliese and Milanese dishes, bread, pasta.

Location: Near Monza, a 20-minute train ride from Milan.

Contact: Giovanna Passeri, , Via V. Foppa. 1, Monticello (Lecco), 22068 Italy; (39) 09202928, Fax (39) 09202928.

GIULIANO BUGIALLI'S COOKING IN FLORENCE

Florence/April-July, September-October, December *(See display ad page 321)*

Cookbook author Giuliano Bugialli offers 1-week hands-on vacation programs. Established in 1973. Maximum class/group size: 18. 6 programs per year. Facilities: Large newly-equipped kitchens in a Chianti villa, wood-burning brick oven and hearth. Also featured: Dining in fine restaurants and trattorias, gastronomic and oenologic trips, tastings.

Emphasis: Italian authentic cooking of all regions.

Faculty: Giuliano Bugialli is author of The Fine Art of Italian Cooking, Tastemaker Award winners Giuliano Bugialli's Classic Techniques of Italian Cooking, Giuliano Bugialli's Foods of Italy, Julia Child Award winner Foods of Sicily & Sardinia.

Costs: $3,500, which includes most meals, planned excursions, and first class or superior hotel lodging in central Florence.

Location: In Florence with classes in a Chianti country villa.

Contact: Giuliano Bugialli's Cooking in Florence, 60 Sutton Place South, #1KS, New York, NY 10022 US; 212-813-9552, Fax 212-486-5518.

GRITTI PALACE SCHOOL OF FINE COOKING

Venice/November-March, July

This resort hotel offers 2-, 3-, and 5-day vacation demonstration courses. Established in 1974. Maximum class/group size: 22-25. Facilities: Specially-equipped mirrored room fitted with a stove.

Emphasis: Regional Italian cuisine, seasonal ingredients, wine selection, setting of a table, flower arrangements.

Faculty: Gritti's chef Celestino Giacomello.

Location: The hotel Gritti Palace, palace of Doge Andrea Gritti in the 15th century, overlooking the Grand Canal, is 30 minutes by boat from Venice Airport.

Contact: Elizabeth Viliani, Gritti Palace School of Fine Cooking, Campo Santa Maria del Giglio, Venice, 2467 Italy; 800-325-3589, Fax 512-834-7598. In Canada: 800-325-3589.

HAMILTON FITZJAMES

Tuscany/April, October-December

This travel company offers 1-week trips that feature cooking demonstrations, truffle hunting, visits to wineries, fine restaurants, cheesemakers, prosciutto maker, butcher, flour mill, vinegar and olive oil producers. Maximum class/group size: 16. 2-3 programs per year. Facilities: Restaurant kitchens.

Emphasis: Food and wine of Tuscany.

Faculty: Burton Anderson, author of The Treasures of the Italian Table and Guide to the Wines of Italy; art historian Athlyn Fitz-James.

Costs: $4,600, includes deluxe lodging, most meals, planned activities.

Location: Milan, Neive, Parma, Florence.

Contact: Dale Gregorczyk, Hamilton Fitzjames, 1011 Upper Middle Rd. E., Oakville, ON, L6H 5Z9 Canada; 800-801-6147/905-842-1845, Fax 905-842-2196, E-mail HamiltonFitzjamesAmerica@compuserve.com, URL http://www.hamiltonfitzjames.com. Also: 22, ave. de la Republique, 21200, Beaune, France; (33) (0)3-80-22-02-62, Fax (33) (0)3-80-22-04-67, E-mail 106134.566@compuserve.com.

HOTEL CIPRIANI COOKING SCHOOL

Venice/April, October-November

This resort hotel offers 5-day demonstration programs taught by noted culinarians with visits and dining in private palazzos and homes. Established in 1978. Maximum class/group size: 24-40. Facilities: Demonstration kitchen in a large meeting room overlooking the Venetian lagoon, video, 3 gas burners and oven. Also featured: Visit to the Rialto market, a lagoon or mainland excursion, wine presentations, dining in fine restaurants, concluding banquet.

Emphasis: Italian and international cuisines utilizing the foods and wines of Venice.

Faculty: Well-known instructors, including Julia Child, Marcella Hazan, Renato Piccolotto, Giuliano Hazan, and Faith Willinger.

Costs: $3,385, which includes double occupancy deluxe lodging at the Hotel Cipriani, meals, planned activities. Hotel amenities include a heated pool, tennis, sauna.

Location: Approximately 30 minutes by water-taxi from the airport.

Contact: Dr. Natale Rusconi, Managing Director, Hotel Cipriani Cooking School: Meet the Stars of Gastronomy, Giudecca 10, Venice, 30133 Italy; (39) 041-520-7744, Fax (39) 041-520-3930, E-mail cipriani@gpnet.it. In the U.S.: Orient Express Hotels 800-237-1236 or 212-838-7874.

IL BORGHETTO COOKING SCHOOL

Florence/April-May, September-October

This country villa offers 6-day vacation programs. Established in 1995. Maximum class/group size: 8. Facilities: Newly constructed restaurant kitchen with 4 work areas. Also featured: Visits to local markets and castles, guided tours, shopping in Florence and Siena.

Emphasis: The theoretical and practical aspects of Tuscan cuisine, including ingredient selection and techniques.

Faculty: Francesca Cianchi, former chef at Mezzaluna in New York City.

Costs: The $3,000 fee includes meals and lodging with private bath at the 15th-century Il Borghetto; $1,000 additional for nonparticipant in double room, $5,000 for 2 participants sharing a room.

Location: The central Chianti region, 10 miles south of Florence. Il Borghetto is on an estate with olive grove, medieval grain silos, swimming pool, library.

Contact: Francesca Cianchi, Il Borghetto Cooking School, San Casciano Val di Pesa, Florence, 50020 Italy; (39) 055-8244442, Fax (39) 055-8244247. In the U.S.: Elaine Muoio, Italian Rentals, 3801 Ingomar St., NW, Washington, DC 20015; 202-244-5345, Fax 202-362-0520.

IL CHIOSTRO DI TOSCANA

Vagliagli, Siena/June, October

This private school offers 1-week programs that feature demonstrations by master chefs and gourmet meals, winery tours, visits to olive oil producers, butchers, and bakers, sightseeing. Established in 1995. Maximum class/group size: 9. 1-2 programs per year. Facilities: Tuscan farmhouse. Also featured: Workshops in drawing and painting, portrait photography, Italian opera appreciation.

Emphasis: Tuscan cuisine.

Faculty: 4 chefs from New York and Tuscany.

Costs: $900-$1,585 includes 2 meals daily, shared lodging at a villa outside of Siena, planned activities.

Location: Vagliagli, a hamlet on the Chiantigiana 15 minutes from Siena, 45 minutes from Florence, in Tuscany.

Contact: Michael Mele, Director, Il Chiostro di Toscana, 241 W. 97th St., #13N, New York, NY 10025 US; 800-990-3506, Fax 800-990-3506, E-mail mmele@msn.com, URL http://www.costar.net/tuscany.

INTERNATIONAL COOKING SCHOOL OF ITALIAN FOOD AND WINE

(See also page 237) (See display ad above) **Bologna/May-June, September-October**

This private school established by culinary professional Mary Beth Clark offers participation courses that include the 6-day Basic certificate course and the 7-day Piedmont Truffle Festival. Established in 1987. Maximum class/group size: 10. 5 programs per year. Facilities: Renaissance palazzo with professional kitchen and individual work stations. Also featured: Visits to food producers and outdoor food markets, olive oil tastings, private winery tours, truffle hunt, dining in Michelin-star restaurants, private demonstrations.

Emphasis: Traditional and contemporary regional light Italian cuisines with fresh ingredients, including specialties from Bologna, Parma and Emilia-Romagna, Tuscany, Rome, Piedmont, Sicily.

Faculty: Chef Mary Beth Clark, cooking teacher and cookbook author; a pasta chef; a Neapolitan pizza chef; Executive Chef Andrea Merlini.

Costs: $3,200-$3,600 includes most meals, first class to deluxe lodging, ground transport, planned activities.

Location: Bologna's historic center, a block from the outdoor food market.

Contact: Mary Beth Clark, International Cooking School Of Italian Food And Wine, 201 E. 28th St., #15B, New York, NY 10016-8538 US; 212-779-1921, Fax 212-779-3248, E-mail MaryBethClark@worldnet.att.net, URL http://www.shawguides.com/cook/mbc.

INTERNATIONAL DINING ADVENTURES (See page 350)

ISTITUTO ZAMBLER VENEZIA

Venice/Year-round

This private school offering courses in Italian language, culture and cooking offers 3 lessons/week or 6 lessons/2 weeks with a choice of 12 menus. Established in 1933. Facilities: Private apartments

or residences.
Emphasis: Italian cuisine.
Costs: 300.000 lire for 3 lessons, 600.000 lire for 6 lessons.
Location: Central Venice.
Contact: Giulia Battaglia, Istituto Zambler Venezia, Dorsoduro 3116/A (Camp S. Margherita), Venice, 30123 Italy; (39) 041-5224331, Fax (39) 041-5285628, E-mail zambler@tin.it, URL http://virtualvenice.net/zambler.

ITALIAN COOKERY WEEKS

Orvieto and Ostuni *(See also page 290, Cuisine International)*

Culinary professional Susanna Gelmetti offers weekly 6-day hands-on courses in Orvieto and Ostuni. Established in 1990. Maximum class/group size: 20. Also featured: Shopping at the local market, sightseeing in Assisi and Perugia (Orvieto) and Lecce and Alberobello (Ostuni), truffle hunt. Also available: tailor-made group classes, swimming, golf, horseback riding.
Emphasis: Italian regional cuisine.
Faculty: Susanna Gelmetti, who was chef at London's Accademia Italiana delle Arti and author of Italian Country Cooking; well-known Italian chefs and guest English chefs.
Costs: Cost of £950 includes meals, lodging at 15th and 16th century farm estates with en-suite baths, and planned excursions.
Location: Montebello, a farm estate in Orvieto, and Lo Spagnulo, a converted 15th century castle near Ostuni.
Contact: Susanna Gelmetti, Italian Cookery Weeks, Box 2482, London, NW10 1HW England; (44) (0)181-208-0112, Fax (44) (0)171-401-8763. In the U.S.: Judy Ebrey, Cuisine International, P.O. Box 25228, Dallas, TX 75225; 214-373-1161, Fax 214-373-1162, E-mail CuisineInt@aol.com, URL http://www.cuisineinternational.com.

ITALIAN COUNTRY COOKING CLASSES WITH DIANA FOLONARI

Positano/May-June, September-October

Culinary professional Diana Folonari offers 1-week participation courses. Established in 1980. Maximum class/group size: 12. Facilities: Ms. Folonari's home.
Emphasis: Italian cuisine.
Faculty: Diana and Vic Folonari.
Costs: Fee for classes is $1,500; cost of 8 nights at the Villa Franca Hotel (other hotels available) is $690-$840 per person; one-way transfer from Naples is $115.
Location: Via del Canovaccio 10 in Positano, which is accessible by train and limousine from airports in Rome and Naples.
Contact: Martha Morano, President, Italian Country Cooking Classes with Diana Folonari, 72 Madison Ave., New York, NY 10017 US; 800-223-9832, Fax 212-252-1818, E-mail E-erom.com@worldnet.att.net, URL http://www.lesromantiques.com.

ITALIAN CUISINE IN FLORENCE

Florence/Year-round

Culinary professional Masha Innocenti CCP offers 3- to 5-day hands-on courses. Special 1- or 2-day demonstration classes for larger groups with travel agencies (in Japan, U.S., etc.). Established in 1983. Maximum class/group size: 8 hands-on/18 demo. 12 programs per year. Facilities: 300-square-foot kitchen with modern equipment. Also featured: Demonstrations for groups, private instruction, food lectures.
Emphasis: Regional and nouvelle Italian cuisine, Italian desserts.
Faculty: Masha Innocenti, CCP, a member of the IACP, holds a diploma from Scuola di Arte Culinaria Cordon Bleu and is a member of the Associazione Italiana Sommeliers and the Commanderie des Cordons Bleus de France.
Costs: $850 for the 5-day gourmet cuisine courses, $530 for the 3-day intensive course. Rates include meals. Private classes are $185/day.

Contact: Masha Innocenti, Italian Cuisine in Florence, Via Trieste 1, Florence, 50139 Italy; (39) 055-499503, Fax (39) 055-480041, E-mail mirel@box1.tin.it. U.S. Contact: William Grossi, 182 Four Corners Rd., Ancramdale, NY 12503; 518-329-1141.

ITALIAN GOURMET COOKING CLASSES IN PORTOVENERE

Portovenere/October-April, July

The Grand Hotel Portovenere offers 7-day courses consisting of 5 classes that cover sauces, soups, salads, antipasti, pasta, main dishes, and desserts. 5 programs per year. Facilities: The hotel restaurant kitchen. Also featured: Visit to a vineyard, olive oil mill, Cinque Terre, Portofino. Shorter courses and specific menus are available. Classes taught in English, French, German.

Emphasis: Gourmet Italian (Mediterranean) cuisine using fresh herbs and ingredients.

Faculty: Grand Hotel Portovenere Chef Paolo Monti.

Costs: 1,800,000 Lira, which includes shared lodging at the 4-star Grand Hotel Portovenere, single room surcharge is 350,000 Lira.

Location: On the Italian Riviera.

Contact: Paolo Monti, Chef, Grand Hotel Portovenere, Via Garibaldi, 5, Portovenere, 19025 Italy; (39) 0187-79-26-10, Fax (39) 0187-79-06-61, E-mail Paolo.Monti@agora.stm.it, URL http://hella.stm.it/market/cucina_italiana/home.htm.

LA BOTTEGA DEL 30 COOKING SCHOOL

Villa a Sesta, Siena/April-June, September-November

This restaurant in Chianti, recognized by Michelin, offers weekly 5-day hands-on, full participation, cooking courses in spring and fall. Year round requests may be honored. Established in 1995. Maximum class/group size: 8-10. Facilities: Dining room, library, wine cellar, wood-burning oven, large modern kitchen with 10 work areas/stove tops. Also featured: Wine instruction, day trips, market visits, winery tours, dining in fine restaurants, sightseeing, visits to food producers.

Emphasis: Tuscan cooking, using fresh local ingredients. Includes appetizers, soups and sauces, home-made pasta, main courses, desserts. Olive oil and wine are pressed on the premises.

Faculty: Helen Stoquelet, owner and chef of La Bottega del 30 restaurant, and her husband, Franco Camelia, Director; Translator, guide, both specialists in cuisine. U.S. representative Diana Place.

Costs: $2,350, including 7 nights double occupancy apartment lodging, breakfast, lunch, 4 dinners, 2 excursions. Apartments are newly refurbished with pool, surrounded by hills of vineyards.

Location: School is separate from restaurant but in same Tuscan village. 20 minutes outside of Siena, 90 minutes south of Florence.

Contact: Franco Camelia, La Bottega del 30 Cooking School, via S. Caterina 2, Villa a Sesta, Siena, NO 53019 Italy; (39) 0577-359226, Fax (39) 0577-359226, E-mail labottegadel30@novamedia.it. In the U.S.: Diana Place, Essence of Italy, Box 956, Boca Raton, FL 33429; 888-213-5678, 561-361-0301, E-mail EssenceofItaly@worldnet.att.net, URL http://home.att.net/~essenceofitaly.

LA CUCINA COOKING SCHOOL (See page 221)

LA CUCINA AL FOCOLARE

(See also page 329) (See display ad page 326) **Tuscany/April-June, September-November**

Fattoria Degli Usignoli, a converted 15th-century friary, offers 1-week hands-on culinary vacations. Established in 1992. Maximum class/group size: 10-15. 10 programs per year. Facilities: Professional kitchen with the latest equipment, including a wood-burning oven, rotisserie, and individual work stations. Also featured: Tuscan bread bakery, grape picking, outdoor market tour, sightseeing in Florence, Siena, and San Gimignano, wine tastings.

Emphasis: Tuscan specialties, pizza, breads, grill and rotisserie dishes, culinary history, wine appreciation.

Faculty: Piero Ferrini, chef-professor from Florence; Paolo Blasi, sommelier; Peggy Markel, program founder and director.

Costs: $3,200 ($2,995, $2,750), including single (double, triple) apartment with private kitchen, meals, wine, excursions, ground transport. Amenities include tennis, horseback riding, swimming.

Location: On 55 acres overlooking the Valdarno Valley, 18 miles southeast of Florence. The Fattoria produces its own Chianti and extra virgin olive oil.

Contact: Peggy Markel, Director, La Cucina al Focolare, P.O. Box 54, Boulder, CO 80306-0646 US; 800-988-2851, Fax 303-440-8598, E-mail tuscancook@earthlink.net, URL http://www.cookinitaly.com.

LA CUCINA DEL CASTELLO

Siena/Year-round *(See display ad page 327)*

Ivory Isle Travel offers a 7-day program featuring 4 demonstration classes, shopping at San Lorenzo Central Market, visits to food producers, sightseeing in Tuscany, wine tours and tastings, and dining in local restaurants. Established in 1997. Maximum class/group size: 10-14. 12 programs per year. Facilities: Professional kitchen featuring a large communal oven. Also featured: Horseback riding, cycling, hiking, weeks for private groups.

Emphasis: Authentic Tuscan culture experienced through the exploration of its culinary traditions.

Faculty: Chef Giancarlo Giannelli, owner of L'oste Poeta Restaurant in Tocchi; anthropologist Vittorio Cambria, an expert on the culture of the region.

Costs: $2,250 all-inclusive for land portion. Lodging in a restored 1,000-year-old medieval village with views of the countryside.

Location: 25 minutes south of Siena.

Contact: Roz Katz, Senior Travel Consultant, Ivory Isle Travel Culinary Adventures, 519 Park Dr., Kenilworth, IL 60043 US; 800-245-9335/847-256-2108 x139, Fax 847-256-0102, E-mail tuscany@tipc.com, URL http://www.tipc.com/tuscany. Contact in Italy: Vittorio Cambria, Podera La Ripa, 53010 Tocchi, Monticiano, SI Italy; Fax (39) 0577-757-100.

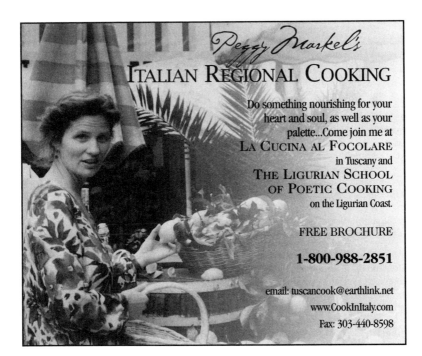

LA SETTIMANA DELLA CUCINA

(See also page 290, Cuisine International) **Bologna/Year-round**

This travel company offers 7-day programs that include daily hands-on classes. Established in 1996. Maximum class/group size: Up to 16. Facilities: I Notai restaurant just off Bologna's main square. Also featured: Winery tours and instruction, tour of ceramic factory, visits to markets and producers of cheese, balsamic vinegar, and prosciutto, dinner and demonstration by Valentino Marcatilli, chef of San Domenico in Imola.

Emphasis: Italian cuisine and traditional Bolognese and Emilia Romagna region cuisine, including tortellini, green lasagne, tagliatelle.

Faculty: Cookbook author Monica Cesari Sartoni, who has over 15 years experience and writes for gastronomy publications and the Gambero Rosso restaurant guide.

Costs: 3,900,000 Lira, which includes double occupancy lodging at a 4-star hotel in city center, meals and regional wines in fine restaurants and trattorias, and guided tours.

Location: Central Bologna.

Contact: Marcello and Raffaella Tori, Owners, Bluone Tour Operator, Via Parigi, 11, Bologna, 40121 Italy; (39) 051-263546, Fax (39) 051-267774, E-mail ra.ra@posta.alinet.it, URL http://www.alinet.it/bluone. In the U.S.: Judy Ebrey, Cuisine International, P.O. Box 25228, Dallas, TX 75225; 214-373-1161, Fax 214-373-1162, E-mail CuisineInt@aol.com, URL http://www.cuisineinternational.com.

LA VILLA CUCINA

Tuscany/May, September-October

This private school offers 1-week vacations that combine a villa holiday with a cooking vacation. Included are 6 hands-on classes, visits to markets, wineries, and food producers, sightseeing, dining in typical regional restaurants. Established in 1998. Maximum class/group size: 12 hands-on, 18 travel. 4+ programs per year. Facilities: Professional kitchens in private villas. Also featured: Private instruction.

Emphasis: Authentic regional Italian cooking.

Faculty: Daniel C. Rosati, a member of the IACP and N.Y. Association of Cooking Teachers.

Costs: $3,300 includes lodging, most meals, ground transport, planned activities.

Location: Western Tuscany, 55 minutes south of Pisa.

Contact: Daniel Rosati and Gina Russomanno, Owners, La Villa Cucina, 326 Broad St., Bloomfield, NJ 07003 US; 973-566-9727, Fax 973-743-6096, E-mail DanRosati@msn.com.

L'AMORE DI CUCINA ITALIANA

(See also page 301) (See display ad page 292) **Pomino/May, September, October**

This travel company offers 1-week hands-on culinary vacations. Established in 1992. Facilities: Kitchen of the Locanda di Praticino. Also featured: Shopping expeditions; visits to outdoor markets, winery, cheese and olive oil producers; the museum of a noted shoe manufacturer; dining at fine restaurants.

Emphasis: Tuscan and regional Italian cuisine.

Faculty: Cristina Blasi and Gabriela Mari, owners of a cooking school in Florence and authors of a book about the cooking of ancient Rome; wine and olive oil experts.
Costs: $2,195 ($1,995 for non-cook guest) includes meals, shared lodging ($350 single supplement) at the Locanda di Praticino, planned activities.
Location: Pomino, about a half hour drive from Florence.
Contact: Ralph P. Slone, L'Amore di Cucina Italiana, Inland Services, Inc., 360 Lexington Ave., New York, NY 10017 US; 212-687-9898, E-mail incook@earthlink.net, URL http://home.earthlink.net/~incook/.

LAURA NICCOLAI COOKING SCHOOL

Naples, Siena/March-December *(See also page 292, International Kitchen)*

Culinary professional Laura Niccolai offers 1-week demonstration courses (4 classes) in S'Agata sui Due Golfi, Naples, and Locanda Dell'Amorosa in Sinalunga, Siena. Established in 1987. Maximum class/group size: 20. Facilities: Well-equipped, professional kitchens. Also featured: Naples: dinners in selected restaurants, visits to Limoncello liqueur and mozzarella factories, guided tours to Capri and Positano. Siena: wine and olive oil tastings, cultural and gastronomic tours.
Emphasis: Traditional and modern Italian, Tuscan, and Neapolitan cuisine with emphasis on light and healthy recipes.
Faculty: IACP-member Laura Niccolai studied with Michelin 3-star chef Gualtiero Marchesi and French pastry chef Jain Bellouet.
Costs: $3,000 ($3,500) for the Naples (Siena) program, which includes double occupancy lodging, meals, planned activities. Lodging at luxury hotels in Sorrento (Naples) and the Locanda dell'Amorosa in Siena.
Location: S. Agata sui Due Golfi is 30 miles south of Naples; Locanda dell'Amorosa is 50 miles south of Florence.
Contact: Laura Niccolai, Laura Niccolai Cooking School, Via Termine 9, S. Agata sui Due Golfi-NA, 80064 Italy; (39) 081-8780152, Fax (39) 081-8780152, E-mail LNCooking@aol.com. In the U.S.: The International Kitchen, Inc., 1209 N. Astor St., #11N, Chicago, IL 60610; 800-945-8606, Fax 847-295-0945/312-654-8446, E-mail info@intl-kitchen.com, http://www.intl-kitchen.com.

LAURA'S COOKING CLASSES

Tuscania/March-June, August-December

This private school offers monthly 1-week cooking vacations that feature hands-on preparation, demonstrations, cultural excursions, sightseeing, shopping, and some restaurant dining. Established in 1994. Maximum class/group size: 8. 12 programs per year. Facilities: Home plan kitchen in a country house on a 120-acre estate with an Etruscan necropolis, olive grove, Arabian horse breeding. Also featured: Visits to food producers, dining in private homes.
Emphasis: Easy and up-to-date Italian cooking, from the Romans' cookery through Renaissance dishes and traditional recipes. Home grown products.
Faculty: Laura Caponetti Brezzi has a degree in History of the Italian Renaissance Gastronomy, was consultant to Italian hotels, is a journalist and cookbook author
Costs: $1,310 includes double room, meals, airport transport, planned activities.
Location: 2 km from the medieval hill town of Tuscania, near the Popes Spa thermal bath, 45 miles from Orvieto, 1 hour from Rome.
Contact: Laura Caponetti Brezzi, Owner, Laura's Cooking Classes, Tenuta del Guado Antico, Tuscania, 01017 Italy; (39) 0761 435792, Fax (39) 0761 435792, E-mail caponetti@iol.it, URL http://www.casacaponetti.com.

A LESSON IN FLAVORS – DONNA FRANCA TOURS

May-October

This tour operator offers eleven-day hands-on culinary tours. Established in 1967. Maximum class/group size: 20. 8-10 programs per year. Facilities: Kitchens of private villas. Also featured: Winery tours, private visits to Venice gardens, visits to prosciutto balsamic vinegar producers, medieval banquet, excursions to Lake Como, Ferrara, Siena, San Gimignano, Florence, Perugia, Spoleto.

Emphasis: Traditional, nuova cucina, medieval cuisine.
Faculty: Donna Franca studied at the Hotel Cipriani and has operated culinary tours since 1972. A member of Les Dames d'Escoffier, she conducts lessons in her private villa in Cetona, Siena and organizes customized culinary courses.
Costs: Include double occupancy lodging in castles and villas, most meals, land transport, and planned excursions. Single supplement ranges from $150-$300.
Location: Tuscan countryside outside of Siena.
Contact: Franca Franzaroli, President, Donna Franca Tours, 470 Commonwealth Ave., Boston, MA 02215-2795 US; 617-375-9400/800-225-6290, Fax 617-266-1062, E-mail dtours2156@aol.com, URL http://www.donnafranca.com.

LIGURIAN SCHOOL OF POETIC COOKING

(See also page 325) (See display ad page 326) **Tellaro/March-May, September-November**

Peggy Markel of La Cucina al Focolare offers history, simple ingredients, new techniques, fish, sauces, fresh vegetables, and herbs. Established in 1996. Maximum class/group size: 6-8. 2 programs per year. Facilities: Professional kitchen, copper utensils. Also featured: Exploration of wines of the Cinque-Terre.
Emphasis: Fish, sauces, Ligurian cuisine.
Faculty: Chef Angelo Cabani, owner of the Michelin 1-star Locanda Miranda Inn.
Costs: $2,750 ($2,250), including meals, shared (single) lodging at the Inn, and planned excursions.
Location: Overlooking the Bay of Poets (La Spezia).
Contact: Peggy Markel, Director, Ligurian School of Poetic Cooking, P.O. Box 54, Boulder, CO 80306-0646 US; 800-988-2851, Fax 303-440-8598, E-mail tuscancook@earthlink.net, URL http://www.cookinitaly.com.

LORETTA PAGANINI SCHOOL OF COOKING (See page 245)

LUNA CONVENTO COOKING CLASSES WITH ENRICO FRANZESE

(See also page 290. Cuisine International) **Amalfi**

The Luna Convento Hotel offers 1-week culinary vacations that include 4 morning demonstration and participation classes. Established in 1991. Maximum class/group size: 15-18. Facilities: Luna Convento Hotel's Saracen Tower, overlooking the sea. Also featured: Guided excursions to Sorrento, Ravello, Pompeii, and Amalfi; dinner at Don Alfonso, a Michelin 3-star restaurant owned by Alfonso and Livio Iccharino.
Emphasis: Regional Neapolitan cuisine.
Faculty: Enrico Franzese, who trained at the Cipriani in Venice and the Hassler in Rome, won the 1990 Parma Ham Chef's Competition in Bologna, and appears on Italian television; interpreter Rosemary Anastasio.
Costs: $2,300 ($2,600), which includes meals, planned excursions, transportation from Naples, and first class double (single) occupancy lodging and private bath at the 4-star Luna Convento Hotel, a restored 13th-century convent.
Location: Amalfi, a resort area on Italy's west coast, is about 150 miles south of Rome and 40 miles south of Naples.
Contact: Andrea Milone, Luna Convento, Amalfi, SA, NO 84011 Italy; (39) 089-871-002. In the U.S.: Judy Ebrey, Cuisine International, P.O. Box 25228, Dallas, TX 75225; 214-373-1161, Fax 214-373-1162, E-mail CuisineInt@aol.com, URL http://www.cuisineinternational.com.

MAMMA AGATA COOKING CLASSES

Ravello/Year-round

This travel company owned by Chiara Lima, daughter of Mamma Agata, offers 1-week courses, daily lessons, and private dinners by Mamma Agata. Established in 1997. Maximum class/group size: 4. 12 programs per year. Facilities: Mamma Agata's simple, well-equipped kitchen in her cliff-top home. Also featured: Visits to cheese and lemon factories, boat excursion, sightseeing in Pompeii, Paestum, Capri, Ischia, Sorrento and Amalfi coasts.

Emphasis: Mediterranean and Italian cuisine including pasta, veal, fish, chicken, vegetables, pizza, cakes, traditional marmalade and conserve.
Faculty: Mama Agata has cooked for Humphrey Bogart, Audrey Hepburn, Jacqueline Kennedy, and Gore Vidal.
Costs: $200/day, $1,000/week includes morning class.
Location: The Amalfi coast.
Contact: Chiara Lima, Proprietor, Amalfi Coast Service Centre, Via Trinità, 31, Ravello, Salerno, 84010 Italy; (39) 089-858386, Fax (39) 089-858386, E-mail chiaralm@amalficoast.it, URL http://www.webworld.co.uk/mall/coastline.

MANGIA FIRENZE "A TASTE OF TUSCANY"
Florence/April-November

This private school offers half-day to week-long custom hands-on cooking classes, walking tours of Florence, Giro Gastronomico; Tuscany programs. Established in 1984. Maximum class/group size: 6 hands-on, 12 travel. 100+ programs per year. Facilities: Fully equipped apartment kitchen overlooking Florence's Mercato Centrale. Also featured: Winery tours, day trips, private instruction, visits to markets and food producers, dining in fine restaurants and private homes, sightseeing.
Emphasis: Tuscan cuisine.
Faculty: Judy Witts Francini, CCP.
Costs: One day $200, 3 days $650, 5 days $850. Longer programs range from $250-$1,250. Monday walking tours $100.
Location: Central Florence.
Contact: Judy Witts Francini, Owner, Mangia Firenze "A Taste of Tuscany", Via Taddea, 31, Florence, 50123 Italy; (39) 055-29-25-78, Fax (39) 055-29-25-78, E-mail info@mangiafirenze.com, URL http://www.mangiafirenze.com.

MARGHERITA AND VALERIA SIMILI'S COOKING COURSES
Bologna/September-May

Culinary professionals Margherita and Valeria Simili offer hands-on classes. Established in 1986. 100+ programs per year. Also featured: 1-week intensives for groups of 10 or less.
Emphasis: Most classes focus on breads. Other topics include pasta, traditional desserts.
Faculty: Margherita and Valeria Simili.
Costs: $80-$100 per class.
Contact: Margherita Simili, , 116 Via San Felice, Bologna, 40122 Italy; (39) 051-52-37-71, Fax (39) 051-52-37-71.

MARIA BATTAGLIA – LA CUCINA ITALIANA
Verona/September-May

Culinary professional Maria Battaglia offers 4-day cooking programs each season at La Foresteria Serego Alighieri, a 14th century villa in Valpolicella. Established in 1981. Also featured: A trip to the Verona market, a demonstration at the Ferron Rice Mill, and tours of the Masi winery and Serego Alighieri Estate.
Emphasis: Northern, central, and southern Italian cuisine.
Faculty: Maria Battaglia studied Italian cooking in Bologna, Florence, Messina, Sardinia, and Milan. She was a recipe consultant and spokesperson for Contadina Foods and was awarded the Diploma di Merito by the Federazione Italiana Cuochi in Milan and Verona.
Costs: $2,160 per person, including breakfast, double occupancy lodging at the La Foresteria villa apartments, which have private kitchens, and all cooking classes and excursions. $1,620 for noncook.
Location: A 25-minute drive north of Verona, 10 minutes from Catullo airport in Villafranca.
Contact: Maria Battaglia, President, Maria Battaglia-La Cucina Italiana, P.O. Box 6528, Evanston, IL 60204 US; 847-328-1144, Fax 847-328-1787, E-mail mbcucina@aol.com. Italy contact: La Foresteria Serego Alighieri, 37020 Gargagnago di Valpolicella, Verona, Italy, (39) 045-770-36-22, Fax (39) 045-770-35-23.

MEDITERRANEAN AND VEGETARIAN COOKING

San Miniato/July-October

Culinary professional Deborah Gravelle offers 2-week hands-on vacation programs that consist of daily lessons in a specific subject, menu planning, and wine instruction. Established in 1995. Maximum class/group size: 8. 4 programs per year. Facilities: Large kitchen in a 16th-century house, Podere di Agliatone, that was originally a nun's monastery and was rebuilt as a farm house. Also featured: Excursions in the Chianti area, visits to winemakers, dining in typical Tuscan restaurants and private homes, visits to markets and food producers.

Emphasis: Vegetarian and Mediterranean foods.

Faculty: Two instructors.

Costs: $1,450, which includes shared lodging at Podere di Agliatone, meals, excursions.

Location: The Tuscan countryside. The closest airports are in Rome and Milan.

Contact: Deborah Gravelle, Mediterranean and Vegetarian Cooking, 2332 Cristwood Ct., Santa Rosa, CA 95401 US; 707-523-3140, Fax 707-579-6229, E-mail wkind@wco.com.

THE MIRABELLA SCHOOL

Mirabella Eclano

This travel provider offers 1-week cooking vacation programs that include four days of hands-on morning and afternoon classes. Maximum class/group size: 6. Also featured: Wednesday excursion to the Isle of Capri in the Bay of Naples.

Emphasis: Heart-healthy, low-fat Mediterranean cuisine, including antipasti, risotto, pasta sauces, meat and vegetable dishes.

Faculty: Carla Dora, hostess/instructor, who grows the produce, olives, walnuts, and grapes that are used in recipes and her private-label wine and olive oil.

Location: In the Apennines 90 minutes from Naples and 3 hours from Rome.

Contact: Carolyn B. Shinkle, The Mirabella School, 1300 Sugar Hill Lane, Xenia, OH 45385 US; 937-445-1150, Fax 937-445-0523.

MOLISE, ITALY CULINARY ADVENTURE

Isernia/May-June, September

This private school offers a 6-day cooking vacation that features morning hands-on preparation of 35+ Neopolitan and Ligurian specialties and afternoon tours of museums, cathedrals, and archeological sites. Established in 1995. Maximum class/group size: 10. 2 programs per year. Facilities: The kitchen of Casino del Barone (House of Barons). Also featured: Optional activities include walking, horseback riding, cycling, hiking, and painting.

Emphasis: Regional Italian cuisine.

Faculty: Laura Cimorelli, Course Director, is author of Le Ricette Del Casino del Barone.

Costs: $1,050 double, $1,250 single occupancy, includes lodging at Casino del Barone, a 17th century villa owned by the Cimorelli family, meals, planned activities.

Location: South central Italy, 60 miles from Naples, 120 miles from Rome, south of the Abruzzi.

Contact: Laura Cimorelli, Molise, Italy Culinary Adventure, Studio Elle, Sal. Pollaiuoli 13/7A, Genoa, 16123 Italy; E-mail stuelle@box1.tin.it. In Italy: Laura Cimorelli (39) 10-246-8667, Fax (39) 10-246-8555, E-mail stuelle@mbox.vol.it. In U.S.: Dorothy Duffy Price 650-948-0596, Fax 650-949-1341.

PANE, VINO E LINGUA

Florence/April-October

Enoteca de'Giraldi wine shop and tavern offers 1- and 2-week hands-on programs that combine exposure to the Italian language with cooking classes and Tuscan food and wine activities, includes excursions to the Chianti region. Established in 1994. Maximum class/group size: 12. 6 programs per year. Facilities: Facilities of the Enoteca de'Giraldi. Also featured: Visits to markets, food producers, and farms, winery tours.

Emphasis: Tuscan food, wine and language.

Faculty: Andrea Moradei, owner and Tuscan food and wine expert.
Costs: Lire 1.800.000 (990.000) with language instruction, Lire 1.340.000 (740.000) without language instruction for two (one) weeks.
Contact: Andrea Moradei, Owner, Enoteca de'Giraldi, Via de'Giraldi, 4r, Florence, 50122 Italy; (39) 055-216518, Fax (39) 055-216518, E-mail info@koinecenter.com, URL http://www.koinecenter.com/code/enoteca.html.

PROVENCAL GETAWAY VACATIONS (See page 306)

THE RHODE SCHOOL OF CUISINE
Vorno, Tuscany/March-November *(See also page 307) (See display ad above)*

This Italian villa offers weekly 7-day hands-on vacation programs. Established in 1993. Maximum class/group size: 10. Also featured: Cultural excursions, dining at Michelin-star restaurants, market visits, mushroom hunting in season, activities for non-cooking guests.
Emphasis: Italian cuisine, sauces, wine and cheese.
Faculty: Giuseppe Iasevoli, local Tuscan chefs.
Costs: $2,295 ($2,595), which includes shared (single) lodging, meals, and ground transport. Non-cook rate is $1,795 ($1,995). Villa Michaela, a 19th-century modernized 12-bedroom/12-bath villa on 50 acres with swimming pool and tennis court.
Location: 30 minutes from Pisa, 50 minutes from Florence.
Contact: Tim Stone, The Rhode School of Cuisine, 800-447-1311, Fax 415-388-4658, URL http://www.to-gastronomy.com.

ROBERTO'S ITALIAN TABLE
Venice, Portofino, Florence, Rome/March-May, September-November

Culinary professional Robert Wilk offers 6-day culinary and cultural holidays combining daily hands-on cooking lessons and demos with cultural events. The Art of the Venetian Table, The Magic of Spring in Portofino, Autumn Harvest in Tuscany, and The Lusty Table of Rome. Established in 1995. Maximum class/group size: 14. 6-10 programs per year. Facilities: Kitchens of Hotel Cipriani (Venice), Hotel Splendido (Portofino), Villa San Michele (Florence), kitchens of private homes and palaces. Also featured: Tours of countryside, wineries and vineyards, banquet, cultural presentations and performances, market visits, dining in fine restaurants and private homes, sightseeing.
Emphasis: Culinary and cultural holiday in Italy.
Faculty: Chef Renato Piccoloto (Hotel Cipriani), Chef Carmine Giuliani (Hotel Splendido),chef Attilio de Fabrizio (Villa San Michele), Cultural Director Dr. Joseph A. Precker, cooking teacher Valentina Sforza Harris, host Roberto Wilk.
Costs: $3,500-$4,000, includes lodging, meals, wines, and all planned activities. Lodging: deluxe Relais & Châteaux hotels or family owned hotel.
Location: Venice, Portofino, Florence, Rome.
Contact: Robert Wilk, Roberto's Italian Table, 3441 Dorsoduro, Venice, 30123 Italy; (39) 041-715-197, Fax (39) 041-714-571, E-mail r.wilk@ve.nettuno.it.

SCUOLA DI ARTE CULINARIA "CORDON BLEU"

(See also page 159)
Tuscany/Year-round

This private school offers 1- to 9-session courses at the school in central Florence and 7-day cooking and art vacations at a farmhouse in the Chianti countryside. Maximum class/group size: 12. Facilities: 40-sq-meter teaching kitchen, 30-sq-meter professional kitchen with brick oven in Chianti. Also featured: Visits to markets, food producers, wineries, artisans' workshops; dining in trattorias and fine restaurants, cultural programs, sightseeing.

Emphasis: Basic, advanced, Tuscan, new Italian cuisine; bread, history and nutrition, wines, olive oil.

Faculty: Cristina Blasi and Gabriella Mari, 12 years teaching, sommeliers and olive oil experts, authored a book on ancient Roman cooking and a book on Mustard, members Commanderie de Cordon Bleus de France, IACP, AICI, Compagnia del Cioccolato and Arcigola.

Costs: Classes in Florence begin at 100,000 Lira/session. 7-day Tuscany program is $2,200, which includes lodging, meals, ground transport, planned activities. Lodging in farmhouse is double occupancy with private bath.

Location: School is in central Florence, near the Duomo Cathedral, 15 minutes from Florence airport. Farmhouse is 10 minutes from Greve, the heart of Chianti.

Contact: Gabriella Mari, Co-Director, Scuola di Arte Culinaria "Cordon Bleu", Via di Mezzo, 55/R, Florence, 50121 Italy; (39) 055-2345468, Fax (39) 055-2345468, E-mail cordonbleu@aspide.it, URL http://www.cordonbleu-it.com. Other contact: URL: http://www.aspide.it/piazza/cordonbleu.

SICILIAN COOKING ADVENTURES

Catania, Sicily/Year-round

Culinary professional Marina Tudisco offers 4- and 7-day hands-on cooking vacation programs, professional program during the winter, short course for English-speaking foreigners living in Italy. Maximum class/group size: 8-12 hands-on. Also featured: Visits to the fish market and sherbet, marzipan and biscuit makers; trip to Mt. Aetna, archaeological tours, sightseeing, visit to Taormina, dining at specialty restaurants.

Emphasis: Basic, advanced, new Italian cuisine, desserts, Old Sicilian cooking with Grandma recipes, history of gastronomy.

Faculty: Marina Tudisco is a member of the IACP and Accademico della Cucina Italiana, representative for Sicily of Commanderie des Cordons Bleus de France, studied in Rome with Enrica Jarratt, owner of Cordon Bleu Culinary Arts School, teaches in North America.

Costs: $720 ($1,210) for 4 (7) days, includes meals and lodging in 4-star hotels. $100/class + $30/day for touring. $300 nonrefundable deposit, balance 45 days prior.

Location: Catania, Siracusa, and Taormina.

Contact: Marina Tudisco Maggini, Director, Sicilian Cooking Adventures-CucinArte Culinary Arts School, Via Barriera del Bosco, 16/B, 95030 S. Agata li Battiati, Catania, 95100 Italy; (39) 095-411444. Also: Davide Ciancio, c/o Nicober Viaggi, Via Androne, 43, 95100 Catania, Italy; (39) (0)95-312164, Fax (39) (0)95-327936.

SPOLETO COOKING SCHOOL

Spoleto/July-August

The Spoleto Arts Symposia offers 1-week hands-on courses that include dining and demonstrations at fine restaurants, visits to a truffle factory, olive oil producers, and wine makers, shopping at local food market. Established in 1997. Maximum class/group size: 10. 4 programs per year. Facilities: The kitchens of La Scuola Alberghiero. Also featured: Italian language class with culinary focus, visit to opera master class of the Spoleto Vocal Arts Symposium, guest speakers.

Emphasis: Regional cooking of Umbria.

Faculty: Chefs of La Scuola Alberghiero dello Statodi Spoleto, a national chef training school; restaurant owners and chefs.

Costs: $1,600, which includes most meals, lodging at the 2-star Hotel Aurora, planned activities. More luxurious lodging is available at additional cost.

Location: Spoleto, in the Umbrian hills, is 75 miles north of Rome, 12 miles south of Assisi.

Contact: Clinton J. Everett, Executive Director, Spoleto Cooking School, 760 West End Ave. #3A, New York, NY 10025 US; 212-663-4440, Fax 212-663-4440, E-mail cjeveret@nightingale.org, URL http://www.spoletoarts.com.

STOVETOP CULINARY TOURS

Torino and Florence/May-June

These culinary professionals offer 2-week programs: 1 week in Piedmonte, 1 week in Veneto to Trentino region. Established in 1990. Maximum class/group size: 20. 2 programs per year. Facilities: Fully-equipped professional kitchens. Also featured: Tours and tastings to wineries and cheese, olive oil, chocolate producers, sightseeing.

Emphasis: Professional-level teaching suited to up-grading chef skills and serious amateurs.

Faculty: Ivano Zambotti and Rod Donne, culinary professors at George Brown College; Italian professional chefs for demos; sommeliers for wine tastings.

Costs: C$3,750, includes shared lodging in 4-star hotels w private bath, meals, ground transport, planned activities, airfare Toronto/Italy (Torino). $30/night single supplement.

Location: Torino and Florence.

Contact: Rod Donne, Professor, George Brown College, School of Hospitality, 300 Adelaide St., East, Toronto, ON, M5A 1N1 Canada; 416-415-2247, E-mail rdonne@gbrownc.on.ca.

TASTING PLACES

Sicily, Tuscany, Venice, Umbria/May-July, September-October

This private school offers one-week hands-on cookery courses in different regions of Italy. Established in 1992. Maximum class/group size: 16-24. 37 programs per year. Facilities: Well-equipped kitchens with pizza ovens, grills, and open fire cooking. Also featured: Visits to markets, vineyards, restaurants, and wine tastings.

Emphasis: Italian regional cooking; wine tastings, celebrity guest chefs.

Faculty: Restaurateurs Mauro Bregoli, Alvaro Maccioni, Carla Tomasi (author and proprietor of Turnaround Cooks), Maxine Clark, Thane Prince, Sebastian Snow, Alastair Little.

Costs: From £950, which includes shared 3- and 4-star lodging, meals and excursions. Lodging: 18th century palazzo (Sicily); hotel Fattoria Montelucci (Tuscany); Foresteria Serego Alighieri (Venice), La Cacciata (Umbria).

Location: Italy: Tuscany, near Arezzo; Umbria, in the hilltop city of Orvieto; Sicily, southern coast; Veneto.

Contact: Sara Schwartz & Sarah Robson, Tasting Places, Unit 40, Buspace Studios, Conlan St., London, W10 5AP England; (44) (0)171-460-0077, Fax (44) (0)171-460-0029, E-mail ss@tasting-places.com, URL http://www.tastingplaces.com.

TOSCANA SAPORITA TUSCAN COOKING SCHOOL

Massarosa, Lucca, Tuscany/April-May, Sept.-Nov. *(See also page 292, Intl. Kitchen)*

Culinary professional Anne Bianchi offers programs that include 7-day traditional Tuscan cuisine and 7-day celebrity chef tours. Each features 17-20 hours of hands-on instruction and afternoon tours. Established in 1995. Maximum class/group size: 8-12. 16-18 programs per year. Facilities: Toscana Saporita has a professional kitchen, wood-beamed dining room, formal herb gardens. Also featured: Market shopping, fine dining, wine country tours and tastings, demos at food producers, tours of Lucca, Torre del Lago Puccini, Massa Carrasa marble quarries, Cinque Terre.

Emphasis: Tuscan cuisine.

Faculty: Anne Bianchi, author of 5 cookbooks, including Solo Verdura and The Complete Guide to Cooking Tuscan Vegetables; Sandra Lotti, chef and author; Nancy Allen, teacher at Peter Kump's New York Cooking School. Celebrity chefs include Reed Hearon (SF) and Mario Batali (NYC).

Costs: $1,800 includes lodging, meals, ground transport, escorted tours. $55/day for extended stay. Lodging at Toscana Saporita, an olive oil producing 60-acre estate with swimming pool, hiking trails.

Location: 15 miles west of Pisa International Airport, 35 miles west of Florence International Airport.

Contact: Anne Bianchi, Founder, Toscana Saporita, 265 Lafayette St., #A-22, New York, NY 10012 US; 212-219-8791, Fax 212-219-8791, E-mail toscana@compuserve.com, URL http://www.cyberstudio.it/saporita. Additional contact: The International Kitchen, Inc., 1209 N. Astor St., #11N, Chicago, IL 60610; 800-945-8606, Fax 847-295-0945/312-654-8446, E-mail info@intl-kitchen.com, http://www.intl-kitchen.com.

TUTTI A TAVOLA

(See also page 292, International Kitchen)) **Radda in Chianti/Year-round**

This private school offers 1- and 3-day cooking vacations that include visits to markets and food producers and sightseeing. Maximum class/group size: 8. Facilities: Farmhouse kitchens. Also featured: Private instruction.

Emphasis: Regional Tuscan cuisine.

Faculty: Women of Radda.

Costs: $500 for three days, $185 for one day. Includes market visit and dinner.

Location: Radda in Chianti, about halfway between Florence and Siena.

Contact: Mimma Ferrando, Tutti A Tavola, 1-Radda in Chianti, Muricciaglia, Italy; (39) 0577-742919, Fax (39) 0577-742807, E-mail ferrando@chiantinet.it. In the U.S.: The International Kitchen, Inc., 1209 N. Astor St., #11N, Chicago, IL 60610; 800-945-8606, Fax 847-295-0945/312-654-8446, E-mail info@intl-kitchen.com, http://www.intl-kitchen.com.

TWO FOR COOKING – TUSCAN COUNTRY EXPERIENCE

Argenina, Chianti/June-July, September-November

Tuscan native Julia Scartozzoni offers one-week private cooking vacation courses with lodging in a private country home. Established in 1994. Maximum class/group size: 2 (one couple only). Also featured: Private tours of the Chianti region, private castles, wineries, food producers; dining in private homes and fine restaurants.

Emphasis: Private Tuscan vacation experiences.

Faculty: Julia Scartozzoni has studied with noted chefs, co-owned a restaurant in Tuscany, and is an artist of interior designs.

Costs: $8,500/couple includes meals, private lodging at the Village of Argenina and attendants, and planned activities. Non-cook couple rate is $6,900.

Location: The Chianti region, 15 miles north of Siena, 30 miles south of Florence.

Contact: Julia Scartozzoni, Two for Cooking-Tuscan Country Experience, Argenina, Gaiole in Chianti, Siena, 53013 Italy; (39) 0337-79-0032, Fax (39) 0577-73-1100.

VENETIAN COOKING IN A VENETIAN PALACE

(See also page 290, Cuisine Intl.) **Venice/January-March, June, September-October**

Culinary professional Fulvia Sesani offers cooking classes in her 13th-century Venetian palace. Established in 1984. Maximum class/group size: 10. Facilities: Modern, fully-equipped kitchen. Also featured: Shopping in the Rialto market, visits to the Ducal palace, museums, and private homes, dinner at Harry's Bar, and the Palazzo Morosini. Also available: day classes and private lessons.

Emphasis: Traditional Venetian cooking, edible works of art.

Faculty: Fulvia Sesani.

Costs: All-inclusive land costs range from $3,300 to $3,700.

Location: The Palazzo Morosini is in the Santa Maria Formosa area of Venice.

Contact: Fulvia Sesani, Venetian Cooking in a Venetian Palace, Castello 6140, Venezia, NO 30122 Italy; (39) 041-522-8923. In the U.S.: Judy Ebrey, Cuisine International, P.O. Box 25228, Dallas, TX 75225; 214-373-1161, Fax 214-373-1162, E-mail CuisineInt@aol.com, URL http://www.cuisineinternational.com.

VILLA CENNINA/THE ART OF ITALIAN CUISINE

Siena/May-August

Culinary Director and Chef Gianluca Pardini offers 7-day program of hands-on culinary and wine instruction and cultural immersion in a 16th century villa. Established in 1983. Maximum class/group size: 15. 6 programs per year. Facilities: Professionally-equipped teaching kitchen on the Villa Cennina estate. Also featured: Visits to wineries, food producers, markets, restaurants, tours to culinary and cultural arts centers.

Emphasis: Regional cuisines and Italian wines.

Faculty: Gianluca Pardini, chef and educator, also serves as a consultant to international restaurant associations and culinary schools.

Costs: $2,100 includes lodging, meals, wine with lunch and dinner, ground transport, planned excursions, and use of Villa amenities, which include a 9-hole golf course, swimming pool, tennis courts, hiking trails, mountain bikes. Horseback riding nearby.

Location: In the Tuscan hills surrounded by medieval towns and villages, 30 miles from Florence.

Contact: Pat Kuh, Villa Cennina/The Art of Italian Cuisine, Food for Thought, 140 N. La Grange Rd., La Grange, IL 60525 US; 708-482-3737, Fax 708-482-3445, E-mail thinkfood@juno.com.

VILLA CROCIALONI COOKING SCHOOL

Fucecchio, Tuscany/Year-round *(See also page 290, Cuisine International)*

This private villa on 35 acres offers 5-day programs every other week that include hands-on classes daily and an excursion to Viareggio (seashore) fish market and Central Market in Florence. Established in 1996. Maximum class/group size: 4-5. 25 programs per year. Facilities: Family kitchen with large oven and 6 burners. The villa produces olive oil, vegetables, herbs, and raises farm animals. Also featured: Pool, jogging, visits to Santa Croce leather factories and Prada factory, and to Lucca, on request.

Emphasis: Tuscan-American cuisine.

Faculty: Buncky Pezzini, who trained in New York City, has 43+ years experience in Italian cuisine.

Costs: $1,800 includes lodging at the villa, meals, excursions, ground transport.

Location: 50 minutes from Florence, 40 minutes from Pisa, 20 minutes from Lucca, less than 2 hours from Siena.

Contact: Patricia (Buncky) Pezzini, Villa Crocialoni Cooking School, Via Delle Cerbaie #60, Fucecchio, Florence, 50054 Italy; (39) 0571-296237, Fax (39) 0571-296237, E-mail houseview@flownet.it, URL http://www.flownet.it/tuscan-cooking-school. In the U.S.: Judy Ebrey, Cuisine International, P.O. Box 25228, Dallas, TX 75225; 214-373-1161, Fax 214-373-1162, E-mail CuisineInt@aol.com, URL http://www.cuisineinternational.com.

VILLA DELIA TUSCANY COOKING SCHOOL

Ripoli di Lari/April-July, September-November

This Tuscan villa offers 10-day vacation packages that include 7 hands-on morning cooking classes. Established in 1995. Maximum class/group size: 20. Also featured: Wine instruction, day trips, private instruction, market visits, winery tours, dining in fine restaurants, sightseeing in cultural centers.

Emphasis: Tuscan cooking.

Faculty: Marietta, the resident chef.

Costs: $3,500 ($3,900), which includes double (single) occupancy lodging, deluxe lodging with private bath, meals, and excursions. Lodging at a modernized 16th-century Tuscan villa with swimming pool and tennis courts.

Location: 20 minutes from Pisa.

Contact: Kim Lloyd, Director of Sales, Umberto Management Ltd., 1380 Hornby Street, Vancouver, BC, V6Z 1W5 Canada; 604-669-3732, Fax 604-669-9723, E-mail inquire@umberto.com, URL http://www.umberto.com.

VILLA PAMBUFFETTI

April-June, September-November *(See also page 292, International Kitchen)*

This hotel offers five-day cooking vacations, including 5 hands-on classes, visits to markets and food producers, winery tours, sightseeing. Maximum class/group size: 8. Facilities: Hotel restaurant kitchen.

Emphasis: Umbrian regional cuisine.

Faculty: Alessandra Angelucci, who appears on the Italian food channel.
Costs: $2,100 includes lodging, meals, excursions.
Location: Perugia, Assisi.
Contact: Alessandra Angelucci, Villa Pambuffetti, Via della Vittoria, 20, Montefalco, 06036 Italy; (39) 0742-379417. In the U.S.: The International Kitchen, Inc., 1209 N. Astor St., #11N, Chicago, IL 60610; 800-945-8606, Fax 847-295-0945/312-654-8446, E-mail info@intl-kitchen.com, http://www.intl-kitchen.com.

VILLA UBALDINI

Florence/April-October

This villa originally built in the 13th century offers 1-week hands-on programs devoted to kitchen-tested recipes of Italian cuisine, focusing on nutrition and the Mediterranean diet, as well as recipes of the 12th-17th centuries. Instruction can be in English, Italian, Spanish, French, or German. Established in 1995. Maximum class/group size: 16. 10 programs per year. Facilities: Large teaching kitchen. Also featured: Winery tours, visits to food producers, guided tours of lesser known Tuscany, dining in private homes and Villa Ubaldini, cultural activities, including art history lectures and tours.
Emphasis: Medieval, Renaissance and Italian cuisine; cooking with flowers; table decoration; eating culture through the centuries.
Faculty: Margherita Vitali, manager of the courses and a founder of the Italian Assn. of Cooking Teachers, and Chef Janet Hansen, author of Medieval fires....Renaissance stoves.
Costs: $2,900-$3,500, which includes shared lodging at Villa Ubaldini, private bath, most meals, planned excursions. $500 deposit, $150 nonrefundable. Amenities: pool, billiards. Golf, horses, gliders available at additional cost.
Location: 20 minutes north of Florence in an environmentally protected area.
Contact: Margherita Vitale, Villa Ubaldini, via Genova 10, Grosseto, 58100 Italy; (39) 0564-451754, Fax (39) 0564-452779.

WALKING AND COOKING IN TUSCANY

(See also page 292, International Kitchen) **San Gimignano, Tuscany/March-November**

This private school offers 5-day trips that combine hands-on cooking lessons with guided walking tours of the region. Maximum class/group size: 8. Facilities: Renovated farmhouse hotel. Also featured: Visits to markets and food producers, winery tours, sightseeing.
Emphasis: Tuscan cuisine.
Faculty: Local chefs.
Costs: $1,900 includes lodging, meals, excursions.
Location: San Gimignano, Tuscany.
Contact: Guido Fratini, Perimetri Nuovi Percorsi, Via Puccini, 51, Pistoia, 51100 Italy; (39) 0573-358184, Fax (39) 0573-358183. In the U.S.: The International Kitchen, Inc., 1209 N. Astor St., #11N, Chicago, IL 60610; 800-945-8606, Fax 847-295-0945/312-654-8446, E-mail info@intl-kitchen.com, http://www.intl-kitchen.com.

WORLD OF REGALEALI

(See also page 290, Cuisine International) **Sicily/April-May, October-November**

Cookbook author Marchesa Anna Tasca Lanza offers weekly 2- and 5-day demonstration courses in Regaleali, Ms. Lanza's ancestral family home. Established in 1989. Maximum class/group size: 12. Facilities: Large professional kitchen with wood-burning oven in an 18th century farm house; adjoining estate and winery. Also featured: Visits to archeological sites, markets, and programs on the estate's agricultural enterprises: bread-baking, vegetable garden tour, tastings of estate olive oil and wines, cheese-making demonstration.
Emphasis: Sicilian cooking utilizing meats, cheeses, vegetables, and wines from the estate.
Faculty: Anna Tasca Lanza, author of The Heart of Sicily; Mario Lo Menzo, the family chef; other local and guest chefs.

Costs: $1,200 ($2,200), which includes all meals and 2 (5) days lodging at Regaleali.
Location: 2 hours by train from Palermo, in central Sicily.
Contact: Anna Tasca Lanza, World of Regaleali, Viale Principessa Giovanna, 9, Palermo, Mondello, NO 90149 Italy; (39) 091-450-727. In the U.S.: Judy Ebrey, Cuisine International, P.O. Box 25228, Dallas, TX 75225; 214-373-1161, Fax 214-373-1162, E-mail CuisineInt@aol.com, URL http://www.cuisineinternational.com.

JAPAN

INTERNATIONAL DINING ADVENTURES (See page 350)

KONISHI JAPANESE COOKING CLASS

Tokyo/September-July

Culinary professional Kiyoko Konishi offers hands-on classes, flexible schedule, home-style Chinese dishes also included. Established in 1969. Maximum class/group size: 8. 40+ programs per year. Facilities: 300-square-foot kitchen with Japanese and Chinese utensils and Japanese tableware. Also featured: Classes for youngsters, market visits, private lessons.
Emphasis: Japanese cooking basics, healthy, seasonal and traditional menus for family and entertaining with artistic presentation, sushi, nabemono (sukiyaki and shabushabu), raw &/or cooked seafood and vegetable dishes.
Faculty: Mrs. Kiyoko Konishi has taught to foreigners in English for 30+ years and is author of Japanese Cooking for Health and Fitness, Entertaining with a Japanese Flavor, and three bilingual cooking videos.
Costs: 5,000 yen/class includes tax, payable on arrival. A list of nearby hotels is available.
Location: Central Tokyo.
Contact: Kiyoko Konishi, Principal, Konishi Japanese Cooking Class, 3-1-7-1405, Meguro, Meguro-ku, Tokyo, 153-0063 Japan; (81) 3-3714-8859, Fax (81) 3-3714-8859.

LE CORDON BLEU – TOKYO

Tokyo/Year-round *(See also pages 154 and 301) (See display ad page 149)*

This private school acquired by Le Cordon Bleu Paris offers half-day to 1-year courses, daily demonstrations, gourmet sessions, cuisine and pastry courses, flower arranging, and Introduction to Cuisine, Pastry, Catering and Bread Baking. Established in 1991. Maximum class/group size: 8-16. 2-10 programs per year. Facilities: Classrooms resemble professional working kitchens; individual work spaces with refrigerated marble tables, convection ovens; specialty appliances. Also featured: Flower arrangement, evening wine classes, guest chef demonstrations.
Emphasis: Diploma and certificate program.
Faculty: 7 full-time French and English Master Chefs from Michelin-star restaurants and fine hotels.
Costs: Range from 10,000 yen (full day course) to 550,000 yen (12 weeks).
Location: Tokyo's Daikanyama district.
Contact: Taeko Okabe, Student Service and Sales Manager, Le Cordon Bleu-Tokyo, Daikanyama, Shibuya-ku, Tokyo, 150 Japan; (81) 3-548901-41, Fax (81) 3-548901-45, URL http://www.cordon-bleu.net. Toll free in the U.S. and Canada: 800-457-CHEF.

A TASTE OF CULTURE

Tokyo/Year-round

Culinary professional Elizabeth Andoh offers participation classes that focus on a specific theme, tasting programs devoted to traditional Japanese ingredients, market tours to Tokyo neighborhoods, includes Shibuya, Shinjuku, Kitchijoji, Jiyugaoka, and Kappabashi. Established in 1970. Maximum class/group size: 6 classes and tours/12-15 tastings. 4-6 programs per year. Facilities: Home kitchen fully equipped for teaching Japanese cooking. Also featured: 3-to-5-day intensive courses, private instruction, dining opportunities.
Emphasis: Practical, skill-building sessions on how to prepare Japanese food combined with

information about Japan's ancient and modern foodways.

Faculty: Bi-lingual American, longterm Japan resident, Elizabeth Andoh's formal culinary training was at Yanagihara Kinsaryu School of Traditional Japanese Cuisine, Tokyo. A member of the Japan Food Journalists Assn and IACP, she contributes to The New York Times.

Costs: $55/class, $40/tasting or market tour, $90/class + tasting or tour. Reduced rates for 2 persons enrolling in the same session. Periodic adjustments for currency fluctuations. Hotel list available.

Location: Central Tokyo.

Contact: Elizabeth Andoh, Director and Instructor, A Taste of Culture, 1-22-18-401 Seta, Setagaya-ku, Tokyo, 158-0095 Japan; (81) 3-5716-5751, Fax (81) 3-5716-5751, E-mail aeli@gol.com.

MEXICO

CULINARY ADVENTURES OF MEXICO

San Miguel de Allende/February-April, June, August, October-November

This culinary travel company offers 1-week culinary tours that include hands-on classes, market visits, visits to food producers, dining in private homes, sightseeing, 2-day trips, optional language classes. Established in 1996. Maximum class/group size: 8.

Emphasis: Traditional Mexican cuisine and Mexican fusion.

Faculty: Local prominent chefs.

Costs: $1,800, which includes luxury lodging, meals, ground transport, planned activities. $400 single supplement.

Contact: Kristen Rudolph, Director, Culinary Adventures of Mexico, Jesus 23, San Miguel de Allende, 37700 Mexico; (52) 4117-9276, Fax (52) 4117-8228, E-mail culadv@unisono.ciateq.mx.

FLAVORS OF MEXICO – CULINARY ADVENTURES, INC.

Oaxaca, Veracruz/Year-round

Marilyn Tausend's Culinary Adventures offers 7- to 10-day cooking vacations to different regions of Mexico and the US Southwest. Established in 1988. Maximum class/group size: 6-16. Facilities: Typically indoor and outdoor home and restaurant kitchens. Varies by locale. Also featured: Visits to food markets and cottage industries, cheese/mescal/pit-roasted pigs meals and demonstrations in homes of local cooks, tours of historical and archaeological sites and artisans' workshops.

Emphasis: Regional Mexican and Southwestern cuisine with an emphasis on ingredients, techniques, and the cultural background behind the dishes.

Faculty: Marilyn Tausend, co-author of Mexico the Beautiful Cookbook; Mexican cooking authority Diana Kennedy; Maria Dolores Torres Yzabál, culinary consultant, teacher, and author Ricardo Munoz Zurita, Executive Chef, Univ. Nacional Autonoma de Mexico.

Costs: Approximately $2,550, which includes meals, double occupancy lodging, planned excursions, and local transport. Lodging in small hotels popular with Mexican families.

Location: Oaxaca and Veracruz in southern Mexico, the San Luis Valley in southern Colorado and northern New Mexico. Sites vary each year.

Contact: Marilyn Tausend, Culinary Adventures, Inc., 6023 Reid Dr. N.W., Gig Harbor, WA 98335 US; 253-851-7676, Fax 253-851-9532.

FOOD OF THE GODS FESTIVAL

Oaxaca/October

Zapotec Tours offers an annual 1-week program that features market and village tours, cooking classes, lectures and demonstrations, visits to food producers, dining around. Established in 1994.

Emphasis: Traditional Oaxacan cuisine.

Costs: From $529, includes hotel lodging and meals.

Location: City of Oaxaca.

Contact: Vezire Adili, Reservation Manager, Food of the Gods Festival, 5121 N. Ravenswood

Ave., Ste. B, Chicago, IL 60640 US; 800-446-2922, Fax 773-506-2445, E-mail zapotectours@oaxacainfo.com, URL http://www.oaxacainfo.com.

FRONTERA GRILL/FRONTERA INSTITUTE (See page 211)

JALISCO CULINARY ARTS

Chapala/February-April, September-November

This private school offers 1-week vacation programs that feature hands-on classes in Mexican cuisine. Established in 1997. Maximum class/group size: 18. Facilities: Separate classroom at a quality hotel in Chapala, with additional demonstrations at local restaurants and food vendor facilities. Also featured: Visits to markets and area villages.

Emphasis: Jalisco regional cooking, the traditional and modern dishes of Mexico.

Faculty: Cultural activities conducted by Andrew Bosworth, Ph.D. Culinary team led by Bruce Jensen, food service consultant and restaurateur for 20+ years. Local and guest chefs supplement the teaching team.

Costs: Start at $1,195, includes shared lodging, all meals except Sunday, planned excursions and cultural activities.

Location: Lake Chapala, just 30 miles south of Guadalajara Mexico.

Contact: University Travel Consultants, Inc., 922 University City Blvd., Blacksburg, VA 24060 US; 800-638-2701, Fax 540-951-2921, E-mail bill@universitytravel.com, URL http://www.fspronet.com/jca/jca.html.

MEXICAN CUISINE SEMINARS WITH LULA BERTRAN

Mexico City/Year-round

Culinary professional Lula Bertran offers Be my Guest for One Day, a program for visitors. Hands-on seminars and personalized programs for individuals and groups by appointment. Established in 1980. Maximum class/group size: 2-8. Facilities: Conference room and fully-equipped kitchen. Also featured: Guided market visits, dining at traditional restaurants, short trips to nearby towns.

Emphasis: Mexican cuisine, food history, eating customs; introduction to Mexico City's food world.

Faculty: Food writer and teacher Lula Bertran is founding member of three food associations in Mexico and international advisor of the board of the IACP Foundation.

Costs: ~$150 for day-long session may include market visit, cooking class, food shops-city tour, lunch or dinner. Suggestions for hotel lodging in moderate ($60/night) or deluxe hotels.

Location: Mexico City.

Contact: Lula Bertran, Mexican Cuisine Seminars, Diego Fernandez de Cordoba, #135, Mexico City, 11000 Mexico; (52) 5-202-7251, Fax (52) 5-540-3633, E-mail lbertran@hotmail.com.

MEXICAN HOME COOKING

Tlaxcala/Year-round

Culinary professional Estela Salas Silva offers monthly 5- to 10-day and customized hands-on cooking vacations. Established in 1996. Maximum class/group size: 6. Facilities: Fully-equipped 500-square-foot kitchen, on-site gardens. Also featured: Day trips, market visits, sightseeing.

Emphasis: Mexican home cooking, pre-Hispanic to modern.

Faculty: Estela Salas Silva, chef in Mexico City and San Francisco since 1974; Rojelio Salas Silva, chef and restaurateur since 1989 in Mexico City; local guest cooks.

Costs: Land cost begins at approximately $780, which includes meals and single occupancy lodging in the Silva family home with use of kitchen and grounds.

Location: 10 minutes outside the city of Tlaxcala, 30 minutes north of Puebla Airport.

Contact: Estela Salas Silva, Mexican Home Cooking, Apartado 64, Tlaxcala, Tlaxcala, 90000 Mexico; (52) 246-80978.

SEASONS OF MY HEART COOKING SCHOOL

Oaxaca/Year-round *(See also page 160)*

This private school offers 1-week course twice a year, 4-day long weekend courses, day classes, all hands-on. 1993. Maximum class/group size: 4-20. Facilities: Rancho Aurora, a working farm, has a

handmade kitchen with 5 stations, outdoor kitchen with parilla, wood-fire and pre-Hispanic cooking utensils. Also featured: Visits to corn and chocolate mills, mezcal factory, markets, archaeological sites, farms, weavers, pottery makers, private homes. Private group tours and lectures.
Emphasis: Native and pre-Hispanic foods, chiles, wild plants and herbs, contemporary dishes.
Faculty: Susana Trilling, teacher, writer, lecturer, chef, caterer, IACP member. Part-time teachers, registered tour guides, cheese makers, bread bakers, chocolate makers, chefs, farmers, herbal healers.
Costs: Day classes begin at $65; 4-day long weekend course $650, 1-week courses $1,495, includes meals, lodging, planned activities. Lodging in Oaxaca hotel (Casa Colonial or Posada de Chencho) or casita on ranch.
Location: Ranch surrounded by archaeological sites in the mountains a half hour outside Oaxaca City in southern Mexico.
Contact: Susana Trilling, Director, Seasons of My Heart Cooking School, Apdo. 42, Admon 3, Oaxaca, Oaxaca, CP, 68101 Mexico; (52) 951-87726/954-83115, Fax (52) 951-87726/65280, E-mail seasons@antequera.com.

VILLA DE LA ROCA COOKING HOLIDAYS
Zihuatanejo, Gro./May-October

This resort offers 1-week cooking holidays with 3-hour hands-on classes daily. Established in 1996. Maximum class/group size: 10. 24 programs per year. Facilities: 8,000-square-foot bed & breakfast with 1,000-square-foot kitchen overlooking bay. 30-seat restaurant on premises. Also featured: Private instruction, market visits, sightseeing, tours of food producers.
Emphasis: Interior Mexican cuisine featuring fresh seafood and local fruits and vegetables.
Faculty: Chefs Cameron Graham and Linda Fox have 25+ yrs experience in catering and restaurant management. Chef Graham was named one of America's best caterers by Town & Country magazine.
Costs: $1,500/week double, $1,850/week single, includes lodging and meals at the villa, Mexican tax and staff gratuity.
Location: On La Ropa beach, a 10-min walk to town.
Contact: Linda Fox, President, Villa de la Roca Cooking Holidays, 3839 Dry Creek Dr., #122, Austin, TX 78731 US; 512-459-9232, Fax 512-459-8005, E-mail LRFOX2@aol.com.

MOROCCO

ART OF LIVING – CULINARY TOURS WITH SARA MONICK (See page 289)

COOKING AT THE KASBAH ★
Casablanca, Marrakech and others/May

Cookbook author Kitty Morse offers a fully-escorted 2-week tour emphasizing local cuisine and culture, including 2 days of seminars with Kitty Morse and local experts. Established in 1983. Maximum class/group size: 24 maximum. 1 program per year. Facilities: Morse's home, a restored pasha's residence; private homes, deluxe or first class hotel kitchens. Also featured: Demonstrations in Morse's home and visits with personal friends; excursions to historic locales and marketplaces; special events; golf and tennis available.
Emphasis: Moroccan cuisine and culture.
Faculty: Native of Casablanca, Kitty Morse, is a member of the IACP, and the So. Cal. Culinary Guild. She is author of The Vegetarian Table: North Africa, A Biblical Feast, and Cooking at the Kasbah!.
Costs: Approximately $4,000, including roundtrip airfare from New York to Casablanca, land transport, 2 meals daily, planned activities. Double occupancy lodging in first class or deluxe hotels.
Location: Coastal cities of Casablanca, El Jadida, Azemmour, Essaouira; Imperial cities of Fez, Marrakech, Meknes, Rabat. Also Ouarzazate, and Tinehrir.
Contact: Natalie Tuomi, Carlsbad Travel, 2727 Roosevelt St., Carlsbad, CA 92008 US; 800-533-2779, Fax 760-729-2482, E-mail carlsbadtravel@juno.com, URL http://www.kittymorse.com.

NETHERLANDS

LA CUISINE FRANCAISE
Amsterdam/September-June

Culinary professional Patricia I. van den Wall Bake-Thompson offers 1- and 4-session demonstration and participation courses. Established in 1980. Maximum class/group size: 25 demo/16 hands-on. 60 programs per year. Facilities: 90-sq-meter kitchen rebuilt in 1994, private dining room. Also featured: Sessions in English for groups, private classes, and market visits.

Emphasis: French, Italian, and English/Dutch cuisines.

Faculty: School owner and instructor Patricia I. van den Wall Bake-Thompson was born in Great Britain and studied home economics at Harrow Technical College.

Costs: Each session is $45 (75-105 Dutch Guilder), payable in advance. Lodging by arrangement in nearby hotels.

Location: In an 18th century canal house, which also houses a private restaurant, in central Amsterdam and 20 minutes from the airport.

Contact: Pat van den Wall Bake-Thompson, La Cuisine Francaise, Herengracht 314, Amsterdam CD, 1016 Netherlands; (31) 20-6278725, Fax (31) 20-620-3491, E-mail pat-cuis@euronet.nl.

NEW ZEALAND

AN EPICUREAN AFFAIR
Blenheim, Marlborough/Year-round

Stone Aerie Estate offers customized half-day programs that include a buffet lunch or dinner, guided tour of the estate, and cooking and/or wine demonstration. Established in 1995. Maximum class/group size: 10-20. Facilities: Designed by Jeremy Jones, Stone Aerie was featured in New Zealand House and Garden. The custom-designed kitchen is Gaggenau equipped.

Emphasis: Food and wine of the Marlborough region.

Faculty: Francie Shagin, owner of Stone Aerie, teaches Italian cooking. Jeremy Jones was chef/owner of Sycamore Tree in Wellington and Peppertree Restaurant outside Blenheim. He runs an interior design firm.

Costs: From A$92 (wine or cooking) to A$150 (wine and cooking).

Location: A 15-minute drive from Marlborough.

Contact: Jeremy Jones, An Epicurean Affair, Dog Point Road, R.D.2, Blenheim, Marlborough, New Zealand; (64) 3-572-9639, Fax (64) 3-572-9634.

EPICUREAN WORKSHOP
Auckland/March-December

This cookware store and school offers demonstration and participation classes, 1-hour Gourmet on the Run classes twice weekly. Established in 1989. Maximum class/group size: 35-40 demo/8 hands-on. 80-100 programs per year. Facilities: Teaching kitchen with overhead mirrors. Also featured: Children's and young adults classes, Taste of New Zealand classes for visitors, private classes, and 5-day vacation programs.

Emphasis: Seasonal themes, classics, technique and method, ethnic cuisines, local and overseas guest chef specialties.

Faculty: Director Catherine Bell, CCP is a graduate of Leith's School in London and an IACP member. Local chefs include Ray McVinnie, Greg Heffernan, and other prominent chefs.

Costs: Demonstrations range from NZ$7.50-NZ$65, hands-on classes from NZ$100, young chef classes from NZ$45.

Location: Newmarket, a major shopping center in Auckland.

Contact: Catherine Bell, Epicurean Workshop, 27 Morrow St., P.O. Box 9255, Newmarket, Auckland, New Zealand; (64) 9 524-0906, Fax (64) 9-524-2017, E-mail CatherineB@xtra.co.nz.

RUTH PRETTY COOKING SCHOOL
Te Horo/February-October

Ruth Pretty Catering offers full-day weekend classes in a rural setting. Established in 1994. Maximum class/group size: 32 demo, 10 hands-on. 40 programs per year. Facilities: Commercial catering kitchen. Also featured: Group classes that include a talk and focus on such topics as gardening or wine (no cooking).

Emphasis: Each class offers a menu with a topic heading.

Faculty: Caterer Ruth Pretty and a variety of New Zealand and overseas instructors.

Costs: A$140/class.

Location: One hour from Wellington (capital city).

Contact: Ruth Pretty, Principal, Ruth Pretty Cooking School, P.O. Box 41, Te Horo, 6470 New Zealand; (64) (0)6-3643161, Fax (64) (0)6-3643262, E-mail pretty@freemail.co.nz.

SMALL KITCHEN SCHOOL
Christchurch/Year-round

Caterer Michael Lee-Richards offers demonstrations, lunch-and-learn classes, participation sessions. Maximum class/group size: 12-80. Facilities: Newly-designed kitchen. Also featured: Private lessons for groups, tours, live-in classes at Gov Bay.

Emphasis: Various topics and guest chef specialties, workshops.

Faculty: Owner-instructor Michael Lee-Richards is chef at Michaels of Canterbury, host to the Canterbury Gourmet Society, director of Catering, and oversees The Lonsdale, his private catering firm.

Costs: NZ$43 for 1-day class to NZ$280 for 3-day class. Bed & breakfast and hotels available.

Contact: Michael Lee-Richards, Small Kitchen School, P.O. Box 22-543, Christchurch, New Zealand; (61) 3-365-2837, Fax (61) 3-365-1621.

SINGAPORE

RAFFLES CULINARY ACADEMY
Singapore/Year-round

Raffles Hotel offers one or two classes daily, both demonstration and hands-on. Established in 1995. Maximum class/group size: 24. Facilities: Well-equipped residential-type kitchen with demonstration area. Also featured: Wine instruction, classes for youngsters, market visits, dessert and pastry classes, social etiquette classes.

Emphasis: Singapore and Asian ethnic cuisines, Raffles Hotel signature dishes.

Faculty: Raffles Hotel chefs.

Costs: $60-$100/class. Specialty classes range from $35-$75.

Location: Singapore's civic district.

Contact: Antonio Daroya, Manager, Raffles Culinary Academy, 1 Beach Rd., 189673 Singapore; (65) 337-1886, Fax (65) 339-7013.

VIOLET OON'S KITCHEN
Singapore/January-November

Culinary professional Violet Oon offers ad-hoc cooking programs cum culinary experiences. Established in 1980. Maximum class/group size: 10-30 demo, 5-15 hands-on. 100 programs per year. Facilities: Kitchen shop. Also featured: Programs for children and teens, wine instruction, day trips, market visits, private instruction, visits to food producers.

Emphasis: Asian/Singaporean, Peranakan/Nonya cuisine.

Faculty: Violet Oon.

Costs: S$65-S$85.

Location: Chinatown, 30 minutes from Changi Airport.

Contact: Violet Oon, Director, Violet Oon's Kitchen, Blk 80 Chay Yan St., Unit 01-08, 160080 Singapore; (65) 323-7379, Fax (65) 323-5009, E-mail vok@pacific.net.sg.

SPAIN

ALAMBIQUE SCHOOL, S.A.

Madrid/October-June

This school in a cookware store established by Clara Maria Amezua offers 1- to 10-session demonstration and participation courses for beginners and professionals. Established in 1973. Maximum class/group size: 15-20 demo/10-12 hands-on. Facilities: Teaching kitchen equipped with microwave, gas, and electric ovens. Also featured: Classes for children, tailor-made courses.

Emphasis: Classic and regional Spanish cuisines, French, Mediterranean, Chinese, Italian, Japanese.

Faculty: Includes Isabel Maestre, Enrique Alvarez de Bastearccheci, Hiroko Bertchman, Natalia Martinez Arroyo, Claude Maison d'Arbray, Gabriella Llamas, Salvador Gallello, Francisco Tejero.

Costs: Single sessions range from $22-$37, courses from $100-$700.

Contact: Clara Maria Gonzalez Amezua, Alambique School, S.A., Calle de la Encarnacion, 2, Madrid, 28013 Spain; (34) 547-88-27, Fax (34) 559-78-02.

ART OF LIVING – CULINARY TOURS WITH SARA MONICK (See page 289)

COOKING VACATION IN AN OLD WATERMILL

Jalon/October-February

Restaurant Molino de Vino, an 18th century watermill, offers 6-day cooking vacations. Established 1987. Facilities: Kitchen of the restaurant Molino de Vino. Also featured: Tour of Valencia's vegetable and fish market, medieval night, vineyard tour, tour to Castell de Guardeleste, wine tasting.

Emphasis: Mediterranean cooking, Spanish specialty meals, farmhouse-style cooking.

Faculty: Hendric Schroeder, German chef and proprietor, and his wife, Olivea, host the course in English. Teaching also available in German and Spanish.

Costs: $2,100 includes shared lodging (private villa with 4 bedrooms and 3 baths, 2 ensuite), 1 meal daily, ground transport, planned activities. $200 single supplement, $850 for noncooking guest.

Location: At the Costa Blanca between Valencia and Alicante, 15 miles from the coast and the fishing villages of Javea, Denia, and Benidorm.

Contact: Hendric and Olivea Schroeder, Proprietors, Restaurant Molino de Vino C.B., Camino Ferra Baja 1, Jalon (Alicante), 03727 Spain; 34-6-648-0354, Fax 34-6-648-0354.

EL BULLI

Gerona/February-March

This Michelin 3-star restaurant offers a 3-day demonstration that features the chef's meat and seafood combinations and recipes planned for the coming season. Discussion in Spanish and Catalan. Maximum class/group size: 18. Facilities: El Bulli restaurant.

Emphasis: Chef specialties.

Faculty: Catalan Chef Ferran Adria , named Best European Chef for 1995.

Costs: $1,200.

Location: Catalonia.

Contact: Chef Ferran Adria, El Bulli Restaurant, Cala Montjoi, Gerona, 17480 Spain; (34) 72-15-04-57, Fax (34) 72-15-07-17, E-mail bulli@grn.es.

EL CENADOR DE SALVADOR

Madrid/January-May

This restaurant offers one-month hands-on classes that meet for four hours daily, Tuesday to Friday, and include the preparation of four different dishes each day. Instruction in Spanish. Maximum class/group size: 10.

Emphasis: Creative cuisine from traditional dishes.

Faculty: Chef Salvador Gallego.

Costs: $400/course.

Contact: Salvador Gallego, El Cenador de Salvador, Av. de Espana, 30, Moralzarzal (Madrid), 28411 Spain; (34) 1-857-7710, Fax (34) 1-857-7780.

EL TXOKO DEL GOURMET

(See also page 163)

Basque Country/Year-round

This specialty food shop offers short courses and seminars (14-21 hours) structured around a region or food category. Facilities: Basement workroom. Also featured: Private courses for a minimum of 10 students (3 mos advance notice).

Emphasis: Theme topics and Basque specialties such as pinchos (skewered multi-layer tapas).

Faculty: Pepa Armendariz.

Location: Opposite San Sebastian's covered market.

Contact: Pepa Armendariz, El Txoko del Gourmet, Aldamar, 4 Bajo, San Sebastian, 20003 Spain; (34) 943-422-218, Fax (34) 943-427-641.

ESCUELA TELVA

Madrid, Vigo, Seville, Cordoba/Year-round

This private school offers a 9-month demonstration course that meets once weekly and covers 3-5 dishes per class; 2- to 10-session demonstration and hands-on theme courses. Instruction in Spanish. Established in 1988. Maximum class/group size: 5-10 hands-on, 20-30 demo.

Emphasis: A variety of topics.

Costs: $900 for the 9-month course, $30/class.

Location: Madrid, Galicia (Vigo), Andalucia (Seville), Cordoba.

Contact: Lorena San Martin, Escuela Telva, Crucero 25 de Mayo, 10, Madrid, 28016 Spain; (34) 91-345-7290, Fax (34) 91-350-7347.

ESCUELA DE COCINA LUIS IRÍZAR

(See also page 164)

Basque Country/July-September

This private school offers 15-day courses for professionals and amateurs that include visits to markets and food producers, winery tours, sightseeing, and dining in Michelin-star restaurants. Instruction in Spanish. Established in 1992. Maximum class/group size: 15. Facilities: Fully-equipped kitchen, separate classroom, TV and video.

Emphasis: Basque and French cooking.

Faculty: Founder Luis Irízar has served as chef in Spain's leading restaurants. His staff includes 3 full-time instructors plus part-time teachers for continuing ed.

Costs: $250.

Location: Old part of San Sebastian, facing the port and beach.

Contact: Virginia Irízar, Escuela de Cocina Luis Irizar, c/Mari, #5, Bajo, San Sebastian, 20003 Spain; (34) 43-431540, Fax (34) 43-423553.

ESCUELA DE COCINA MENCHU

Madrid/Year-round

This private school offers 4-session (once weekly) beginner and advanced demonstration courses, special classes on demand. Instruction in Spanish. Established in 1994. Maximum class/group size: 15.

Emphasis: A variety of topics.

Costs: $60 for beginner course, $75 for advanced course.

Contact: Menchu Puerto, Escuela de Cocina Menchu, Anastasio Aroca 3, lo 2a, Madrid, 28001 Spain; (34) 915-496-792.

ESCUELA DE HOSTELERIA ARNADI

(See also page 164)

Barcelona/Year-round

This private school offers 2-day seminars (Oct-Dec) on tapas and Mediterranean cooking, 18-month (6 hours weekly) pastry courses (start Sept and Mar), holiday food (turrones, marzipan) courses (Dec). Established in 1982. Also featured: Month-long summer courses (June-July) focus on summer specialties such as Spanish/Catalan cooking and Spain's regional dishes.

Emphasis: Basque and Mediterranean cuisine and other topics.
Costs: $120-$280.
Contact: Mey Hoffman, Escuela de Hosteleria Arnadi, C/Argenteria 74-78, Barcelona, 08003 Spain; (34) 933-195-889/882, Fax (34) 933-195-859.

ESCUELA DE HOSTELERIA JOSE LUIS

Madrid/September-July

This private school offers 1- to 3-week demonstration courses in Basic Cuisine (36 hours), Advanced Cuisine (24 hours), Pastry (36 hours), and Cocktails (8 hours). Classes meet for 3 hours daily, Monday-Thursday. Instruction in Spanish. Established in 1990. Maximum class/group size: 6-10.
Emphasis: Basic and advanced cuisine and pastry, cocktails.
Costs: Basic Cuisine $230, Advanced Cuisine $260, Pastry $240, Cocktails $120.
Contact: Juan Guitart, Escuela de Hosteleria Jose Luis, Sambara, 153, Madrid, 28027 Spain; (34) 913-772-766, Fax (34) 914-070-854.

INTERNATIONAL DINING ADVENTURES (See page 350)

LA TABERNA DEL ALABARDERO

Seville/Year-round *(See also page 165)*

This private school offers 5-session evening courses on such topics as game, pastry, holiday cuisine, international specialties. Instruction in Spanish. Established in 1993. Maximum class/group size: 15.
Emphasis: A variety of topics.
Costs: $150-$220/course.
Contact: Pedro Olives, La Taberna del Alabardero, Pza. de Molviedro, 4, Seville, 41001 Spain; (34) 954-560-637, Fax (34) 954-215-555.

SWEDEN

CUISINE ECLAIRÉE (See page 290)

TAIWAN

TAIPEI CHINESE FOOD FESTIVAL

Taipei/August

The Taipei Tourist Hotel Association and co-sponsor, American Express International, sponsor an annual 3-day event that includes lectures on culinary culture, demonstrations by noted chefs, and professional cooking competitions. 1 program per year. Facilities: TWTC Exhibition Hall A (Hsinyi Rd.) Taipei. Also featured: Taipei travel agencies organize gourmet tours that include visits to the National Palace Museum, the Fu Hsing Dramatic Arts Academy, the Tsushih Temple at Sanhsia, and the pottery kilns at Yingko.
Emphasis: Theme changes each year.
Costs: Ticket prices $7.70 for adults, $5.80 for children.
Contact: Alex Hsiao, Deputy Director, Taiwan Visitors Assn., One World Trade Ctr., #7953, New York, NY 10048 US; 212-466-0691, Fax 212-432-6436, URL http://www.tbroc.gov.tw.

THAILAND

THE ART OF THAI COOKING (See page 185)

ASIA TRANSPACIFIC JOURNEYS

Bangkok, Chiang Mai/Year-round

This special interest travel provider (formerly Bolder Adventures) offers 3- to 6-day programs at 3 schools: Thai House in Nontaburi, The Oriental Hotel in Bangkok, Chiang Mai Thai Cookery School in Chiang Mai. Schedule includes market visits and daily classes. Custom designed tours in

Viet Nam. Established in 1991. Maximum class/group size: 15. Facilities: Varies. Also featured: Tours of Bangkok, Chiang Mai and environs, Thailand and region. Departures are flexible and available as stand-alone trips or part of a longer itinerary.

Emphasis: Regional Thai and Vietnamese cuisines.

Faculty: Thai cooks include Peep Chinsanaboom, Pip Fargrajang. Somphon and Elizabeth Nabnian. Various cooks in Viet Nam.

Costs: Thai House: $795 for 4 days includes guest house lodging, meals, market trip. Oriental Hotel: $1,490 for 6 days includes lodging, $450 classes only, $100/class. Chiang Mai: $150 for 3 days of classes, no lodging. Viet Nam: $100/class.

Location: Thai House: Nontaburi outside Bangkok. Oriental Hotel, Bangkok. Chiang Mai: Thailand's northern hill country. Viet Nam: Hanoi, Saigon, Hoi An.

Contact: Marilyn Staff, Sala Siam-Asia Transpacific Journeys, 3055 Center Green Dr., Boulder, CO 80301 US; 800-642-2742, Fax 303-443-7078, E-mail travel@SoutheastAsia.com, URL http://www.SoutheastAsia.com.

BAAN RATA CENTER FOR FAR EAST CULTURE

Kata Beach, Phuket/Year-round

This restaurant, private school, travel company offers classes and 3-day courses in Asian cuisines with the stress on authenticity as well as modern adaptations. Established in 1998. Maximum class/group size: 12 hands-on, 25 demo, 20 travel. 20-25 programs per year. Facilities: The Gallery Grill's professional restaurant kitchen with views of garden and sea. Also featured: Visits to food producers and markets, dining in private homes, sightseeing.

Emphasis: Asian cuisines and fusion foods.

Faculty: Chefs of the Gallery Grill and Boathouse, visiting chefs from Hong Kong, Singapore, Vietnam, and Australia.

Costs: $25 per class, $150 for a 3-day course, which includes continental breakfast and lunch.

Location: One hour by air from Bangkok.

Contact: Sue Farley, Director, Baan Rata Center for Far East Culture, 3/2 Moo 2 Patak Rd., A. Muang Kata Beach, Phuket, 83100 Thailand; (66) (0)76-330123, Fax (66) (0)76-330482, E-mail sue@baankata.com, URL http://www.baankata.com.

CHIANG MAI THAI COOKERY SCHOOL

Chiang Mai/Year-round

This family business run by Somphon and Elizabeth Nabnian offers daily 1-, 2-, and 3-day courses that include intro to Thai ingredients, paste making, or a market tour, preparation of 7 dishes, herb garden tour, fruit tasting. Established in 1993. Maximum class/group size: 3-18 hands-on. Facilities: Purpose-built kitchen.

Emphasis: Northern and traditional Thai cuisine.

Faculty: Experienced Thai chef Somphon Nabnian.

Costs: 800B (1,600B, 2,300B) for 1 (2, 3) days.

Location: Outside Chiang Mai.

Contact: Elizabeth Nabnian, The Chiang Mai Thai Cookery School, 1-3 Moon Muang Rd., Chiang Mai, 50200 Thailand; (66-53) 206388, Fax (66-53) 206387, E-mail cmcook@infothai.com, URL http://www.infothai.com/s/cmcook.htm.

INTERNATIONAL DINING ADVENTURES (See page 350)

ROYAL THAI SCHOOL OF CULINARY ARTS

(See also page 165)

Bang Saen, Chonburi/Year-round

This private school offers 1- to 9-week beginning to advanced hands-on courses that emphasize the feel, taste, and presentation of Thai, Burmese and Fusion cuisines. Established in 1997. Maximum class/group size: 8/class. 120+ programs per year. Facilities: Demonstration and 2 hands-on kitchens each for 8 students. Also featured: Visits to markets and food producers, sightseeing, private instruction.

Emphasis: Royal Thai, regional Thai, Burmese, and Fusion cooking.
Faculty: Le Cordon Bleu Grand Diplome graduates.
Costs: $1,700-$2,800 includes single lodging in ocean-view room, two meals daily.
Location: On the beach in Bang Saen about 88 km from the center of Bangkok.
Contact: Chris Kridakorn-Odbratt, Head chef, Royal Thai School of Culinary Arts, 53 Sukhumvit 33 Rd., Bangkok, 10110 Thailand; 66 - 38 - 748 404, Fax 66-38-748 405, E-mail lechef@chef.net, URL http://www.rtsca.com.

THAI COOKING SCHOOL AT THE ORIENTAL, BANGKOK
Bangkok/Year-round

The Oriental resort hotel offers weekly 4-day demonstration courses (Monday-Thursday). Established in 1986. Maximum class/group size: 15. Facilities: Classroom, participation/demonstration room, kitchen, eating area.
Emphasis: Introduction, ingredients, snacks, salads/soups, desserts, fruit and vegetable, carving/curries, condiments, side/dishes, steam, stir fry, fry and grill menu preparation and how to order.
Costs: $1,788 ($2,500), including double (single) occupancy lodging at The Oriental, most meals, planned activities. Amenities: swimming pools, sports center, tennis, squash, spa and health center.
Location: Overlooking the Chao Phraya River, 15 minutes from the main business district.
Contact: Chanida Srimanoj, Secretary, Thai Cooking School at the Oriental, 48 Oriental Ave., Bangkok, 10500 Thailand; (662) 236-0400, Fax (662) 439-7587, E-mail bscorbkk@loxinfo.co.th. In the U.S.: Mandarin Hotel Group 800-526-6566.

THAI HOUSE COOKING SCHOOL
Nontaburi/Year-round

This private school offers 1-, 2- and 4-day courses that include lectures, preparation of ingredients, hands-on cooking. Maximum class/group size: 2-10. Facilities: Open-air kitchen, farm and orchard. Also featured: 2- and 4-day courses include an excursion to an open-air market.
Faculty: Pip Fargrajang.
Costs: $100 for 1-day course. 2- (4-) day courses are $255 ($530) which includes lodging at Thai House. Price includes transport from/to Bangkok.
Location: 22 km from Bangkok.
Contact: The Thai House Cooking School, 22 Pra-athit Rd., Bangkok, 10200 Thailand; (662) 2800740-41, Fax (662) 2800741.

TURKEY

CULINARY EXPEDITIONS IN TURKEY
May-June, September-October

Kathleen O'Neill, food researcher, resident in Turkey, offers 9- to 11-day trips. Culinary cruises along the Aegean and Mediterranean; local market tours; culinary explorations of Istanbul; land expedition featuring the regional home cuisine of Gaziantep and southeast Turkey. Established in 1996. Maximum class/group size: 10. 5 programs per year.
Emphasis: Turkish regional cuisine and home cuisine.
Costs: $2,100-$2,400 includes lodging, meals, planned activities.
Location: Turkey, the Aegean and Mediterranean.
Contact: Kathleen O'Neill, Owner, Culinary Expeditions in Turkey, P.O. Box 1913, Sausalito, CA 94966 US; 415-437-5700, Fax 925-210-1337, E-mail koneill@evocative.com, URL http://www.evocative.com/~turkey.

THE GOURMET TOUR: EAT SMART IN TURKEY
Istanbul and other cities/June, October

This travel company specializing in tours of Turkey offers 2-week trips that include cooking demonstrations, visits to markets and wineries, dining at neighborhood and gourmet restaurants, lectures by food authorities, shopping at Istanbul's Covered Bazaar, and sightseeing excursions. Maximum class/group size: 10-18. 2 programs per year.

Emphasis: Food, wine, and culture of Turkey.

Faculty: Noted food historians and cookbook writers.

Costs: $2,990 ($3,490), which includes shared (single) deluxe or first class hotel lodging, meals, transportation in Turkey, and planned activities.

Location: Itinerary includes Istanbul, Kayseri, Cappadocia, Ankara, Antalya, Pamukkale, Aphrodisias, Marmaris, Seláuk (Ephesus).

Contact: Elizabeth Kurumlu, Vice President, Innovations In Travel, 211 S. Beverly Dr., #101, Beverly Hills, CA 90212 US; 800-2-TURKEY/310-276-7167, Fax 310-276-7315, E-mail yavuz@ix.netcom.com. Additional contact: Bagdat Cad. Tanueri Apt. No: 509/3, 81070 Suadiye, Istanbul, Turkey; (901) 216-4119692, Fax (901) 216-4119473.

VIETNAM

DISCOVER VIETNAM

3 historical cities/April, June-August, November

This travel company specializing in cycling, culinary and custom adventure tours of Vietnam offers 16-day tours that feature classes in the cuisine of three regions of Vietnam. Established in 1997. Maximum class/group size: 15. Facilities: Hotel restaurant. Also featured: Lectures on the country's history and culture, dining in fine restaurants, sightseeing.

Faculty: Master chef Nguyen Thi Hanh of the Huong Giang Hotel in Hue.

Costs: $2,895 includes luxury hotel lodging, meals, ground transport, planned activities.

Location: Hanoi, Hue, Hoi An, Nha Trang, Ho Chi Minh City.

Contact: Hans Krausche, Discover Vietnam, 1088 Clarendon Crescent, PO Box 10096, Oakland, CA 94610 US; 510-839-4019, Fax 510-268-0119, E-mail HKrausche@aol.com, URL http://www.discovervietnam.com/culinary.html.

WALES

WALES CULINARY CENTRE – CULINARY ADVENTURES

Cardiff/Year-round

This private school offers weekly courses and daily demonstrations, leisure and activity weekends, Kids in the Kitchen, Global Gourmand travel holidays, as well as certificate courses. Established in 1997. Maximum class/group size: 25 demo/8 hands-on/14 travel. Facilities: Cooking theatre seating 12-30, newly-equipped purpose-built kitchen. Lodging in the 7-bedroom St. Fagans Country Guest House.

Emphasis: International cuisines, including Celtic, Welsh, and Austrian.

Faculty: Margaret Rees, IACP SOFHT, and international experts in food and wine.

Location: The historic village of St. Fagans, on the outskirts of Cardiff, the capital city of Wales.

Contact: Margaret Rees, Director, Culinary Adventures, St. Fagans, Cardiff, CF5 6DS Wales; (44) (0)1222-565400, Fax (44) (0)1222-565400, E-mail ca@culinary-adventures.com, URL http://www.culinary-adventures.com.

WEST INDIES

COOKING IN PARADISE

St. Barthelemy, Gustavia/April

Cookbook author Steven Raichlen offers Cooking in Paradise, a 1-week hands-on culinary vacation in the Caribbean. Established in 1979. Maximum class/group size: 10. 1 programs per year. Facilities: Open-air kitchen. Also featured: Visits to top island restaurants, gourmet picnic, sailing cruise; swimming, fishing, scuba diving, shopping in the capital of Gustavia.

Emphasis: Healthy Caribbean cuisine.

Faculty: Cooking teacher and syndicated columnist Steven Raichlen writes for Food & Wine and the LA Times Syndicate and is author of IACP award-winner Miami Spice and the New Barbecue Bible.
Costs: $2,995 ($2,495 for non-cook guest), which includes meals, activities, and lodging at the Yuana Hotel, overlooking the ocean.
Location: St. Barthelemy's beaches and hills are 8 miles from the island of St. Maarten.
Contact: Barbara Raichlen, Cooking in Paradise, 1746 Espanola Dr., Miami, FL 33133 US; 305-854-9550, Fax 305-854-2232, E-mail 102333.1603@compuserve.com.

WORLDWIDE

BUTTERFIELD & ROBINSON

Year-round

This organizer of biking and walking trips worldwide offers 4- to 9-day trips that include daily bike rides and walks, cooking classes at restaurants, trattorias and culinary schools, culinary lectures, tastings, fine dining. Established in 1966. Also featured: Market tours, museum visits, sightseeing.
Emphasis: French and Italian cuisines, New England cuisine.
Faculty: Sarah Leah Chase, author of The Nantucket Open-House Cookbook, Cold-Weather Cooking, Pedaling through Provence & Pedaling through Tuscany.
Costs: 8-day Provence biking trip $6,750, 9-day Tuscany walking trip $6,850, 4-day Nantucket biking trip $3,975 includes deluxe lodging, meals, planned activities. Single supplement available.
Location: Provence , Tuscany, Nantucket.
Contact: Butterfield & Robinson, 70 Bond St., Toronto, ON, M5B 1X3 Canada; 800-678-1147/416-864-1354, Fax 416-864-0541, E-mail info@butterfield.com, URL http://www.butterfield.com.

COOKING LIGHT'S SHIP SHAPE CRUISE ADVENTURE

6 locations/October-November

Cooking Light magazine offers a 1-week cruise featuring cooking demonstrations, fitness programs, shore excursions, wine seminars. Established in 1996. Maximum class/group size: 100+. 1 programs per year. Facilities: Celebrity Cruises' Galaxy offers a full-service spa, 10,000-square-foot fitness center, 2 swimming pools.
Emphasis: Healthy living with emphasis on food, fitness, and the good life.
Faculty: Cooking Light editors.
Costs: From $1,124 includes shared cabin.
Location: Ports include Antigua, Barbados, Catalina, Martinique, St. Thomas, San Juan.
Contact: HMS Travel, 800-303-4800.

INTERNATIONAL DINING ADVENTURES

Florence/Year-round *(See also page 290, Cuisine International)*

This travel company offers 5- to 15-day tours that include 2 or 3 cooking lessons and visits to farms, markets, wineries, and other sites of interest to food enthusiasts. Established in 1996. Maximum class/group size: 20 maximum. Also featured: Dining in fine restaurants, visits to artisans' studios, sightseeing.
Emphasis: Authentic regional cuisine, role of food in culture.
Faculty: Judith Witts Francini, former pastry chef at San Francisco's Stanford Court Hotel; Marina Tudisco, owner of a cooking school in Sicily; Ernesto Pino, who studied with Wolfgang Puck; and other food experts.
Costs: $2,490-$4,190, which includes double occupancy lodging in 4-star hotels, ground transportation, most meals, planned activities.
Location: Includes Florence, Taormina, Barcelona, San Sebastian, Bangkok, Chiang Mai, Tokyo, New Orleans, Seattle.
Contact: Judy Ebrey, Director, International Dining Adventures, PO Box 25228, Dallas, TX 75225 US; 214-373-1161, Fax 214-373-1162, E-mail CuisineInt@aol.com, URL http://www.CuisineInternational.com.

3

Wine Courses

Taught by Members of the American Wine Society (AWS) and the Society of Wine Educators (SWE)

CALIFORNIA

MICHAEL A. AMOROSE
San Francisco/Year-round

First offered 1974. Twenty-five single 2-hour sessions per year. Enrollment 40-50 students per class. Specialties: California and Pacific Northwest. 9 wines per session. Vintages: Current. Price range $10-$40 per bottle. Source of wines: Local wine merchants. Instructor credentials: Instructor is author of 10 books on wine. Tuition $25-50/session. Class location: Meetings and conventions.

Contact: Michael A. Amorose, 555 California St., #1700, San Francisco, CA 94104; 415-951-3377, Fax 415-951-3296, E-mail michael.amorose@ey.com.

SAM (ROSARIO) ARMATO
Manhattan Beach/Fall and Spring

First offered 1975. Four 7-session courses per year. Enrollment 15 students per class. Specialties: Principal varieties of Europe and America, wine tasting for varietal identification and evaluation. 6-8 wines per session. Vintages: Vary according to availability. Price range $8-$30 per bottle. Source of wines: Wine shops. Instructor credentials: 35 years of tasting and evaluating wines, member of many wine societies including SWE and Les Amis du Vin, wine consultant to restaurants, wine importer. Tuition $125 per course. Class location: Varies.

Contact: Sam Armato, California Wine Productions, PO Box 3637, Manhattan Beach, CA 90266; 310-545-3877, Fax 310-545-5955, E-mail sarmato@earthlink.net.

MARIAN W. BALDY, PH.D.
Chico/Fall

First offered 1972. 45-session 3-credit course. Enrollment 80 students per class. Specialties: Sensory evaluation of California varietal wines. 30 wines per course. Vintages: 1984-current. Price range $8-$70 per bottle. Source of wines: The Wine Club and Safeway stores. Instructor credentials: Doctorate degree in genetics, winemaker, author of The University Wine Course, certified wine educator, SWE. Tuition $300 per course. Class location: California State University, Chico.

Contact: Marian W. Baldy, Ph.D., California State University, School of Agriculture, First and Normal Sts., Chico, CA 95929-0310; 916-895-5844, Fax 916-898-5845, E-mail mbaldy@oavax.csuchico.edu, URL http://www.csuchico.edu/agr/mbaldy/index.html.

JOHN BUECHSENSTEIN
Northern California sites/Year-round

First offered 1978. Eight to ten 1- or 2-day courses per year, several week-long seminars, special seminars by arrangement. Enrollment 50 students per class. Specialties: Sensory evaluation, winemaking; wine tasting tours of France. 30-100 wines per session. Vintages: Current to early 1980's, occasional rare old. Price range $5-$80 per bottle. Source of wines: Retailers, importers, donations. Instructor credentials: Winemaker, B.S. from UC-Davis, member AWS, IFT, SWE, ASEV. Tuition $150-$300 per course. Class location: University of California-Davis, Mendocino College, Culinary Institute of America at Greystone; also private seminars.

Contact: John Buechsenstein, Wine Education & Consultation, 309 Hillview Ave., Ukiah, CA 95482; 707-468-8245, Fax 707-468-8245, E-mail johnb@pacific.net.

ANN CIERLEY
Bakersfield/Year-round

First offered 1984. Two 6-week courses per year, 2- to 3-hour short course upon request. Enrollment 15-30 students class (short course unlimited). Specialties: California wines. 6-8 wines per session. Vintages: Current-1990. Price range $7.50-$50 per bottle. Source of wines: Stars Theatre Restaurant cellar, private cellars. Instructor credentials: Credentialed college instructor, wine consultant and wine buyer for restaurants and private cellars. Member SWE. 30 years experience in wine. Tuition $75/complete course. Individual group sessions by arrangement. Class location: Stars Theatre Restaurant for complete course. Your site for single session.

Contact: Ann Cierley, 5509 Muirfield Dr., Bakersfield, CA 93306; 805-871-8950.

DRAEGER'S CULINARY CENTER – PETER MARKS, M.W.

San Mateo/Year-round

First offered 1984. One to two 10-session courses per year. Enrollment 32 students per class. Specialties: Wines of the World: survey on important wines and wine-growing regions of the world. 8-12 wines per session. Vintages: Current. Price range $6.99-$120 per bottle. Source of wines: Worldwide. Instructor credentials: Master of Wine, Wine Director of Draeger's Marketplaces, member S.W.E.. Tuition $475 per course. Class location: Draeger's Culinary Center.

Contact: Draeger's Culinary Center, Registration, 222 E. 4th Ave., San Mateo, CA 94401; 650-685-3704, Fax 650-685-3728.

BARRY C. LAWRENCE, C.F.I.

Healdsburg/Monthly except June and July

First offered 1983. 14 courses per year; 6-day sessions, 4-day wine classes. Enrollment 48 students per class. Specialties: California wine history, North Coast Wine Country Sonoma, Napa, Lake and Marin. Lecture and tasting, videos, films and slides. 4-7 wines per session. Vintages: Current. Price range $4.75-$35 per bottle. Source of wines: Local retailers. Instructor credentials: Credentialed California Dept. of Education, President Nor-Cal. S.W.E., winery and vineyard owner. Tuition $425 includes meals, lodging and bus tour at local wineries and Redwood Forrest. Class location: Elderhostel, Bishops Ranch in California.

Contact: Barbara Crossland, Elderhostel, North America, 75 Federal St., Boston, MA 02110-1941; 617-426-8056 (catalog), Fax 707-433-3431, E-mail crosslan@juno.com, URL http://www.elderhostel.org.

FRED McMILLIN

San Francisco/Spring, Summer, Fall

First offered 1965. Five to six 3-session courses per year. Enrollment 15 students per class. Specialties: Wine history, California wine history, ranking the great varietals. 20 wines per session. Vintages: Current to 1980. Price range $5-$50 per bottle. Source of wines: Wineries, retail, privately made. Instructor credentials: Northern California editor for American Wine on the Web, two degrees in chemical engineering; offered teaching position in philosophy. Tuition $80-$100 per course. Class location: San Francisco State University-College of Extended Learning, San Francisco Community College-Ft. Mason Campus.

Contact: Fred McMillin, 2121 Broadway, #6, San Francisco, CA 94115; 415-563-5712, Fax 415-567-4468.

BOB MILLER

San Jose/Year-round

First offered 1974. Three 10- to 15-session courses per year. Enrollment 16 students per class. Specialties: Western US, French, German, Italian, Australian. 8-10 wines per session. Vintages: Current to 1960's, occasionally a pre-1960 vintage. Price range $3.50-$100+ per bottle. Source of wines: Retail wine stores, wineries, private cellars. Instructor credentials: California Community Colleges Teaching Credential for Food & Wine area, member SWE, started in wine business in 1960, past WINO Regional Director for over 16 years, M.S. in Counseling. Tuition $14-$200/person/meeting. Class location: San Jose and occasionally at Mission College in Santa Clara.

Contact: Bob Miller, 195 Castillon Way, San Jose, CA 95119-1502; 408-953-1441/225-4084, E-mail millerr@earthlink.net.

MR. STOX FOOD & WINE CLASSES – SCOTT RACZAC & RON MARSHALL

Anaheim/Spring and Fall

First offered 1980. 3-session classes/wine classes, 1-session classes/food classes. Enrollment 20 students/wine, 30 students/food/class. Specialties: California and French wines, food classes on all regions. 12 wines per session. Vintages: Vintages vary-more emphasis on varietals. Price range $7-$50 per bottle. Source of wines: Wine Spectator Grand Award Cellar. Instructor credentials: Chef Scott Raczak and wine expert Ron Marshall. Manage 22,000-bottle cellar, extensive independent study. Tuition $100/three wine classes, $40/food classes with lunch, $75/food classes with dinner. Class location: Private room at Mr. Stox Restaurant.

Contact: Debbie Marshall, Mr. Stox Restaurant, 1105 E. Katella Ave., Anaheim, CA 92805; 714-634-2994, Fax 714-634-0561, E-mail mrstox@mrstox.com, URL http://www.mrstox.com.

G.M. "POOCH" PUCILOWSKI

Sacramento/Year-round

First offered 1973. Ten to twelve 1 to 6-session courses per year. Enrollment 15-40 students per class. Specialties: California. 6-9 wines per session. Vintages: Current. Price range $5-$40 per bottle. Source of wines: Local wineries and wine stores. Instructor credentials: Past president of SWE; offers commercial courses to restaurants, wine retailers and wholesalers; chief judge and consultant California State Fair Wine Competition. Tuition $50-$100 per course. Class location: Local restaurants.

Contact: G.M. "Pooch" Pucilowski, 4595 College Oak Dr., Sacramento, CA 95041; 916-485-5550, Fax 916-485-5553, E-mail gmpooch@pacbell.net.

UNIVERSITY OF CALIFORNIA – DAVIS

Davis/September-June

First offered 1940. Ten 2 to 3-hr. classes/qtr. Enrollment 20-30 students per class. Specialties: Enology and viticulture. /session. Price range . Instructor credentials: Instructors all have Ph.D.'s. Tuition $4,500 per year in-state, $14,000 per year out-of-state. Class location: University of California-Davis.

Contact: Judy Hendrickson, University of California, Davis, Dept. of Viticulture & Enology, One Shields Ave., Davis, CA 95616; 530-752-8035, Fax 530-752-0382, E-mail jlhendrickson@ucdavis.edu, URL http://wineserver.ucdavis.edu.

DR. ALAN YOUNG

San Francisco/Year-round

First offered 1975. Home study programs. Enrollment 15 students per class. Specialties: Worldwide varieties; winemaking and growing; winery visits; 3-day seminars; study tours to wine areas of world. 8 wines per session. Vintages: Current to 1975. Price range $5-$150 per bottle. Source of wines: International. Instructor credentials: Australian wine consultant and author of 15 books, international faculty. Tuition Home study $250. Class location: Home, and various international sites.

Contact: Dr. Alan Young, International Wine Academy, 38 Portola Dr., San Francisco, CA 94131-1518; 415-641-4767, 800-345-8466, Fax 415-641-7348, E-mail ayoung@sirius.com, URL http://www.wineacademy.com.

COLORADO

B&C WINE ADEPT PROGRAM

Aspen/October, November, May, June

First offered 1990. Two 1-week sessions 4 times per year. Enrollment 10 students per class. Specialties: Broad general knowledge with emphasis on tasting. 6 wines per session. Vintages: Primarily current releases. Price range $10-$200 per bottle. Source of wines: Retail. Instructor credentials: Certified Wine Educator, 20+ years in restaurants. Tuition $1200/one session, $2000/both sessions. Class location: Client's location in Aspen, and nationwide.

Contact: Stephen Reiss, Ph.D., Buyers & Cellars Wine Consultants, Box 10206, Aspen, CO 81612; 970-923-6172, Fax 970-923-6670, E-mail wineguy@wineeducation.com, URL http://www.WineEducation.com.

THE VINEYARD WINE SEMINARS

Denver/Spring and Fall

First offered 1972. Six 3-session courses per year. Enrollment 40 students per class. Specialties: Basic Seminar, American Seminar, European Seminar. 5-8 wines per session. Vintages: Younger vintages. Price range $6-$20 per bottle. Source of wines: Current store inventory. Instructor credentials: 4-person staff has over 50 years cumulative wine experience. Tuition $60/person, $115/two people. Class location: Local restaurant.

Contact: Cheryl Lopez, The Vineyard, 261 Fillmore St., Denver, CO 80206; 303-355-8324, Fax 303-355-1413, E-mail viney158@vineyardwineshop.com, URL http://www.vineyardwineshop.com.

CONNECTICUT

COLONEL JACK E. DANIELS

Norwalk, Wilton, Ridgefield/September-June

First offered 1972. 8 sessions introductory, 7 sessions intermediate, 6 sessions advanced. Enrollment 29 students per class. Specialties: Introductory is based on Kevin Zraly and Steve Spurrier texts, Intermediate involves country by country analysis, Advanced features grape varieties. 5-8 wines per session. Vintages: Within past 10 years with exceptions. Price range $8-$35/ bottle with exceptions. Source of wines: Local wine shops. Instructor credentials: Former Executive Director, SWE, Regional Director Les Amis du Vin, Officer Chaine des Rotisseurs. Lived in California, Germany, France. Visited vineyards

throughout the world. Tuition $160. Class location: Varies from year to year. High schools, universities, community centers.
Contact: Jack E. Daniels, Wine Information, 561 Ridgebury Rd., Ridgefield, CT 06877; 203-744-WINE, Fax 203-743-2890, E-mail JEDani@aol.com.

ANDREW V. RANDI
New Haven/Fall and Spring

First offered 1981. Two 6 to 8-session courses per year. Enrollment 18 students per class. Specialties: French champagnes, intro to wine, US vs France, identify varietals (blind tastings). 6 wines per session. Vintages: Currently available. Price range $15-25/session. Source of wines: Local wine warehouses. Instructor credentials: . Tuition . Class location: Gateway Community Technical College.
Contact: Andrew V. Randi, Gateway Community Technical College, 60 Sargent Dr., New Haven, CT 06510; 203-867-6013.

RICK ROSS/ALL ABOUT WINE
Hartford/Year-round

First offered 1988. Varies. Enrollment 20 students per class. Specialties: Various. 5-8 wines per session. Vintages: Generally young, current vintages. Price range Varies. Source of wines: Varies. Instructor credentials: Member SWE, 20+ years of study. Tuition Varies. Class location: Mostly state community colleges, some town-sponsored adult-ed.
Contact: Rick Ross, All About Wine, 45 Rushford Meade, Granby, CT 06035-2325; 860-653-6057, Fax 860-653-4399, E-mail redwine5@earthlink.net.

EUGENE J. SPAZIANI – AMENTI DEL VINO
New Haven, Manchester, Norwich, Mystic, Darien/Spring and Fall

First offered 1973. Four to ten 1- to 5-session courses per year. Enrollment 20 students per class. Specialties: Basic wine knowledge, history, wine fundamentals, all wine producing regions of the world. 6-12 wines per session. Vintages: Nouveau to 1956. Price range $5-$200 per bottle. Source of wines: Retail outlet. Instructor credentials: College graduates, usually M.S. degrees and a Ph.D. Tuition $30-$45/session. Class location: Community colleges, hotels, elks, community centers.
Contact: Eugene J. Spaziani, Amenti del Vino, 57 East Main Street, Mystic, CT 06355; 860-536-0249, Fax 860-536-7224, E-mail genespaz@aol.com.

WINE WANDERINGS, INC./LOU CAMPOLI AND CATHI CARROLL
Norwalk/Spring and Fall

First offered 1993. 5-8 classes per year. Enrollment 36 students per class. Specialties: All countries/regions of the world, viticulture, vinification, food and wine, wine travel. 10-14 wines per session. Vintages: All, but generally current vintages. Price range $6-$100 per bottle. Source of wines: Retail stores and wineries. Instructor credentials: Both instructors: 15 years wine education, enrolled in diploma level of Wine & Spirit Education Trust (UK), Board of CT chap. AIWF, member IACP, member SWE, wine column journalists, 30+ yr corp. career. Tuition $30-$50/class. Class location: Norwalk Community-Technical College.
Contact: Lou Campoli and Cathi Carroll, Wine Wanderings, Inc., 192 Gillies Lane, Norwalk, CT 06854; 203-853-9550, Fax 203-853-9550.

DELAWARE

UNIVERSITY OF DELAWARE
Newark/Spring

First offered 1995. One 30-session course per year. Enrollment 30 students per class. Specialties: General focus. /session. Vintages: Varies. Price range Varies. Source of wines: Local distributors. Instructor credentials: Varies per instructor. Tuition $515. Class location: University of Delaware.
Contact: Paul Wise, University of Delaware, Hotel, Restaurant & Institutional Management, Amy Rextrew House, 321 S. College Ave., Newark, DE 19716-3365; 302-831-6077, Fax 302-831-6395, E-mail Paul.Wise@mvs.udel.edu, URL http://www.udel.edu/HRIM.

FLORIDA

LAURENCE J. SAUTER, PH.D

Ormond Beach/Fall, Winter, Spring

First offered 1978. Two 12-session courses per year. Enrollment 40-50 students per class. Specialties: . 2-3 wines per session. Vintages: Current to 1970's. Price range $5-$25 per bottle. Instructor credentials: PhD Bus.& Public Administration, Professor of Engineering, member SWE. Tuition $50 per course.

Contact: Laurence J. Sauter, Ph.D, 3639 Conifer Ln., Ormond Beach, FL 32174; 904-677-3488.

GEORGIA

YVES DURAND

Roswell/Winter

First offered 1984. One session/week on-going; European wine trips. Enrollment 15 students per class. Specialties: European varieties. 10-12 wines per session. Vintages: 1985-1960's. Price range $70-$250 per bottle. Source of wines: Own cellar. Instructor credentials: Best Sommelier USA 1984, 3rd Best Sommelier World 1985, former president Sommelier Soc. America, TV show host, Emmy Award Winner for Best Performer for TV serial Wining & Dining with Yves Durand. Tuition $60-$100/class. Class location: Own home.

Contact: Yves Durand, 400 Hollyberry Dr., Roswell, GA 30076; 770-993-4337, Fax 770-992-8484.

ANITA LOUISE LA RAIA

Atlanta/January, March, May, September

First offered 1978. Four 6-session Basic Diploma courses per year. plus advanced classes; weekend classes for hospitality industry/culinary schools for CEUs, private classes. Enrollment 70 students per class. Specialties: Worldwide varieties, home study course book, audio and video casettes. 6 wines per session. Vintages: Current to 1970. Price range $6-$80 per bottle. Source of wines: Local wine stores. Instructor credentials: Certficate from the Wine & Spirits Guild of Great Britain, member SWE. Tuition $225/Basic six-session course. Class location: Kennesaw/State/University, Dept. of Cont. Educ., Wyndham Garden Hotel Buckhead, and the Wine Center.

Contact: Anita Louise LaRaia, The Wine School, P.O. Box 52723, Atlanta, GA 30355; 770-901-9433, E-mail anitalaraia@mindspring.com.

DONALD REDDICKS

Atlanta/Fall, Winter, Spring

First offered 1979. Three 3-session courses per year. Enrollment 20 students per class. Specialties: German wines. 8-10 wines per session. Vintages: Current. Price range $8-$20 per bottle. Source of wines: Retail stores. Instructor credentials: Certified instructor SWE, member AWS and GWS. Tuition $100 per course. Class location: Goethe Institute, local hotels and restaurants.

Contact: Donald Reddicks, Harry's Farmers Market, Alpharetta, 7055 Hunters Branch Dr., Atlanta, GA 30328; 770-393-4584, 770-664-6300, Fax 770-772-9050.

ILLINOIS

CUISINE COOKING SCHOOL

Moline/September-June

First offered 1985. Varies; 4-hr classes may be taken separately. Enrollment 10 students per class. Specialties: Chardonnay, Riesling, Cabernet Sauvignon and Pinot matched with food. 5-6 wines per session. Vintages: About a 6-yr range, varies with grape. Price range $7-$30 per bottle. Source of wines: Wine shop, collected in travels. Instructor credentials: Member SWE, Internal Assoc. of Cooking Professionals, Iowa State University Graduate in Food Science. Tuition $45/class. Class location: Cuisine Cooking School.

Contact: Mary Sue Salmon, Cuisine Cooking School, 1100 23rd Ave., Moline, IL 61265; 309-797-8613, Fax 309-797-8641.

PATRICK W. FEGAN

Chicago/Year-round

First offered 1975. Twenty-two 5-week courses, forty-five single evening seminars. Enrollment 6-42 students per class. Specialties: Worldwide varieties. 6-10 wines per session. Vintages: Current to 1986, older for advanced classes. Price range $4-$75 per bottle. Source of wines: Area retail shops. Instructor credentials: Member SWE. Tuition $140-$250/five-week session; $40-$60/single evening seminar. Class location: Chicago Wine School.

Contact: Patrick W. Fegan, Chicago Wine School, 312-266-9463, Fax 312-266-9769, E-mail pwfegan@aol.com, URL http://www.wineschool.com.

GREAT LAKES WINE

Chicago/Year-round

First offered 1985. 6 courses per year. Enrollment 10 students per class. Specialties: Germany, France. 12 wines per session. Vintages: Varies. Price range Varies. Source of wines: Great Lakes Wine Co.. Instructor credentials: . Tuition No charge. Class location: Great Lakes Wine Co..

Contact: Gregory D. Miller, Great Lakes Wine, 1927 N. Milwaukee Ave., Chicago, IL 60647; 773-489-2112, Fax 773-489-5801, E-mail auslese@aol.com.

IRENE HUFFMAN

Moline, Rock Island/Fall, Spring

First offered 1984. Two 4-session courses per year. Six 1-subject sessions per year. Enrollment 10-15 students per class. Specialties: Worldwide varieties. 8-10 wines per session. Vintages: Range of vintages. Price range $5-$50 per bottle. Source of wines: Purchased. Instructor credentials: Wine judge for CA State Fair, wine columnist for 3 local newspapers. Tuition $95 per course, $25-$40/single subject session. Class location: Blackhawk College Outreach Center and Plaza One Hotel, Rock Island.

Contact: Irene Huffman, 1225 W. 5th St., Moline, IL 61264; 309-787-6941, Fax 309-787-0962.

KENTUCKY

SCOTT HARPER

Louisville/Year-round

First offered 1988. Numerous sessions and courses available. Enrollment 12 students per class. Specialties: Tasting (sensory evaluation), Italy, California, France, food and wine pairing. 6-8 wines per session. Vintages: Wide range. Price range $8-$50 per bottle. Source of wines: Local wholesalers. Instructor credentials: 12 years as wine buyer in restaurants, Sterling Vineyards School of Hospitality, state fair wine judge. Tuition $20-$100 per course. Class location: Bristol Bar & Grille.

Contact: Scott Harper, Bristol Bar & Grille Inc., 300 N. Hurstbourne Pkwy., Louisville, KY 40222; 502-426-9318, E-mail wscottharper@msn.com.

LOUISIANA

FORREST K. DOWTY

Lafayette/Fall and Spring

First offered 1978. 8-session beginner course in spring, 6-session advanced wine and food course in fall. Enrollment 50 students per class. Specialties: Worldwide varieties, occasional wine trips. 8-10 wines per session. Vintages: Wide range. Price range $2-$100 per bottle. Source of wines: Wholesaler. Instructor credentials: Owns Magnolia and Reliable Marketing of Lafayette (wine wholesaler), SWE Certificate of Proficiency. Tuition $100-$150.

Contact: Forrest K. Dowty, Box 3587, Lafayette, LA 70502; 318-233-9244, Fax 318-261-3570.

MARYLAND

THE FLAVORS IN THE GLASS AND HOW THEY GET THERE

Baltimore/March-May

First offered 1997. One 10-session seminar held each Monday. Enrollment 35 students per class. Specialties: Global wines. Components, aromas, wine making, grape growing, barrels, bottles and corks,

flavor continuum, varietals and blends, sherry, port and champagne. 6-20+ wines per session. Vintages: 1982 to recent releases and barrel samples. Price range $7.50-$70 per bottle. Source of wines: Barrel samples and library wines not available in stores. Instructor credentials: Wine Manager of F.P. Winner (wholesaler), AWS Certified Wine Judge, Certfied Wine Educator. Tuition $350. Class location: F.P. Winner Ltd.
Contact: Lisa M. Airey, F.P. Winner Ltd., 1101 Desoto Rd., Baltimore, MD 21223; 410-646-5500, x103, Fax 410-646-6464, E-mail winnerwine@aol.com.

MASSACHUSETTS

MICHAEL APSTEIN, M.D.

Boston/Year-round except summer

First offered 1980. Five 6-session courses per year. Enrollment 30 students per class. Specialties: Worldwide varieties. 5 wines per session. Vintages: Current to 1983. Price range $5-$30 per bottle. Source of wines: Retail stores. Instructor credentials: Wine writer and educator for 16 years, wine editor of Grand Diplome Cooking Course, freelance wine writer The Boston Globe, national competition judge, gastroenterologist specializing in liver disease. Tuition $142 per course. Class location: Boston Center for Adult Education.
Contact: Michael Apstein, M.D., Boston Center for Adult Education, 5 Commonwealth Ave., Boston, MA 02116; 617-267-4430.

ROGER ORMON AT BROOKLINE LIQUOR MART, INC.

Allston (Boston)/Year-round

First offered 1993. Weekly (every Saturday). Enrollment No limit/class. Specialties: All regions, imported (Bordeaux and Burgundy) are a specialty. 8 wines per session. Vintages: All. Price range $5-$100 per bottle. Source of wines: Our stock. Instructor credentials: Certified Wine Educator, Instructor of Wine since since 1981 (Harvard and Boston U. extension courses). Tuition Free. Class location: Brookline Liquor Mart.
Contact: Roger Ormon, Brookline Liquor Mart, Inc., 1354 Commonwealth Ave., Allston (Boston), MA 02134; 617-734-7700/800-BLM-WINE, Fax 617-232-5725, E-mail Roger@BLMWine.com, URL http://www.BLMWine.com.

GARY SANDMAN

Springfield/Year-round

First offered 1988. 12 monthly 1-night sessions per year. Enrollment Open/class. Specialties: Topics vary each month. 10-12 wines per session. Vintages: Mostly current. Price range $5-$40 per bottle. Source of wines: Kappy's Liquors. Instructor credentials: 15+ years as a wine retailer and educator. Tuition $20/session. Class location: Springfield Marriott.
Contact: Gary Sandman, Kappy's Liquors, 1755 Boston Rd., Springfield, MA 01129; 413-543-4495, Fax 413-543-4414.

WESTPORT RIVERS LONG ACRE HOUSE

Westport/Year-round

First offered 1996. 10 wine and food classes, 4-6 beer and food classes. Classes can be customized for private groups. Enrollment 35 students per class. Specialties: Four Seasons, a series highlighting vineyard and winery activities, paired with foods for each season. Guest chefs and speakers, cooking demonstrations. 2-3 wines per session. Vintages: Current, library selections, vertical tastings. Price range $10-$35 per bottle. Source of wines: All Westport Rivers' wines. Instructor credentials: Exec. Chef, New England Culinary Institute degree, grad. of Madeleine Kamman's School for American Chefs at Beringer Vineyards, former Sous Chef at Beringer Vineyards, food writer and cookbook author. Tuition $40/class. Class location: Westport Rivers Long Acre House (the Wine & Food Education Center at Westport Rivers).
Contact: Kerry Downey Romaniello, Westport Rivers Vineyard & Winery, 417 Hixbridge Rd., Westport, MA 02790; 508-636-3423, x7, Fax 508-636-4133, URL http://www.westportrivers.com.

MICHIGAN

TERRENCE G. DUNN

Grand Rapids/Fall

First offered 1984. One 16-week course per year. Enrollment 20 students per class. Specialties: France, Germany, Italy, East and West Coast United States. 5 wines per session. Vintages: Current. Price range $8-$25 per bottle. Source of wines: Local wholesalers. Instructor credentials: Member SWE, Certified Wine Educator, F & B Instructor Grand Rapids Community College. Tuition $55/credit hour plus $20 lab fee. Class location: Grand Rapids Community College, Hospitality Education Department.

Contact: Terrence G. Dunn, Grand Rapids Community College, 151 Fountain NE, Grand Rapids, MI 49503; 616-771-3690/3705, Fax 616-771-3698.

MINNESOTA

GEORGE RENIER

Duluth/Fall and Spring

First offered 1979. Two 8 to 10-session courses per year. Enrollment 18-35 students per class. Specialties: Winemaking at home; dandelion, sweet potato, grape (concord) and others. 3 wines per session. Vintages: 1978, 1983, 1987, 1995 and others. Price range . Source of wines: Over 400 gallons in private wine cellar. Instructor credentials: Judge AWS, started making wine in 1955. Tuition $15-$20 per course + equipment fee. Class location: Woodland Jr. High School.

Contact: George Renier, 2418 E. 4th St., Duluth, MN 55812; 218-724-2558.

MISSOURI

MILLARD S. COHEN

St. Louis/Year-round

First offered 1981. 24 courses per year. Enrollment 18 students per class. Specialties: Wine appreciation. 6 wines per session. Price range $10+ per bottle. Source of wines: Varied. Instructor credentials: Member SWE, Missouri Wine Advisory Board, International Wine & Food Society. Tuition $25/couple. Class location: Dierberg's Cooking Schools in Creve Coeur, Southroads, Chesterfield, and Mid-Rivers.

Contact: Millard S. Cohen, P.O. Box 419050, St. Louis, MO 63141-9050; 314-872-8500, Fax 314-872-8517, E-mail milco@bigfoot.com.

CORKDORK U. – THE "NO ATTITUDE" WINE SCHOOL

St. Louis/Year-round

First offered 1998. Single-session and multiple-session programs. Enrollment 20 students per class. Specialties: Wine classes from beginner to connoisseur. Emphasis on food and wine matching, and making wine choices. 6-8 wines per session. Vintages: Current-1978. Price range $6-$200 per bottle. Source of wines: Retail stores. Instructor credentials: Instructor, Ed Deutch is a member of the SWE, has taught wine classes at Univ. of Minnesota's Open U., and is a regular wine speaker for banquets, corporate programs and social groups for past 20 years. Tuition $30-$60/session. Class location: A turn of the century house in St. Louis' Central West End.

Contact: Ed Deutch, Corkdork U.-The Vintage Room Wine Store, 4736 McPherson Ave., St. Louis, MO 63108; 314-FOR-WINE/367-9463, E-mail eddie@corkdork.com, URL http://www.corkdork.com.

NEW JERSEY

ADULT SCHOOL OF THE CHATHAMS, MADISON, AND FLORHAM PARK

Morristown/Spring and Fall

2 courses per year. Enrollment 25 students per class. Specialties: Wine examined from a varietal perspective. 8 wines per session. Price range $10-$20 per bottle. Source of wines: Purchased at retail stores. Instructor credentials: Walter A. Kotrba, retailer/wine writer for 12 years and member SWE. Tuition . Class location: Pierre's Restaurant.

Contact: Walter Kotrba, Shop Rite Wines & Spirits, 120 Rt. 206 South, Chester, NJ 07930; 908-879-5352, Fax 908-879-7295, E-mail wakotrba@interpow.net, URL http://www.users.interpow.net/~wakotrba.

L.B. BRATTSTEN
New Brunswick/Spring and Fall

First offered 1992. Two 14-session courses. Enrollment 65 students per class. Specialties: All regions. 36 wines per session. Vintages: 1998-1978. Price range $6-$60 per bottle. Source of wines: Retail stores. Instructor credentials: AWS Certified Judge. Tuition . Class location: Rutgers University.

Contact: L.B. Brattsten, Rutgers University, Dept. of Entomology, New Brunswick, NJ 08901-8536; 732-932-8166, Fax 732-932-7229, E-mail brattsten@aesop.rutgers.edu.

JOSEPH FIOLA, PH.D
Cream Ridge/Fall, Spring, Summer

First offered 1989. Two to three 1-session courses per year. Enrollment 50-100 students per class. Specialties: Northeastern U.S.. 3-8 wines per session. Price range . Instructor credentials: PhD. horticulture, won AWS gold and bronze awards for wines produced. Tuition $50 per course.

Contact: Joseph Fiola, Ph.D, Rutgers Fruit R&E Center, 283 Rte. 539, Cream Ridge, NJ 08514; 609-758-7311, x17, Fax 609-758-7085, E-mail fiola@aesop.rutgers.edu.

CONNIE FOWLER
Summit/Spring, Fall, Winter

First offered 1995. Four to six 3- to 6-session courses per year. Enrollment 35 students per class. Specialties: Introduction to Wine Appreciation, Food & Wine Pairing, Chardonnay, Cabernet Sauvignon, California, Italy. 6 wines per session. Vintages: Recent (what is available on the retail market). Price range $6-$60 per bottle. Source of wines: Retail shops. Instructor credentials: Member SWE, 10 years wine industry experience. Tuition Varies. Class location: Summit, NJ area restaurants. Some classes are held through Summit Area Adult School.

Contact: Connie Fowler, 16 Sherman Ave., Summit, NJ 17901; 908-277-1348, Fax 908-277-1348, E-mail rsfowler@eclipse.net.

BOB LEVINE AND LINDSEY CHURCHILL
Princeton/Fall and Spring

First offered 1971. Two 5-session courses per year. Enrollment 32 students per class. Specialties: Introduction to Wine Appreciation or specialized course on a particular region such as New Wines of Italy or Wines of the Languedoc and Rousselin. 5-8 wines per session. Vintages: Various. Price range From modest to very high. Source of wines: Wine stores. Instructor credentials: Founder/1st Pres. of Society of Wine Educators, teacher at Princeton Adult School for 25+ years, seminars for wine professionals. Lindsey Churchill, student of Bob Levine's and co-instructor past 5 years. Tuition $150 for most courses. Class location: Princeton Adult School.

Contact: Bob Levine, Princeton Adult School, Box 701, Princeton, NJ 08540; E-mail rjl@gurus.com.

JOHN J. MAHONEY/TRI-STATE WINE COLLEGE
Hammonton/October, April, May

First offered 1980. 6-session intro courses, 2-session advanced courses and wine tours. Enrollment 40 students, 15 students/advanced course/class. Specialties: Wine making, tasting techniques, faulty wines, every country's wines. 9-12 wines, 10-15 wines/advanced course/session. Vintages: 10 years for intro courses, vertical seminars. Price range $7.50-$50/bottle (intro), $50-$200/bottle (advanced). Source of wines: Retail for intro, private cellars for advanced. Instructor credentials: B.A., M.A., Ph.D., Certified Wine Educator, Professional Wine Judge in many states and Spain, wine instructor for 20 years. Tuition $189/regular course including graduation dinner, $30-$90/advanced course. Class location: Tri-State Wine College at Tomasello's Winery. Lecture at New Jersey universities on wine.

Contact: John J. Mahoney, Tri-State Wine College, Box 100, High Street, Milmay, NJ 08340; 609-476-2728, Fax 609-476-2728, E-mail j.mahoney@juno.com.

STAGE LEFT WINE COURSE – FRANCIS SCHOTT
New Brunswick/January

First offered 1990. One or two 8-session courses. Enrollment 50 students per class. Specialties: Estate bottled wines from small producers. 12 wines per session. Vintages: Current to 1961. Price range $10-$600 per bottle. Source of wines: Purchased from wholesalers. Instructor credentials: Member SWE, member American Sommelier Society, Owner/Wine Director of NJ Restaurant and wine catalog. Tuition $450. Class location: Stage Left: An American Cafe.

Contact: Francis Schott, Stage Left: An American Cafe, 5 Livingston Ave., New Brunswick, NJ 08901; 722-828-4444, Fax 722-828-6228, E-mail CabFrancis@aol.com, URL http://www.stageleft.com.

NEW YORK

SILVI FORREST

New York/Year-round

First offered 1982. Wine tasting classes by arrangement for professional organizations, corporate groups and graduate students at Universities. Enrollment 15-30 students per class. Specialties: Presented by grape variety, region or country. 6-8 wines per session. Price range $8-$50 per bottle. Source of wines: Vintners, distributors, retailers. Instructor credentials: Member SWE, writer food and wine articles/catalogs, advertising/marketing for leading wine retailers and wine associations. Tuition Tuition priced by event. Class location: NY, NJ, CT.

Contact: S. Forrest, 303 West 66th St., New York, NY 10023; 212-874-0683.

GREG GIORGIO

Altamont/Year-round

First offered 1987. Five to six single session seminars per year. Enrollment 20 students per class. Specialties: Rhone/Provence, Red Zinfandel. 6-7 wines per session. Vintages: Current. Price range $5-$30 per bottle. Source of wines: Local wine shops. Instructor credentials: Member SWE. Tuition $10-$20/session. Class location: Local restaurants.

Contact: Greg Giorgio, P.O. Box 74, Altamonte, NY 12009; 518-861-5627.

INTERNATIONAL WINE CENTER

New York/Year-round

First offered 1981. Beginner 7-session Certificate Course, intermediate 14-session Higher Certificate Course, advanced 2-yr Diploma Course. Enrollment 30-50 students per class. Specialties: Worldwide. /session. Vintages: Various. Price range . Instructor credentials: All instructors with Wine & Spirit Education Trust (WSET) Diploma; Master of Wine Mary Ewing Mulligan. Tuition $388-$1,550. Class location: Chelsea Wine Vault.

Contact: Susan Rusbasan, 1133 Broadway, Ste. 520, New York, NY 10010; 212-627-7170, Fax 212-627-7116, E-mail IWCNY@aol.com.

RONALD A. KAPON

New York/Fall, Spring, Summer

First offered 1969. Two 6-session courses per year, 25-30 dinners and wine tastings per year. Enrollment 20-100 students per class. Specialties: Worldwide varieties. 6-12 wines per session. Price range $6-$150 per bottle. Source of wines: Purchased. Instructor credentials: Ph.D. Econ., Sr. Bureau Chief East-West News Bureau; grad. German Wine Academy; vp/co-dir. Tasters Guild Intl., NY Chapter; co-founder Travel World newsletter; prof. FDU Hotel, Rest. & Tourism School. Tuition $100 per course, $20-$100 for dinners and wine tastings. Class location: Queens College Continuing Education Dept., Flushing, (718) 997-5700; Gramercy Park Hotel, NYC; Wine Workshop, NYC.

Contact: Ronald A. Kapon, 230 W. 79th St., #42N, New York, NY 10024-6210; 212-799-6311, Fax 212-799-0245.

HARRIET LEMBECK

New York/Fall and Spring

First offered 1975. Two 10-session courses per year and 4-session spirits course in fall; over thirty 1-4-session courses per year through New School. Enrollment 32 students per class. Specialties: Worldwide varieties, advanced courses on specific subjects. 6-10 wines per session. Vintages: Current, some older. Price range $7.99-$150. Source of wines: Retailers, importers. Instructor credentials: Author of Grossman's Guide to Wines, Beers, & Spirits, 6th & 7th Eds.; Director of New School wine program; Charter Director, SWE. Tuition $525/10-session course; $700 with spirits ($200 spirits alone); $60-$200 per course through New School. Class location: New School affiliation; all classes taught in a new wine classroom in a historic building at 203 E. 29th St., between 2nd and 3rd Ave.

Contact: Harriet Lembeck, Wine & Spirits Program, 54 Continental Ave., Forest Hills, NY 11375-5229; 718-263-3134, Fax 718-263-3750.

HARVARD LYMAN
Stony Brook/Fall

First offered 1972. One 14-session course per year. Enrollment 25-30 students per class. Specialties: All regions, wine making, appreciation. 4-6 wines per session. Vintages: Current to late 1990's. Price range $5-$55 per bottle. Source of wines: Local cooperating merchants. Instructor credentials: Member AWS and SWE. Tuition $270/three credits plus $40-$45 lab fee; $90/one credit option. Class location: SUNY-Stony Brook.

Contact: Harvard Lyman, SUNY-Stony Brook, Biochemistry Dept., Stony Brook, NY 11794-5215; 516-632-8534, Fax 516-632-9780, E-mail hlyman@allinl.cc.sunysb.edu.

DAVID G. MALE
Williamsville/Fall and Spring

First offered 1986. Two 10-session courses per year. Enrollment 30 students per class. Specialties: New and old world wines. 8-10 wines per session. Vintages: Current to 1970. Price range $15-$60 per bottle. Source of wines: Local retailer. Instructor credentials: Graduate of German Wine Acad. and Scholarship of Wine Program, member AWS and SWE, president InterVin Intl. Tuition $235 per course. Class location: Eagle House restaurant in affiliation with Amherst Continuing Ed.

Contact: David G. Male, Vintage House, 441 Sprucewood Terr., Williamsville, NY 14221-3910; 716-634-2456, Fax 716-634-7061, E-mail vinehous@aol.com.

NEW YORK WINE TASTING SCHOOL
New York/Spring, Fall, Winter

First offered 1975. 8-week courses. Enrollment 12 students per class. Specialties: All wines tasted blind, plus cordials, liqueurs, and spirits appropriate to the area. country wines, Italy, US, Bordeaux, Burgundy, sparkling wines. 12 wines per session. Vintages: 1997-1979. Price range $7-$135 per bottle. Source of wines: Purchased from merchants or from own cellar. Instructor credentials: Ms. Rundel and Mr. Sheldon are SWE members. Mr. Sheldon is a CWE (Certified Wine Educator). Tuition $350 per course. Class location: Michael's of Broadway Restaurant, (Wall Street area, NYC).

Contact: John Sheldon or Judy Rundel, New York Wine Tasting School, 32 Bergen St., Brooklyn, NY 11201; 718-852-4839, Fax 718-852-4839, E-mail jrundel@aol.com.

DAVID PERRY – CHARMER INDUSTRIES
New York metro area/Year-round

First offered 1993. Four 7-session courses per year. Enrollment 24 students per class. Specialties: General. 10 wines per session. Vintages: Varies. Price range $6-$60 per bottle. Source of wines: Suppliers. Instructor credentials: Master of Arts in Teaching, Certified Wine Educator (C.W.E.), WSET diploma candidate. Tuition $195-may vary. Class location: Various locations in the metropolitan NY area.

Contact: David Perry, Charmer Industries, 48-11 20th Ave., Astoria, NY 11105; 718-545-7400, Fax 718-726-2385, E-mail dperry@charmer.com.

TAO PORCHON-LYNCH
White Plains/Fall and Winter

First offered 1982. Two 8-session courses per year. Enrollment 24 students per class. Specialties: Worldwide varieties, wine and food seminars; wine tours to France, Italy, Spain/Portugal, Australia, Argentina/Chile. 8 wines per session. Vintages: Range of vintages. Price range $10-$150/bottle (avg. $30). Source of wines: Importers, wine stores, own cellar. Instructor credentials: Ed. of The Beverage Communicator; Regional V.P. and founder AWS; family owned vineyard in France; Cert. wine judge, member SWE. Wine/food tours to France, Italy, Australia, S. America. Tuition $280 per course. $55-$75/five-course dinner and wine. Class location: Westchester Marriott Residential Inn.

Contact: Tao Porchon-Lynch, Les Amoureux du Vin, 5 Barker Ave., #501, White Plains, NY 10601; 914-761-7700 #2501, 997-0949, Fax 914-997-2617.

HERBERT F. SPASSER, D.D.S.
New York/Summer, Winter

First offered 1976. Two 4-session courses per year. Enrollment 14 students per class. Specialties: An Adventure into Wine. 5-6 wines per session. Vintages: Current to 1960's. Price range $10-$75. Source of wines: Purchased and samples. Instructor credentials: Wine advisor to Restaurant Society of N.Y., member of Confrerie de la Chaine des Rotisseurs; SWE Certified wine educator. Tuition $230 per course. Class location: Office in Manhattan and home in Westfield, NJ.

Contact: Herbert F. Spasser, D.D.S., 116 Central Park South, #8, New York, NY 10019; 212-765-1877.

TASTING AND SERVING WINE – JANE UTZ

Long Island/Fall

First offered 1989. One 8-session course per year. Enrollment 20 students per class. Specialties: History, geography, fermentation, world wine producers as countries, wine and food, wine and health, grape varieties, viticulture, tasting techniques. 6-10 wines per session. Vintages: Current to 1980s. Price range $4-$60 per bottle. Source of wines: Vineyard visits, liquor stores, gifts, travel. Instructor credentials: Certified teacher in NY State (25 years in public school system), member SWE, American Wine Society, LI wine industry since 1976 (Hargrave Vineyard). Tuition $200 per course. Class location: Use of a different vineyard site each year. 1998: Macari Vineyard.

Contact: Jane Utz, 2705 Laurel Ave., Southold, NY 11971; 516-765-5731, E-mail JanePake@aol.com.

THE WINE SCHOOL AT WINDOWS ON THE WORLD

New York/Spring, Summer, Fall

First offered 1976. Four 8-session courses per year. Enrollment 100 students per class. Specialties: U.S., European. 10 wines per session. Vintages: Wide range. Price range $7-$150 per bottle. Source of wines: Windows on the World wine cellar. Instructor credentials: Wine Director of Windows on the World; author of Windows on the World Complete Wine Course, 1993 James Beard Wine & Spirits Professional of the Year, co-host TV Food Network Show Wines A to Z. Tuition $650 per course. Class location: World Trade Center.

Contact: Kevin Charles Zraly, The Wine School at Windows on the World, 16 Woodstock Lane, New Paltz, NY 12561; 914-255-1456, Fax 914-255-2041.

WINE WORKSHOP

New York/September-December, February-June

First offered 1991. 100 classes per year. Enrollment 18-40 students per class. Specialties: Essentials of Wine, regionals, hard-to-find wines. 12-15 wines per session. Vintages: Current to 1961. Price range $40-$400 per bottle. Source of wines: Own cellars, auctions, confirmed sources. Instructor credentials: Michael Capon, second generation wine merchant, and his son John; guest winemakers and importers. Tuition $45-$295/class. Class location: The Culinary Loft, 515 Broadway in SoHo. Also: Luxury dinners (18 persons max) at top restaurants range from $195 to $1,500/person.

Contact: Jana Kravitz, Acker Merrall and Condit, 160 W. 72nd St., 2nd Flr., New York, NY 10023; 212-875-0222, Fax 212-799-1984, E-mail winewkshop@aol.com, URL http://www.ackerbids.com.

WINES FOR FOOD

New York/Year-round

First offered 1981. Four 1-session courses rotate every six weeks. Enrollment 50-100 students per class. Specialties: White and red wines, wines of Italy and France; The Ultimate Champagne Class (Nov-Dec), private and corporate events, 7-course wine/food demos. 33 wines per session. Vintages: Various. Price range $3.99-$50 per bottle. Source of wines: Purchased. Instructor credentials: Willie Gluckstern has written wine lists for 200+ Manhattan restaurants, teaches at Peter Kump's New York Cooking School, is author of The Wine Avenger (Simon & Schuster). Tuition $55/class. Class location: Park Central Hotel, 870-7th Ave. at 56th St..

Contact: Willie Gluckstern, Wines for Food, 158 W. 76th St., New York, NY 10023; 212-724-3030, Fax 212-501-0717, URL http://www.wineavenger.com.

OHIO

MATTHEW CITRIGLIA

Columbus/Year-round

First offered 1988. Two 1 to 5-session courses per year. Enrollment 5 students minimum/class. Specialties: Basic grape varieties and where grown; Italy, France, Spain; fortified, sparkling. 5-6 wines per session. Vintages: Varies. Price range $8-$35 per bottle. Instructor credentials: AWS certified judge, 14 years in wine business, Segrams School of Hospitality graduate. Tuition $35-$45 per course. Class loca-

tion: Vintage Wine in Columbus.
Contact: Matthew Citriglia, The Wine Mentor, 222 Hanford St., Columbus, OH 43206; 888-846-8023, x161, Fax 614-876-1038, E-mail winegeek@via.net.

JOHN F. KEEGAN – MIAMI UNIVERSITY BOTANY DEPARTMENT
Oxford/September-December, January-May

First offered 1995. 1 section in Fall and 2 sections in Spring. Enrollment 90 students per class. Specialties: History, viticulture and enology, health effects, major varieties, oak and cork, overview of the major wine making regions. 3-4 wines after 1st few sessions/session. Vintages: Usually recent. Price range $3-$50 per bottle. Source of wines: Worldwide. Instructor credentials: Certified Wine Educator, Masters in Horticulture, German Wine Academy. Tuition $90 for wines, college credit available.
Contact: John F. Keegan, Miami University, Botany Dept., 316 Pearson Hall, Oxford, OH 45056; 513-529-4200, Fax 513-529-4243, E-mail Keeganjf@muohio.edu.

JAMES R. MIHALOEW – THE CLEVELAND WINE LINE
Cleveland/Strongsville area/Year-round

First offered 1984. Four 5- to 12-session courses. Enrollment 5-15 students per class. Specialties: The Wine Course: basic apprec. Sensations in Wine: advanced sensory eval. Explorations in Wine with Food: wine-food pairing, interactions. Tastings. 4-8 wines per session. Vintages: Currently available. Price range $9-$30 per bottle. Source of wines: Retail stores, occasionally own cellar. Over 500 titles. Instructor credentials: Wine Judge Training (AWS) first year course coordinator and instructor, AWS Life Member, member SWE, teaching experience and wine judging. Tuition $100-$240 per course. Class location: Various locations.
Contact: James R. Mihaloew, 13463 Atlantic Road, Strongsville, OH 44136; 440-238-4184.

GARY L. TWINING
Bainbridge, Euclid, Chesterland/Spring, Fall, Winter

First offered 1983. Thirty 1- to 4-session courses per year. Enrollment Open/class. Specialties: Aging wine, general usage, specific wine styles and origins, vertical wine tastings, component tastings. Focus on strong basics. 6-8 wines per session. Vintages: Current to 1984 and older. Price range Value selections to $80 per bottle. Source of wines: Retail outlets. Instructor credentials: Taught wine appreciation courses for Creative Activities Program at Ohio State Univ. for 7 years, guest lecturer for SWE's national convention, certified member of SWE, 14 years in wholesale wine trade. Tuition Varies. Class location: Loretta Paganini School of Cooking (Chesterland), Kenston Community Education (Bainbridge), Euclid Continuing Education (Euclid).
Contact: Gary L. Twining, 301 Greenwood Ct., Elyria, OH 44035; 440-458-6912, Fax 216-771-3919.

MARIO VITALE
Moreland Hills/Year-round

First offered 1990. 10-15 evening and 3 to 4-session courses per year. Enrollment 20 students per class. Specialties: European, especially Italy. 8 wines per session. Vintages: Current to 1960's. Price range . Instructor credentials: Member AWS. Tuition $50-$100 per course, $15-$50/one-night.
Contact: Mario Vitale, Western Reserve Wines, 34101 Chagrin Blvd., Cleveland, OH 44023; 216-831-2116, Fax 216-831-6369.

OREGON

ROBERT A. LINER/MATTHEW ELSEN – WINE MERCHANTS
Portland/Spring, Summer, Fall

First offered 1990. Three 3-session courses per year. Enrollment 20 students per class. Specialties: France and West Coast, U.S.. 8 wines per session. Vintages: Current. Price range $8-$100 per bottle. Source of wines: Local wholesalers. Instructor credentials: 25 years wine business, Food & Wine Magazine's Top % Wine Shops, Portland's Best Wine Shop - Willamette Weekly. Tuition $85 per course. Class location: Liner & Elsen Wine Merchants.
Contact: Robert A. Liner & Matthew Elsen, Liner & Elsen Wine Merchants, 202 N.W. 21st Ave., Portland, OR 97209; 503-241-9463, Fax 503-243-6706, E-mail LINERELSEN@aol.com, URL http://www.citysearch.com/pdx/linerelsen.

WINE TASTING AND STUDY COURSE BY BOB SOGGE

Eugene/Spring and Fall

First offered 1970. Four full-length sessions, special requests on demand. Enrollment 20 students/basic, 100/specific sessions/class. Specialties: Oregon, California, Washington, Germany, France, Italy, Portugal, Australia, New Zealand. 6-8 wines per session. Vintages: 20-25 years. Price range $4-$150 per bottle. Source of wines: Personal cellar and the market. Instructor credentials: Member SWE, Napa Valley Wine Library School, Bordeaux Wine Soc., Oregon Wine Growers Assoc. Visits to major growing areas of the world. 28 years teaching, 15 years in wine wholesale and importing business. Tuition $125/basic 6-week course. Class location: Chanterelle Restaurant.

Contact: Bob Sogge, Wine Tasting & Study Course, 3620 Donald St., Eugene, OR 97405; 541-484-9848.

HEIDI YORKSHIRE

Portland

First offered 1994. On demand. Enrollment /class. Specialties: Customized interactive wine classes. French for Food & Wine Lovers, and Italian for Food & Wine Lovers seminars, emphasizing labels and food and wine terms. /session. Price range . Instructor credentials: Wine Columnist for The Oregonian, author of Wine Savvy:The Simple Guide to Buying and Enjoying Wine Anytime, Anywhere; contributor to Bon Appetit and other magazines. Tuition .

Contact: Heidi Yorkshire, Wine Savvy, PO Box 12081, Portland, OR 97212; 503-335-3155, Fax 503-280-8964, E-mail heidiyorkshire@juno.com.

PENNSYLVANIA

WILLIAM H. CLARK

Langhorne/Fall and Spring

First offered 1976. Two 10-session courses per year. Enrollment 50 students per class. Specialties: U.S., European, Australian, and South American. 5 wines per session. Price range $15-$50 per bottle. Instructor credentials: Member SWE. Tuition $125 per course. Class location: Neshaminy Adult School.

Contact: William H. Clark, 74 Hollybrooke Dr., Langhorne, PA 19047; 215-752-4895, Fax 215-752-4895, E-mail bill_clark@prodigy.com, URL http://pages.prodigy.com/pa/bordeaux/bordeaux.html.

LOUIS J. DIGIACOMO

Berwyn/Fall and Spring

First offered 1978. Four 8-session courses per year. Enrollment 42 students per class. Specialties: All major wine countries. 6 wines per session. Vintages: Current. Price range $8-$20 per bottle. Source of wines: Pennsylvania state stores. Instructor credentials: Author of The Clear & Simple Wine Guide, member AWS. Tuition $60 per course. Class location: Local high schools-Main Line School Night Assn. (PO Box 8175, 260 Gulph Creek Rd., Radnor, PA 19087).

Contact: Louis J. DiGiacomo, 204 Country Rd., Berwyn, PA 19312; 215-644-6233, Fax 215-563-3337.

INDEPENDENT WINE CLUB

Philadelphia/Year-round

First offered 1992. 40 nights of classes per year, 50 nights of tastings per year. Most classes are 1-night sessions, intro courses are 3 nights, some intermediate/advanced are 2 weeks. Enrollment 15 students per class, 70 students/tasting/class. Specialties: Individual regions, grape varieties, viticulture, vinification and more. Classes focus on a wine book provided. 8-15 wines per class, 8-55 wines/tasting/session. Vintages: Mostly recent. Older wines have dated back to 1865. Price range $10-$700 per bottle. Source of wines: Retail, instructors' personal cellars, wineries. Instructor credentials: Neal Ewing, Certified Wine Educator from SWE, has Higher Certification and partial completion of Diploma from England's Wine & Spirit Education Trust, and extensive wine writing experience. Tuition Full members ($225 per year): $35/one night, $75/three nights. Select members ($60 per year): $50/one night, $100/three nights. Non-members: $75/one night, $145/three nights. Fees include wine book. Class location: Jack's Firehouse Restaurant.

Contact: Neal Ewing, IWC, PO Box 1478, Havertown, PA 19083; 610-649-9936, Fax 610-649-9936, E-mail IWCWine@aol.com.

SHIRLEY MARTIN COUNTRY WINES
Pittsburgh/September-October

First offered 1975. Two 3-session courses per year. Enrollment 5-20 students per class. Specialties: Winemaking, beermaking offered. 6-8 wines (final)/session. Price range . Source of wines: Mostly home-made, some commercial. Instructor credentials: Member of AWS and a graduate of Penn. State University with a major in home economics education; AWS certified judge. Tuition $25 per course. Class location: Country Wines, Pittsburgh.

Contact: Shirley Martin, Country Wines, 3333 Babcock Blvd., Ste. 2, Pittsburgh, PA 15237-2421; 412-366-0151, Fax 412-366-9809, E-mail info@countrywines.com, URL http://www.countrywines.com.

DICK NAYLOR
Stewartstown/Year-round

First offered 1980. Two 6-session courses per year. Enrollment 20-30 students per class. Specialties: Dry red wines. 4-6 wines per session. Vintages: Current to late 1980's. Price range $7-$25 per bottle. Source of wines: Naylor and private collections. Instructor credentials: Articles in wine-related magazines, member AWS, wine maker. Tuition $65-$75 per course. Class location: Naylor Wine Cellars.

Contact: Dick Naylor, Naylor Wine Cellars, 4069 Vineyard Rd., Stewartstown, PA 17363; 717-993-2431, 800-292-3370, Fax 717-993-9460.

THE RESTAURANT SCHOOL
Philadelphia/October-March

First offered 1998. Two 7-session semesters and one 3-session semester. Enrollment 50 students per class. Specialties: Professional level course. Wines of the world, wine making, sparkling, fortified and dessert, wine list writing, customer psychology, buying and selling. 6-9 wines per session. Vintages: Currently available plus a few older vintages. Price range $5-$100 per bottle. Source of wines: Private cellar and retail. Instructor credentials: . Tuition $385/seven-session semesters, $165/three-session semester, $60/individual session. Class location: The Restaurant School.

Contact: Michelle Gambino, The Restaurant School, 4207 Walnut St., Philadelphia, PA 19104; 215-222-4200, Fax 215-222-4219.

SYLVIA SCHRAFF
Altoona/Fall and Spring

First offered 1989. Two 4-session courses per year. Enrollment 30 students per class. Specialties: Worldwide varieties. 2+ wines per session. Price range $6-$15 per bottle. Instructor credentials: Masters degree in Nursing, retired CEO of home nurse agency; Certified Wine Judge. Tuition $45 per course. Class location: Penn State University-Altoona.

Contact: Sylvia Schraff, Oak Spring Winery, R.D. 1, Box 612, Altoona, PA 16601; 814-946-3799.

EDWARD TURBA
Pittsburgh/Year-round

First offered 1988. 8 sessions per year. Enrollment 10+ students per class. Specialties: Worldwide varieties. 5 wines per session. Vintages: 7 year spread. Price range $10-$40 per bottle. Source of wines: Individual vendors and state stores. Instructor credentials: Sommelier LeMont Restaurant (teaches waiters); Bronze Medal in Pittsburgh Magazine wine competition, member SWE, Award of Excellence by the Wine Spectator. Tuition $250 per 3 hr. session plus wine. Class location: Privately arranged.

Contact: Edward Turba, LeMont Restaurant, 1114 Grandview Ave., Pittsburgh, PA 15211; 412-488-8499/431-3100, Fax 412-431-1204, E-mail ETurba@aol.com, URL http://www.le-mont.com.

WOODEN ANGEL RESTAURANT/ALEX SEBASTIAN
Beaver/Year-round

First offered 1980. Monthly tasting at the restaurant and custom courses for student group or sponsor. Enrollment Multiples of 12/class. Specialties: American only. 12-24 wines per session. Vintages: Custom to the group. Price range . Source of wines: Liquor Control Board or Wooden Angel's Grand Award Wine Cellar. Instructor credentials: California State Fair/Intervin qualified judge, 30 years in the wine business, first California Wine Wizard. Tuition Varies. Class location: Wooden Angel Restaurant. Custom programs available at a sponsor's location in Western Pennsylvania or Eastern Ohio.

Contact: Alex Sebastian, 308 Leopard Lane, Bridgewater, Beaver, PA 15009-3096; 724-774-7880, Fax 724-774-7994, E-mail woodangl@ccia.com.

PUERTO RICO

PEDRO J. BORAS

Guaynabo/August-June

First offered 1978. Four 5 to 8-session courses per year; four to five 1-day seminars per year. Enrollment 25-30 students per class. Specialties: Wine history, production, blind tastings. 5-8 wines per session. Vintages: Current to late 1970's. Price range $8-$60 per bottle. Source of wines: Local distributors. Instructor credentials: Member SWE. Tuition $185-$225 per course; $60/one-day seminar. Class location: Local cooking school.

Contact: Pedro J. Borras, Colegio del Vino, Green Hill, G-2, Garden Hill, Guaynabo, PR 00966; 787-783-6099, Fax 787-758-6105, E-mail borras@pol.net.

RHODE ISLAND

EDWARD KORRY – JOHNSON & WALES UNIVERSITY

Providence/September-June

First offered 1994. Eight 1- to 4-session courses. Enrollment 15 students per class. Specialties: Wine and food pairing, France, Spain, Portugal, Italy, Germany, U.K., Chile, Australia. 6-8 wines per session. Vintages: 1997-1988. Price range $6-$50 per bottle. Source of wines: Local distributors. Instructor credentials: Edward Korry, Ass. Professor, member SWE, wine judge, faculty member of Johnson & Wales University, articles in wine-related magazine. Tuition $45/session. Class location: Johnson & Wales University College of Culinary Arts.

Contact: Bill Day, Johnson & Wales University, 1 Washington Ave., Providence, RI 02905; 401-598-1130, Fax 401-598-1856, E-mail Bday@JWU.edu, URL http://www.JWU.edu.

TENNESSEE

SHIELDS HOOD WINE CLASSES

Memphis/September, October, November, February, March, April

First offered 1978. 3 sessions/month. Enrollment 40 students per class. Specialties: Wines of the world, varietals, wines of California. 10 wines per session. Vintages: 10 years. Price range $8-$40 per bottle. Source of wines: Wines of the world. Instructor credentials: Member SWE, Certified Wine Educator. Tuition $25/session. Class location: Local restaurants.

Contact: Shields T. Hood, 905 James St., Memphis, TN 38106; 901-774-8888/901-853-2693, Fax 901-946-4751, E-mail HoodWine52@aol.com.

JOHN IACOVINO

Oak Ridge/Fall, Spring, Summer

First offered 1989. Three 4 to 6-session courses per year. Enrollment 25-30 students per class. Specialties: Burgundy. 6-9 wines per session. Vintages: Current to early 1980's. Price range $8-$80 per bottle. Instructor credentials: Member AWS, Grand Seneschal of Ducal Ordes of the Cross of Burgundy. Tuition $100-$120 per course. Class location: Local restaurants, homes.

Contact: John Iacovino, 120 Westlook Circle, Oak Ridge, TN 37830; 423-483-8330, Fax 423-482-2495, E-mail JAIacovino@aol.com.

UT COMMUNITY PROGRAMS

Knoxville/Year-round

First offered 1978. Six 7-session courses per year. Enrollment 18 students per class. Specialties: French, Italian, Australian, Chilean, California, and eastern U.S.. 7-8 wines per session. Vintages: 1990-1970. Price range $6-$35 per bottle. Source of wines: Retail. Instructor credentials: Member AWS & Cross of Burgundy Wine Society with 20 years teaching experience. Tuition $145 per course. Class location: University of Tennessee.

Contact: University of Tennessee, Community Programs, 600 Henley St., #105, Knoxville, TN 37996-4110; 423-974-0150, Fax 423-974-0154, E-mail utcommunity@gateway.ce.utk.edu.

VIRGINIA

JOSEPH V. FORMICA

Richmond/Fall

First offered 1974. One 15-session course per year. Enrollment 20 students per class. Specialties: Old and New World. 6 wines per session. Vintages: Current to 1985. Price range $8-$120 per bottle. Source of wines: Local wine shops. Instructor credentials: Certified by Wine & Spirit Education Trust of G.B. and SWE, member SWE and AWS. Tuition $150 plus $95 lab fee. Class location: J. Sargeant Reynolds Comm. College.

Contact: Joseph V. Formica, The Wine School, 8402 Gaylord Rd., Richmond, VA 23229; 804-786-9730, Fax 804-828-9946, E-mail formica@hsc.vcu.edu.

STEFAN GRABINSKI

Richmond/Fall, Winter, Spring

First offered 1970. 3-6 sessions per year. Enrollment 16 students per class. Specialties: Wine country and regions, wine making techniques, grape varieties, and wine evaluation. 6 wines per session. Vintages: Current to 1980. Price range $8-$20 per bottle. Source of wines: Local wine shops. Instructor credentials: Graduate Walden School of Wine (Canada), member SWE and AWS, winery consultant, Founding Member Richmond Wine Society, President of Hanover Wine Guild, former vineyard owner. Tuition $100 plus wine fee. Class location: Local college and restaurants.

Contact: Stefan Grabinski, 4944 Farrell Ct., Richmond, VA 23228; 804-270-6255, Fax 804-270-0163, E-mail stefin@juno.com.

ARGENTINA

LA SOCIEDAD DEL CATADOR

Buenos Aires/Year-round

First offered 1992. 12 courses per year. Enrollment 16 students per class. Specialties: Taste and elaboration. 5-8 wines per session. Vintages: Wide range. Price range $6-$15 per bottle. Source of wines: Local and imported. Instructor credentials: Wine Director Sociedad Del Catador. Tuition $80 per course. Class location: La Sociedad Del Catador.

Contact: Ricardo Ianne, Sociedad Del Catador, Av. Libertador 276, (1638) Vte Lopez, Argentina; (54) 718-0232, Fax (54) 718-0231, E-mail catador@overnet.com.ar.

GUSTAVO JORGE PRECEDO

Buenos Aires, Cordoba, La Plate, Santa Fe/March-November

First offered 1992. 16 courses per year in Buenos Aires, 2 courses per year in Montevideo. 4 sessions per course for 1st, 2nd and 3rd levels. 3 sessions per course for sparkling wines. 3-4 days/trip. Enrollment 25 students per class. Specialties: Wine-making, tastings, fortified wines and Blend vs varietal, alcohol and acid levels, aromas, sparkling wines, trips in Argentine, Uraguay, Brazil, Chile. 3-4 wines per session. Price range $7-$15 per bottle. Source of wines: Local wineries and wine shops. Instructor credentials: Member SWE, professor of wine tasting classes in School of Hotel Administration. Tuition $100-$150 per course, $400-$600/trip. Class location: Club del Vino & Savoy wine shop in Buenos Aires. Cavas Privadas wine club in Montevideo.

Contact: Gustavo Jorge Precedo, Club del Vino, Cabrera 4737, (1414) Buenos Aires, Argentina; (54) 1-833-0050, Fax (54) 1-833-0045, E-mail clubdelvino@interar.com.ar or gustavoprecedo@interar.com.ar.

CANADA

MICHAEL BOTNER – ACCOUNTING FOR TASTE

Ottawa/Fall, Winter, Spring

First offered 1996. Four 4-session courses, one 8-session course per year. Enrollment 24 students per class. Specialties: The Wine Course: How to taste, choose, serve, match; Discovery Series: wines of old and new world. 6 wines per session. Vintages: Current to 1978. Price range $8-$100 per bottle. Source of wines: Ontario Liquor Control Board. Instructor credentials: Governor, National Capital Sommelier Guild, writer, chair of the Cellars of the World International Wine Competition, member SWE. Tuition $150 per course, $275/series.

Contact: Michael Botner, Accounting for Taste, 195 Rodney Crescent, Ottawa, Ontario, K1H 5J8, Canada; 613-523-3389, Fax 613-523-3397, E-mail acctaste@synapse.net.

LINDA BRAMBLE, PH.D. – BROCK UNIVERSITY

St. Catharines/September-December

First offered 1998. 1 course/term. Enrollment 45 students per class. Specialties: Varietal approach, studying 12 major int'l varietals from Chardonnay to Tempranillo. Sensory evaluation experiments and background on each wine. 6 wines per session. Vintages: 10 years. Price range $15-$20. Source of wines: LCBO and private agents. Instructor credentials: Third level on IWEG Dipoma, university courses on oenology, wine marketing and viticulture. Tuition $275 Canadian. Class location: Brock University.

Contact: Barbara Smart, Cool Climate Oenology & Viticulture Institute, Brock University, St. Catharines, Ontario, L2S 3A1, Canada; 905-688-5550, x4652, Fax 905-641-0406, E-mail lbramble@ican.net.

KENSINGTON WINE MARKET

Calgary/Spring, Fall, Winter

First offered 1992. Approximately 30 single-session courses per year. Enrollment 20 students per class. Specialties: All regions. 6 wines per session. Vintages: Fairly recent. Price range $10-$50 per bottle. Instructor credentials: All self-taught store employees. Tuition Average $25/class. Class location: In store.

Contact: Nancy Carten, Kensington Wine Market, 1257 Kensington Rd., N.W., Calgary, Alberta, Canada; 403-283-8000, Fax 403-283-4283, E-mail kwinemkt@bmlive.com.

MARGARET SWAINE

Toronto/Year-round

First offered 1996. 8 courses per year. Enrollment 90 students per class. Specialties: Wine and food pairings themed to different regions. 9 wines per session. Vintages: All. Price range $8-$90 per bottle. Source of wines: All countries. Instructor credentials: Wine columnist since 1979, Toronto Life and other magazines. Tuition $125/session includes dinner. Class location: Top Toronto restaurants.

Contact: Margaret Swaine, 80 Alcorn Ave., Toronto, Ontario, M4V 1E4, Canada; 416-961-5328, Fax 416-961-4251, E-mail mswaine@netcom.ca.

ENGLAND

PATRICIA STEFANOWICZ

London/Fall and Spring

First offered 1988. Two 1- to 6-session courses per year. Enrollment 5-15 students per class. Specialties: Grape varieties around the world, winemaking, wine faults, Italy, France, Germany/Austria, New World, Wine & Art, Wine and Music tasting events. 4-12 wines per session. Vintages: Wide range. Price range L 5-100 per bottle. Source of wines: Wine merchants, private collections. Instructor credentials: Masters in Architecture/Fine Arts, Higher Nat'l Diploma in Horticulture (Oenology), Wine & Spirit Education Trust Diploma with Honors, member SWE. Tuition L 15-100 per course. Class location: Varies. Privately arranged.

Contact: Patricia Stefanowicz, Vino Vino, 31 Randolph Crescent, Little Venice, London, W9 1DP, England; (44) 171 286 4505, Fax (44) 171 266 3166.

JAPAN

ASSOCIATION DE SOMMELIER JAL (A.S.J.)

Tokyo/April-September

First offered 1991. One main Preparatory Course of 12-14 sessions per year. Enrollment 80-100 students per class. Specialties: All subjects and regions around the world. 16-22 wines per tasting session. Vintages: 1997-1990. Price range $9-$100 per bottle. Source of wines: Mainly supplied by Mercian Wine Corporation. Instructor credentials: Pres./Dir. Assoc. de Sommelier JAL, Senior Sommelier of Japan since 1987, Wine Judge of Japanese Wine Magazine, author/supervisor of JAL Flying Sommelier Book, Cert. Japanese Sake, Cert. Beer Judge. Tuition 120.000yen (about $900)/twelve sessions of Main Course. Class location: Headquarter of Mercian Wine Corporation in Kyobashi.

Contact: Hayato KOJIMA, Association de Sommelier JAL, 1-2-1 Oyamadai, Setagaya, Tokyo, 158-0086, Japan; (81) 3-3704-0722.

4

Food & Wine Organizations

AMERICAN CULINARY FEDERATION (ACF)
St. Augustine, Florida

Founded in 1929. Membership 25,000. Oldest nationwide professional cooks' association. Objectives: to advance the culinary profession and offer training, education, and fellowship. Over 295 local chapters in the U.S. and Caribbean. Membership benefits include: educational seminars at national convention and regional meetings; subscription to monthly magazine, The *National Culinary Review,* competitions for medals in ACF-approved culinary arts shows sponsored by local chapters. The ACF and the NRA sponsor the U.S. Culinary "Olympic" Team, which competes in the Culinary "Olympics" every 4 years in Germany.

The American Culinary Federation Educational Institute, an ACF subsidiary, accredits culinary programs through its Accreditation Commission, awards loans and scholarships to students, and provides a U.S. Department of Labor recognized 3-year National Apprenticeship Training Program for Cooks. The ACF Certification Department certifies chefs on the basis of knowledge and experience. Certification categories include Certified Cook/Pastry Cook (CC, CPC), Certified Sous Chef/Chef de Cuisine and/or Pastry Chef (CSC, CCC, CPC), Certified Culinary Educator (CCE), Executive Chef and/or Executive Pastry Chef (CEC, CEPC), and the highest level, Certified Master Chef/Pastry Chef (CMC, CMPC). The American Academy of Chefs, another ACF subsidiary, is the honor society of American chefs.

Contact: ACF, P.O. Box 3466, St. Augustine, FL 32085 US; 800-624-9458/904-824-4468, Fax 904-825-4758, E-mail acf@aug.com, URL http://www.acfchefs.org/acf.html.

AMERICAN DIETETIC ASSOCIATION (ADA)
Chicago, Illinois

Founded in 1917. Membership 69,000, of which 75% are registered dietitians in diverse areas of practice. Promotes optimal nutrition through activities, publications, educational meetings, media and marketing. Establishes and enforces quality standards for 600+ educational programs and internships. Maintains a legislative affairs office in Washington, DC, and has a nonprofit foundation for research and scholarship purposes.

Contact: ADA, 216 W. Jackson Blvd., Chicago, IL 60606-6995 US; 312-899-0040, Fax 312-899-1979, E-mail webmaster@eatright.org, URL http://www.eatright.com

AMERICAN INSTITUTE OF BAKING (AIB)
Manhattan, Kansas

Founded in 1919. This non-profit organization's objective is to promote the cause of education in nutrition, baking, and bakery management. Employs about 135 personnel and is supported by the contributions of 600+ member companies. Programs include Baking Science and Technology and Bakery Maintenance Engineering. Correspondence courses include Science of Applied Baking, Bakery Maintenance Engineering, and Maintenance Engineering (page 50). The Certified Baker Program provides companies with training.

Contact: AIB, 1213 Bakers Way, Manhattan, KS 66502 US; 800-633-5137/913-537-4750, Fax 913-537-1493, URL http://www.aibonline.org.

AMERICAN INSTITUTE OF WINE & FOOD (AIWF)
San Francisco, California

Founded in 1981. Membership 10,000+, 30+ chapters. This non-profit educational organization's objectives are to advance the appreciation of wine and food and stimulate greater scholarly education in gastronomy. Membership is open to all and benefits include a discounted invitation to conferences, discounts on national and chapter programs, wine and food publications, and six copies a year of *American Wine & Food* newsletter. Annual dues range from $45 for students to $550 for corporations.

Contact: AIWF, 1550 Bryant St., #700, San Francisco, CA 94103 US; 415-255-3000/800-274-AIWF, E-mail webmanagement@aiwf.org, URL http://www.aiwf.org.

AMERICAN VEGAN SOCIETY(AVS)
Malaga, New Jersey

Founded in 1960. Nonprofit educational organization dedicated to teaching a compassionate way of living and abstinence from animal products. Lectures, discussions, and live-in weekend classes in vegan cooking are available. Member services: books & videos, annual convention, quarterly newsletter. Annual dues are $18.

Contact: AVS, 56 Dinshah Ln., P.O. Box H, Malaga, NJ 08328-0908; 609-694-2887.

AMERICAN WINE SOCIETY (AWS)
Rochester, New York

Founded in 1967. Membership 5,000, 90 chapters. Non-profit consumer organization dedicated to bringing together wine lovers and promoting wine education. Membership benefits include an annual conference, quarterly journal, and technical manuals. Members with at least 2 years of chapter comparative tastings or equivalent are eligible for the AWS Wine Judge Certification Program. Annual dues are $36 per person or couple. Professional Memberships are $58 per year, Lifetime Memberships, for ages 60 and over, are $280.

Contact: AWS, 3006 Latta Rd., Rochester, NY 14612-3298 US; 716-225-7613, Fax 716-225-7613, E-mail Angel910@aol.com, http://www.vicon.net/~aws.

THE BREAD BAKERS GUILD OF AMERICA
Pittsburgh, Pennsylvania

Founded in 1993. Membership 1,300. Objectives are to bring together individuals involved in the production of high quality bread products, to raise professional standards, and to encourage the education of people interested in careers as bread baking professionals. The Guild publishes a newsletter and sponsors seminars and baking competitions. Membership is open to anyone but the focus of the Guild is professional bread bakers. Business membership yearly dues start at $100; non-business dues: $35 students, $50 educators, $65 all others.

Contact: Gina Piccolino, Director of Membership Services, The Bread Bakers Guild of America, Box 22254, Pittsburgh, PA 15222; 412-322-8275, Fax 412-322-3412, E-mail bbguild@sgi.net, URL http://www.bbga.org.

CAREERS THROUGH CULINARY ARTS PROGRAM, INC. (C-CAP)
New York, New York

Founded in 1990 by cookbook author Richard Grausman. Membership 10,000 students in 190 schools in Arizona, California, Illinois, New York, Pennsylvania, Washington, DC, Virginia. Nonprofit corporation whose mission is to promote and provide career opportunities in the foodservice industry for inner-city youth through culinary arts education and apprenticeship. Provides public high school home economics teachers with basic culinary techniques that are taught to students, who can serve internships with local foodservice establishments. Matches each teacher with an industry professional who serves as a guest speaker and instructor/role model. Conducts annual culinary competitions and awards scholarships.

Contact: C-CAP, 155 W. 68th St., New York, NY 10023 US; 212-873-2434, Fax 212-873-1514.

CONFRERIE DE LA CHAINE DES ROTISSEURS
New York, New York

Founded in 1950. Membership 7,200 (143 U.S. chapters). International gastronomic organization, purpose is to encourage educational functions and promote fellowship among individuals with a serious interest in wine and cuisine. The nonprofit, tax-exempt Chaine Foundation supports culinary educational programs. Membership benefits at the local level include gastronomic functions. On a regional and national level, members can join in Chaine-sponsored excursions and attend the national convention. Professionals make up approximately 30% of the membership. Membership is normally by invitation only. Interested individuals who do not know a member should contact the National Office for information.

Contact: Confrerie de la Chaine des Rotisseurs, 444 Park Ave. So., New York, NY 10016 US; 212-683-3770, Fax 212-683-3882.

COOKING CLUB OF AMERICA
Minnetonka, Minnesota

Established in 1998. Membership organization for people who enjoy cooking. Benefits include a subscription to Cooking Pleasures magazine. Annual dues are $24.

Contact: Cooking Club of America, North American Outdoor Group, Inc., 12301 Whitewater Dr., Minnetonka, MN 55343 US; 888-850-8202.

COUNCIL ON HOTEL, REST., AND INSTITUTIONAL EDUCATION (CHRIE)
Washington, D.C.

Founded in 1946. 2,000 members from 52 countries. This trade and professional organization's mission is to foster the international advancement of teaching and training in the field of hospitality and tourism management and facilitate the professional development of its members. Membership benefits include an annual conference and several publications. CHRIE also publishes *A Guide to College Programs in Hospitality & Tourism*, which describes curricula, admission requirements, scholarships, and internships.

Contact: CHRIE, 1200-17th St., NW, Washington, DC 20036-3097 US; 202-331-5990, Fax 202-785-2511, E-mail alliance@access.digex.net, URL http://www.chrie.org.

EDUCATIONAL FOUNDATION OF THE NATIONAL RESTAURANT ASSOCIATION

Chicago, Illinois

Founded in 1987. Nonprofit organization created to advance the professional standards of foodservice management through education. The Foundation develops courses, video training, seminars, and other programs. The Professional Management Program (ProMgmt.), for undergraduate hospitality students, covers: unit revenue/cost management, risk management, human resources/diversity management, operations, and marketing. The organization also offers ProMgmt. scholarships.

Contact: Educational Foundation of the National Restaurant Association, 250 S. Wacker Dr., #1400, Chicago, IL 60606-5834 US; 312-715-1010, Fax 312-715-0807, URL http://www.restaurant.org/educate/educate.htm.

FOODSERVICE EDUCATORS NETWORK INTERNATIONAL (FENI)

Chicago, Illinois

Established in 1998. Nonprofit network of foodservice educators whose mission is to advance professional growth through seminars and workshops, an annual Educators' Summit, summer institutes, awards to top programs, consulting services, resource guides, a continuing education scholarship program, and an electronic publication, Network News.

Contact: Mary Petersen, Executive Director, FENI, 959 Melvin Rd., Annapolis, MD 21403; 410-268-5542, Fax 410-263-3110, E-mail EAHX15A@Prodigy.com.

INSTITUTE OF FOOD TECHNOLOGISTS

Chicago, Illinois

Established in 1939. 28,000 members. Nonprofit scientific society dedicated to supporting the improvement of the food supply and its use through science, technology, and education. Networks worldwide through 23 divisions of expertise, sponsors student chapters, schedules regional meetings, affiliates with food science and technology associations, provides scientific lecturers. Membership benefits include the publications *Food Technology* and *Journal of Food Science* and an annual meeting and food expo.

Contact: Institute of Food Technologists, 221 N. LaSalle St., #300, Chicago, IL 60601-1291 US; 312-782-8424, Fax 312-782-8348, E-mail info@ift.org, URL http://www.ift.org.

INTERNATIONAL ASSOCIATION OF CULINARY PROFESSIONALS (IACP)

Louisville, Kentucky

Founded in 1978. Membership 4,000+ representing over 35 countries. This not-for-profit professional association's objectives include: providing continuing education and professional development, sponsoring of the annual IACP Julia Child Cookbook Awards, promoting the exchange of culinary information among members of the professional food community, establishing professional and ethical standards, and funding scholarships. Membership benefits include the annual spring and regional conferences, newsletters and research reports, the annual *IACP Membership Directory*, the Certified Culinary Professional (CCP) certification program. Annual dues are $175 (plus $50 one-time fee) for Professional Members, $300 (plus $50) for Cooking School Members, $350 (plus $50) for Business Members; $875 (plus $100) for Corporate Members, and $50 for Student Members.

Contact: IACP, 304 W. Liberty St., #201, Louisville, KY 40202 US; 502-581-9786/800-928-4227, Fax 502-589-3602, E-mail iacp@hqtrs.com, URL http://www.iacp-online.org.

INTL. ASSOC. OF CULINARY PROFESSIONALS (IACP) FOUNDATION

(See display ad page 376) **Louisville, Kentucky**

Founded in 1984. Supports the IACP by soliciting, managing, and distributing funds for educational and charitable work related to the culinary profession. Maintains programming in four emphasis areas: library funding, research, scholarship, and world hunger. Provides and administers tuition-credit and tuition-cash awards for professional training at both the primary and continuing education level. Publications: the annual *Perspectives*. Committee members meet during the annual IACP conference.

Contact: Ellen McKnight, Director of Development, IACP Foundation, 304 W. Liberty St., #201, Louisville, KY 40202 US; 502-587-7953, Fax 502-589-3602, E-mail ellenm@hqtrs.com, URL http://www.gstis.net/~epicure.

INTERNATIONAL FOOD SERVICE EXECUTIVES ASSOCIATION

Margate, Florida

Founded 1901. Membership 5,000. Nonprofit educational and community service organization. Services include scholarships, monthly meetings, *Hotline Magazine*., and certification program. Annual dues are $130 for certification only, $150 for certification and IFSEA membership.

Contact: IFSEA, 1100 S. State Rd. 7, #103, Margate, Florida 33068; 954-977-0767, Fax 954-977-0874, E-mail hq@ifsea.org, URL http://www.ifsea.org.

INTERNATIONAL FOODSERVICE EDITORIAL COUNCIL (IFEC)

Hyde Park, New York

Founded in 1956. Membership 250. This nonprofit association is dedicated to improving the quality of media communications in the foodservice industry. Membership benefits include an annual directory and conference, newsletter, and networking. Scholarships are awarded to students whose career aspirations combine foodservice and communications. Membership is open to individuals employed in editorial functions within the industry.

Contact: Carol Metz, Executive Director, IFEC, P.O. Box 491, Hyde Park, NY 12538 US; 914-452-4345, Fax 914-452-0532, E-mail ifec@aol.com.

INTERNATIONAL WINE & FOOD SOCIETY

London, England

Founded in 1933. Educational and social organization devoted to bringing together wine and food lovers and promoting a wider knowledge of the wines of the world. Membership benefits include yearly festivals, monthly wine events, dinners, seminars, day trips, and an annual review. Branches in 40 countries operate autonomously with admission at the discretion of each branch.

Contact: International Wine & Food Society, 9 Fitzmaurice Place, Berkeley Square. London, W1X 6JD, England; (44) 171-495 4191, Fax (44) 171-495 4172.

JAMES BEARD FOUNDATION, INC.

New York, New York

Established in 1986. Membership 5,000. Nonprofit organization whose mission is to maintain the ideals that made James Beard the "Father of American Cooking" and to maintain his home as the first historical culinary center in North America. Membership benefits include discounts on more than 200 events (workshops and dinners featuring well-known American chefs) each year, a subscription to *Beard House* magazine, member directory, and the annual James Beard Awards, which include cookbook, journalism, chef, and restaurant categories. Tax-deductible annual dues begin at $60 for nonprofessionals and $125 for professionals.

Contact: James Beard Foundation, Inc., 167 W. 12th St., New York, NY 10011; 212-675-4984, 800)-36-BEARD, Fax 212-645-1438, E-mail jbeard@pipeline.com, URL http://www.jamesbeard.org.

NATIONAL RESTAURANT ASSOCIATION (NRA)

Washington, D.C.

Established in 1919. Membership 30,000+. This national trade association for the foodservice industry provides educational, research, communications, convention, and government services; interacts with legislators and political leaders; offers a media relations program and speech bank; has a toll-free information hotline. Publications: the monthly *Restaurants USA* magazine, *Washington Weekly* political report, operations manuals. Sponsors Restaurant, Hotel-Motel Show in Chicago each May. Membership is open to any entity that operates facilities and/or supplies meal service to others regularly. Dues are revenue-based and begin at $140.

Contact: NRA, 1200-17th St., NW, Washington, DC 20036-3097 US; 202-331-5900, Fax 202-331-2429, E-mail isal@restaurant.org, URL http://www.restaurant.org.

NATIONAL ASSOCIATION FOR THE SPECIALTY FOOD TRADE, INC.

New York, New York

Established in 1952. Membership 2,000 companies in the U.S. and overseas. Nonprofit trade association that fosters trade, commerce, and interest in the specialty food industry. Sponsors the semi-annual International Fancy Food and Confection Show every winter (West Coast) and summer (East Coast). Other services include the Annual Product Awards, the Scholarship and Research Fund, and a bimonthly magazine. Applicants must be in business for at least one year. Annual dues start at $200.

Contact: NASFT, 120 Wall St., 27th Floor, New York, NY 10005-4001; 212-482-6440, Fax 212-482-6455, http://www.specialty-food.com/Client/NASFT.asp.

NORTH AMERICAN VEGETARIAN SOCIETY (NAVS)

Dolgeville, New York

Established in 1974. Nonprofit educational organization dedicated to promoting the vegetarian way of life. Affiliated with the International Vegetarian Union. Sponsors an annual Vegetarian Summerfest conference and World Vegetarian Day. Members receive a quarterly magazine. Dues begin at $20 annually.

Contact: NAVS, P.O. Box 72, Dolgeville, NY 13329 US; 518-568-7970, Fax 518-568-7979, E-mail navs@telenet.net, URL http://www.cyberveg.org/navs.

OLDWAYS PRESERVATION AND EXCHANGE TRUST

Boston, Massachusetts

Established 1990. Nonprofit educational institution dedicated to preserving healthy food traditions and fostering cultural exchange in the fields of food, cooking, and agriculture. The Chefs Collaborative: 2000, an educational initiative launched in 1993, promotes sustainable cuisine by teaching children, supporting local farmers, and educating consumers. Membership in Oldways and the Collaborative is open to all.

Contact: Oldways Preservation and Exchange Trust, 25 First St., Cambridge, MA 02141 US; 617-621-3000, Fax 617-621-1230, E-mail oldways@tiac.net, URL http://www.oldwayspt.org.

THE RETAILER'S BAKERY ASSOCIATION (RBA)

Laurel, Maryland

Established 1918. Trade association representing 3,500 member companies who bring consumers bakery foods from bakery departments, independent bakeries, and foodservice facilities. Objective is to create training programs, and connect retailers with suppliers and experts to help build profitable bakeries. Grants certification as Certified Master Baker (CMB) and Certified Journey Baker (CJB). Provides a baking curriculum for 30+ post-secondary schools, hosts an annual trade show and a baking competition for students, publishes career guides, videos, and a newsletter.

Contact: Bernie Reynolds, Education Director, RBA, 14239 Park Center Dr., Laurel, MD 20707-5261; 800-638-0924 or 301-725-2149, Fax 301-725-2187, E-mail breynolds@rbanet.com, URL http://www.rbanet.com.

SOCIETY OF WINE EDUCATORS (SWE)

East Longmeadow, Massachusetts

Formed in 1977. Membership 1,500. Nonprofit organization dedicated to improving information about wine making, wine service, wine and food pairing, wine and health. Membership services include the annual conference, educational programs, and trips to wine regions worldwide. Administers an annual test of wine knowledge and awards a Certificate of Proficiency. Publications include the quarterly *SWE Chronicle*. Annual dues are $55 (single), $82.50 (couple), plus a $15 application fee the first year. Industry membership is $200 annually.

Contact: SWE, 8600 Foundry St., Mill Box 2044, Savage, MD 20763 US; 301-776-8569, Fax 301-776-8578, E-mail vintage@erols.com, URL http://www.wine.gurus.com.

SOMMELIER SOCIETY OF AMERICA

New York, New York

Founded in 1954. Not-for-profit organization for industry professionals dedicated to wine education from basic to internationally competitive levels. Represented internationally through the Association de le Sommellerie Internationale. Offers 6-month Sommelier Certificate Course and 8-session Wine Service Program. Chapters in New York City, Chicago, Atlanta, Los Angeles. Membership benefits include competitions, social functions, quarterly newsletter, job networking. Annual dues are $75 (professional), $25 (student), $100 (club), $250 (corporation), $500 (trade association).

Contact: The Sommelier Society of America, Box 1770, New York, NY 10159 US; 212-679-4190.

TASTERS GUILD INTERNATIONAL

Ft. Lauderdale, Florida

Founded in 1985. 78 chapters in the U.S. Objective is to promote the appreciation, understanding, and moderate use of wine and food through education, dinner seminars, tastings, and travel opportunities. Conducts an annual international wine judging each spring and a wine competition in the fall. Publishes *Tasters Guild Journal*, and sponsors an annual Food and Wine Cruise. Local chapters sponsor wine and food events and discounts are offered by affiliated wine and gourmet establishments. Annual dues are $35 per family.

Contact: Tasters Guild International, 1451 W. Cypress Creek Rd., #300-78, Ft. Lauderdale, FL 33309 US; 954-928-2823, Fax 954-928-2824, E-mail jjschagrin@aol.com, URL http://www.tastersguild.com.

UNITED STATES PERSONAL CHEF ASSOCIATION (USPCA)

Albuquerque, New Mexico

Founded in 1991. Membership 1,300+. For-profit association that trains personal chefs. Member services: continuing education, business support, regional training, referral system. Publications: bi-monthly magazine. Meals are customized to the client's taste and prepared in the client's home.

Contact: USPCA, 3615 Hwy. 528, #107, Albuquerque, NM 87114 US; 800-995-2138, Fax 505-899-4097, E-mail uspcainc@uspca.com, URL http://www.uspca.com.

WINE BRATS

San Jose, California

Founded in 1993. Membership 11,000+, 42 chapters in the U.S. Organization of (mostly young) wine enthusiasts dedicated to providing education, encouraging responsible drinking, and organizing social events. Membership is free.

Contact: Joel Quigley, Executive Director, Santa Rosa, CA; 707-545-4699, E-mail info@wine.brats.org, URL http://www.wine.brats.org.

WOMEN CHEFS AND RESTAURATEURS (WCR)

Louisville, Kentucky

Founded in 1993. Promotes the education and advancement of women in the restaurant industry and the betterment of the industry as a whole. The WCR publishes the quarterly newsletter Entrez!, and conducts an annual convention and regional events to promote networking and discussion of such issues as flexible working arrangements, job sharing, and child care. Provides nationwide job networking, publishes membership and service directory, and awards grants ande scholarships. Membership categories/annual dues include Professional (employed in the restaurant industry)/$100, Student/$35, Small Business/$250, Corporate/$1,500.

Contact: Melissa Mershon, WCR, 304 W. Liberty St., #201, Louisville, KY 40202; 502-581-0300, Fax 502-589-3602, E-mail wcr@hqtrs.com, URL http://www.culinary.net/cgi-bin/iccentry.cgi.

5

Appendix

AMERICAN CULINARY FEDERATION EDUCATIONAL INSTITUTE (ACFEI) ACCREDITING COMMISSION

St. Augustine, Florida

Accreditation by the ACFEI Accrediting Commission, the educational arm of the American Culinary Federation, evaluates the quality of an educationally-accredited post-secondary institution's program in culinary arts and foodservice management. Objectives, staff, facilities, policies, curriculum, instructional methods, and procedures are examined to determine if they meet ACFEI standards for entry-level culinarians. To be eligible, a program must contain a majority of required competencies; be offered by a school that is accredited by an agency recognized by the U.S. Dept. of Education; be full-time, include at least 1,000 contact hours, and result in a certificate, diploma, or degree; have a full-time coordinator who qualifies as a Certified Culinary Educator, Executive Chef, or Executive Pastry Chef, or has an appropriate master's degree; and have existed continuously for at least two years and graduated a sufficient number of students. Accreditation application must be authorized by the department Dean and 50% of full-time faculty must have credentials equivalent to an ACFEI Certified Culinary Educator, Sous Chef, or Pastry Chef.

Contact: For a current list of accredited programs: The Educational Institute, ACF, P.O. Box 3466, St. Augustine, FL 32085; 904-824-4468.

ACFEI-Accredited schools as of July, 1998:

ALABAMA

BISHOP STATE COMMUNITY COLLEGE
414 Stanton Rd., Mobile, AL 36617 **Contact:** Levi Ezell 334-473-8692

FAULKNER STATE COMMUNITY COLLEGE
3301 Gulf Shores Pkwy., Gulf Shores, AL 36542 **Contact:** Ron Koetter 334-968-3108

JEFFERSON STATE COMMUNITY COLLEGE
Pinson Valley Pkwy. at 2601 Carson Rd., Birmingham, AL 35215 **Contact:** Joe Morris 205-853-1200

ARIZONA

SCOTTSDALE CULINARY INSTITUTE
8100 Camelback Rd., Scottsdale, AZ 85251 **Contact:** Elizabeth Leite 602-990-3773

CALIFORNIA

CALIFORNIA CULINARY ACADEMY
625 Polk St., San Francisco, CA 94102 **Contact:** Admissions 415-771-3555

CITY COLLEGE OF SAN FRANCISCO
50 Phelan Ave., San Francisco, CA 94112 **Contact:** Frank Ambrosic 415-239-3154

DIABLO VALLEY COLLEGE
321 Golf Club Rd., Pleasant Hill, CA 94523 **Contact:** Jack Hendrickson 510-685-1230 #556

LOS ANGELES TRADE-TECHNICAL COLLEGE
400 W. Washington Blvd., Los Angeles, CA 90015 **Contact:** Steven Kasmar 213-744-9480

ORANGE COAST COLLEGE
2710 Fairview Blvd., Costa Mesa, CA 92625-5005 **Contact:** Daniel Beard 714-432-5835

SANTA BARBARA CITY COLLEGE
721 Cliff Dr., Santa Barbara, CA 93100 **Contact:** John Dunn 805-965-0581

COLORADO

COLORADO INSTITUTE OF ART-THE SCHOOL OF CULINARY ARTS
200 E. 9th Ave., Denver, CO 80203 Contact: Gary Prell 303-778-8300

PUEBLO COMMUNITY COLLEGE
900 W. Orman Ave., Pueblo, CO 81004 **Contact:** Carol Himes 719-549-3071

CONNECTICUT

MANCHESTER COMMUNITY COLLEGE
P.O. Box 1046, 60 Bidwell St., Manchester, CT 06040 **Contact:** Glen Lemaire 203-647-6121

FLORIDA

ART INSTITUTE OF FORT LAUDERDALE
1799 S.E. 17th St., Fort Lauderdale, FL 33316 **Contact:** Klaus Friedenreich 305-463-3000 #208

ATLANTIC VOCATIONAL TECHNICAL CENTER
4700 Coconut Creek Pkwy., Coconut Creek, FL 33066 **Contact:** Moses Ball 305-977-2066

FLORIDA COMMUNITY COLLEGE AT JAX
4501 Capper Rd., Jacksonville, FL 32218 **Contact:** 904-766-6652

FLORIDA CULINARY INSTITUTE
1126 53rd Ct., W. Palm Beach, FL 33407 **Contact:** David Pantone 561-688-2001/800-826-9986

GULF COAST COMMUNITY COLLEGE
5230 W. U.S. Hwy. 98, Panama City, FL 32401 **Contact:** Travis Herr 904-769-1551 #3850

PINELLAS TECHNICAL EDUCATION CENTER-CLEARWATER CAMPUS
6100 154th Ave., N. Clearwater, FL 33516 **Contact:** Vincent Calandra 813-538-7167

SOUTHEAST INSTITUTE OF CULINARY ARTS
Collins at Del Monte Ave., St. Augustine, FL 32084 **Contact:** David Bearl 904-824-4401

GEORGIA

ART INSTITUTE OF ATLANTA-THE SCHOOL OF CULINARY ARTS
3376 Peachtree Rd., NE, Atlanta, GA 30326 **Contact:** Jim Morris 404-266-1341 #227, 800-275-4242,

SAVANNAH TECHNICAL INSTITUTE
5717 White Bluff Rd., Savannah, GA 31499 **Contact:** Marvis Hinson 912-351-4553

HAWAII

KAPIOLANI COMMUNITY COLLEGE-UNIVERSITY OF HAWAII
4303 Diamond Head Rd., Honolulu, HI 96816 **Contact:** Frank Leake 808-734-9483, Fax 808-734-9212

LEEWARD COMMUNITY COLLEGE
96045 Ala Ike, Pearl City, HI 96782 **Contact:** Fern Tomisato 808-455-0375

MAUI COMMUNITY COLLEGE
310 Kaahumanu Ave., Kahului, HI 96732 **Contact:** Karen Tanaka 808-984-3225, Fax 808-984-3314

IDAHO

BOISE STATE UNIVERSITY
1910 University Dr., Boise, ID 83725 **Contact:** Julie Kulm, CEC, CCE 208-385-1532/1957

ILLINOIS

BELLEVILLE AREA COLLEGE
24950 Maryville Rd., Granite City, IL 62040 **Contact:** 618-931-0600

COLLEGE OF DUPAGE
22nd St. and Lambert Rd., Glen Ellyn, IL 60137 **Contact:** George Macht 630-942-3663

COOKING AND HOSPITALITY INSTITUTE OF CHICAGO, INC.
361 W. Chestnut, Chicago, IL 60610 **Contact:** Linda Calafiore 312-944-0882

ELGIN COMMUNITY COLLEGE
1700 Spartan Dr., Elgin, IL 60120 **Contact:** Mike Zema 708-697-1000 #7461

JOLIET JUNIOR COLLEGE
1215 Houbolt Rd., Joliet, IL 60431 **Contact:** Patrick Hegarty 815-729-9020 #2448, Fax 815-744-5507

KENDALL COLLEGE
2408 Orrington Ave., Evanston, IL 60201 **Contact:** Michael Artlip 847-866-1362, Fax 847-866-1320

INDIANA

IVY TECH STATE COLLEGE
3800 N. Anthony Blvd., Fort Wayne, IN 46805 **Contact:** Alan Eyler 219-480-4240

IVY TECH STATE COLLEGE
1440 E. 35th Ave., Gary IN 46409 **Contact:** Sharon Matusik 219-981-1111

IVY TECH STATE COLLEGE
One W. 26th St., Indianapolis, IN 46208 **Contact:** Vincent Kinkade 317-921-4619

IOWA

DES MOINES AREA COMMUNITY COLLEGE
2006 S. Ankeny Blvd., Bldg. 7, Ankeny, IA 50021 **Contact:** Robert Anderson 515-964-6532, Fax 515-964-6486

KIRKWOOD COMMUNITY COLLEGE
6301 Kirkwood Blvd., P.O. Box 2068, Cedar Rapids, IA 52406 **Contact:** Mary Jane Germann 319-398-5468

KANSAS

JOHNSON COUNTY COMMUNITY COLLEGE
12345 College at Quivira, Overland Park, KS 66210 **Contact:** Jerry Vincent 913-469-8500

KENTUCKY

JEFFERSON COMMUNITY COLLEGE
109 E. Broadway, Louisville, KY 40202 **Contact:** Gail Crawford 502-584-0181 #317

NATIONAL CENTER FOR HOSPITALITY STUDIES-SULLIVAN COLLEGE
3101 Bardstown Rd., Louisville, KY 40232 **Contact:** Walter ìSpudî Rhea 502-456-6504 #123

LOUISIANA

BOSSIER PARISH COMMUNITY COLLEGE
2719 Airline Dr., N., Bossier City, LA 71111 **Contact:** Elizabeth Dickson 318-746-6120

DELGADO COMMUNITY COLLEGE
615 City Park Ave., New Orleans, LA 70119 **Contact:** Iva Bergeron 504-483-4208

LOUISIANA TECHNICAL COLLEGE-LAFAYETTE CAMPUS
1101 Bertrand Dr., Lafayette, LA 70502 **Contact:** Jerry Sonnier 318-262-5962 #232

LOUISIANA TECHNICAL COLLEGE-NEW ORLEANS CAMPUS
9800 Navarre Ave., New Orleans, LA 70124 **Contact:** Christina Nicosia 504-483-4626

MICHIGAN

GRAND RAPIDS COMMUNITY COLLEGE
151 Fountain, NE, Grand Rapids, MI 49503 **Contact:** Robert Garlough 616-771-3690

HENRY FORD COMMUNITY COLLEGE
5101 Evergreen Rd., Dearborn, MI 48128 **Contact:** Dennis Konarski 313-845-6390

MACOMB COMMUNITY COLLEGE
44575 Garfield Rd., Clinton Twnshp., MI 48038-1139 **Contact:** David Schneider 810-286-2088

MONROE COUNTY COMMUNITY COLLEGE
1555 S. Raisinville Rd., Monroe, MI 48161 **Contact:** Kevin Thomas 313-242-7300

NORTHWESTERN MICHIGAN COLLEGE
1701 E. Front St., Traverse City, MI 49684 **Contact:** Fred Laughlin 616-922-1197

OAKLAND COMMUNITY COLLEGE
27055 Orchard Lake Rd., Farmington Hills, MI 48018 **Contact:** Susan Baier 810-471-7786

MINNESOTA

HENNEPIN TECHNICAL COLLEGE-BROOKLYN PARK CAMPUS
9000 Brooklyn Blvd., Brooklyn Park, MN 55445 **Contact:** Robert Menne 612-425-3800 #2553

ST. PAUL TECHNICAL COLLEGE
235 Marshall Ave., St. Paul, MN 55102 **Contact:** Manfred Krug and Marilyn Krasowski 612-221-1300

MONTANA

UNIVERSITY OF MONTANA COLLEGE OF TECHNOLOGY-MISSOULA
909 S. Ave. West, Missoula, MT 59801 **Contact:** Dennis Lerum 406-243-7811

NEBRASKA

METROPOLITAN COMMUNITY COLLEGE
P.O. Box 3777, Omaha, NE 68103 **Contact:** Dana Goodrich 402-449-8309

NEVADA

COMMUNITY COLLEGE OF SOUTHERN NEVADA
3200 E. Cheyenne Ave. S2D, N. Las Vegas, NV 89030-4296 **Contact:** Joe DelRosario 702-651-4192

NEW HAMPSHIRE

NEW HAMPSHIRE COLLEGE-THE CULINARY INSTITUTE
2500 N. River Rd., Manchester, NH 03106 **Contact:** Bill Petersen 603-644-3128

NEW JERSEY

HUDSON COUNTY COMMUNITY COLLEGE
161 Newkirk St., Jersey City, NJ 07306 **Contact:** Siroun Meguerditchian 201-714-2193

NEW MEXICO

ALBUQUERQUE TVI COMMUNITY COLLEGE
525 Buena Vista SE, Albuquerque, NM 87106 **Contact:** Carmine Russo 505-224-3755

NEW YORK

NEW YORK INSTITUTE OF TECHNOLOGY
300 Carlton Ave., P.O. Box 9029, Central Islip, NY 11722 **Contact:** Susan Sykes Hendee 516-348-3247

PAUL SMITH'S COLLEGE
Paul Smiths, NY 12970 **Contact:** Paul Sorgule 518-327-6218

SCHENECTADY COUNTY COMMUNITY COLLEGE
78 Washington Ave., Schenectady, NY 12035 **Contact:** Anthony ìTobyî Strianese 518-346-1390

SULLIVAN COUNTY COMMUNITY COLLEGE
Hospitality Dept., Box 4002 Le Roy Rd., Lake Sheldrake, NY 12759 **Contact:** Ed Nadeau 914-434-5750

SUNY/COBLESKILL AGRICULTURAL AND TECHNICAL COLLEGE
P.O. Box 4002, Cobleskill, NY 12043 **Contact:** Alan Roer 518-234-5425

NORTH CAROLINA

GUILFORD TECHNICAL COMMUNITY COLLEGE
P.O. Box 309, Jamestown, NC 27282 **Contact:** Keith Gardiner 910-334-4822 #2302, Fax 910-454-2510

OHIO

CINCINNATI TECHNICAL COLLEGE
3520 Central Pkwy., Cincinnati, OH 45223 **Contact:** Richard Hendrix 513-569-1662

COLUMBUS STATE COMMUNITY COLLEGE
550 E. Spring St., Columbus, OH 43215 **Contact:** Carol Kizer 614-227-2579

HOCKING TECHNICAL COLLEGE
3301 Hocking Pkwy., Nelsonville, OH 45764 **Contact:** Thomas Landusky, CEC 614-753-3591, Fax 614-753-5286

SINCLAIR COMMUNITY COLLEGE
44 W. Third St., Dayton, OH 45402-1460 **Contact:** Steven Cornelius 513-449-5197

OREGON

WESTERN CULINARY INSTITUTE
1316 S.W. 13th Ave., Portland, OR 97201 **Contact:** Larry Lewis 503-223-2245/800-666-0312

PENNSYLVANIA

INDIANA UNIVERSITY OF PENNSYLVANIA
125 S. Gilpin St., Punxsutawney, PA 15767 **Contact:** Al Wutsch 800-438-6424

INTERNATIONAL CULINARY ACADEMY
107 Sixth St., Fulton Bldg., Pittsburgh, PA 15222 **Contact:** Joseph D. Parrotto, Jr. 412-471-9330

PENNSYLVANIA INSTITUTE OF CULINARY ARTS
1200 Clark Bldg., 717 Liberty Ave., Pittsburgh, PA 15222 **Contact:** Pauline Geraci 412-566-2433 #230

PENNSYLVANIA COLLEGE OF TECHNOLOGY
One College Ave., Williamsport, PA 17701 **Contact:** Bill Butler 717-326-3761/717-327-4505

WESTMORELAND COUNTY COMMUNITY COLLEGE
Armbrust Rd., College Stn., Youngwood, PA 15697 **Contact:** Dr. Paul Lonigro 412-925-4016

SOUTH CAROLINA

GREENVILLE TECHNICAL COLLEGE
P.O. Box 5616, Stn. B, Greenville, SC 29606 **Contact:** Alan Scheidhauer 803-250-8404

HORRY-GEORGETOWN TECHNICAL COLLEGE
2050 Hwy. 501 E., P.O. Box 261966, Conway, SC 29528-6066 **Contact:** Carmen Catino 803-347-3186

TRIDENT TECHNICAL COLLEGE
P.O. Box 118067, Charleston, SC 29423-8067 **Contact:** Scott Roark 803-722-5571, Fax 803-720-5614

TENNESSEE

OPRYLAND HOTEL CULINARY INSTITUTE
2800 Opryland Dr., Nashville, TN 37214 **Contact:** Dina Starks 615-871-7765

TEXAS

ART INSTITUTE OF HOUSTON
1900 Yorktown, Houston, TX 77056 **Contact:** Michael Nenes 713-623-2040

ST. PHILIP'S COLLEGE
1801 Martin Luther King, San Antonio, TX 78284 **Contact:** Mary A. Kunz 210-531-3315

UTAH

SALT LAKE COMMUNITY COLLEGE
4600 S. Redwood Rd., Salt Lake City, UT 84130 **Contact:** Ricco Renzetti 801-957-4066

VIRGINIA

ATI CAREER INSTITUTE
7777 Leesburg Pike #100S, Falls Church, VA 22043 **Contact:** Sabrina Taffer/John Martin 800-444-0804

WASHINGTON

BELLINGHAM TECHNICAL COLLEGE
3028 Lindbergh Ave., Bellingham, WA 98225 **Contact:** Patricia McKeown 206-676-7761

RENTON TECHNICAL COLLEGE
3000 Northeast 4th St., Renton, WA 98056 **Contact:** Karl Hommer 206-235-7863

SEATTLE CENTRAL COMMUNITY COLLEGE
1701 Broadway, Seattle, WA 98122 **Contact:** Dr. Melissa Dallas 206-344-4331

SKAGIT VALLEY COLLEGE
2405 E. College Way, Mount Vernon, WA 98273 **Contact:** Lyle Hildahl 360-428-7618

SOUTH SEATTLE COMMUNITY COLLEGE
6000 16th Ave., SW, Seattle, WA 98106 **Contact:** Dan Cassidy 206-764-5344

SPOKANE COMMUNITY COLLEGE
North 1810 Greene St., Spokane, WA 99207-5399 **Contact:** Doug Fisher 509-533-7284

WISCONSIN

BLACKHAWK TECHNICAL COLLEGE
6004 Prairie Rd., P.O. Box 5009, Janesville, WI 53547 **Contact:** Joe Wollinger 608-757-7730

FOX VALLEY TECHNICAL COLLEGE
1825 N. Bluemound Dr., Appleton, WI 54913-2277 **Contact:** Albert Exenberger 414-831-5491

MADISON AREA TECHNICAL COLLEGE
3550 Anderson St., Madison, WI 53704 **Contact:** Mary Hill 608-243-4455

MILWAUKEE AREA TECHNICAL COLLEGE
700 W. State St., Milwaukee, WI 53233 **Contact:** Marietta Advincula 414-278-6507, Stephen Kissler 414-297-6255, Fax 414-297-7733

WAUKESHA COUNTY TECHNICAL COLLEGE
800 Main St., Pewaukee, WI 53072 **Contact:** William Griesemer 414-691-5254, Fax 414-691-5155

RECOMMENDED READING

THE ART OF EATING
Quarterly newsletter. Address: Box 242, Peacham, VT 05862; 800-495-3944, 802-479-3033, Fax 802-592-3400, E-mail circ@artofeating.com, http://www.artofeating.com. Annual subscription: $30. Essays about food and wine, emphasizing tradition and the relationship of food to place.

ART CULINAIRE
Quarterly magazine. Address: 40 Mills St., Morristown, NJ 07960; 973-993-5500, Fax 973-993-8779, E-mail getartc@aol.com. Per issue (annual subscription) price: $18 ($59) US,, $22 ($75) other countries. Subscriptions: Box 9268, Morristown, NJ 07963. *Hardcover magazine with 80 pages of color photographs, industry-related articles, recipes.*

THE ASIAN FOODBOOKERY
Quarterly magazine. Address: Box 15947, Seattle, WA 98115; 206-523-3575, E-mail lucky8rice@aol.com. Annual subscription price: $16 US, Canada, Mexico; $20 other countries. *Food and cooking of Asia through book reviews, essays, and travel accounts.*

AUSTRALIAN GOURMET TRAVELLER
Monthly magazine. Address: 54 Park St., Sydney NSW, 2000, Australia; (61) 2-282-8000, Fax (61) 2-267-8037. Per issue (annual) price A$5.95 (A$59 in Australia, A$128.65 overseas air). *Color glossy devoted to food, wine and travel, including restaurants, recipes, and travel.*

BBC GOOD FOOD
Monthly magazine. Address: 80 Wood Ln., London W12 OTT, UK; (44) 181-576-2000, Fax (44) 181-576-3824. Annual subscription £49.35. Subscriptions: Box 425, Woking GU21 1GP, UK; (44) 1483-733724, Fax (44) 1483-756792. *Recipes, celebrity chefs, food issues, wine, shopping news.*

BBC VEGETARIAN GOOD FOOD
Monthly magazine. Address: 80 Wood Ln., London W12 OTT, UK; (44) 181-576-3767, Fax (44) 181-576-3825. Annual subscription £45.60. Subscriptions: Box 425, Woking GU21 1GP, UK; (44) 1483-733712. *Promotes vegetarian food. Includes recipes, food news, nutrition, health, and environment updates.*

BECOMING A CHEF
320-page paperback. Published 1995. Authors: Andrew Dornenburg and Karen Page. Price $29.95. Orders: International Thomson Publishing, Inc. 800-842-3636 or 606-525-6600, Fax 606-525-7778. *Interviews with over 60 noted U.S. chefs about their early influences, training, personal and career experiences, restaurants; includes recipes.*

BON APPÉTIT
Monthly magazine. Address: 360 Madison Ave., New York, NY 10017; 212-880-8800. Per issue (annual subscription) price: $2.95 ($18) in U.S.; ($30) in Canada and abroad. Subscriptions: Box 59191, Boulder, CO 80322; 800-765-9419; URL http://www.epicurious.com. *Kitchen & tableware design, chefs, travel & restaurants, recipes for home cooks, wine & spirits tasting panel, wine reviews, articles about vineyards.*

CAREER OPPORTUNITIES IN THE FOOD & BEVERAGE INDUSTRY
240-page paperback. Published 1994. Author: Barbara Sims-Bell. Price $14.95. Orders: 800-322-8755 or 212-683-2244, Fax 212-213-4578. *Provides career profiles of 70 jobs, including duties, salary range, employment and advancement prospects, prerequisites, best locations.*

CHEF
Monthly magazine. Address: 20 North Wacker Dr., Chicago, IL 60606; 312-849-2220, Fax 312-849-2174, E-mail chefmag@aol.com, URL http://www.chefmagazine.com. Per issue (annual subscription) price: $5.00 ($24) US, ($35) Canada, ($60) other. *News and trends, chef interviews, columns on marketing, management, career mobility, menu and restaurant design.*

CHILE PEPPER
Bimonthly magazine. Address: 1400 Two Tandy Center, Ft. Worth, TX 76102; 888-SPICYHOT, 817-215-9000, Fax 817-214-9010, E-mail jgregory@mmgweb.com, URL http://www.chilepeppermag.com. Per issue (annual subscription) price: $3.99 ($18.95). Subscriptions: Box 2940, Ft. Worth, TX 76102; 800-767-9377. *Devoted to spicy foods from around the world.*

CHOCOLATIER
Six issues per year. Address: 45 W. 34th St., New York, NY 10001; 212-239-0855, Fax 212-967-4184, URL http://www2.godiva.com/chocolatier. Editor: Michael Schneider. Per issue (annual subscription) price: $4.95 ($23.95). *Chocolate and elegant desserts, their preparation and presentation, with photographs and recipes.*

COFFEE JOURNAL
Quarterly magazine. Address: 123 N. 3rd St., Minneapolis, MN 55401; 612-338-4125, Fax 612-338-0532, E-mail coffeejrnl@aol.com, URL http://www.coffeejournal.com Editor: Susan Bonne. Per issue (annual subscription) price: $3.95 ($12.97)US, $4.95 Canada. Subscriptions: Box 3000, Denville, NJ 07834-9479; 800-783-4903. *Lifestyle magazine featuring gourmet coffees and teas, new blends, brewing techniques, exotic travel, recipes, fiction, recommendations.*

COOK'S ILLUSTRATED
Bimonthly magazine. Address: 17 Station St., Brookline, MA 02147; 617-232-1000, Fax 617-232-1572, E-mail cooksill@aol.com, URL http://www.cooksillustrated.com. Per issue (annual subscription) price: $4 ($19.95) US, $4.95 Canada. Subscriptions: Box 7444, Red Oak, IA 51591-0444; 800-526-8442. *Cooking technique; comparisons of kitchen-tested methods and products.*

COOKING LIGHT
Ten issues per year. Address: 2100 Lakeshore Drive, Birmingham, AL 35209; 205-877-6000, Fax 205-877-6469, E-mail letters@cookinglight.com, URL http://cookinglight.com. Per issue (annual subscription) price: $2.95 ($20) US, $3.50 ($24) Canada, ($24) other countries. Subscriptions: Box 830656, Birmingham, AL 35282-9086; 800-999-1750, Fax 205- 877-6504. *Healthy lifestyle magazine; 65% devoted to food & preparation, 35% to personal care and fitness.*

CUISINE
Bimonthly magazine. Address: 2200 Grand Ave., Des Moines, IA 50312; 515-282-7000, Fax 515-283-0447, E-mail Cuisine@cuisinemag.com, URL http://www.augusthome.com/cuisine.htm. Per issue (annual subscription) price: $4.99 ($19.95). Subscriptions: Box 400807, Des Moines, IA 50340; 800-311-3995. *For beginning cooks. Step-by-step format illustrated with color photographs.*

THE CULINARY SLEUTH
Bimonthly newsletter. Address: Box 156, Spring City, PA 19475; 610-948-6031, Fax 610-948-6081, E-mail FoodSleuth@aol.com. Annual subscription $15. *Food history, folklore, unusual food tidbits, healthy eating, free recipe offers, food-related resources.*

CULINARY TRENDS
Quarterly magazine. Address: 6285 East Spring St., Ste. 107, Long Beach, CA 90808-9927; 714-826-9188, Fax 714-826-0333. Subscription price: $29/yr US, $58/yr foreign. *Stories and recipes of interest to the career culinarian and home chefs.*

EATING WELL
Bimonthly magazine. Address: 823A Ferry Rd., Charlotte, VT 05445; 802-425-3961, Fax 802-425-3675, E-mail EWellEdit@aol.com, URL http://www.eatingwell.com. Per issue (annual subscription) price: $2.99 ($18) US, $3.99 ($24) Canada, Subscriptions: Box 52919, Boulder, CO 80322; 800-678-0541. *Reports and comments on the dietary movement in America; nutrition reports, food and cooking articles and recipes.*

EPICURIOUS
On-line food and wine publication. Address: 140 E. 45th St., New York, NY 10017; 212-697-3132. URL http://www.epicurious.com. *Articles and a database of recipes from Gourmet and Bon AppÈtit magazines, a food dictionary, a restaurant database, and forums on a dozen topics.*

FINE COOKING
Bimonthly magazine. Address: 63 South Main St., Newtown, CT 06470; 203-426-8171, Fax 203-426-3434; E-mail fc@taunton.com, URL http://www.taunton.com/fc. Per issue (annual subscription) price: $4.95 ($26) US, $5.95 ($32) other. Subscriptions: Box 5507, Newtown, CT 06470; 800-888-8286. *Hows and whys of cooking technique, complemented by detailed information on food, preparation, and principles of good cooking.*

FOOD ARTS
Ten issues per year. Address: 387 Park Ave. S., New York, NY 10016; 212-684-4224, Fax 212-684-5424. Per issue (annual subscription) price: $4 ($30) US, $5 Canada, $7 other. Subscriptions: Box 7808, Riverton, NJ 08077; Fax 212-481-0722. *For chefs, restaurateurs, food & beverage directors, caterers; editorial on various aspects of the food business.*

FOOD ILLUSTRATED
Monthly magazine. Address: 136-142 Bramley Rd., London W10 6SR, UK; (44) (0)171-565-3000, Fax (44) (0)171-565-3056, E-mail food@johnbrown.co.uk. Per issue (annual subscription) price: $5.95 US ($71.40), $6.95 Canada (£35.40 UK, £45 Europe, £60 rest of the world). Subscriptions: 330 Pacific Ave., #404, Virginia Beach, VA 23451-2983; 888-428-6676. *Oversized illustrated lifestyle magazine emphasizing seasonal food and organic produce.*

FOOD AND TRAVEL
Monthly magazine. Address: 135 Greenford Rd., Sudbury Hill, Harrow, Middlesex HA1 3YD, UK; (44) (0)81-869-8410, Fax (44) (0)81-869-8411, URL http://www.foodandtravel.com. Per issue (annual subscription) price: $6.95 US, £2.50 UK (£30 UK, £40 Europe & Eire, £55 rest of the world). *Lifestyle magazine with balanced coverage of food and drink, travel, and entertaining.*

FOOD & WINE
Monthly magazine. Address: 1120 6th Ave., New York, NY 10036; 212-382-5618, Fax 212-764-2177, E-mail foodandwine@amexpub.com, URL http://www.foodwinemag.com. Per issue (annual subscription) price: $3.50 ($29). Subscriptions: Box 3003, Harlan, IA, 51593-0022, 800- 333-6569. *Focuses on food stories, recipes, tabletop design stories, travel.*

FOOD WRITER
Bimonthly newsletter. Address: Box 156, Spring City, PA 19475. 610-948-6031. Fax 610-948-608. E-mail Foodwriter@aol.com, URL http://www.food-journalist.net. Annual subscription: $69. *Covers food writing, editorial and freelance contacts, food industry news, education, interviews with industry leaders, marketing resources.*

GOURMET
Monthly magazine. Address: 560 Lexington Ave., New York, NY 10022; 212-880-8800, Fax 212-753-2596, URL http://www.epicurious.com. Per issue (annual subscription) price: $2.95 ($20) US, $3.50 ($34) Canada. Subscriptions: 800- 365-2454. *Balanced coverage of travel, food, culture, and entertainment.*

THE JOURNAL OF ITALIAN FOOD, WINE, & TRAVEL
Bimonthly magazine. Address: 609 W. 114th St., #77, New York, NY 10025; 212-316-3026, Fax 212-316-3476, E-mail bibendi@aol.com. Per issue (annual subscription) price: $2.95 ($17) US, $5 Canada. Subscriptions: 800)-438-2385. *Covers food, wine & travel in Italy, North America, and other countries; recipes, book reviews, restaurant and chef profiles, regional Italian articles, wine reviews, commentary.*

KITCHEN GARDEN
Bimonthly magazine. Address: 63 S. Main St., Newtown, CT 06470; 800-283-7252, 203-426-8171, Fax 203-426-3434, E-mail kg@taunton.com, URL http://www.taunton.com/kg. Per issue (annual subscription) price: $4.95 ($24) US, $5.95 ($30) other countries. Subscriptions 800-888-8286. *For gardeners who grow and prepare their own herbs, fruits, and vegetables. Combines gardening articles and tips with detailed cooking instructions and recipes.*

THE MAGAZINE OF LA CUCINA ITALIANA
Bimonthly magazine. Address: 230 Fifth Ave., New York, NY 10001; 212-725-8764, Fax 212-889-3907, E-mail piacere@earthlink.net, URL http://www.piacere.com. Per issue (annual subscription) price: $4.95 ($24) US, $5.95 ($32) Canada, ($40) other. *Emphasis on Italian cooking and culture.*

NORTH CAROLINA'S TASTE FULL
Bimonthly magazine. Address: 1202 S. 16th St., Wilmington, NC 28402; 910-763-1601, Fax 910-763-0321, E-mail nctf@aol.com. Annual subscription: $19.95 US, $27.50 Canada, $32 other. *Food, travel, and entertaining in N.C.*

NW PALATE
Bimonthly magazine. Address: Box 10860, Portland, OR 97296; 503-224-6039, Fax 503-222-5312, E-mail nwpalate@teleport.com. Per issue (annual subscription) price: $3.95 ($21) US, $5.95 ($33) Canada. *Food, wine, and travel of the Pacific Northwest; includes recipes, wine & food personalities, getaways, wine reviews, news events.*

PASTRY ART & DESIGN
Quarterly magazine. Address: 45 W. 34th St., New York, NY 10001; 212-239-0855, Fax 212-967-4184. Per issue (annual subscription) price: $5.95 ($30). *Pastry kitchen/bakery profitability, kitchen equipment, tabletop presentations, pastry school curricula; includes interviews, techniques, trends, competition updates.*

SAVEUR
Eight issues per year. Address: 100 Sixth Ave., New York, NY 10013; 212-334-1212, Fax 212-334-1260, E-mail saveur@here.com. Per issue (annual subscription) price: $5 ($24) U.S. ($38) other. Subscriptions: 800-462-0209. *Examines the gourmet world of food with emphasis on its origins, regional & cultural diversities, and the people who create it.*

SHAPE COOKS
Quarterly. Address: 21100 Erwin St., Woodland Hills, CA 91367; 818-884-6800, Fax 818-716-5626. Per issue (9 issue) price: $2.95 ($16.97) U.S. $3.50 ($27.79) Canada ($28.22) other countries. Subscriptions: 800-493-4337. *Techniques, tools, and products for preparing nutritious, tasty, well-balanced meals for active women and their families. Features on gardening, travel, entertainment.*

SIMPLE COOKING
Bimonthly newsletter. Address: Box 8, Steuben, ME 04680; E-mail outlawcook@earthlink.net, http://home.earthlink.net/~outlawcook/. Per issue (annual subscription) price: $4 ($24) US, $4.25 Canada, $5 other. *Essays on food, cooking and the culinary life; includes recipes, food book reviews, product notes.*

SIMPLY SEAFOOD
Quarterly magazine. Address: 5305 Shilshole Ave. NW, Seattle, WA 98107; 206-789-6506, Fax 206- 789-9193, E-mail editor@simplyseafood.com, http://www.simplyseafood.com. Per issue (annual subscription) price: $2.95 ($11.85) US, $3.55 ($14) Canada. *Informational writing, practical instruction, recipes related to handling and preparation of seafood.*

SMART WINE MAGAZINE
Monthly magazine. Address: 867 W. Napa St., Sonoma, CA 95476; 707-939-0822, Fax 707-939-0833. Per issue (annual subscription) price: $3.95 ($29.95). Subscriptions: 800-895-WINE. *Covers wine, food and lifestyle with an irreverent attitude. Includes recipes, reviews, wine country reports, wine events.*

VEGETARIAN JOURNAL
Bimonthly magazine. Address: Box 1463, Baltimore, MD 21203; 410-366-8343, E-mail TheVRG@aol.com, URL http://www.vrg.org. Per issue (annual subscription) price: $3.50 ($20) US, ($30) Canada. *Vegetarian meal planning, nutrition, recipes, and natural food product reviews.*

VEGETARIAN TIMES
Monthly magazine. Address: 1140 Lake St., Oak Park, IL 60301; 708-848-8100, Fax 708- 848-8175, E-mail 74651.215@compuserve.com, URL http://www.vegetariantimes.com. Per issue (annual subscription) price: $2.99 ($29.95) US, $4.50 ($41.95) Canada, ($54.91) other. Subscriptions: Box 420235, Palm Coast, FL 32142; 800-829-3340. *Vegetarian recipes, nutritional breakdowns, photography; articles on health, nutrition, fitness; profiles, product news and reviews.*

VEGGIE LIFE
Bimonthly magazine. Address: 1041 Shary Circle, Concord, CA 94518; 800-777-1164, Fax 510-671-0692, URL http://www.veggielife.com. Per issue price: $3.95 US, $4.95 Canada. *Growing green cooking lean, feeling good.*

VOGUE ENTERTAINING & TRAVEL
Bimonthly magazine. Address: Locked Bag 2550, Crows Nest, NSW 2065, Australia; (61) 2-9964-3888, Fax (61) 2-9964-3882. Per issue price $7.50, A$5.50 in Australia. Annual subscription: A$33 in Australia. *Lifestyle magazine featuring food, entertaining, travel.*

WEIGHT WATCHERS MAGAZINE
Monthly magazine. Address: 360 Lexington Ave., New York, NY 10017; 212-370-0644, Fax 212-687-4398. Per issue price: $1.95 US, $2.50 Canada. Subscriptions: 800-876-8441. *Information and news on women's health, wellness, nutrition, and fitness; low-fat cooking techniques and recipes.*

THE WINE ADVOCATE
Bimonthly newsletter. Address: Box 311, Monkton, MD 21111; 410-329-6477, Fax 410-357-4504, URL http://www.wineadvocate.com. Annual subscription: $40 US, $60 Canada, $85 other. *Consumer guide to fine wine.*

THE WINE ENTHUSIAST
Fourteen issues per year. Address: 8 Saw Mill River Rd., Hawthorne, NY 10532; 800-356-8466, Fax 914-345-3028, E-mail wineenth@aol.com, URL http://www.wineenthusiastmag.com. Annual subscription: $24.95 US, $49.95 other. Subscriptions: 800-356-8466. *Focuses on wine for both new and experienced wine drinkers.*

WINE SPECTATOR
Nineteen issues per year. Address: 387 Park Ave. S., New York, NY 10016; 212-684-4224, Fax 212- 684-5424. URL http://www.winespectator.com. Per issue price: $2.95 ($40) US, $3.95 ($53.50) Canada, 39FF, £2.50 ($125) other countries. Subscriptions: Box 50463, Boulder, CO 80323, 800-752-7799. *Wine features, ratings and tasting notes, fine restaurants, cooking and entertaining, world travel and the arts, shopping and collectibles.*

WINE & SPIRITS
Eight issues per year. Address: 2 W. 32nd St., #601, New York, NY 10001; 212-695-4660, Fax 212-695-2920, E-Mail winespir@aol.com. Per issue price $2.95 ($4.95) US, $3.50 ($5.95) Canada. Subscriptions: Box 50463, Boulder, CO 80323-0463, 800-395-3364. *Guide to wine, for consumers and buyers.*

WINE TIDINGS
Eight issues per year. Address: 5165 Sherbrooke St. W., #414, Montreal, QB, Canada; 514-481-5892, Fax 514-481-9699. Per issue price $3.25 US, $3.75 Canada. *Rates wines in Canada and abroad, winemaker profiles.*

CAREER SCHOOL TUITION RANKINGS

Programs are indexed by total tuition for those less than nine months and by annual tuition (in-state) for those nine months or more. Additional costs, e.g., housing, meals, supplies are not included.

TOTAL TUITION FOR PROGRAMS OF LESS THAN NINE MONTHS

ANNUAL TUITION FOR PROGRAMS OF NINE MONTHS OR LONGER

$1,000-$1,999

$2,000-$2,999

$3,000-$4,999

$5,000-$7,499

$7,500-$9,999

$10,000 or over

SCHOOLS & ORGANIZATIONS THAT OFFER SCHOLARSHIPS

SCHOOLS THAT OFFER CLASSES FOR CHILDREN AND TEENS

INDEX OF ADVERTISERS

ACCREDITING AGENCIES

ACCSCT Accrediting Commission of Career Schools/Colleges of Technology
ACFEI American Culinary Federation Educational Institute
ACICS Accrediting Council for Independent Colleges and Schools
COE Council on Occupational Education
MSA Middle States Association of Colleges and Schools
NASC Northwest Association of Schools and Colleges
NCA North Central Association of Colleges and Schools
NEASC New England Association of Schools and Colleges
SACS Southern Association of Colleges and Schools
WASC Western Association of Schools and Colleges

CURRENCY CONVERSION TABLE

(as of August 21, 1998)

Country	*Currency per U.S. $1*	*Country*	*Currency per U.S. $1*
Australia (A$)	1.70	Israel (Shekel)	3.72
Canada (C$)	1.53	Italy (Lira)	1775.00
France (FF)	6.03	New Zealand (NZ$)	2.02
Great Britain (£)	1.63	South Africa (R)	6.35
Ireland (IR£)	1.40	Spain (Pts)	152.65

Cooking School Index

A

B

C

D

E

F

G

H

I

J

K

L

M

N

T

U

V

W

Y

Z